Numerical 'experiments' now play a key role in astrophysical research. From the origin and growth of large-scale structure in the Universe, and the evolution of galaxies and galaxy clusters, through to the hydrodynamics of interstellar flows and the formation of stars and the solar system, numerical simulations often provide the most fruitful means of furthering our understanding.

An international conference held by UNAM, in Mexico, was dedicated to reviewing the exciting and important results of a decade of numerical simulations of particle and fluid dynamics and to identifying key areas for future research. Presented in this opportune volume are the proceedings of this dynamic meeting.

The review articles and research papers gathered together here constitute a comprehensive and up-to-date account of modelling our Universe through dynamical and hydrodynamical numerical simulations. They provide an excellent introduction and overview for graduate students as well as a critical update for researchers.

Numerical Simulations in Astrophysics

Numerical Simulations in Astrophysics

Proceedings of the First UNAM-CRAY Supercomputing Workshop,
*"Numerical Simulations in Astrophysics: Modelling
the Dynamics of the Universe"*,
held in Mexico City, México
July 26–30, 1993

Edited by

José Franco
Instituto de Astronomía–UNAM, México

Susana Lizano
Instituto de Astronomía–UNAM, México

Luis Aguilar
Observatorio Astronómico Nacional–UNAM, México

Enrique Daltabuit
Dirección Gral. de Cómputo Académico–UNAM, México

Published by the Press Syndicate of the University of Cambridge
The Pitt Building, Trumpington Street, Cambridge CB2 1RP
40 West 20th Street, New York, NY 10011–4211, USA
10 Stamford Road, Oakleigh, Melbourne 3166, Australia

© Cambridge University Press 1994

First published 1994

Printed in Great Britain at the University Press, Cambridge

A catalogue record for this book is available from the British Library

Library of Congress cataloguing in publication data available

ISBN 0 521 46238 X hardback

Contents

Participants   xi
Preface   xiv

Large Scale Structure of the Universe and the Formation of Galaxies

The Large Scale Structure of the Universe
    C. S. Frenk 1

The Evolution of Superhorizon-Sized Voids in the Early Universe
    S. L. Vadas 13

Perturbation Solution to the Linearized Einstein Constraint Equations in Spherical Symmetry as a Test for Numerical General–Relativistic Codes
    H. Harleston L. 18

Nested Grid Methods for Cosmological Hydrodynamics and N-body Systems
    P. Anninos & M. L. Norman 24

Dynamics of Clusters of Galaxies
    A. E. Evrard 28

Large Scale Structure and Motions from Simulated Galaxy Clusters
    R. A. C. Croft & G. Efstathiou 40

Adaptive Smoothed Particle Hydrodynamics and Galaxy Formation
    P. R. Shapiro, H. Martel & J. V. Villumsen 45

Mergers and Galaxy Formation
    J. F. Navarro 53

A Model for Galaxy Formation in CDM Cosmology
    M. B. Mosconi, P. B. Tissera & D. G. Lambas 63

The Formation of a Cluster of Galaxies
    S. N. Dutta & D. N. Spergel 68

A Hydrodynamical Simulation of Cluster Formation
    G. L. Bryan & M. L. Norman 73

Simulations of the Formation of Dwarf Ellipticals: Models with no Halo
    E. Athanassoula 77

Nonlinear Evolution of Elliptical Perturbations
    A. S. González 80

Particle-Mesh Simulations of the Formation of Binary Galaxies and the Spins of Local Group Members
    S. Gelato, D. F. Chernoff & I. Wasserman 85

Galaxy Evolution and Stellar Clusters

Instabilities of Stellar Disks
    J. A. Sellwood 90

Stellar Cluster Dynamics
  S. J. Aarseth 101

Mapping the Galaxy Using Three Different Galactic Potentials
  F. Valera, L. Aguilar & W. Schuster 111

A Quiet Start for an Integrable Non-Axisymmetric Potential
  S. E. Levine 117

Dissipation in Dynamical Models of Galaxies
  J. Palouš & B. Jungwiert 122

Numerical Simulations of Collisions of Spherical Galaxies
  M. M. Vergne & J. C. Muzzio 127

Modeling the Dynamics of the Interacting Galaxy Pair NGC 4676
  J. Gilbert & J. A. Sellwood 131

Radial Orbit Instability in a Hubble–Expanding Universe
  D. D. Carpintero & J. C. Muzzio 135

Dynamics of Massive Black Holes as Possible Candidates of Dark Matter
  G. Xu & J. P. Ostriker 140

Galactic Orbits for 280 Halo Stars
  F. Valera, W. Schuster & L. Aguilar 145

Galactic Winds from Starburst Galaxies
  A. Habe 151

Shocks and Dust in Active Galactic Nuclei
  S. M. Viegas & M. Contini 157

The Physics of Self-Propagating Star Formation
  G. Tenorio-Tagle 159

Supernova Explosions and Interstellar Hydrodynamics

Supernova Explosion Calculations
  J. R. Wilson 165

Convection in Supernova Cores
  T. Shimizu, S. Yamada & K. Sato 170

Instabilities in Supernova Explosions
  B. Fryxell 175

Superbubbles and Supernova Remnants in Magnetized Interstellar Media
  K. Tomisaka 184

3-D Models for Supershells in a Cloudy Medium
  S. A. Silich, J. Franco, J. Palouš & G. Tenorio-Tagle 193

Producing the Soft X-ray Background with Multiple Supernova Remnants
  R. K. Smith 198

Formation of Molecular Clouds in Expanding Supershells: 3-D Models
  S. Ya. Mashchenko & S. A. Silich 202

2D Simulations of SN Remnants with a Moving Precursor
  F. Brighenti & A. D'Ercole 206

Mass-Loaded Astrophysical Flows
  S. J. Arthur, J. E. Dyson & T. W. Hartquist 210

The Effects of a Magnetic Field in the Evolution of Cosmic Ray Mediated Shocks
  B. Jun, D. A. Clarke & M. L. Norman 216

Particle Acceleration with Spontaneous Excitation of Alfvén Waves in the Magnetospheres of Neutron Stars
  H. Hanami 220

Instability of C-shocks in the ISM
  G. Tóth 224

MHD Experiments on the Thick Galactic Disk of Gas
  M. A. Martos & D. P. Cox 229

Turbulence in the Interstellar Medium
  A. Pouquet 237

A Turbulent Model for the Interstellar Medium
  T. Passot, E. C. Vázquez-Semadeni & A. Pouquet 246

The Hydrodynamics of Cloud Interactions
  R. I. Klein & C. F. McKee 251

Dynamical Evolution of HII Regions Powered by Stellar Winds
  F. Comerón 267

Line Formation During the Disruption of Cloud Cores by Photoionization
  J. A. Rodríguez-Gaspar, G. Tenorio-Tagle & J. Franco 271

Highly Collimated Outflows in Two and Three Dimensions
  P. E. Hardee & D. A. Clarke 275

Numerical Simulations of Stellar Outflows
  F. Rubini, G. Manzini, S. Lizano & C. Giovanardi 280

Variable Velocity Jets: Internal Working Surfaces
  S. Biro & A. C. Raga 284

Accretion Disks, Star Formation and Planetary Growth

Accretion Disks in Astrophysics
  S. K. Chakrabarti 288

Orbitally-Modulated Emission Line Profiles from Non-Keplerian Accretion Disks
  I. G. Martínez-Pais 297

Numerical Simulation of Co-planar Star-Disk Interaction
  D. Molteni, G. Gerardi & S. K. Chakrabarti 301

The Stability of Circumstellar Disks
  S. M. Miyama, T. Nakamoto, N. Kikuchi, S. Inutsuka, K. Kobayashi & T. Takeuchi 305

Instabilities in Protostellar Disks
  G. Laughlin & P. Bodenheimer 313
The Dynamics of Massive Protostars and their Photoionized Disks
  H. W. Yorke & A. Welz 318
Three-Dimensional Fragmentation Calculations of Protostellar Collapse
  P. Bodenheimer 327
Collapse and Fragmentation of Magnetized Cylindrical Clouds
  K. Tomisaka 336
Effect of Deceleration on the Gravitational Instability of Shocked Gas Layer
  T. Yoshida & A. Habe 341
The Formation of a "Protocluster"
  J. A. Turner, A. S. Bhattal, S. J. Chapman, M. J. Disney, H. Pongracic & A.
  P. Whitworth 345
The Formation of Hierarchical Binary Systems in Turbulent GMCs
  S. J. Chapman, A. S. Bhattal, M. J. Disney, H. Pongracic, J. A. Turner & A.
  P. Whitworth 348
Two Formation Mechanisms for Binary (and Multiple) Protostars in Shocked Interstellar Gas Layers
  S. J. Chapman, A. S. Bhattal, M. J. Disney, H. Pongracic, J. A. Turner & A.
  P. Whitworth 351
Star Formation via Interactions of Shocks with Molecular Clouds
  H. Pongracic, S. J. Chapman, M. J. Disney, A. H. Nelson, J. A. Turner & P.
  A. Whitworth 355
The Formation of OB Subgroups
  S. J. Chapman, A. S. Bhattal, M. J. Disney, J. A. Turner & A. P. Whitworth 360
Numerical Simulations of Planetary Growth
  D. M. Kary & J. J. Lissauer 364

Participants

Sverre Aarseth	U. Cambridge, UK (sja@ast-star.cam.ac.uk)
Luis Aguilar	UNAM, México (aguilar@bufadora.astrosen.unam.mex)
Peter Anninos	U. Illinois, USA (panninos@ncsa.uiuc.edu)
Anabel Arrieta	UNAM, México (anabel@astroscu.unam.mx)
Jane Arthur	UNAM, México (jane@astroscu.unam.mx)
E. Athanassoula	Obs. Marseille, France (lia@obmara.cnrs-mrs.fr)
Sandra Ayala	UNAM, México (sayala@astroscu.unam.mx)
Javier Ballesteros	UNAM, México (javier@astroscu.unam.mx)
Erika Benitez	UNAM, México (erika@astroscu.unam.mx)
Susana Biro	UNAM, México (susana@ast-star.man.ac.uk)
Peter Bodenheimer	UC Santa Cruz, USA (peter@helios.ucsc.edu)
Fabrizio Brighenti	U. Bologna, Italy (brighenti@alma02.cineca.it)
Greg Bryan	UI Urbana-Champaign, USA (gbryan@ncsa.uiuc.edu)
Jorge Cantó	UNAM, México (canto@astroscu.unam.mx)
Daniel D. Carpintero	Obs. Astron. La Plata, Argentina (ddc@caglp.edu.ar)
René Carrillo	UNAM, México (rene@astroscu.unam.mx)
Sandip Chakrabarti	Tata Insitute, India (chakraba@tifrvax.bitnet)
Simon J. Chapman	U. Wales, UK (sjc@vax1.astronomy.cardiff.ac.uk)
Carlos Chavarría	UNAM, México (chavarri@astroscu.unam.mx)
Enrique Chavira	INAOE, México (echavira@tonali.inaoep.mx)
Gerardo Cisneros	CRAY Research Inc., México (gerardo@cray.com)
Pedro Colín	UNAM, México (colin@astroscu.unam.mx)
Fernado Comerón Tejero	U. Barcelona, Spain (fcomeron@mizar.ub.es)
Maria Eugenia Contreras	UNAM, México (maru@astroscu.unam.mx)
Rafael Costero	UNAM, México (costero@astroscu.unam.mx)
Rupert A. C. Croft,	U. Oxford, UK (racc@oxds02.astro.ox.ac.uk)
Enrique Cruz Martínez	UNAM, México (craymail@ds5000.dgsca.unam.mx)
Fidel Cruz Peregrino	UNAM, México (fidel@astroscu.unam.mx)
Paola D'alessio	UNAM, México (dalessio@astroscu.unam.mx)
Enrique Daltabuit	UNAM, México (enrique@pcdaltabuit.dgsca.unam.mx)
Durruty Jesús De Alba Martínez	U. Guadalajara, México (durruty@redudg.udg.mx)
Rosa Izela Díaz	UNAM, México (rosa@astroscu.unam.mx)
Deborah Dultzin	UNAM, México (deborah@astroscu.unam.mx)
Suvendra Dutta	Princeton U., USA (dutta@astro.princeton.edu)
Vladimir Escalante	UNAM, México (vladimir@astroscu.unam.mx)
August Evrard	U. Michigan, USA (evrard@pablo.physics.lsa.umich.edu)
Tatiana Fetisova	Space Research Institute, Russia
Julieta Fierro	UNAM, México (julieta@astroscu.unam.mx)
Daniel Flores	UNAM, México (daniel@astroscu.unam.mx)
José Franco	UNAM, México (pepe@astroscu.unam.mx)
Carlos Frenk	Durham U., UK (csf@starlink.durham.ac.uk)
Bruce Fryxell	NASA/Goddard, USA (fryxell@neutrino.gsfc.nasa.gov)
Sergio Gelato	Cornell U., USA (gelato@astrosun.tn.cornell.edu)
Alejandro González	U. Sussex, UK (mafa1@cluster.sussex.ac.uk)
Ignacio González Martínez-Pais	Ins. Astrofís. Canarias, Spain (igm@iac.dnet.nasa.gov)
Violeta Guzmán Jiménez	U. Guadalajara, México
Asao Habe	Hokkaido U., Japan (habe@astro1.phys.hokudai.ac.jp)
Hitoshi Hanami	Iwate U., Japan (d1269@jpnkudpc.bitnet)

Participants

Philip Hardee	U. Alabama, USA (hardee@venus.astr.ua.edu)
Hugo Harleston	UNAM, México (hugo@roxanne.nuclecu.unam.mx)
Joaquín Hernández	UNAM, México (craymail@ds5000.dgsca.unam.mx)
Armando Hernández Portilla	UNAM, México (armando@aleph.cinstrum.unam.mx)
Byung-Il Jun	U. Illinois/NCSA, USA (bjun@ncsa.uiuc.edu)
Richard Klein	UC Berkeley/Livermore, USA (klein@radhydro.berkeley.edu)
Gloria Koenigsberger	UNAM, México (gloria@astroscu.unam.mx)
Dimitri Kouznetsov	UNAM, México (kusnecov@aleph.cinstrum.unam.mx)
Rafael Lacambra	UNAM, México (rlm@ds5000.dgsca.unam.mx)
Gregory Laughlin	UC Santa Cruz, USA (laugh@helios.ucsc.edu)
Stephen Levine	UNAM, México (levine@bufadora.astrosen.unam.mx)
Jack Lissauer	SUNY-Stony Brook, USA (jlissauer@sbast1.ess.sunysb.edu)
Susana Lizano	UNAM, México (lizano@astroscu.unam.mx)
Gabriel López Walle	UNAM, México (gabriel@ds5000.dgsca.unam.mx)
Mordecaik-Mark Mac Low	U. Chicago, USA (mordecai@jets.uchicago.edu)
Roel Martínez	UNAM, México (roel@astroscu.unam.mx)
Marco Antonio Martos	U. Arizona, USA (martos@quark.physics.arizona.edu)
Andres Meza	U. Chile, Chile (nzamora@cecvx1.cec.uchile.cl)
Maripaz Miralles	UNAM, México (miralles@astroscu.unam.mx)
Shoken Miyama	Nat. Astrophys. Obs., Japan (miyama@yso.mtk.nao.ac.jp)
Mirta Beatriz Mosconi	Obs. Astron. Cordova, Argentina (atrim@amaf.edu.ar)
Julio Navarro	U. Durham, UK (jfn@star.dur.ac.uk)
Michael Norman	U. Illinois, USA (norman@ncsa.uiuc.edu)
Lorenzo Olguin	UNAM, México (lorenzo@astroscu.unam.mx)
Jan Palouš	Charles U., Czech Republik (palousj@csearn.bitnet)
Manuel Peimbert	UNAM, México (peimbert@astroscu.unam.mx)
Miriam Peña	UNAM, México (miriam@astroscu.unam.mx)
Gabriela Piccinelli	UNAM, México (gabi@astroscu.unam.mx)
Helen Pongracic	U. Sydney, Australia (helenp@physics.su.oz.au)
Annick Pouquet	Obs. Côte d'Azur, France (pouquet@obs-nice.fr)
Ivanio Puerari	Obs. Marseille, France (puerari@fromrs51.bitnet)
Margarita Rosado	UNAM, México (margarit@astroscu.unam.mx)
José Angel Rodríguez-Gaspar	Ins. Astrofís. Canarias, Spain (jrg@iac.es)
Francesco Rubini	U. Firenze, Italy (rubini@sisifo.arcetri.astro.it)
Martha Adriana Sánchez	UNAM, México (martha@ds5000.dgsca.unam.mx)
Alfredo Santillán	UNAM, México (alfredo@astroscu.unam.mx)
Antonio Sarmiento	UNAM, México (ansar@astroscu.unam.mx)
William Schuster	UNAM, México (schuster@bufadora.astrosen.unam.mex)
Jerry Sellwood	Rutgers U., USA (sellwood@physics.rutgers.ed)
Paul Shapiro	UT Austin, USA (shapiro@castro.as.utexas.edu)
Sergey Silich	Main Astron. Obs., Ukraine (galaxy@gao.kiev.ua)
Tetsuya Shimizu	U. Tokyo, Japan (shimizut@tkyvax.phys.s.u-tokyo.ac.jp)
Randall Smith	U. Wisconsin, USA (rsmith@wisp4.physics.wisc.edu)
Charles Swanson	CRAY Research Inc., USA (cds@cray.com)
Guillermo Tenorio-Tagle	Ins. Astrofís. Canarias, Spain (gtt@iac.es)
Patricia Beatriz Tissera	Obs. Astron. Cordova, Argentina (tissera@famaf.edu.ar)
Kohji Tomisaka	Niigata U., Japan (tomisaka@ed.niigata-u.ac.jp)
Silvia Torres-Peimbert	UNAM, México (silvia@astroscu.unam.mx)
Gabor Toth	Princeton U., USA (toth@astro.princeton.edu)
Sharon L. Vadas	Fermilab, USA (vasha@fnas11.fnal.gov)

Fabian Valera — INAOE, México (fvalera@tonali.inaoep.mx)
Enrique Vazquez-Semadeni — UNAM, México (enro@astroscu.unam.mx)
Octavio Valenzuela — UNAM, México
Shoba Veeraraghavan — U. Arizona, USA (shoba@as.arizona.edu)
Maria Marcela Vergne — Obs. Astron. La Plata, Argentina (mvergne@fcaglp.edu.ar)
Sueli M. Viegas — U. Saõ Paolo, Brazil (viegas@iag.usp.ansp.br)
Felipe C. Wachlin — Obs. Astron. La Plata, Argentina (fcw@caglp.edu.ar)
Antony Peter Whitworth — U. Wales, UK (apw@vi.astro.cardiff.ac.uk)
James Wilson — Lawrence Livermore Natl. Lab., USA (wilson@ricker.llnl.gov)
Guohong Xu — Princeton U., USA (xu@astro.princeton.edu)
Harold Yorke — U. Wuerzburg, Germany (yorke@astro.uni-wuerzburg.dbp.de)
Tatsuo Yoshida — Ibaraki U., Japan (iyoshida@tansei.cc.u-tokyo.ac.jp)

Preface

The First UNAM-CRAY Supercomputing Workshop, **Numerical Simulations in Astrophysics: Modelling the Dynamics of the Universe**, was held in the beautiful setting of the *Palacio de Minería*, a historical building located in downtown Mexico City. The goal of this meeting was to bring together scientists working on dynamical and hydrodynamical models in astrophysics (at all scales, from the large scale structure of the Universe to planetary systems), to exchange ideas and to stimulate new directions of research. A total of 109 participants, from 16 different countries, gathered together during one week to discuss their most recent results, and the outcome of this fruitful week is reported in this book.

The organization of these proceedings is as follows. First, it starts with the dynamics of galaxy clusters and the large scale structure of the Universe, and moves into the details of galaxy structure and evolution. Then, it goes on to the hydrodynamics of the general interstellar medium and supernova explosions, and ends at smaller scale astrophysical phenomena with the rich activity of protostellar and accretion disks and planetary growth. The review papers provide a well balanced overview of all these topics, and the peer reviewed contributed papers illustrate the current activity in each field. The final result is a comprehensive, but concise, compendium of the most recent trends in astrophysical simulations, which will allow both the expert and the novice to keep up with this fascinating and fast changing branch of theoretical astrophysics.

We would like to acknowledge Drs. P. Bodenheimer, C. Frenk, C. McKee, S. Miyama, J. Ostriker, J. Scalo, G. Tenorio-Tagle, and K. Tomisaka, of the Scientific Organizing Committee, and Drs. G. Cisneros, G. Koenigsberger, J. Thomas, and E. Vazquez-Semadeni, of the Local Organizing Committee, for their excellent work in the design and organization of the meeting. We also thank our budget managers Lics. Manuel Comi, from IA-UNAM, and Daniel Jiménez, from DGSCA-UNAM, for sparing us from the usual book-keeping headaches with their very efficient financial organization, and to Toña Zimerman for her wonderful art work and poster design. Very special thanks to our talented executive secretaries Lila Perillat, Luz María Gisquet and Faride Estevané, for maintaining the spirit of the meeting with their shining smiles and very efficient work, and to our fast-moving crew Reynaldo Hernández, Guadalupe Valdelamar, and Juan Torres for keeping the meeting "trouble-free". They certainly created a joyful atmosphere during the coffee breaks.

The editing of these proceedings required the help of a large number of persons. In particular, we thank our good friends Jane Arthur, Will Henney, Stan Kurtz and Warren Miller for their invaluable help. The fine work done by Alberto García with the original art and postcript files is greatly appreciated. Also thanks to Juana Orta and Bertha Hernández for handling the manuscripts at the early stages of the editing process, and Rossy Díaz for her help with some of the typing.

Last, but obviously not least, we are very grateful to the generous grants provided by Cray Research Inc., Coordinación de la Investigación Científica de la UNAM, Dirección General de Asuntos del Personal Académico de la UNAM, and the financial support given by the Instituto de Astronomía–UNAM and Dirección General de Servicios y Cómputo Académico–UNAM.

José Franco, Susana Lizano, Luis Aguilar, Enrique Daltabuit
México D. F., México
January, 1994

The Large-scale Structure of the Universe

By Carlos S. Frenk

Physics Department, University of Durham, Durham DH1 3LE, United Kingdom.

Theoretical ideas bearing on the physics of the early universe and the identity of the dark matter lead to fully specified initial conditions for the formation of cosmic structure. Primordial density fluctuations grow by gravitational amplification in the expanding universe. The resulting distribution of *mass* at the present epoch can be calculated reliably using large N-body simulations. The distribution and properties of *galaxies* are considerably more difficult to predict because hydrodynamic, radiative and stellar effects are involved. Although these uncertainties weaken tests of specific models, numerical work over the past decade has led to a number of important conclusions. One is the exclusion of the model in which the dark matter consists of massive neutrinos and the initial fluctuations are gaussian. Another is the growing acceptance of the inflationary cold dark matter cosmogony as the basic framework for understanding the formation of cosmic structure. The simplest version of this theory accounts for many observed properties of galaxies and clusters but appears to underestimate the strength of galaxy clustering on large scales. Variants of this model, involving a non-zero cosmological constant or a small contamination from neutrino dark matter, provide a good match to the large-scale data, including the measured anisotropies in the microwave background radiation.

---

1. Introduction

At early times, the Universe was remarkably uniform. Direct evidence for this was provided last year by the COBE satellite which mapped the structure of the microwave background – the relic radiation emitted when the Universe was only a few hundred thousand years old. The departures from homogeneity detected by COBE amount to an *rms* temperature fluctuation of only about 1 part in 10^5 on angular scales greater than 10° [57]. On smaller scales, a variety of experiments have set upper limits to the temperature fluctuations of comparable amplitude, e.g. [15,35]. Yet, the nearby universe is patently far from uniform – maps of the galaxy distribution clearly show a wide variety of structures, from aggregates of a few tens to elongated superclusters of several thousand galaxies. Understanding how the Universe evolved from its remarkable early simplicity to its present, highly structured state is the central problem of physical cosmology.

The 1980's saw considerable progress in attempts to understand the growth of cosmic structure. This resulted from a timely combination of new theoretical ideas, computational methods and observational discoveries. Some of the most influential ideas came from the particle physics community: the notion of an inflationary universe and the suggestion that the dynamically dominant dark matter may consist of weakly interacting elementary particles. According to the inflationary paradigm, the universe underwent a brief period of exponential expansion at early times, driven by the vacuum energy of a scalar field [1,37,42]. This idea explains why the universe is as old as it is, why it appears nearly homogeneous on large scales, and why it is so clumpy on small scales. The rapid expansion gives rise to a flat geometry and allows a long-lived universe. Quantum fluctuations in the field that drives the inflation are amplified to macroscopic scales and give rise to seeds which may subsequently grow by gravitational instability into the clustering pattern characteristic of the present day distribution of galaxies, e.g. [48]. The amplitudes of primordial density fluctuations produced by inflation are gaussian distributed and scale-invariant.

The temperature anisotropies detected by COBE appear to have just the properties expected in the inflationary model: a power spectrum corresponding to scale-invariant fluctuations ($|\delta_k|^2 \propto k^n$, with $n = 1.1 \pm 0.5$, where $k \equiv 2\pi/\lambda$ denotes spatial frequency) and a Gaussian distribution. There are, however, alternative mechanisms to generate primordial seeds which are also consistent with the COBE data. One idea posits the existence of topological defects in the early universe arising from field ordering phase transitions associated with the spontaneous breaking of global symmetries, e.g. [10,60] and references therein. Defects such as strings or textures give rise to local density perturbations which may subsequently grow, as in the inflationary case, through gravity.

Whether produced by quantum processes or by topological defects, primordial density perturbations develop at a rate which is determined by the amount and nature of the dark matter, known to dominate the dynamics of the Universe on scales larger than the cores of galaxies. Although the identity of the dark matter remains a mystery, several candiates have been proposed and investigated over the past decade. Perhaps the most popular ones are weakly interacting elementary particles – neutrinos, supersymmetric particles or axions – but baryonic candidates such as brown dwarfs or Jupiters are by no means discounted [11].

In this article, I review current ideas regarding the origin of the cosmic large-scale structure. In Section 2 I discuss the main argument in favour of non-baryonic dark matter. The general theory for the gravitational origin of large-scale structure is summarised in Sections 3 and 4 in which specific models are briefly discussed. The shortcomings of the standard cold dark matter model and various ways to overcome them are discussed in Section 5. A summary and conclusions are given in Section 6.

2. The mean density of the Universe

One of the most dramatic successes of the Hot Big Bang theory has been its ability to explain the relative abundances of the light elements. Thus, current estimates of the "primordial" abundances of H, ^{2}H, ^{3}He, ^{4}He, and ^{7}Li are all consistent with those expected a few minutes after the Big Bang in a Friedman-Robertson-Walker universe with present baryon density in the range:

$$0.01 \leq \Omega_b h^2 \leq 0.015. \tag{2.1}$$

where h is the present value of Hubble's constant in units of 100 km/s/Mpc, Ω_b is given in units of the critical density needed to close the universe and the range quoted represents a 95% confidence limit [60]. Over the last decade the cosmology community has increasingly accepted the plausibility of the inflation paradigm and of its associated prediction that the mean matter density of the universe, Ω, should be very close to the critical value for closure. The discrepancy between $\Omega = 1$ and $\Omega_b h^2 \sim 0.0125$ is the main motivation for the hypothesis that most of the mass in the universe is in some nonbaryonic and invisible form.

For a long time direct dynamical evidence for such large amounts of dark matter was lacking. However, recent comparisons of large-scale motions of galaxies and clusters with the density field that induces them have consistently led to large estimates of Ω [4,40,49,53,58] For example, analysis of the "QDOT" redshift survey of IRAS galaxies [30,40] gives:

$$\Omega^{-0.6} b_I = 1.3 \pm 0.20, \qquad (2.2)$$

where the "biasing parameter", b_I, is defined as the ratio of QDOT galaxy fluctuations to mass fluctuations in the same region, $b_I = (\delta N/N)/(\delta\rho/\rho)$. Note that dynamical analyses do not allow the determination of Ω independently of b_I. However, in most models of galaxy formation we expect galaxies to form preferentially in high density regions and so to be at least as, and probably more strongly clustered than the mass, corresponding to $b_I \gtrsim 1$. Thus, the most straightforward interpretation of the QDOT result is that $\Omega = 1$ and $b_I = 1.3 \pm 0.2$.

The conclusion that $\Omega = 1$ is of such fundamental importance that caution is well exercised. The samples used in the dynamical analyses are still small and may be subject to as yet ill-understood systematic effects. A further reason for caution was recently put forward by White et al. [69]. These authors argue that the large mass fraction in baryons seen in rich galaxy clusters (about 17 % for $h = 0.5$) compared with the low baryon density allowed by Big Bang nucleosynthesis (equation 1) can only be explained if either Ω is low or the nucleosynthesis limit is incorrect. For most of this article, I will continue to assume that $\Omega = 1$, but we must bear in mind that this remains a controversial assumption.

3. The linear evolution of density fluctuations

There are many excellent reviews of this topic, e.g. [20,63], so I will not dwell on it here. The primordial density and velocity fields are the initial conditions required to model the formation of large scale structure. Density fluctuations can be of two types, adiabatic if the total energy density is perturbed, or isocurvature if only the energy density of some species of matter is perturbed. The fluctuations generated by inflation are adiabatic since they arise from fluctuations in the energy density of the quantum field which contains most of the energy of the universe. The structure of the density field is usually expressed in terms of its power spectrum, $|\delta_k|$, and a prescription for the phases of the fluctuations. The phases can be either random, in which case the mass in randomly placed spheres has a gaussian distribution with dispersion $\sim k^3 |\delta_k|^2$, or correlated as in the case of strings or textures. Since quantum fluctuations in a free field have random phases, the perturbations that come out of inflation are gaussian. Once a density field has been specified, the associated velocities in linear theory are easily derived using Zel'dovich's formalism [71].

As the universe evolves, the primordial spectrum is distorted since different waves evolve in different ways. In the linear regime, i.e. while the amplitude of the fluctuations remains small, $|\delta_k| \propto (1+z)^{-1}$, where z is the redshift, and each wave evolves independently of the others according to a transfer function, $T(k,t)$, which describes the temporal evolution of each mode. In this way, the primordial spectrum is propagated forwards in time:

$$\delta_k(t) = T(k,t) \delta_k^p, \qquad (3.3)$$

where δ_k^p is the primordial power spectrum, for example, the $n = 1$ Harrison-Zeldovich spectrum from inflation.

The form of the transfer function depends on the contents of the universe and their interactions which together determine the damping mechanisms that operate. Here, I briefly summarize the case in which the dark matter is made of weakly interacting elementary particles and the initial density fluctuations are as predicted in inflation: gaussian, adiabatic, and scale-invariant. There are two possibilities to consider.

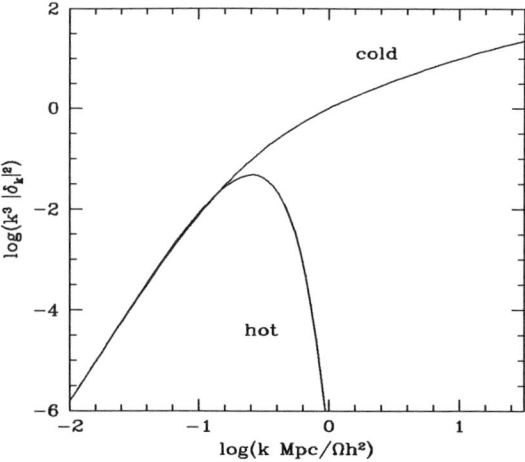

FIGURE 1. Power per decade as a function of spatial frequency for density fluctuations in universes dominated by weakly interacting elementary particles. These are linear power spectra evolved from the adiabatic, constant curvature fluctuations predicted by inflation. Hot and cold dark matter differ in the magnitude of the random velocities at early times. The curves are taken from [6,8].

3.1. Hot dark matter.

Hot dark matter consists of light particles which retain significant thermal motions for an extended period. The best example are neutrinos with a mass of ~ 30 ev. Their abundance is essentially fixed at the time when the neutrinos cease to be in thermal equilibrium with the radiation bath: $n_\nu = 4n_\gamma/11 \sim 100$ cm^{-3}. While the temperature of the Universe remains above $\sim (30/K)$ ev, the neutrinos are relativistic. Any fluctuations that come within the horizon during this epoch are wiped out by free-streaming, a relativistic version of Landau damping in collisionless fluids. There is a critical wavelength,

$$\lambda_\nu = \frac{2\pi}{k_c} = 41\left(\frac{m_\nu}{30\mathrm{ev}}\right)^{-1}\mathrm{Mpc} = \frac{13}{\Omega h^2}\mathrm{Mpc}, \qquad (3.4)$$

shortwards of which fluctuations are damped by thermal motions [7,8]. If the primordial spectrum had the inflationary shape, the postrecombination spectrum would be like that shown in Figure 1.

3.2. Cold dark matter.

In this case, thermal motions are never important. This would occur if, as in the case of photinos, the dark matter particles are very heavy ($m_x \gtrsim 1$ Gev) or if, as in the case of axions, the particles are created with negligible momentum. After the temperature of the Universe has fallen below m_x/K, and while the particles remain in equilibrium, their abundance drops relative to that of photons by the Boltzmann factor $e^{m_x/KT}$. This process continues until the universe reaches the "freeze out" temperature for the particle at which time the annihilation rate drops below the expansion rate. Since thermal motions are never important, free streaming is negligible. The main damping mechanism is the Mézáros effect whereby the growth of matter fluctuations is stiffled during the radiation era when the dominant photon-baryon fluid undergoes acoustic oscillations [38,47]. This effect produces a bend in the spectrum from the initial power-law index n to $n-4$ at a characteristic scale $\lambda_c \simeq 13(\Omega h^2)^{-1}$, corresponding to the horizon size at the epoch when the energy density in matter becomes equal to that in radation. The cold dark matter spectrum is shown in Figure 1.

3.3. *Normalisation of the power spectrum*

For a given cosmological model, there is only one free parameter in Figure 1: the fluctuation amplitude at some fiducial epoch. Until recently this had to be fixed empirically, by reference to the amplitude of the galaxy clustering pattern. The COBE results have changed this. Temperature fluctuations due to adiabatic perturbations on large angular scales are produced by the Sachs-Wolfe effect [54], a general relativistic change in wavelength as photons traverse evolving regions of varying mass density. The amplitude of these fluctuations can be readily related to the amplitude of the mass fluctuations on the largest scales shown in Figure 1, see e.g. [21,39,70]. Thus, in principle, the quadrupole anistropy measured by COBE may be used to normalise the power spectra of Figure 1. There is a potential complication: in many inflationary models, tensor modes associated with long-wavelength gravitational waves, produce a quadrupole anisotropy indistinguishable from that produced by density perturbations via the Sachs-Wolfe effect [16,55]. Neglecting tensor modes can lead to an overstimate of the amplitude of mass fluctuations.

Figure 1 summarizes the main impact of particle physics on studies of large-scale structure. Together with the assumption of random phases, it provides a set of definite predictions for the density field at the recombination era. These predictions still need to be propagated to the present in order to be confronted with observations, but this is mostly a matter of computation, not of principle.

4. The non-linear evolution of density fluctuations

From Figure 1 it is clear that we expect cosmic structures to form in a different order depending on whether the dark matter is hot or cold. In the first case, all fluctuations with mass smaller than $\sim 10^{15}(\Omega h^2)^{-2} M_\odot$ are erased by the time of recombination, so there are no seeds from which galaxies can grow. Instead, superclusters are the first objects to decouple from the overall expansion and collapse, generally into flattened objects or "pancakes". The baryons trapped in these sheets will shock and must subsequently cool and fragment in order for galaxies to form. By contrast, in a cold dark matter universe there is power on all scales, with amplitude increasing with frequency. Subgalactic units are the first to collapse and they subsequently cluster hierarchically, accreting gas in the process, to produce large scale structure. The non-linear phases of the growth of structure in the mass distribution are best studied using N-body simulations.

4.1. *The neutrino-dominated cosmogony*

When they were first discussed in the early 1980s, neutrino models seemed very attractive. A mass measurement near the value needed to close the universe had just been reported [45]; the excessive microwave background anisotropies predicted in purely baryonic models could be avoided [7]; and, for good measure, a preferred scale of $\sim 40\,h^{-1}$ Mpc, reminiscent of the largest superclusters known, was singled out by fundamental physics (equation 4). It therefore came as a great disappointment that on closer examination the idea did not hold out. The arguments against neutrino dark matter have been reviewed by [25,29,63], so I will limit this discussion to a brief summary.

The formation of large-scale structure in a neutrino universe is driven by the sharp cutoff in the fluctuation spectrum at a present-day scale of a few tens of magaparsecs (Figure 1, [31,64,66]). Gravitational evolution leads to a characteristic "beehive" pattern in the mass distribution delineated by filaments roughly the size of the cutoff length. Matter flows along these filaments and accumulates at their intersections. Eventually, the filaments break up and the distribution becomes completely dominated by a few large blobs which contain most of the mass. More recent and bigger simulations show much the same behaviour [13,14,72].

The problem with a neutrino dominated universe is that if we normalize the initial fluctuation spectrum according to the COBE quadrupole anisotropy, then wavelengths of order a few tens of megaparsecs would have only began to collapse very recently. However, collapse of such wavelengths is a prerequisite for galaxy formation in a neutrino universe where there are no primordial fluctuations on galactic scales. In the simulations described in [29], galaxies begin to form only at redshift, $z = 0.4$, much too late since we observe galaxies at redshifts $z \simeq 2$ and quasars at redshifts $z \simeq 5$. Thus, neutrino-dominated models are ruled out. (If gravity waves contribute to the COBE quadrupole anisotropy, the initial fluctuation amplitude would be even smaller than assumed here and the epoch of galaxy formation would be even more recent.)

4.2. The standard cold dark matter cosmogony

Immediately after the demise of the neutrino model, the alternative proposition that the dark matter consists of cold, weakly interacting particles began to be explored in detail [5,17,51]. In due course, the CDM model became the best studied and, in many ways, the most successful cosmogonic theory developed to date. This model has been reviewed extensively, e.g. [18,26-28], so I will restrict attention here to a few general remarks.

The post-recombination fluctuation spectrum in the CDM model bends gradually from an almost flat shape on subgalactic scales to the inflationary scale invariant form on large scales (Figure 1). Structure builds up in a quasi-hierachical fashion from small scales upwards with substantial cross-talk between different scales. The general properties of the *mass* distribution on scales ranging from those of galactic halos to those of clusters and superclusters were calculated in a series of N-body studies [17,22,26,32-34,36,67,68]. It is important to recognize that the CDM theory only specifies the distribution of mass. However, in the real world we do not observe mass, but galaxies. It is therefore necessary to add a new ingredient into the basic theory to relate galaxies and mass. A simple, mathematically tractable model is the "high peak ansatz" in which galaxies are assumed to form only near the peaks of a suitably smoothed version of the linear density field [3,17]. What many refer to as the "standard CDM cosmogony" (SCDM) is the combination of the CDM mass distribution plus the high peak ansatz. We must bear in mind, however, that while the statistical properties of the distribution of mass are known quite well, there is little, if any, physical justification for the high peak model.

The CDM power spectrum can be normalised using the COBE quadrupole anisotropy. It is customary to express the result in terms of σ_8, the present, linearly extrapolated rms mass fluctuation on the fiducial scale, 8 h^{-1} Mpc. Efstathiou et al. [21] find $0.8 < \sigma_8 < 1.33$. The central value is about a factor of 2 higher than the pre-COBE normalisation favoured by Davis et al. [17] from galaxy clustering considerations. Note, however, that any contribution to the quadrupole from gravity waves would reduce the value of σ_8 required for consistency with COBE. The standard CDM model accounts for a wide variety of observed properties of galaxies, clusters, and their distribution on small and intermediate scales, e.g. [26]. Amongst its most spectacular successes are the prediction of the correct abundance of non-linear structures, from galactic to cluster scales and the correct internal structure for galactic halos. On the scale of clusters and below, the model accords well with most quantitative statistical measures of galaxy clustering (two- and three-point correlation functions, cluster-galaxy cross-correlations, etc). As White et al. [67] first showed and Cen & Ostriker [13] later verified, the sites of galaxy formation are biased *ab initio* in the sense

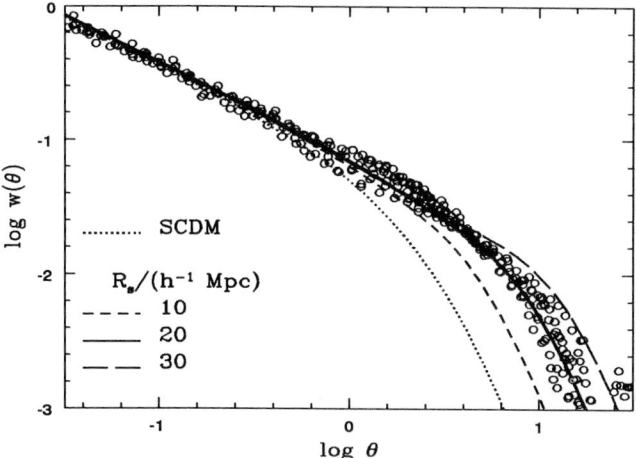

FIGURE 2. Angular autocorrelation function for galaxies in the APM survey [46] and in cold dark matter models. The circles give the APM results for galaxies in the magnitude range $17 \leq b_j \leq 20$, split into six disjoint bins in apparent magnitude, all scaled to the magnitude limit of the Lick catalogue, $b_j = 18.4$. The dotted line shows correlations in the standard CDM model. The remaining three lines show correlations in models of cooperative galaxy formation assuming that the threshold for galaxy formation is modulated by a Gaussian of characteristic scale R_s, with $R_s = 10h^{-1}$ Mpc (short dashes), $R_s = 20h^{-1}$ Mpc (solid line), and $R_s = 30h^{-1}$ Mpc (long dashes). See [9] for further details.

required to reconcile the assumption that $\Omega = 1$, with the lower estimates of Ω obtained from dynamical studies in which galaxies are assumed to trace the mass.

On scales larger than ~ 10 h^{-1} Mpc, the standard CDM model is noticeably less successful than on smaller scales. There are now a number of studies which demonstrate that the distribution of galaxies is about a factor of 2 more strongly clustered than expected in the model, see Figure 2 and [21,24,43,46,48,56,61]. Several ideas have been proposed to modify or replace the standard CDM model and I now turn to a review of some of them.

5. Beyond the standard model

The observed pattern of galaxy clustering seems to have more power on scales $\gtrsim 20$h^{-1}Mpc than expected in the SCDM model. Proposals to overcome this deficiency fall into two distinct categories. In one, new physics are invoked to change the fluctuation spectrum; in the other, the original fluctuation spectrum is retained but the relation between galaxies and mass is altered. Within the first category there are models that propose relatively minor modifications to the CDM spectrum, and models which differ fundamentally from it. Amongst the latter, the most popular are models in which non-gaussian primordial fluctuations are seeded by topological defects, such as cosmic strings or textures. These models require specifying the dark matter component as well and so one has, for example, models with *strings + neutrinos*, *textures + CDM*, etc. Generally speaking, specific predictions for these models (particularly in the case of strings) are considerably more difficult to calculate than in the gaussian case and so the viability of many of these models remains an open question. However, if the defect models are normalised to the COBE quadrupole, the resulting spectrum seems to have somewhat less power on large scales than the SCDM model.

Within the gaussian hypothesis, two modifications of the CDM power spectrum have

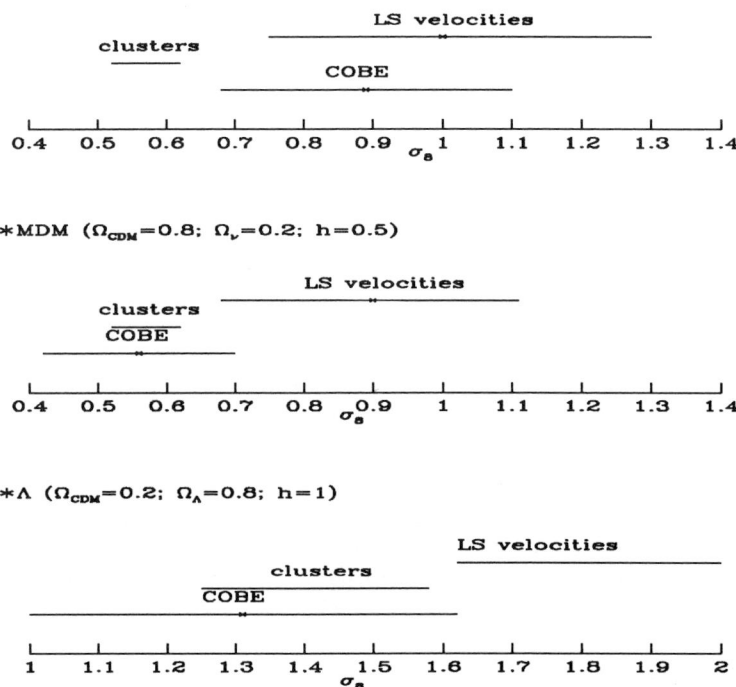

FIGURE 3. Constraints on the amplitude of mass fluctuations in three models of large scale structure. The inferred values of the amplitude at 8 h^{-1} Mpc (σ_8) are shown, as derived from the COBE quadrupole anisotropy, large-scale galaxy flows, and the abundance of rich galaxy clusters. The crosses indicate the best fit value of σ_8 and the lines its $\pm 1\sigma$ uncertainty, except in the "clusters" case where the errors are likely to be dominated by systematics. For the SCDM case, the COBE value includes a 20% downward correction to allow for small departures from the Harrison-Zeldovich spectrum and a small contribution from gravity waves. The three models tested are the standard cold dark matter model (SCDM), a mixed hot (neutrinos) and cold dark matter model (MDM) and a flat, low density cold dark matter model (Λ). The parameters assumed in each model are shown.

proved particularly popular in the past year or so. The first one exploits the fact that the CDM spectrum has a characteristic lengthscale, $\sim 13(\Omega h^2)^{-1}$, corresponding to the size of the horizon at the epoch of matter-radiation equality (see Figure 1). More power on large scales can be realised by shifting this scale towards larger wavelengths, by for example taking $\Omega < 1$. Since inflation only requires the curvature of the universe to be zero, compatibility with inflation can be achieved by adjusting the value of the cosmological constant, Λ. For values of $\Omega h \simeq 0.2$, Efstathiou et al. [21] have shown that a model of this kind gives a good match to the APM clustering data shown in Figure 2. An alternative way of enhancing large-scale power while retaining $\Omega = 1$ is to assume that about 20% of the dark matter density consists of hot particles such as neutrinos with mass $\sim 10ev$ [19,41,52,59]. The net effect of the mixture is to slightly retard the growth of fluctuations on scales on which the neutrino component is too hot to cluster. This occurs for wavelengths smaller than the neutrino Jeans length, $\sim 0.8(1+z)^{1/2}(m_\nu/10ev)h^{-1}$ Mpc, corresponding, roughly, to galaxies today. Thus, for a fixed fluctuation amplitude on 8 h^{-1} Mpc, the flatter spectral shape on small scales translates into additional power on large scales.

The low density CDM and *mixed dark matter* (MDM) models both give an acceptable match to observations of galaxy clustering on large scales. The sceptic might argue that this is not a major triumph since these models contain an additional adjustable parameter – Λ in the first case, Ω_ν in the second– introduced purposely to match these observations. It is therefore important to test these models on other scales, an activity which is now underway, e.g [41].

Modifications to the CDM power spectrum involve new physics. There is, however, a more economical way of achieving the same result – more large scale power in the distribution of galaxies – without affecting the distribution of mass. An example is the *cooperative galaxy formation* model of Bower et al. [9] in which galaxies are related to mass in a more complex fashion than envisaged by the "high peak ansatz." In the "cooperative" model, galaxies also form near high peaks of the mass density field, but the efficiency of galaxy formation is modulated by large-scale density fluctuations. Such a modulation might be produced in a number of ways, for example, by variations in the physical state of pregalactic material caused by the presence of a nearby quasar or bright protogalaxy or by the collapse of a supercluster. The most instructive aspect of the Bower *et al.* calculation is that even a weak modulation can produce significant large-scale clustering. For example, in a CDM universe a $2-3\%$ modulation on a scale exceeding $\sim 10h^{-1}$ Mpc is all that is needed to produce enough additional clustering to fit the APM angular correlation function (see Figure 2). Coherent effects on such large scales may seem *a priori* unlikely but there are a number of observed mechanisms which seem to have the required scales [2,9].

The moral is that while the clustering data, taken at face value, indicate a lack of large-scale power in the CDM model, our ignorance of how galaxies actually formed and the possibility that their distribution may be affected by non-gravitational effects, weakens any conclusions based on such data alone. Fortunately, there are ways to measure the amplitude of *mass* fluctuations directly, with only minor reference to the details of the galaxy distribution. There are two well-studied mass diagnostics, in addition to microwave background fluctuations: the large scale streaming motions of galaxies and the abundance of galaxy clusters. Peculiar velocities [44] measure the contrast relative to the background of the mass lumps that induce them, while the abundance of massive objects, like clusters, is a direct reflection of the inhomogeneity of the mass distribution. (To measure streaming motions, however, one must assume that the intrisinc properties of galaxies such as the relation between their luminosity and circular velocity are universal; this may not be true if environmental effects influence galaxy formation.) The mass scale relevant to the largest observed galaxy clusters, like Coma, is precisely the fiducial scale, $8h^{-1}$Mpc, so the derived fluctuation amplitude on this scale is independent of the shape of the power spectrum but depends strongly on the value of Ω [69]. The way in which these two diagnostics are applied is discussed in [21,34,69].

Figure 3 summarizes the constraints from the mass diagnostics on three currently popular models of large scale structure: the standard CDM model (SCDM), a mixed dark matter model (MDM) and a flat CDM model with a low value of Ω and a non-zero cosmological constant. For each model, the bars show the range in σ_8 (the rms linear fluctuation amplitude on 8 h^{-1} Mpc) allowed by each diagnostic. For a model to be consistent with the measured COBE quadrupole anistropy, large scale peculiar velocities and abundance of massive clusters, all three error bars should overlap. It is clear that the three models shown are about equally consistent with these mass diagnostics, especially if one allows for a plausibly small ($\sim 20\%$) contribution to the COBE quadrupole from gravity waves.

6. Conclusions

New ideas in fundamental physics coupled with technical innovations in theoretical and observational tools during the past decade have given unprecedented impetus to cosmological studies. The long awaited detection of primordial fluctuations in the microwave background radiation represents one of the most important advances in observational cosmology for many years. In a broad brush sense, it confirms the general picture of structure formation from the gravitational growth of small primordial perturbations. At a more detailed level, the COBE detection allows an estimate of the amplitude of primordial density fluctuations.

One of the most influential ideas to have emerged from the particle physics connection is the inflationary model of the early universe which predicts a flat geometry and provides a mechanism to generate primordial density fluctuations which have a gaussian distribution and a power spectrum consistent with the COBE detection. The theoretically appealing idea of a universe with critical density has recently received some empirical support through the comparison of large-scale motions of galaxies with the density field that induces them. This idea, together with the low abundance of baryons allowed by Big Bang nucleosynthesis, leads to the view that the dark matter in our universe consists of weakly interacting, non-baryonic elementary particles. Nevertheless, the key issues of how much dark matter there is and what is it are far from settled. For example, the large fraction of baryons observed in rich galaxy clusters is best understood if the mean cosmic density is much lower than the critical value. Similarly, the possibility that the dark matter is baryonic is by no means ruled out.

One attractive feature of the hypotheses of inflation and non-baryonic dark matter is that they lead to completely specified theories for the gravitational evolution of the *mass* distribution from early times to the present. This evolution can be calculated reliably using large N-body simulations. Understanding the distribution of *galaxies*, on the other hand, requires understanding a number of astrophysical processes such as gas flows, star formation and energy feedback, which are currently beyond reach. N-body/hydrodynamical simulations are beginning to address some of these issues but it will be a long time before a detailed understanding of galaxy formation is achieved. In the meantime, cosmogonic theories can only be tested against observations of galaxies by making a number of simplifying assumptions which considerably weaken the predictive power of the theories. In spite of this, a neutrino-dominated cosmogony can be conclusively ruled – simulations show that this model disagrees with observations for any plausible galaxy formation mechanism. On the other hand, the cold dark matter cosmogony and its variants – a low density model with a flat geometry or a mixed dark matter model – seem to provide an adequate framework for understanding the formation of cosmic structure.

Numerical simulations provide a powerful tool for exploring the consequences of different hypotheses for the initial conditions, geometry, and material content of the Universe. By bringing theoretical predictions onto the observational plane, they enable models to be tested in detail against the data. The 1980s laid the foundations for this new methology which has now become an integral part of cosmology.

REFERENCES

[1] Albrecht, A. & Steinhardt, P. J. 1982, PhysRevLett, 48,
[2] Babul, A. & White, S. D. M. 1991, MNRAS, 253, 31p.
[3] Bardeen, J. M., Bond, J. R., Kaiser, N. & Szalay, A. S. 1986, ApJ, 304, 15.
[4] Bertschinger, E., Dekel, A., Faber, S.M., Dressler, A. & Burstein, D. 1990, ApJ, 364, 370.
[5] Blumenthal, G. R., Faber, S. M., Primack, J. R. & Rees, M. J. 1984, Nature, 311, 517.

[6] Bond, J. R. & Efstathiou, G. 1984, ApJ, 285, L45.
[7] Bond, J. R., Efstathiou, G. & Silk, J. 1980, PhysRevLett, 45, 1980.
[8] Bond, J. R. & Szalay, A. S. 1983, ApJ, 274, 443.
[9] Bower, R., Cole, P., Frenk, C. S. & White, S. D. M. 1993, ApJ, 405, 403.
[10] Brandenberger, R. 1991, *The Birth and Early Evolution of our Universe*, Nobel Symposium No 79, eds J. S. Nilsson, B. Gustafsson, B.-S. Skagerstam, World Scientific, PhysScripta, T36, 114.
[11] Carr, B. J. 1990, CommAp, 14, 257.
[12] Cen, R. & Ostriker, J. P. 1992, ApJ, 399, 331.
[13] Cen, R. & Ostriker, J. P. 1993, ApJ, 414, 407.
[14] Centrella, J. M., Gallagher, J. S. III, Melott, A. L. & Bushouse, H. A. 1988, ApJ, 333, 24.
[15] Cheng, E. S. *et al.* 1993, ApJ, in press.
[16] Crittenden, R., Bond, J. R., Davis, R. L., Efstathiou, G. & Steinhart, P.J. 1993, *Oxford preprint*.
[17] Davis, M., Efstathiou, G., Frenk, C. S. & White, S. D. M. 1985, ApJ, 292, 371.
[18] Davis, M., Efstathiou, G., Frenk, C. S. & White, S. D. M. 1992a, Nature, 365, 489.
[19] Davis, M., Summers, F.J. & Schlegel, M. 1992b, Nature, 359, 393.
[20] Efstathiou, G. 1991, *Physics of the Early Universe*, eds. J. A. Peacock & A. F. Heavens, Kluwer.
[21] Efstathiou, G., Bond, R.J. & White, S. D. M. 1992, MNRAS, 258, 1p.
[22] Efstathiou, G., Davis, M. Frenk, C. S. & White, S. D. M. 1985, ApJS, 57, 241.
[23] Efstathiou, G, Kaiser, N., Saunders, W., Lawrence, A., Rowan-Robinson, M., Ellis, R. S. & Frenk, C. S. 1990, MNRAS, 247, 10p.
[24] Fisher, K. B., Davis, M., Strauss, M. A., Yahil, A. & Huchra, J. P. 1993. ApJ, 402, 42.
[25] Frenk, C. S. 1986, PhilTransRSocLond, A 330, 517.
[26] Frenk, C. S. 1991, *The Birth and Early Evolution of our Universe*, Nobel Symposium No 79, eds J. S. Nilsson, B. Gustafsson, B.-S. Skagerstam, World Scientific, PhysicaScripta, T36, 70.
[27] Frenk, C. S. 1992a, *Current Topics in Astrofundamental Physics: First International School of Astrophysics, D. Challonge*, ed N. Sanchez & A. Zichichi, World Scientific, 345-385.
[28] Frenk, C. S. 1992b, *Texas/ESO-CERN Symposium on Relativistic Astrophysics*, eds. J. Barrow *et al.* , AnnNYAcadSci, 647, 649-658.
[29] Frenk, C. S. 1993, PhilTransRSocLond, in press.
[30] renk, C. S., Kaiser, N. & Lucey, 1994, in preparation.
[31] Frenk, C. S., White, S. D. M. & Davis, M. 1983, ApJ, 271, 417.
[32] Frenk, C. S., White, S. D. M., Efstathiou, G. & Davis, M. 1985, Nature, 317, 595.
[33] Frenk, C. S., White, S. D. M., Efstathiou, G. & Davis, M. 1988, ApJ, 327, 507.
[34] Frenk, C. S., White, S. D. M., Efstathiou, G. & Davis, M. 1990, ApJ, 351, 10.
[35] Gaier, T., Schuster, J., Gunderson, J., Koch, T., Seiffert, M., Meinhold, P. & Lubin, P. 1992, ApJ, 398, L1.
[36] Gelb, J. 1992, Ph.D. Thesis, MIT, USA.
[37] Guth, A. 1981, PhysRevD, 23, 347.
[38] Guyot, & Zel'dovich, Ya. B. 1970,
[39] Holtzman, J. A. 1989, ApJS, 71, 1.
[40] Kaiser, N., Efstathiou, G., Ellis, R. S., Frenk, C. S., Lawrence, A., Rowan-Robinson, M. & Saunders, W. 1991, MNRAS, 252, 1.
[41] Klypin, A., Holtzman, J., Primack, J. & Regos, A. 1993, preprint.
[42] Linde, A. D. 1982, PhysLett, 108B, 389.
[43] Loveday, J. Efstathiou, G., Peterson, B. A. & Maddox, S. J. 1992, ApJ, 400, L43.
[44] Lynden-Bell, D., Faber, S. M., Burstein, D., Davies, R. L., Dressler, A., Terlevich, R. J. & Wegner, G., 1988, ApJ, 326,19.
[45] Lyubimov, V. A., Novikov, E. G., Nozik, V. Z., Tretyakov, E. F. & Kozik, V. F. 1980, PhysLett, 94B, 266.
[46] Maddox, S. J., Efstathiou, G, Sutherland, W. & Loveday, J. 1990, MNRAS, 242, 43p.
[47] Mészáros, P. 1974, AAp, 37, 225.
[48] Moore, B., Frenk, C. S., Weinberg, D. H., Saunders, W., Lawrence, A., Rowan-Robinson, M., Kaiser, N., Efstathiou, G. & Ellis, R. 1992, MNRAS, 256, 477.
[49] Nusser, A. & Dekel, A. 1993, ApJ, 405, 437.
[50] Peebles, P. J. E. 1980, *The large scale structure of the Universe.* (Princeton: Princeton Un.

Press).
[51] Peebles, P. J. E. 1984, ApJ, 284, 439.
[52] Pogosyan, D. Yu. & Starobinsky, A. A. 1993, Cambridge preprint.
[53] Rowan-Robinson, M., Ellis, R. S., Efstathiou, G., Frenk, C. S., Kaiser, N., Lawrence, A., Saunders, W., Parrym I., Allington-Smith, J. & Xiaoyang, X. 1990, MNRAS, 247, 1.
[54] Sachs, R. K. & Wolfe, A. M. 1967, ApJ, 147, 73.
[55] Salopek, D. 1992, PhysRevLett, 69, 3602.
[56] Saunders, W., Frenk, C. S., Rowan-Robinson, M., Efstathiou, G.,Lawrence, A., Kaiser, N., Ellis, R. S. E, Crawford, J., Xiao-Yang Xia & Parry, I. 1991, Nature, 349, 32.
[57] Smoot et al. 1992, ApJ, 396, L1.
[58] Strauss, M. A., Yahil, A., Davis, M., Huchra, J. P. & Fisher, K. 1992, ApJ, 397, 419.
[59] Taylor, A. N. & Rowan-Robinson, M. 1992, Nature, 359, 396.
[60] Turok, N. 1991, *The Birth and Early Evolution of our Universe*, Nobel Symposium No 79, eds J. S. Nilsson, B. Gustafsson, B.-S. Skagerstam, World Scientific, PhysicaScripta, T36, 135.
[61] Vogeley, M. S., Park, C., Geller, M. J. & Huchra, J. P. 1992, ApJ, 391, L5.
[62] Walker, T. P. et al. 1991, ApJ, 376, 51.
[63] White, S. D. M. 1985, *Inner Space, Outer Space*, eds E. W. Kolb, M. S. Turner, D. Lindley, K. A. Olive & D. Seckel (Chicago: University of Chicago Press), 228.
[64] White, S. D. M., Davis, M. & Frenk, C. S. 1984, MNRAS, 209, 27p.
[65] White, S. D. M., Efstathiou, G. & Frenk, C. S. 1993a, MNRAS, 262, 1023.
[66] White, S. D. M., Frenk, C. S. & Davis, M. 1983, ApJ, 274, L1.
[67] White, S. D. M., Frenk, C. S., Davis, M. & Efstathiou, G. 1987a, Nature, 330, 451.
[68] White, S. D. M., Frenk, C. S., Davis, M. & Efstathiou, G. 1987b, ApJ, 313, 505.
[69] White, S. D. M., Navarro, J.F., Evrard, A. E. & Frenk, C. S. 1993b, Nature, in press.
[70] Wright et al. 1992, ApJ, 396, L13.
[71] Zel'dovich, Ya. B. 1970, AAp, 5, 84.
[72] Zeng, N. & White, S. D. M. 1991, ApJ, 374, 1.

The Evolution of Superhorizon-sized Voids in the Early Universe

By Sharon L. Vadas

Center for Particle Astrophysics, University of California, Berkeley, USA.

We study the evolution of a superhorizon-sized void embedded in a radiation-dominated Friedmann-Robertson-Walker (FRW) universe. Voids similar to this would have been formed after first-order inflation. We numerically solve the spherically symmetric general relativistic equations in comoving, synchronous coordinates. When the fluid inside the void is relativistic, we find that radiation diffuses into the void at approximately the speed of light as a strong shock—the void collapses. If the void is empty enough, the collapse occurs in less than the Hubble time outside the void. In this case, for the simple examples we examine here, the void's collapse can be described very well by the special relativistic fluid equations only.

1. Introduction

The Big Bang model is very successful as a late-time cosmology model. However, it has some problems, two of which are the horizon and flatness problems. Inflation provides a natural solution to these problems through rapid expansion and creation of entropy in the early universe. There are two basic types of inflation. At the end of inflation, the inflaton field either rolls or tunnels to the true vacuum. Voids are created from the later type of inflation, so-called "first-order inflation". These voids, dubbed "superhorizon-sized voids", are embedded in the radiation-dominated FRW universe and are larger than the Hubble radius, which is the scale on which microphysics acts.

It has been thought that these superhorizon-sized voids could eventually thermalize on time scales of order the 1st-crossing time, which is the time taken for a photon initially at the inner wall edge to reach the origin [2,5]. This time was estimated to be $\Delta t = t - t_{\text{i}} \simeq H_{\text{out}}^{-1}(t_{\text{i}})(c^{-1}R_{\text{wall}}(t_{\text{i}})/H_{\text{out}}^{-1}(t_{\text{i}}))^2$, where $R_{\text{wall}}(t_{\text{i}})$ is the initial void radius, $cH_{\text{out}}^{-1}(t_{\text{i}})$ is the Hubble radius outside the void and t_{i} is the cosmic time after reheating. However, the 1st-crossing time is actually much smaller than this [6]. This is because the potential difference between the void and the relativistic background spacetime is large and negative. This results in time dilation, and has the following effect: a photon inside the void will travel much farther than a photon outside the void in the same *coordinate time*. One can thus think of a void as creating a warp in spacetime which is opposite from that created by a massive object, like a black hole. We note, however, that when the background spacetime is non-relativistic (i.e. when the void is embedded in a matter-dominated FRW universe), this time-dilation effect is very small.

The 1st-crossing time can be calculated for a relativistic or non-relativistic void embedded in a radiation-dominated FRW universe. (A void is non-relativistic (relativistic) if the fluid inside the void is non-relativistic (relativistic)). It is [6]

$$\Delta t_c = c^{-1}R_{\text{wall}}(t_{\text{i}})\overline{\Phi}_{\text{in}}^{-1}\left[1 + .5(c^{-1}R_{\text{wall}}(t_{\text{i}})/H_{\text{out}}^{-1}(t_{\text{i}}))\overline{\Phi}_{\text{in}}^{-1}\right] \qquad (1.1)$$

for $c^{-1}R_{\text{wall}}(t_i)/H^{-1}_{\text{out}}(t_i) < \sqrt{\rho_{\text{out}}(t_i)/\rho_{\text{in}}(t_i)}$ and $\overline{\Phi}_{\text{in}} \simeq constant$, where the subscripts "in" and "out" refer to quantities inside and outside the void, respectively. In addition, $\overline{\Phi}_{\text{in}} \simeq T_{\text{out}}(t_i)/(c^2\mu)$ for a non-relativistic void, where T is the local temperature and μ is the particle mass, and $\overline{\Phi}_{\text{in}} \simeq (\rho_{\text{out}}(t_i)/\rho_{\text{in}}(t_i))^p$ for a relativistic void, where $1/2 < p < .366$ in the slab-symmetric limit. If $\Delta t_c/H^{-1}_{\text{out}}(t_i) \ll 1$, then $\Delta t_c/H^{-1}_{\text{out}}(t_i) \simeq \overline{\Phi}_{\text{in}}^{-1} c^{-1}R_{\text{wall}}(t_i)/H^{-1}_{\text{out}}(t_i)$, and the time dilation effect is due entirely to the relativistic pressure gradient in the void wall. However, if $\Delta t_c/H^{-1}_{\text{out}}(t_i) \gg 1$, the universe expands outside the void during this time, decreasing the magnitude of the relative potential of the void. This lengthens the collapse time by essentially squaring the previous result: $\Delta t_c/H^{-1}_{\text{out}}(t_i) = .5(c^{-1}\overline{\Phi}_{\text{in}}^{-1} R_{\text{wall}}(t_i)/H^{-1}_{\text{out}}(t_i))^2$.

Thus the idea that a superhorizon-sized void behaves like a small perturbation in spacetime (and thus cannot collapse in less than a Hubble time) must be abandoned. In addition, although estimates of the 1st-crossing can be made from Eqn(1.1), detailed knowledge of the void's evolution can be obtained only through the numerical evolution of the entire spacetime. In this paper, we show the numerical evolution of voids in a radiation-dominated FRW universe. We show that for the simple examples we choose here, voids do collapse in the form of strong, inward shocks, and that if the collapse is quick enough (i.e. less than a Hubble time), the collapse can be described by the special relativistic fluid equations only.

2. The Metric and Equations of Motion

The metric we use to study this problem is the comoving, synchronous metric

$$ds^2 = -c^2\Phi^2(t,r)dt^2 + \Lambda^2(t,r)dr^2 + R^2(t,r)d\Omega^2. \tag{2.2}$$

At time t, $2\pi R(t,r)$ is the spacelike circumference of a sphere centered on the origin which contains all particles with comoving coordinate r. We also take the stress-energy tensor to be that for a perfect fluid with artificial viscosity: $T^{\alpha\beta} = \rho u^\alpha u^\beta + (p+Q)(u^\alpha u^\beta + g^{\alpha\beta})$, where $u^\alpha = (-c\Phi^{-1},0,0,0)$ is the fluid 4-velocity, $\rho = n(1+\epsilon/c^2)$ is the energy density, $p = (\gamma-1)n\epsilon$ is the pressure, n is the mass density, ϵ is the specific energy and γ is a constant. We assume that the total number of particles per comoving volume is constant, $\nabla_\mu(nu^\mu) = 0$, so that $4\pi n R^2 R'/\Gamma = f(r)$, where $' \equiv \partial/\partial r$, $\Gamma = R'/\Lambda$, and $f(r)$ is chosen to be r^2 in order that shocks and explosions are numerically stable at the origin. The fully general relativistic equations can then be written as [3,4]

$$\dot{R} = \Phi U \tag{2.3}$$

$$\dot{U} = -\Phi\left(\frac{G_N M}{R^2} + \frac{4\pi G_N(p+Q)R}{c^2}\right) - \frac{\Gamma\Phi R^2 4\pi(p+Q)'}{wr^2} \tag{2.4}$$

$$\dot{M} = -4\pi(p+Q)R^2\Phi U/c^2 \tag{2.5}$$

$$\dot{n} = -\frac{n\Phi(R^2 U)'}{R^2 R'} \tag{2.6}$$

$$\dot{\epsilon} = -\frac{\Phi 4\pi(p+Q)(R^2 U)'}{\Gamma\, r^2} \tag{2.7}$$

$$\Phi' = -\Phi\frac{(p+Q)'}{nwc^2}, \tag{2.8}$$

where $\dot{} \equiv \partial/\partial t$, $M' \equiv 4\pi c^{-2}\rho R^2 R'$, $\Gamma^2 = 1 + (U/c)^2 - 2G_N M/(Rc^2)$, and $w \equiv 1 + [\epsilon + (p+Q)/n]/c^2$. The relativistic limit is obtained when $\epsilon/c^2 = (\gamma-1)^{-1}T/(c^2\mu) \gg 1$. In addition, the artificial viscosity used here allows for relativistic shocks to penetrate into relativistic fluids: $Q = k^2 n(1+\epsilon/(\Gamma c^2))(U')^2(\Delta r)^2/\Gamma$ for $U' < 0$ (Figure 1) or

$Q = k^2 n w \Gamma^{-2}(U')^2(\Delta r)^2/\Gamma$ for $U' < 0$ (Figures 2,3), and $Q = 0$ otherwise. (These viscosities are only significantly different for discontinuities near the origin).

These equations are solved using the MacCormack predictor-corrector method [1]. Suppose we know all quantities on the ith time slice for all spatial points j. Using Eqn(2.6) as an example, we first predict the new quantities (with forward differencing) for all j: $n_p{}_j^{i+1} = n_j^i - \Delta t \, n_j^i \Phi_j^i \, (R_{j+1}^i{}^2 \, U_{j+1}^i - R_j^i{}^2 \, U_j^i)/[R_j^i{}^2(R_{j+1}^i - R_j^i)]$. After using similar methods to obtain $U_p{}_j^{i+1}$, $R_p{}_j^{i+1}$, $M_p{}_j^{i+1}$ and $\epsilon_p{}_j^{i+1}$, we integrate Eqn(2.8) inwards from the outer boundary using the 4th-order Runge-Kutta method with linear interpolations to determine $\Phi_p{}_j^{i+1}$. We then integrate again (with backward differencing), $n_j^{i+1} = .5(n_j^i + n_p{}_j^{i+1} - \Delta t \, n_p{}_j^i \Phi_p{}_j^i \, (R_p{}_j^i{}^2 \, U_p{}_j^i - R_p{}_{j-1}^i{}^2 \, U_p{}_{j-1}^i)[R_p{}_j^i{}^2(R_p{}_j^i - R_p{}_{j-1}^i)])$, and obtain the other quantities similarly.

We require our time steps to be small enough to satisfy the Courant condition. In addition, we require n, ϵ and M to change as slowly as desired when $G_N \neq 0$. Thus,

$$(\Delta t)^{i+1} = \min\left(\Delta t_{\max}, \, C \, \frac{R_{j+1}^i - R_j^i}{\Gamma_j^i \Phi_j^i (c_S)_j^i}, \, \left[\overline{f}\, \frac{n_j^i}{\dot{n}_j^i}, \, \overline{f}\, \frac{\epsilon_j^i}{\dot{\epsilon}_j^i}, \, \overline{f}\, \frac{M_j^i}{\dot{M}_j^i}\right]_{\text{if } G_N \neq 0}\right), \qquad (2.9)$$

where $\Delta t_{\max}$ is a constant, $(c_S)_j^i = \sqrt{\gamma(p_j^i + Q_j^i)/(n_j^i w_j^i)}$ is the speed of sound, and $\overline{f} < 1$ is a constant.

The initial conditions we employ are as follows. Since we require the inside of a void to be Minkowski spacetime in the limit that the energy density (or "mass-energy", M) goes to zero, we set $U(t_i) = \sqrt{2G_N M(t_i)/R(t_i)}$ or $\Gamma(t_i) = 1$ inside the void. Outside the void, we set $\Phi_{\text{out}}(t) = 1$ and $\Gamma(t_i, R) = 1$, the later of which means that the outside spatial curvature is initially zero: $k = 0$. When $\Gamma = 1$ in the void wall, the gravitational attraction inward is balanced by the outward "kinetic energy" for each Lagrange shell.

The void wall can be either compensated (in the "energy" missing from the void) or uncompensated. If the void is initially uncompensated, then the energy density is given by

$$\rho(t_i, R) = .5c^{-2}4\pi\rho_{\text{out}}(t_i)[(1 + \tanh x) + \alpha(1 - \tanh x)],$$

where $x \equiv (R(t_i) - R_{\text{wall}}(t_i))/\Delta R_{\text{wall}}(t_i)$, $\Delta R_{\text{wall}}(t_i)$ is the wall thickness and $\alpha \simeq \rho_{\text{in}}(t_i)/\rho_{\text{out}}(t_i) \leq 1$ is a constant. If the void is initially compensated, the energy density is $\rho = c^2 M'/(4\pi R^2 R')$, where $M(t_i, R) = .5c^{-2}4\pi\rho_{\text{out}}(t_i)\left[(1 + \tanh x) + \alpha(1 - \tanh x)\right]R^3(t_i)/3$. Note that in this case, the "mass-energy" M reaches its FRW $k = 0$ value outside the void. In addition, the fluid is initially taken to be an isentrope, so that if $\epsilon/c^2 \gg 1$, $\epsilon(t_i, R) = \epsilon_{\text{out}}(t_i)(\rho(t_i, R)/\rho_{\text{out}}(t_i))^{(\gamma-1)/\gamma}$.

3. Numerical Results

In all figures which follow, we set $C = .3$, $c = 1$, $\gamma = 4/3$, $t_i = 1$, and $4\pi\rho_{\text{out}}(t_i) = 3/8$. Thus the initial Hubble time outside the void is $H_{\text{out}}^{-1}(t_i) = 2t_i = 2$.

In Figure 1 we show the collapse of a special relativistic ($G_N = 0$), uncompensated void with $R_{\text{wall}}(t_i) = 20$, $\Delta R_{\text{wall}}(t_i) = .02$, $\Delta R(t_i) = .01$, $\alpha = 10^{-2}$, $\epsilon_{\text{out}}(t_i) = 10^5$, $\Gamma(t_i, R) = 1$ and $k^2 = 3$. Since $\epsilon_{\text{in}}(t_i) = 3.16 \times 10^4$, the fluid is everywhere ultra-relativistic. The triangles (connected by lines) show the pressure and specific energy as a function of R for the Lagrange points at t_i and $t = 2.5$. The void collapses via a shock moving toward the origin at the speed of light. [7] (A radially propagating photon initially at the inner wall edge has position $R(t, r) \simeq R_{\text{wall}}(t_i) - c\overline{\Phi}_{\text{in}}(t - t_i)$. Thus, since $\overline{\Phi}_{\text{in}} \sim \sqrt{10}$, at $t = 2.5$, this photon would be located at $R \simeq 15.3$. This agrees with the location of the shock

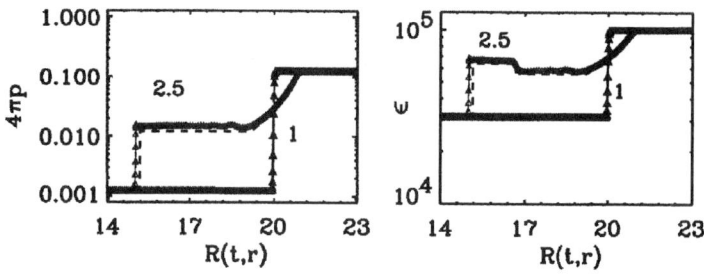

FIGURE 1. Collapse of a special relativistic void (triangles). The dashed line is the slab-symmetric shock tube solution.

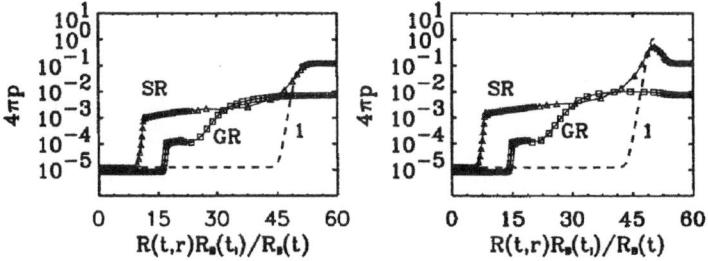

FIGURE 2. Collapse of uncompensated (left) and compensated (right) relativistic voids evolved with (squares) and without (triangles) gravity. Here, $\Delta t_c/H_{\text{out}}^{-1}(t_i) > 1$.

in Figure 1.) It is followed by a (somewhat smeared) contact discontinuity (located at $R \simeq 16.8$) and a rarefaction wave ($R \simeq 19.4 - 20.8$). Since the evolving wall region and $\Delta R_{\text{wall}}(t_i)/R_{\text{wall}}(t_i)$ are relatively small, the solution is approximately the slab-symmetric shock tube solution. (Due to time dilation, the amount of time taken to form the shock is very important. The solution will approach the slab shock tube only if the shock is formed quickly enough. For $\alpha = 10^{-2}$, we find that $\Delta R_{\text{wall}}(t_i)/R_{\text{wall}}(t_i) \leq 10^{-3}$.) The shock tube solution is plotted as dashed lines. We see that the agreement is good. Note that the pressure behind the shock from the code is slightly higher than that from the shock tube. This is due to spherical aberrations which are smaller at earlier times [7].

In Figure 2, we show the collapse of special ($G_N = 0$) and general relativistic voids which are uncompensated (left) and compensated (right). The initial configurations are dashed lines, and the special and general relativistic points are shown at $t = 4$ as boxes ("GR") and triangles ("SR"), respectively. Plotted is $4\pi p$ versus $R\, R_B(t_i)/R_B(t)$, where R_B is the radius of the outer boundary point. In this figure, $R_{\text{wall}}(t_i) = 50$, $\Delta R_{\text{wall}}(t_i) = 1$, $\Delta R(t_i) = .5$, $\alpha = 10^{-4}$, $\epsilon_{\text{out}}(t_i) = 10^6$, $\Gamma(t_i, R) = 1$ and $k^2 = 2$. Note that the GR voids are superhorizon-sized, since $c^{-1}R_{\text{wall}}(t_i)/H_{\text{out}}^{-1}(t_i) = 25$. In addition, the collapse takes longer than an outside Hubble time to occur, since $\overline{\Phi}_{\text{in}}^{-1} c^{-1}R_{\text{wall}}(t_i)/H_{\text{out}}^{-1}(t_i) \simeq 2.5 > 1$ (i.e. $\Delta t_c > H_{\text{out}}^{-1}(t_i)$). For the GR cases, the pressure outside the void redshifts during this time. This causes $\Phi_{\text{in}}(t)$ to decrease, lessening the time dilation effect. This allows SR shocks to reach the origin much more quickly than the GR shocks. These effects make the SR and GR spacetimes very different at this time.

In Figure 3, we show the evolution of voids identical to those from Figure 2, except with $\alpha = 10^{-10}$. Thus, these voids are much "emptier" initially. If the voids collapse, we expect this to occur in less than a Hubble time, since the 1st-crossing time is less than a Hubble

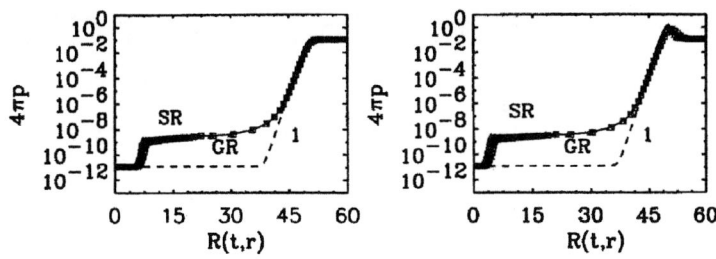

FIGURE 3. Same as for Figure 2, but for initially emptier voids. Here, $\Delta t_c / H_{\text{out}}^{-1}(t_i) < 1$.

time: $\overline{\Phi}_{\text{in}}^{-1} c^{-1} R_{\text{wall}}(t_i) / H_{\text{out}}^{-1}(t_i) \simeq .08 \ll 1$. Shown is the pressure at $t = 1.08$, when the void has nearly collapsed. We see the important result that the SR and GR cases look very similar at this time. In addition, although it is not shown here, ρ, ϵ, Γ and n are very similar at this time also. Thus, the dynamics of the GR void can be approximated at this time by its evolution without gravity for the same initial configuration. This is counter-intuitive, since one would intuitively expect the dynamics of a collapsing *superhorizon-sized* void to depend heavily on gravitational effects.

4. Conclusions

In this paper, we examine the evolution of superhorizon-sized voids in a radiation-dominated FRW universe. For the examples shown here, we find that these voids collapse. In addition, if the collapse occurs in less than an outside Hubble time, the evolution of a void can be approximated by that of an identical initial configuration evolved under zero gravity. This is partly because the pressure gradient force in the void wall acts much more quickly than gravity does, and partly because the time dilation effect "speeds-up" any fluid motion in the void. It is possible however, that if a more realistic initial velocity profile is used to model inflationary voids (i.e. $\Gamma(t_i) \gg 1$ in the void wall), then there will be a significant difference in the evolution with and without gravity.

This work was supported in part by NASA under Grant NAGW-2381 at Fermilab and by the DOE at Chicago and Fermilab.

REFERENCES

[1] Bernstein, D., Hobill, D. & Smarr, L. 1986, *Dynamical Spacetimes and Numerical Relativity*, ed. J. Centrella, Cambridge, New York.
[2] Liddle, A. & Wands, D. 1991, MNRAS, 253, 637.
[3] May, M. & White, R. 1967, MethComputatPhys, 7, 219.
[4] Misner, C. & Sharp, D. 1964, PhysRev, 136, B571.
[5] Turner, M., Weinberg, E. & Widrow, L. 1992, PhysRevD, 46, 2384.
[6] Vadas, S. L. 1993, PhysRevD, in press.
[7] Vadas, S. L. 1993, in preparation.

Perturbation Solution to the Linearized Einstein Constraint Equations in Spherical Symmetry as a Test for Numerical General–Relativistic Codes

By H. Harleston L.

Instituto de Ciencias Nucleares, UNAM; Circuito Exterior, C. U.; Apartado Postal 70–543; 04510 México D. F.; México

A linearized form for the Constraint and Lapse equations in the ADM (or "3 + 1") formalism in spherical symmetry is obtained. This set of (linearized) equations lends itself to an analytical solution when the initial matter and momentum density sources are given. These solutions may then be used as a test for the Constraint, Lapse, and Shift solvers in spherically–symmetric numerical codes for the Einstein Equations. The results from these tests, obtained from a code that has been developed by the author are presented.

1. Introduction

Based on previous work by Holcomb [5] in which a perturbation solution to the Constraint equations is found, we obtain a complete solution to the initial value problem in General Relativity in spherical symmetry including a new analytical solution to the Lapse equation. This solution may be used as a good test for the constraint, lapse, and shift solvers in general relativistic codes.

In this work, a description of this solution is presented together with a comparison between the analytical and numerical results obtained through the use of a general–relativistic spherically–symmetric code that has been developed by the author. In Sections 2 to 4 a brief review of the Einstein constraint, lapse and shift equations in the ADM formalism is presented, in Section 5 the analytical perturbation solution to these equations is described, in Section 6 two example solutions are presented, and, finally, Section 7 includes some comments on the numerical solution. A graphical representation of the results is included.

2. Line Element

The most general line element consistent with spherical symmetry and using an isotropic gauge [1,6,7] may be written as

$$ds^2 = -(\alpha^2 - A^2 \beta^2) c^2 dt^2 + 2 A^2 \beta \, c \, dt \, dr + d\ell^2 \tag{2.1}$$

with

$$d\ell^2 = A^2 \left[dr^2 + r^2 \left(d\theta^2 + \sin^2\theta \, d\varphi^2 \right) \right], \tag{2.2}$$

where, in the standard "3 + 1" notation, α is the Lapse function and β is the only non-vanishing component of the Shift vector. We assume that all functions depend only on the time, t, and the radial coordinate r.

3. Sources

In this work, the sources of gravitational field are modelled as a perfect fluid so that the energy–momentum tensor is given by

$$T^{ab} = \sigma U^a U^b + P g^{ab}, \tag{3.1}$$

where

$$\sigma = \rho + \rho \frac{\varepsilon}{c^2} + \frac{P}{c^2} \tag{3.2}$$

is the relativistic enthalpy. Here, ρ is the rest–mass density, ε is the specific internal energy density, P is the pressure, and $U^a = cU\,\alpha^{-1}(1, U^1/U^0, 0, 0)$ is the bulk four-velocity where U is the boost factor given by $cU \equiv \alpha U^0$.

4. Constraint, Lapse, and Shift Equations

Using the ADM formalism with the line element (2.1) and the fluid energy-momentum tensor given by (3.1), the *Hamiltonian Constraint* equation is

$$\frac{1}{r^2} \frac{\partial}{\partial r}\left[r^2 \frac{\partial}{\partial r}\left(\sqrt{A}\right)\right] = \tfrac{1}{4}\left(\sqrt{A}\right)^5 \left[\tfrac{1}{3} K^2 - \tfrac{3}{4}(K^*)^2 - \kappa \rho_H\right], \tag{4.1}$$

where $\kappa = 8\pi G/c^4$, $K \equiv g^{ab} K_{ab}$ is the trace of the extrinsic curvature tensor and we define an extrinsic curvature variable, K^*, given by $K^* \equiv K^1{}_1 - \tfrac{1}{3} K$. The *Hamiltonian density* is given by $\rho_H \equiv n^a n^b T_{ab}$, where $n^a = \alpha^{-1}(1, -\beta, 0, 0)$ is the normal unit vector to the 3-hypersurfaces in the ADM formalism. The *Momentum Constraint* equation may be written as

$$K^*(r) = \frac{\kappa}{(Ar)^3} \int_0^r (r')^3\, \widetilde{S}_r\, dr', \tag{4.2}$$

where $\widetilde{S}_r \equiv A^3 S_r$ and $S_r \equiv -\gamma_r{}^a n^b T_{ab}$ is the only non-vanishing component of the Momentum Flux vector, S_a. The 3-metric tensor is $\gamma^{ab} = g^{ab} + n^a n^b$.

Combining the Hamiltonian Constraint equation with the original Lapse equation obtained via the ADM formalism [2,5], we can write a "modified" *Lapse* equation which is more amenable to numerical treatment. Thus,

$$\frac{1}{r^2} \frac{\partial}{\partial r}\left[r^2 \frac{\partial}{\partial r}\left(\alpha\sqrt{A}\right)\right] = \tfrac{1}{4}\alpha\left(\sqrt{A}\right)^5 \left[\tfrac{5}{3} K^2 + \tfrac{21}{4}(K^*)^2 + \kappa(2\rho_\alpha - \rho_H)\right] - \left(\sqrt{A}\right)^5 \frac{1}{c}\frac{\partial K}{\partial t} \tag{4.3}$$

where $\rho_\alpha \equiv \gamma_a{}^c \gamma_b{}^d T_{cd} + \rho_H$ is known as the *Lapse Density*. Finally, the *Shift* equation may be integrated to yield

$$\beta(r) = -\tfrac{3}{2} r \int_r^\infty \alpha\, K^*\, (r')^{-1}\, dr'. \tag{4.4}$$

We will now present a way to linearize equations (4.1) and (4.3) in order to obtain an analytical solution for the metric variable A and the Lapse function α. Once these are known, we can integrate equation (4.2) to obtain the extrinsic curvature variable K^* with which we can then solve equation (4.4) for the radial component of the shift vector, β.

5. Analytical Solution

Based on previous work performed by Holcomb [5], we linearize equations (4.1) and (4.3) by assuming that the variables involved in these equations can be written as

$$\sqrt{A} = \Psi_0 + \Psi_1, \quad \alpha = \alpha_0 + \alpha_1, \quad \beta = \beta_0 + \beta_1, \quad U = U_0 + U_1,$$

$$K = K_0 + K_1, \quad K^* = K_0^* + K_1^*,$$

and

$$\rho_H = \rho_{H0} + \rho_{H1}, \quad \rho_\alpha = \rho_{\alpha 0} + \rho_{\alpha 1},$$

with $\Psi_0 = 1$, $\alpha_0 = 1$, $\beta_0 = 0$, $K_1 = 0$, $K_0^* = 0$, and $U_0 = 1$. We also take $\varepsilon = P = 0$ and define a density variable $D \equiv U\rho = D_0 + D_1$ so that $d \equiv A^3 D = d_0 + d_1$. With these choices we obtain, to first order:

$$d_0 = D_0, \quad d_1 = D_1 + 6\Psi_1 D_0, \quad D = d_0(1 - 6\Psi_1) + d_1,$$

$$U_1 = (\tilde{S}_r)^2 / 2(c^2 D_0)^2 \equiv \tfrac{1}{2}\mu^2,$$

$$\rho_{H0} = \rho_{\alpha 0} = c^2 D_0,$$

$$\rho_{H1} = c^2 D_0 \left[(d_1/d_0) - 6\Psi_1 + \tfrac{1}{2}\mu^2\right] \quad \text{and} \quad \rho_{\alpha 1} = c^2 D_0 \left[(d_1/d_0) - 6\Psi_1 + \tfrac{3}{2}\mu^2\right].$$

With these values and results, and after some algebra, both the Hamiltonian Constraint, (4.1), and the Lapse equation, (4.3), may be written in the form

$$\partial_r \partial_r \vartheta = h(t)\vartheta + f_s(t, r), \tag{5.1}$$

where ϑ stands for either $(r\Psi_1)$ or $(r\alpha_1\sqrt{A})$ depending on whether equation (5.1) represents the Hamiltonian constraint or the Lapse equation. The function $h(t)$ is given by

$$h(t) \equiv \tfrac{3}{2}\kappa c^2 d_0 \geq 0, \tag{5.2}$$

and

$$f_s(t, r) = \begin{cases} -rh\left[\tfrac{1}{6}(d_1/d_0) + \tfrac{1}{12}\mu^2\right]; & \text{for Hamiltonian Constraint,} \\ rh\left[\tfrac{1}{6}(d_1/d_0) - \Psi_1 + \tfrac{5}{12}\mu^2\right]; & \text{for Lapse equation.} \end{cases} \tag{5.3}$$

In the process of linearizing these equations we also obtain

$$(K_0)^2 = 3\kappa \rho_{H0} \tag{5.4}$$

and

$$\dot{K}_0 = h\left[1 - \Psi_1 + \tfrac{1}{6}(d_1/d_0) + \tfrac{1}{12}\mu^2\right]. \tag{5.5}$$

The general formal solution to equation (5.1) has the form

$$\vartheta = pe^R + qe^{-R} + \frac{e^R}{2h}\int_0^R e^{-R'} f_s\, dR' - \frac{e^{-R}}{2h}\int_0^R e^{R'} f_s\, dR' \tag{5.6}$$

where $p = p(t)$, $q = q(t)$, and $R = r\sqrt{h}$.

In order to obtain an explicit solution to equation (5.1), we need information about the initial source configuration, i.e. we need to know d_0, d_1, and $\tilde{S}_r$. This allows one to solve (5.1) for Ψ_1 and thus obtain the metric variable $A(t_0, r)$; once we have A we can solve equation (5.1) again for α_1 and hence obtain $\alpha(t_0, r)$. Equation (4.2) can then be solved for $K^*(t_0, r)$ and finally, with K^* and α, equation (4.4) yields $\beta(t_0, r)$ thus completing the analytical solution to the initial value problem.

6. Examples

6.1. Matter Perturbation

As a first example, we present a "Matter Perturbation" solution. For this case we assume that
$$d_0 = 4/(3\,\kappa\,c^4\,t_0{}^2),$$
with $t_0 = 1$,
$$d_1 = \begin{cases} 0, & 0 \le r \le r_1, \\ \delta/r, & r_1 < r < r_2, \\ 0, & r_2 \le r < \infty, \end{cases}$$
where r_1 and r_2 are arbitrary values of the radial coordinate, and
$$\delta = \epsilon\,\frac{8\,\pi}{\kappa\,c^2}\,(c\,t_0)^{-1}.$$

Here ϵ, not to be confused with the specific internal energy ε, is a "small" perturbation parameter; typically, $\epsilon = 0.05$ to 0.001. We also set $\widetilde{S}_r = 0$.

Using these values for d_0, d_1, and $\widetilde{S}_r$, the Hamiltonian constraint equation yields:
$$\Psi_1(0 \le r \le r_1) = C_0\,r^{-1}\,[C_1\,\sinh R],$$
$$\Psi_1(r_1 < r < r_2) = C_0\,r^{-1}\,[C_1\,\sinh R - \cosh\delta_1 + 1],$$
and
$$\Psi_1(r_2 \le r < \infty) = C_0\,r^{-1}\,[C_1\,\sinh R - \cosh\delta_1 + \cosh\delta_2],$$
where $C_0 = \tfrac{1}{6}(\delta/d_0)$, $C_1 = e^{-R_1} - e^{-R_2}$, $\delta_1 = R - R_1$, and $\delta_2 = R - R_2$, with $R_1 \equiv R(r_1)$ and $R_2 \equiv R(r_2)$.

Similarly, the Lapse equation yields:
$$\alpha_{1_a} = \tfrac{1}{2}\,C_0\,r^{-1}\,F(\Psi_{1_a})\,[C_2\,\sinh R - C_1\,R\,\cosh R],$$
$$\alpha_{1_b} = \tfrac{1}{2}\,C_0\,r^{-1}\,F(\Psi_{1_b})\,[C_2\,\sinh R - C_1\,R\,\cosh R + \delta_1\,\sinh\delta_1],$$
and
$$\alpha_{1_c} = \tfrac{1}{2}\,C_0\,r^{-1}\,F(\Psi_{1_c})\,[C_2\,\sinh R - C_1\,R\,\cosh R + \delta_1\,\sinh\delta_1 - \delta_2\,\sinh\delta_2],$$
where the indices a, b, and c denote the three radial regions delimited by r_1 and r_2 (see the expressions for Ψ_1 above), while $F(\Psi_1) \equiv (1 + \Psi_1)^{-1}$ and $C_2 = R_1\,e^{-R_1} - R_2\,e^{-R_2}$. Furthermore, since $\widetilde{S}_r = 0$ then $K_1{}^* = 0$ and $\beta_1 = 0$.

6.2. Momentum Perturbation

A second example is a "Momentum Perturbation" solution in which we assume that
$$d_0 = 4/(3\,\kappa\,c^4\,t_0{}^2),$$
with $t_0 = 1$. Also, we pick $d_1 = 0$ and
$$\widetilde{S}_r = \begin{cases} 0, & 0 \le r < r_1, \\ \delta_s\,r^{-1/2}, & r_1 \le r < r_2, \\ -\delta_s\,r^{-1/2}, & r_2 \le r < r_3, \\ 0, & r_3 \le r < \infty. \end{cases}$$
where r_1, r_2, and r_3 are arbitrary values of the radial coordinate, and
$$\delta_s = \epsilon\,\frac{8\,\pi}{\kappa}\,(c\,t_0)^{-3/2}.$$

Here again, as before, ϵ is a "small" perturbation parameter; typically, $\epsilon = 0.01$ to 0.001. Using these values for d_0, d_1, and $\tilde{S}_r$, the Hamiltonian constraint equation yields:

$$\Psi_1(0 \leq r < r_1) = C_0\, r^{-1}\, [C_1 \sinh R],$$

$$\Psi_1(r_1 \leq r < r_3) = C_0\, r^{-1}\, [C_1 \sinh R - \cosh \delta_1 + 1],$$

and

$$\Psi_1(r_3 \leq r < \infty) = C_0\, r^{-1}\, C_2\, e^{-R},$$

where $C_0 = \frac{3}{16}(\kappa\, \delta_s/h)^2$, $C_1 = e^{-R_1} - e^{-R_3}$, $C_2 = \cosh R_3 - \cosh R_1$, with $R_1 \equiv R(r_1)$ and $R_3 \equiv R(r_3)$.

Similarly, the Lapse equation yields:

$$\alpha_{1_a} = \tfrac{1}{2} C_0\, r^{-1}\, F(\Psi_{1_a})\, [C_3 \sinh R - C_1\, R \cosh R],$$

$$\alpha_{1_b} = \tfrac{1}{2} C_0\, r^{-1}\, F(\Psi_{1_b})\, \{C_3 \sinh R - C_1\, R \cosh R + \delta_1 \sinh \delta_1 + 8\, [\cosh \delta_1 - 1]\}$$

and

$$\alpha_{1_c} = \tfrac{1}{2} C_0\, r^{-1}\, F(\Psi_{1_c})\, \{-e^{-R}\, [C_4 - C_2\, (R - 8)]\},$$

where the indices a, b, and c denote the three radial regions delimited by r_1 and r_3 (see the expressions for Ψ_1 above), also, $C_3 = C_1\, (R_3 - 8) - e^{-R_1}\, (R_3 - R_1)$, $C_4 = R_3 \sinh R_3 - R_1 \sinh R_1$, and, as before, $F(\Psi_1) \equiv (1 + \Psi_1)^{-1}$. Once we know A and α, expressions for $K_1{}^*$ and β_1 can be readily obtained from equations (4.2) and (4.4) for the four radial regions delimited by r_1, r_2, and r_3, but they are too cumbersome to be shown here.

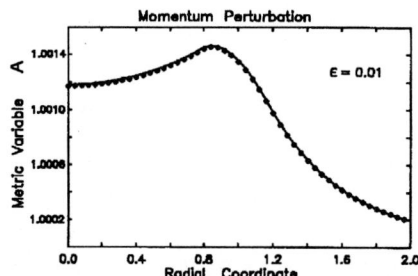

FIGURE 1. Metric variable A versus radial coordinate. Maximum relative percent error: 0.00124%

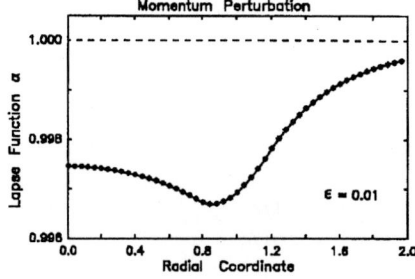

FIGURE 2. Lapse function α versus radial coordinate. Maximum relative percent error: 0.00127%

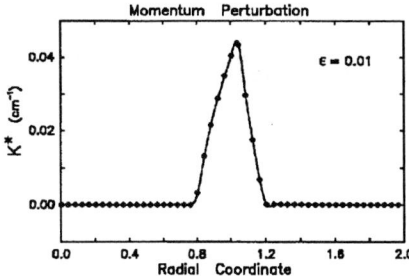

FIGURE 3. Extrinsic Curvature variable K^* versus radial coordinate. Maximum relative percent error: 1.57520%

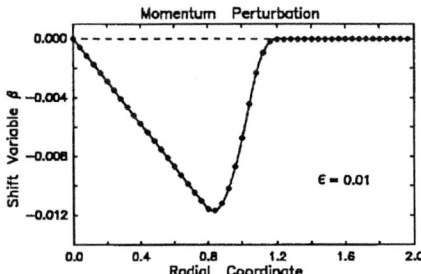

FIGURE 4. Shift variable β versus radial coordinate. Maximum relative percent error: 0.69948%

7. Numerical Solution and Results

The code that we use has been described elsewhere [3,4], so we shall not go into the details here. Naturally, we need to "feed" the code with the initial values of some of the variables relevant to the problem at hand. In this case we need the *outermost* radial zone analytical value of the metric function A and the Lapse function α. The trace of the extrinsic curvature, K_0, and the time derivative of this trace, $\dot{K}_0$, given by equations (5.4) and (5.5) respectively, are also considered as known data to be input into the code. Lastly, we need to feed the code with the values chosen for the sources, i.e. d_0, d_1, and $\widetilde{S}_r$.

The results obtained from the code for the Momentum Perturbation Case are shown in the figures. For this run we used 500 radial zones and the perturbation parameter ϵ was set equal to 0.01. In all the plots the solid line represents the analytical solution, while the dots represent the numerical solution. The excellent agreement between these two solutions is clearly evident. The maximum relative percent errors in our results are also shown on the plots.

REFERENCES

[1] Evans, C. R. 1986, *Dynamical Spacetimes and Numerical Relativity*, ed. J. Centrella, Cambidge U. Press, Cambridge, 3.
[2] Harleston, H. 1990, Ph. D. Thesis, University of Texas at Austin, USA.
[3] Harleston, H. & Holcomb, K. A. 1991, Ap. J., 372, 225.
[4] Harleston, H. & Vishniac, E. T. 1992, Phys. Rev. D15, 45, 4458.
[5] Holcomb, K. A. 1990, G.R.G., 22, 145.
[6] Shapiro, S. L. & Teukolsky, S. A. 1985, Ap. J., 298, 34.
[7] Wilson, J. R. 1979, *Sources of Gravitational Radiation*, ed. L. L. Smarr, Cambridge U. Press, Cambridge, 423.

Nested Grid Methods for Cosmological Hydrodynamic and N-body Systems

By Peter Anninos, Michael L. Norman

National Center for Supercomputing Applications, Beckman Institute, 405 N. Mathews Ave., Urbana, IL 61801, USA.

We describe a three-dimensional hierarchical numerical code designed for applications in the formation and evolution of galaxies and other larger scale structures in an expanding universe. A particle-mesh (PM) code used to simulate the dark matter constituency is coupled with ZEUS-3D, an ideal Eulerian MHD fluid solver to evolve the baryonic matter. We utilize a one-way interface between the coarse and fine grids. Interpolated fields from the coarse grid supply boundary conditions for the subgrid but no allowance is made for a back reaction modification to the coarse grid solution.

1. Introduction

The enormous demand for computer memory and speed sets some fundamental limits on the size and resolution of practical three-dimensional simulations that can be run on current supercomputers. The dynamic range in length and mass scales is severely hampered. One possible way to optimize codes for efficiency is to implement an adaptive mesh refinement (AMR) scheme as a general algorithmic framework. AMR is a programming strategy that dynamically tracks regions of the cube which require additional resolution, introducing refined grids of higher resolution to resolve those regions. For a survey of nested grid algorithms see [6]. Nesting finer scaled meshes within coarse grids is an economical way to improve the resolution and hence the overall accuracy of solutions. The result is a tremendous savings of computer memory and a substantial reduction in execution time over large equally fine grid simulations [5,6,8]. As a first generation predecessor to a more general scheme, we have developed a three-dimensional numerical code with a single level of refinement. The code is designed to model the evolution of structure in an expanding universe containing both dark and baryonic matter. The basis of this code is a standard PM algorithm to evolve the dark matter particles and an ideal nonrelativistic Eulerian MHD fluid solver (ZEUS-3D) to simulate the baryonic matter in an expanding universe. In the following sections we discuss the various components of our code and some tests we have run. Further details can be found in [2,3].

2. The Equations

A list of the basic physical equations begins with the Einstein equation for the expansion factor of evolving homogeneous and isotropic background spacetimes. The dynamical equations describing the behavior of matter within the framework of an expanding cosmology include the geodesic equation for the dark matter comoving particle positions, Poisson's equation for the gravitational potential field in comoving coordinates, and the gas dynamical equations modified to include the expanding universe. Our models treat the gas as a composite of hydrogen and helium in cosmic abundance so that helium is 24% of the total mass of the gas. The equation of state is set by assuming that recombination ($ep \rightarrow H\gamma$) and collisional ionization ($eH \rightarrow eep$) have balanced rates, that the gas is ideal, and that the electrons and ions are in thermal equilibrium. We also model radiative losses with bremstrahlung, Compton and hydrogen and helium line cooling.

3. The Grid Hierarchy

Communications and interactions between parent and child grids can be achieved with either one- or two-way interfaces. One-way schemes allow signals to propagate into the child grids but do not allow the child grid solutions to feed back into the parent cubes and modify the coarse grid solutions. Two-way interfaces are more complicated schemes that allow information to pass from both the parent-to-child and the child-to-parent directions. Information from either grid can be used to influence the solution of the other. Although two-way interfaces are the more appealing choice, they tend to give solutions that contain more noise than the simpler one-way algorithms [6]. Waves leaving the finer grids may generate false waves or other numerical noise when entering the coarse grid, particularly when the coarse/fine grid spacing ratio is large. We have opted to pursue the one-way interface option for the work presented in this article.

We implement two different interpolation schemes to obtain subgrid boundary conditions from the coarse grid data. The first is a linear scheme in which each subgrid boundary inherits the volume weighted average of each of the eight top grid cells that overlap a coarse grid size cell surrounding the subgrid point. A second method is a hybrid linear interpolation perpendicular to the subgrid boundary planes and bicubic interpolation parallel to the planes. This higher order method provides additional smoothness to help preserve the continuity of forces as the gradient of a potential field and results in more stable evolutions when modeling oscillatory or wave-like behavior.

3.1. The Particle Code

We use a standard Cloud-in-Cell PM code [1,4] to model collisionless dark matter. Physical scalar quantities such as the gravitational potential and density are defined on a uniformly spaced cubical grid. Individual particles are given identical masses and volumes equal to a single cubical cell. Forces on each particle are computed by using fourth order centered differences to compute the potential gradients on the zone centers and the same volume weighting scheme used for the mass assignments is used to interpolate forces to individual particles. The Poisson equation is solved using a multidimensional Fourier transform (FFT) on the top grid (which has periodic boundary conditions) and an iterative conjugate gradient method on the fine grid (which has Dirichlet boundary conditions).

Sharp discontinuities in the coarse and fine grid spacings across the grid interface leads to errors with a characteristic length equal to about a coarse grid zone which is significant when compared to the subgrid smoothing length [2,5,8]. Blending the interface boundary over a finite region so that the forces acting on a particle entering the subgrid are not obtained directly from the interpolated forces at the boundary helps to stablize the evolution. We accomplish this by evolving particles with forces that are linear combinations of the coarse and fine grid forces until they cross sufficiently far into the subgrid domain that errors from the interpolated boundary forces are negligible. In this way a particle crosses the interface between parent and child grids in a continous fashion. It is essential for the code to keep track of each particle's position and be able to apply this blending scheme at all times, allowing particles to smoothly enter and exit the subgrid at will.

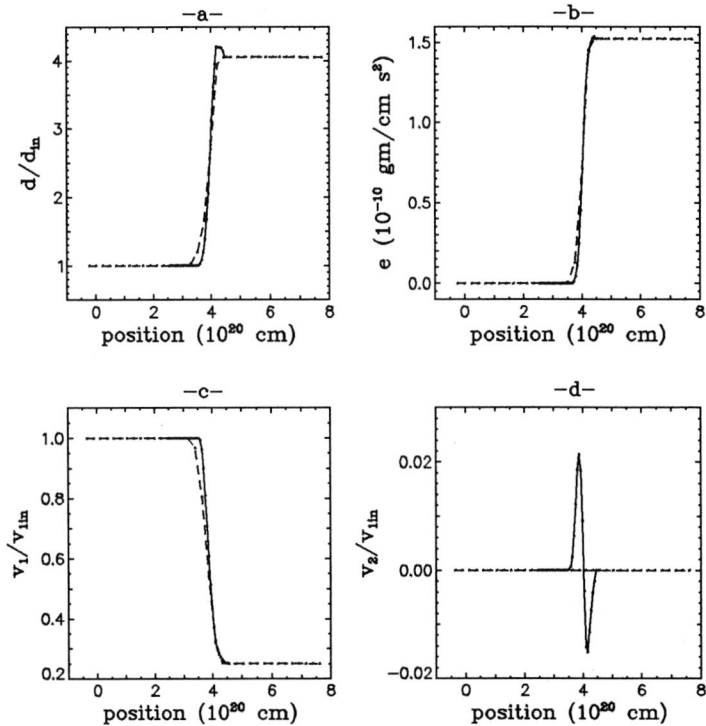

FIGURE 1. Coarse grid (dashed lines) and fine grid (solid lines) solutions projected along the subgrid boundary faces for a one-dimensional strong shock flow along the 1-direction. The solution is resolved with 32^3 cells on both grids.

3.2. The Hydrodynamic Code

ZEUS-3D is a time explicit, finite difference, Eulerian code that solves the ideal (non-resistive, non-viscous and adiabatic) magnetohydrodynamical (MHD) equations. A two-dimensional implementation of the ZEUS algorithms is described in [7]. We have modified the original code to include the cosmological expansion, radiative cooling and refined subgrid evolution.

The time explicit nature of ZEUS-3D will propagate and amplify any errors that may arise from interpolations or other boundary related problems. Smooth grid interfaces can be achieved by implementing some method of noise control [6]. We introduce diffusive terms $\partial f/\partial t = K\nabla^2 f$ near the boundary to act as a 'sponge' and absorb any short wave noise. The diffusion coefficient K is defined as $K = c(x)\Delta x^2/(2\Delta t)$, where $c(x) < 1$ is a function that tends to zero after a few zones turning off the diffusion far from the boundaries. Hence, incompatibilities between the two grid solutions are dissipated by viscous damping. There might be some concern that significant information might at some level be suppressed. For this reason we have limited the range of zones and the value of the diffusion coefficient to be as small as possible and still maintain accurate stable evolutions.

4. A Stringent Code Test - 'The Shock in the Box'

The basic PM and ZEUS-3D codes have been developed and tested independently in previous unrelated work [1,3,7]. We have run a number of additional test-bed computations designed to test the new elements of our code. These include: 1) Jeans instability tests in both an expanding universe and flat Euclidean space, 2) Zel'dovich pancake test in both dark matter and baryon dominated universes, 3) single particle simulations to test for anisotropies and 'ringing' of forces near the grid interfaces, 4) radiative shock tests of the cooling algorithms, and 5) the stability of subgrid shocks.

We cannot present further details of all the tests described above. Instead we will focus on a single important and rather stringent test and refer the reader to [2] for more details of this and other test-bed computations we have performed. The test we detail here is number (5) in the above paragraph which refers to the ability of the subgrid to evolve and maintain a shock across its boundaries. We anticipate that differences in the shock widths at the two grid resolutions will result in pressure gradients at the subgrid interfaces which could generate spurious, purely numerically induced flows. This is indeed what we observe. Fixing the subgrid solution to the coarse grid values prevents the subgrid solution from ringing down to its static post shock value. Outflow from the subgrid arising from post shock oscillations interferes and is obstructed by the coarse grid outflow boundaries. This is most evident in the density shown in figure (1a). A possible remedy to this situation is to smear the subgrid shock over more zones near the boundaries by increasing the linear and quadratic viscosity coefficients. This procedure significantly suppresses noise resulting from shock front gradients.

In the figures (1) we display the shock smeared numerical solutions for a one-dimensional static strong shock configuration after more than 100 time cycles of evolution, giving the solution a chance to relax to its shock jump conditions. Solid (dashed) lines represent the fine (coarse) grid solution. Inflow and outflow boundary conditions appropriate for strong shocks are imposed on the coarse grid. The grid refinement factor is set to 4 so that the subgrid shock which is resolved with the same number of cells as the coarse grid shock is actually thinner in absolute spatial scales by a factor of 4. Figures (1a)-(1d) show the density, energy, v_1 (direction of flow) and v_2 (direction perpendicular to flow) along the 1-direction. The simulation is performed on a 3D grid with a resolution of 32^3 cells and the projected slices along the transverse axes are chosen to lie along the boundary subgrid faces so that figures (1) represent the worst case solutions. As evidenced from figure (1d), errors are of order 2% (v_2 is ideally zero for this 1D flow). Errors continue to drop further from the boundaries so that $v_2 \sim 0.02\%$ at the center of the subgrid.

Acknowledgements

We thank Paul Shapiro for suggesting the 'shock in the box' code test. This work is partially supported by NASA grant NAGW 3152.

REFERENCES

[1] Anninos, P., Matzner, R. A., Tuluie, R. & Centrella, J. M. 1991, ApJ, 382, 71.
[2] Anninos, P., Norman, M. L. & Clarke, D. (in preparation).
[3] Anninos, W. Y. & Norman, M. L. 1993, ApJ (submitted).
[4] Hockney, R. & Eastwood, J. 1981, Computer Simulation Using Particles (New York: McGraw-Hill).
[5] Jessop, C., Duncan, M. & Chau, W. Y. 1993, JCompPhys, in press.
[6] Koch, S. E. & McQueen, J. T. 1987, NASA Technical Memorandum 87808.
[7] Stone, J. M. & Norman, M. L. 1992, ApJS, 80, 753.
[8] Villumsen, J. V. 1989, ApJS, 71, 407.

Dynamics of Clusters of Galaxies

By August E. Evrard

Department of Physics, University of Michigan, Ann Arbor, MI 48109-1120 USA.

The abundance and internal characteristics of rich clusters of galaxies can provide useful constraints on models of large–scale structure formation. The full power of observational constraints will only be realized when theoretical modeling of these complex, multi–component structures reaches a level of detail comparable to observations. This article will review some recent three dimensional, multi–fluid simulations of cluster dynamics and discuss their impact on issues raised from optical and X-ray observations of clusters. In 'bottom–up' formation scenarios (such as the ubiquitous cold dark matter model), galaxies form before rich clusters; hence, cluster formation is intimately linked to galaxy formation which, in turn, is tied to star formation. I will examine two issues which appear relatively insensitive to galaxy/star formation — the baryon fraction in clusters and the connection between X-ray morphology and Ω_o — and end with a topic that is intimately linked to it — dynamical biases in the cluster galaxy population.

1. Introduction

Clusters of galaxies are believed to be the largest collapsed, gravitationally bound structures in the universe. A typical rich cluster (Coma is the usual example) is a multi–component system containing hundreds of bright galaxies, a hot, metal–enriched intracluster medium (ICM) observed in X-rays, and dark matter whose presence has been inferred by application of the virial theorem for over 60 years [39].

Measures of the first two components, the galaxies and gas, are naturally linked by the star formation histories of the galaxies within the cluster. The present relative distributions of all three components reflect their full dynamical and thermal histories, which need not be the same. In particular, an exchange of energy between the different components will result in spatial segregation within the common cluster gravitational potential, with the coolest component being centrally concentrated and the hottest being extended.

Cluster formation must be understood in a cosmological context. Although a standard model of large–scale structure formation does not exist, many popular models envision clusters forming from gravitational amplification of small, initially Gaussian distributed density fluctuations [14, 16]. In nearly all viable models, cluster formation occurs after galaxy formation. Complete understanding of the present distributions of the dark matter, galaxies, and intracluster gas thus requires solving the FOE ('formation of everything') problem. That is, aspects of cluster formation are intimately tied to galaxy formation which, in turn, is linked to fragmentation of gas clouds and star formation [3]. Many unsettled issues regarding clusters (*e.g.*, the origin and distribution of metals in the ICM, whether or not galaxies fairly trace the cluster dark matter) persist because of the uncertainties in modeling galaxy/star formation.

Despite this cautionary tone, there is good reason for optimism, since there are some important aspects of cluster formation which are largely decoupled from star/galaxy formation, at least in a model dependent fashion. In this article, I will briefly outline how one attempts to model cluster formation in a multi–component fashion using a combined N–body and smoothed particle hydrodynamics algorithm (§2). Most of the article is concentrated on applications of this approach to three particular issues — the baryon fraction in clusters (§3), the connection between cosmology and X–ray morphology (§4) and galactic dynamics within clusters (§5). The first two problems are fairly insensitive to galaxy formation while the last is strongly coupled to it.

2. SPH Modelling of Cluster Dynamics

Correctly modeling the dynamics of distinct components within a cluster requires a multi–fluid approach. The simulations discussed in this article all used a combination of an N–body algorithm to provide gravitational forces and smoothed particle hydrodynamics (SPH) [17, 23] to provide gas dynamic forces and thermal energy evolution for the gaseous baryonic component. Details of the P3MSPH code used in Sections 4 and 5 below can be found in [9]. Briefly, SPH is a Lagrangian scheme which uses a smoothing kernel $W(r,h)$ to determine characteristics of the fluid at a given point based on properties associated with the local particle distribution. For example, the density at the position of particle i in the 'gather' interpretation [18] is given by

$$\rho_i = \sum_j m_j W(r_{ij}, h_i) \quad (1)$$

where r_{ij} is the separation of the pair of particles i and j and h_i is the local smoothing scale. The kernel W has compact support on a scale of a few h; hence, h is a measure of the local resolution of the solution. Usually h is adaptively varied both spatially and temporally such that a constant number of neighbors in the range $30 - 100$ is involved in the above sum. Applications to date have employed spherically symmetric kernels; schemes using anisotropic kernels are currently being developed [32].

The gas force on particle i is found with a sum involving the gradient of the kernel

$$\left(\frac{\vec{\nabla} P}{\rho}\right)_i = \sum_j m_j \left(\frac{P_i}{\rho_i^2} + \frac{P_j}{\rho_j^2}\right) \vec{\nabla} W(r_{ij}, h_{ij}) \quad (2)$$

where h_{ij} is an average measure of h_i and h_j used to preserve pairwise symmetry in the force equations. This guarantees conservation of linear and angular momentum (for a spherically symmetric kernel) to machine accuracy. In regions of converging flow, an artificial viscosity is included to prevent free–streaming and increase entropy in a manner satisfying the local shock jump conditions. Radiative cooling terms can be included in the energy equation, as can heating terms due to, for example, photoionizing radiation or other sources.

The Lagrangian nature and wide dynamic range of SPH are well suited to the problem of large–scale structure formation. Schemes using Eulerian finite difference methods have recently come on line [4, 7]. A comparison of several cosmological gas dynamic schemes applied to structure formation in a cold dark matter universe has recently been done [19]. Examined at low resolution, the codes produce very similar results for the thermal and spatial structure of the gas. Higher resolution examination serves to illustrate the relative strengths and weaknesses of each approach. For low density contrasts, the Eulerian codes display superior resolution while at high density contrasts, higher resolution is achieved by the SPH codes. Roughly speaking, the 'break–even' density contrast δ_{eq}, where Eulerian and Lagrangian approaches have comparable resolution, is that at which one particle in the SPH calculation is contained in one cell of the Eulerian code. At densities above δ_{eq}, the Lagrangian method resolves one Eulerian cell with more than one particle (implying 'higher resolution') while for densities less than δ_{eq}, a single Lagrangian particle covers many Eulerian cells ('lower resolution'). It follows then that $\delta_{eq} = N_{cell}/N_{part}$ where N_{cell} is the number of cells in the Eulerian code and N_{part} is the number of particles in the SPH code. In three dimensions, $\delta_{eq} = 256^3/64^3 = 64$ is a realistic value. Since the mass within an Abell radius of rich clusters represents a significant ($\delta > 100$) local density enhancement (see §3 below), a Lagrangian approach is (arguably, of course) currently the most efficient and effective means to model them numerically.

3. The Cluster Baryon Fraction

The combination of X–ray surface brightness and spectral data for a cluster allows a direct estimate of the mass of hot, intracluster gas M_{gas} to be made [31]. The mass in galaxies M_{gal} can be estimated by multiplying the total optical luminosity in cluster galaxies by a representative, galactic mass-to-light ratio. Finally, the cluster binding mass M_{tot} can be inferred by assuming the gas (and/or galaxies) are in hydrostatic equilibrium. From these masses, all determined within some radius r_x where accurate X–ray and optical data exist, one can infer the mass fraction in *observed* baryons $f_b \equiv (M_{gal} + M_{gas})/M_{tot}$. It is interesting to compare this to the global value $<f_b> = \Omega_b/\Omega_o$ by defining the factor

$$\Upsilon = \left(\frac{\Omega_o}{\Omega_b}\right) \left(\frac{M_{gal} + M_{gas}}{M_{tot}}\right) \quad (3)$$

where Ω_o is the present total density and Ω_b the global baryon density, each relative to the critical density.

This exercise has recently been carried out for the Coma cluster [37] and the results for the component masses within an Abell radius $r_A = 1.5\,h^{-1}$ Mpc ($h = H_o/100$ km s^{-1} Mpc^{-1}) are: $M_{gal} = 3.15 \pm 0.66 \times 10^{13}\,h^{-1}M_\odot$, $M_{gas} = 5.66 \pm 1.02 \times 10^{13}\,h^{-5/2}M_\odot$ and $M_{tot} = 1.10 \pm 0.18 \times 10^{15}\,h^{-1}M_\odot$. The quoted 1σ errors are purely statistical, arising from uncertainty in optical photometry (for M_{gal}) and X–ray photon counts (for M_{gas}). In deriving the total mass, several independent estimates were derived which spanned the range $0.67 - 1.10 \times 10^{15}\,h^{-1}M_\odot$. The largest value was adopted, which was derived by scaling a set of 12 N–body/SPH simulations to the temperature of Coma and measuring the mass within r_A. The error is then based on the scatter among the runs. Adopting the largest value is conservative in that it leads to the minimum cluster baryon fraction. The total mass implies a mean overdensity within an Abell radius of $280/\Omega_o$.

The resultant baryon fraction in Coma is $f_b = 0.029 + 0.051\,h^{-3/2}$ with a statistical uncertainty of about 25%. For $h = 0.5$, the baryons represent 17% of the total mass. The limits on the global baryon fraction from nucleosynthesis [33] are very stringent, $\Omega_b h^2 = 0.0125 \pm 0.0025$ at 95% confidence. The data then imply that the baryon fraction within an Abell radius in Coma is enhanced by the factor

$$\Upsilon \in [3.5 - 6.4]\,\Omega_o \quad (4)$$

where the smaller factor assumes $h = 0.5$ and the larger $h = 1.0$.

From here, there are two possible avenues to pursue. One is to assume that the baryon fraction in Coma is representative of the global value ($\Upsilon = 1$) and so the data then constrain Ω_o to be in the range $0.16 - 0.29$. This perfectly valid line of reasoning is anathema to those cosmologists who prefer Ω_o to be unity either on the basis of aesthetics or because it is a natural outcome of inflationary models. (This may be a good time to point out that analysis of cluster X–ray morphology presented in the next section strongly supports this point of view.) The other avenue is thus to assume $\Omega = 1$, which leads to the following possibilities:

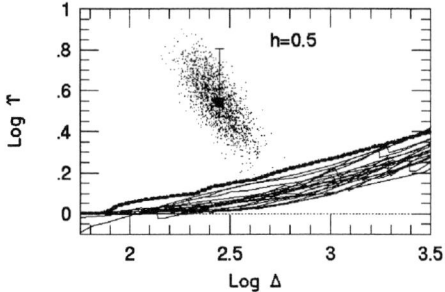

FIGURE 1. The baryon enhancement factor Υ as a function of mean interior overdensity for the spherical, self–similar accretion model (thick, dashed line) and the results of N-body/SPH simulations of 'pressureless' baryon accretion in a CDM universe. The dots represent (Υ, Δ) pairs drawn at random using the uncertainties for Ω_b and for the masses of the different components assuming $h = 0.5$. Arrows show how the value of Υ would change for h in the range of 0.4 to 1.0. From White et al. [37].

(i) the estimated masses are systematically in error by large factors, (ii) the nucleosynthesis bound is in error or (iii) some process packed at least 3.5 times more baryonic matter than dark matter within an Abell radius in Coma. The last possibility may have physical justification in the fact that baryons can dissipate their thermal energy via radiative cooling while the dark matter may not. Classic cooling timescale arguments [38] suggest that dissipation should be modest on the large mass scales contained within an Abell radius. However, dissipation on galaxy scales may be very efficient, such that all the baryons in the universe are packed into very small structures before cluster formation. In that case, galaxies may infall nearly radially into the cluster center, and there deposit via inelastic collisions the baryons which they carried in, leading to a large baryon enhancement.

This possibility — that the baryon fraction in Coma is enhanced through 'known' mechanisms of gravity and dissipation — has recently been ruled out by a combination of semi–analytic and numerical analysis [37]. Treating the cluster as spherically symmetric, one can appeal to Bertschinger's self–similar infall solutions [1] to estimate the post–collapse structure for both dark matter and an 'infinitely dissipative' gas. The former is found by using the solution for infall of a collisionless gas and the latter comes from the solution which assumes accretion onto a central black hole. The solutions presented as a function of the self–similar radius have been re–phrased in terms of the mean interior overdensity Δ and are shown in Figure 1. At an overdensity of 280, an enhancement in the baryon fraction of $\sim 40\%$ is expected, well below the inferred range of equation (4). The cloud of points in Figure 1 are $[\Delta, \Upsilon]$ pairs constructed by random realizations of the quoted uncertainties in all the quantities on the right hand side of equation (3). The observations and model predictions are inconsistent at greater than 99% confidence.

To verify that the spherical model results are relevant to more realistic, fully three dimensional accretion, a series of N–body/SPH runs were performed using CDM cluster initial conditions. The code used was a combined TREE/SPH code written by J. Navarro [27]. To mimic a perfectly dissipative gas, an isothermal equation of state was assumed, using a temperature well below that of any resolvable potential well in the simulation. A set of runs, each with 65536 particles representing gas and dark matter, were performed. The enhancements seen in the simulations, shown as the light lines in Figure 1, all fall below the spherical model solution. That the spherical model should provide an upper limit to the enhancement factor is perhaps not too surprising, since it represents the limiting case of zero angular momentum for all mass elements. In the three dimensional case, sublumps acquire angular momentum and may 'miss' the center on first infall, resulting in a more extended baryon distribution.

These toy models produce clusters which are far from realistic. The models contain no hot, intracluster medium while observations indicate that the majority of baryons end up in this phase. A more realistic treatment, such as the models discussed in the next section, show the ICM to be slightly more *extended* than the underlying dark matter, *i.e.*, $\Upsilon < 1$. This result is due energy exchange between the dark matter and ICM during mergers [28, 29].

The upshot is that a large (factor > 3) enhancement in the baryon fraction within an Abell radius in a rich cluster like Coma is impossible with conventional dynamics. Consistency with $\Omega_o = 1$ requires other explanations for the discrepancy, such as those listed above.

4. A Morphology–Cosmology Connection

There are several ways clusters can be used as cosmological diagnostics. Their abundance as a function of, for example, velocity dispersion σ, is extremely sensitive to the normalization of the fluctuation spectrum [36]. However, small errors in σ can translate into large errors in the normalization [10], and handling this correctly requires both good data and careful analysis. Because of degeneracy between the spectrum normalization and Ω_o, abundances provide little information on Ω_o (unless the spectrum normalization is known by other means).

A more fruitful approach to constraining Ω_o is to examine the structure of the hot, intracluster gas. The motivating idea is that, because the linear growth of perturbations diminishes as Ω decreases, structure formation in a low Ω_o universe should occur earlier than if $\Omega = 1$. An analysis based on a spherical model for cluster collapse yields an age difference between clusters in models with $\Omega_o = 0.2$ and $\Omega = 1$ of $\sim 0.3\ H_o^{-1} \sim 4-6$ billion years [30]. The sound crossing time for 10 keV gas in the central 1 Mpc of a cluster is only 0.6 billion years, so this age difference corresponds to many sound crossing times within the region surveyed by X-ray imaging instruments. This leads to the expectation that clusters in low density models should have more relaxed X-ray isophotes than their critical counterparts.

This effect has now been verified and quantified with a set of 24 P3MSPH simulations [11]. Eight sets of initial conditions, two each in comoving periodic boxes of side 30, 40, 50 and 60 Mpc ($h = 0.5$), were generated in a constrained manner [2] from an initial CDM fluctuation spectrum. Each initial density field was evolved in three different cosmological backgrounds: (i) an unbiased, open universe with $\Omega_o = 0.2$; (ii) an unbiased, vacuum energy dominated universe with $\Omega_o = 0.2$ and $\lambda_o = 0.8$ and (iii) a biased, critical density ($\Omega = 1$) universe with *rms* present, linear mass fluctuations in a sphere of $8\ h^{-1}$ Mpc equal to $\sigma_8 = 0.59$. A baryon content of $\Omega_b = 0.1$ was assumed for the models, with all the baryons in the form of gas. The rest of the mass was assumed to be collisionless dark matter.

Present day X-ray images of half of the simulated clusters are shown in Figure 2, along with a set of *Einstein* IPC images of four Abell clusters. The simulated images show the IPC band-limited flux in X-rays with an angular resolution of about $1'$. They are cleaner than the observations because no noise or background of points sources have been added. The low density models, shown in the first two rows, are much more centrally concentrated and display much less asymmetry than the critical universe clusters. These differences arise because the low Ω_o clusters suffer fewer merging events at late times, a result expected from analytic arguments [20, 22, 30]. We have quantified the differences using statistics measuring the surface brightness fall-off (the familiar β_{fit} parameter), mean isophotal

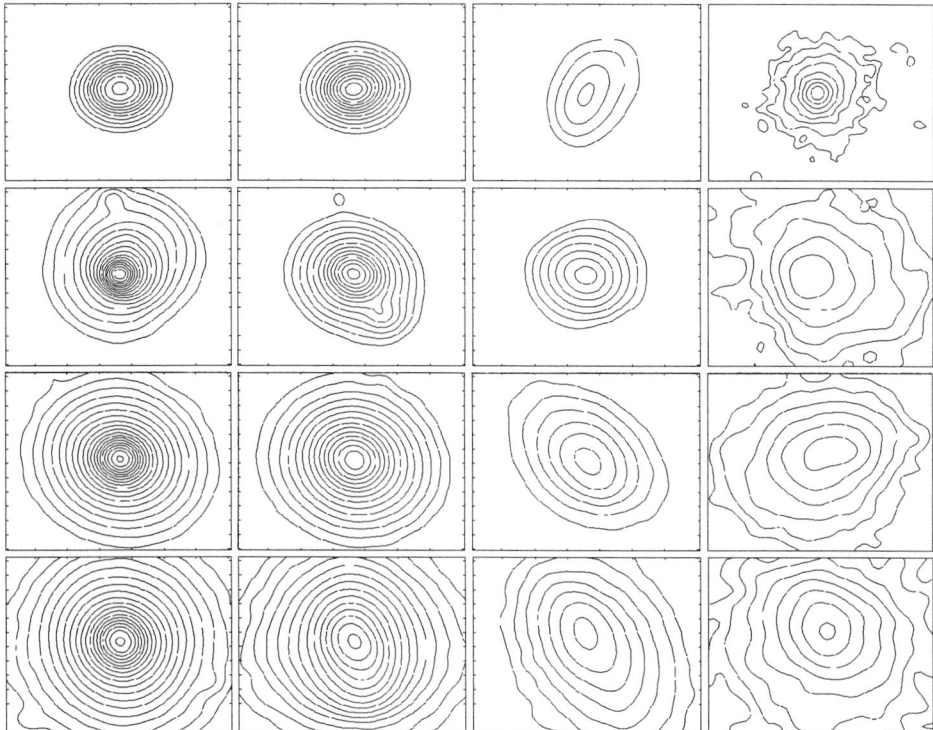

FIGURE 2. Contour maps of cluster X–ray emission showing the dependence of X–ray morphology on the underlying cosmology. The first three columns show simulated clusters 'viewed' at $z=0.04$ evolved in different cosmologies (from left to right): (i) $\Omega_o = 0.2$; (ii) $\Omega_o = 0.2$, $\lambda_o = 0.8$; and (iii) $\Omega = 1$, $\sigma_8 = 0.59$. The fourth column consists of *Einstein* IPC observations of Abell clusters; from top to bottom, they are: A496, A399, A2256 and A401. Each row in the first three columns corresponds to the same initial density field generated in (from top to bottom) 30, 40, 50 and 60 Mpc cubes. The contours in all maps are spaced by factors of 1.8 in surface brightness from 3.5×10^{-4} cts s^{-1} arcmin^{-2}, and the spatial scale for every map is identical, the distance between tic–marks being 188 kpc. From Evrard et al. [11].

center shift [26] and the mean eccentricity. The same measures have been made for both the simulated and observed clusters. Histograms of any of these show a distinct difference between the low and high density models, with the observations strongly favoring $\Omega = 1$ over either of the $\Omega_o = 0.2$ universes [25]. This result is supported by recent analysis of the abundance of rich clusters. In order to reproduce observations, an $\Omega_o = 0.2$ CDM dominated universe requires a very high fluctuation amplitude $\sigma_8 = 1.25 - 1.58$, which requires galaxies to be *less* clustered than the mass distribution [36].

There is a potential conflict between this result and that of the previous section. The large baryon fraction in Coma could be explained by a low Ω_o, but the morphology of X–ray emission from clusters strongly disfavors low values. One possibility is that $\Omega_o \sim 0.6$, and both constraints are satisfied. (This idea can be discounted by appealing to the Principle of Non–Ugliness.) One might instead doubt the robustness of the X–ray morphology constraints. After all, the physics in the models shown in Figure 2 is fairly simplistic, incorporating gravity for both components, as well as shock heating and a thermal pressure gradient for the gas.

Although there is more physics beyond this simple approach, it is difficult to finger a mechanism which would strongly distort in an *anisotropic fashion* the present cluster X-ray morphologies in the low Ω_o models. Feedback due to winds from early–type galaxies would have to occur very recently, and the winds would have to be coherently directed so as to distort the isophotes in a manner similar to that which occurs naturally by merging. Recent simulations incorporating winds in $\Omega = 1$ clusters show little effect on the overall morphology [24]. Adding radiative cooling would produce a large central cooling flow, but there is no reason to suspect this will strongly affect the morphology of the outer regions. Finally, unrealistically large tidal torques would be required to distort the structure of the X–ray gas in the inner ~ 1 Mpc region, where the bulk of the X–rays are observed.

It all boils down to this. Generating anisotropy in the gas distribution at late times requires a directed source of energy input with magnitude comparable to the binding energy of the cluster. The most natural source for such directed energy is the merging of two systems of roughly comparable mass. To save the low Ω_o models, one needs to come up with a mechanism(s) which replaces merging, but produces the same effects. It is not at all clear how to do this.

5. Galactic Dynamics in Clusters

As a final topic, let's turn from X–ray to optical wavelengths. The theme common with the preceding two sections is the determination of Ω_o from clusters. It is well known that dynamical mass estimates based on the virial theorem have yielded mass to light ratios around $150\,h^{-1}$ in solar units, which is about a factor 5 smaller than that required to reach closure density [13,15]. Again, the two possible interpretations are: (i) the estimate is unbiased and $\Omega_o \sim 0.2$ or (ii) there is a systematic bias in the estimate which makes it consistent with $\Omega = 1$. One possibility for the latter is that the dark matter is very weakly clustered on comoving scales $\lesssim 10$ Mpc. Another is that the galaxies are condensed toward the cluster center, and that one is measuring only some inner fraction of the total cluster mass and missing an extended, outer dark envelope. The latter issue has been investigated numerous times with N–body experiments over the past decade. Unfortunately, the interpretation of these simulations is clouded by the rather naive way in which galaxies were represented within the cluster. In some studies, heavy particles meant to represent the luminous parts of galaxies were merely put in 'by hand' in the initial conditions [8, 34]. Others tagged a subset of the collisionless dark matter particles, based on plausible physical arguments, to represent the kinematics of the galaxy population [5].

These experiments generally produced results in the desired direction; that is, the galaxies represented a cooler, condensed population within the dominant dark matter potential well and mass estimates based on them underestimated the total mass of the system. However, the magnitude of the effect varied by rather large factors, depending on the treatment. Furthermore, the physics responsible for the bias remains poorly understood. Dynamical friction [6], incomplete or 'non–violent' relaxation [45, 40], or some combination of the two remain the most viable mechanisms.

The limiting factor in performing such experiments is the uncertainty involved in galaxy formation. Ideally, one would like to form galaxies *in situ* and subsequently follow their hierarchically clustering to the scale of rich clusters. The combined N–body/gas dynamic methods are now making this possible. In the simplest scenario, one allows the baryons to radiatively cool within their parent dark halos. Since local temperatures and densities are known, cooling rates can be calculated. The principle uncertainty is due to lack of resolution — the baryons represented by a single particle or single cell, which is typically $\gtrsim 10^8 M_\odot$, is assumed to be a single phase medium characterised by the given density and temperature. This treatment, though not perfect, has much more physical validity than

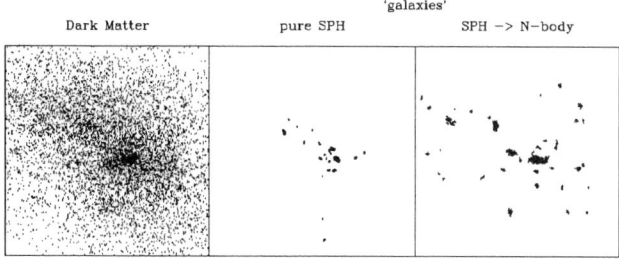

FIGURE 3. Distribution of particles in a 1 Mpc region centered on a rich cluster at $z=0$ from a P3MSPH simulation using CDM initial conditions. The left panel shows a random subset of the dark matter particles (the total number in the cluster is nearly 56,000). The middle panel shows baryonic particles found in 'galaxies' at the end of the simulation, using SPH dynamics throughout the evolution. The right panel shows the final 'galaxy' distribution when particles in galaxies are treated as collisionless 'stars' from $z=0.7$ to $z=0$. All galaxies with more than 32 particles are shown. See the text for further discussion.

tagging particles in an N–body experiment. The end result is a two–phase structure in the baryons, with cold, dense knots one associates with galaxies surrounded by halos of hot, rarefied gas.

Simulations of this sort have been successful recently in producing clusters with anywhere from three [21] to several tens [12] of such 'galaxies' within them. An example is shown in Figure 3. This cluster was modeled with P3MSPH using 2×64^3 particles to represent the dark matter and baryons. The simulation modeled a periodic cube 22.5 Mpc on a side ($h=0.5$). The limiting spatial resolution was ~ 30 kpc and the mass per baryon particle was $3 \times 10^8 M_\odot$. An L_* galaxy would thus be modeled by about 300 particles. Radiative cooling for the baryons based on collisional ionization equilibrium was included. One of the main aims of such a simulation is to address the issue of dynamical biases in the galaxy population.

The results are striking. The galaxies, shown in the middle panel of Figure 3, are much more centrally concentrated than the dark matter. Table 1 gives a summary of the relevant properties. Galaxies were defined to be cluster members if they lie within a radius of 1.6 Mpc from the cluster center. This radius is that within which the mean density, based on the known dark mass distributions, is 180 times the background value. The number of galaxies so found N_{gal} is shown for galaxies with particle counts above $N_{cut}=32$ and 128.

Table 1 : Cluster Parameters from Pure SPH Treatment

N_{cut}	N_{gal}	σ_{gal} (km s^{-1})	σ_{gal}/σ_{DM}	R_{gal}/R_{DM}	M_{vir}/M_{true}
32	29	458	0.84	0.35	0.43
128	10	469	0.86	0.22	0.25

The galaxies represent a cooler population, as witnessed by the values of the 'velocity bias' parameter, the ratio σ_{gal}/σ_{DM} where σ is the one–dimensional galaxy velocity dispersion obtained from an average of the three orthogonal directions. A modest $\sim 15\%$ bias is evident in the velocities. In contrast, the ratio of the half–mass radii R_{gal}/R_{DM} (determined from the known three dimensional positions and using 1.6 Mpc as an outer radius) shows a much more pronounced bias. Galaxies with baryon mass above $10^{10} M_\odot$ (32 particles) are more concentrated than the dark matter by a factor 3, while galaxies above a mass cut a factor 4 larger are even more concentrated. Application of the virial theorem to determine binding masses results in a large (factor $2-4$) underestimate of the total mass of the cluster.

What is worrisome about this treatment is that the galaxies are assumed to be purely gaseous throughout the evolution of the cluster. Their interactions with the surrounding medium and with each other during collisions entail viscous drag, which is unphysical for a galaxy comprised mainly of stars. Of particular concern is the fact that the largest galaxy in the center of the cluster ends up containing *more than half* of the total baryons in cluster galaxies. Although bright, central cD galaxies are not uncommon in rich clusters, it is not the norm for the central cD to be brighter than the sum of all the other galaxies in the cluster.

A simple way to test the effects of this collisional, or 'pure SPH', treatment on the galactic dynamics within the cluster is to turn the SPH gas particles in galaxies into collisionless 'stars'. Ideally, the star formation process should be modeled self–consistently within the code; however, there still exists a large amount of uncertainty in parameterising star formation rates and the associated feedback. For the purposes of examining galactic orbits within the cluster, the key element is that the star formation in galaxies occur before the bulk of the cluster is assembled. With this in mind, the following simple experiment was performed as a variation to the original SPH run. At a redshift $z=0.7$, before cluster collapse but after many of the cluster galaxies were assembled, the particles labeled in galaxies were 'instantaneously' turned into collisionless 'star' particles. At this time, all the remaining gas was removed and the mass associated with it was added to the dark matter particles. A collisionless, two–fluid run, using as initial conditions the dark matter particles and galaxies comprised of the star particles above, was then evolved from $z=0.7$ to $z=0$, a time interval of 7.3 Gyr with $h=0.5$. This run thus follows the evolution of the same cluster as in the pure SPH run, but with the galaxies being treated as collisionless entities during the epoch of cluster formation. This 'SPH $\rightarrow$ N-body' treatment is arguably more realistic, if one believes the observed stellar populations in cluster galaxies pre–date the cluster itself.

The resulting cluster galaxy distribution is shown as the right panel in Figure 3. The galaxies in the SPH $\rightarrow$ N-body treatment are clearly more extended than their pure SPH counterparts. Figure 4 shows the orbits of three galaxies identified at $z=0.7$. The top row shows the orbit under the pure SPH treatment while the lower row shows orbits under the SPH $\rightarrow$ N-body treatment. The predominantly radial infalling orbits take the galaxies very close to the cluster center. In the SPH case, the viscous gas interactions in the high density, central region prove effective at 'braking' the galaxies. This prevents them from completing a full orbit and also enhances the accretion rate onto the large, central galaxy. In contrast, the galaxies comprised of collisionless stars fly through the center relatively undisturbed, as expected if the cluster velocity dispersion is larger than the internal velocities of the galaxies (which is the case here). No extremely large central galaxy forms. Instead, two galaxies separated by 0.5 Mpc, each roughly one–fifth the mass of the largest in the pure SPH run, are the most conspicuous objects in the cluster at $z=0$.

A listing of the relevant cluster properties inferred from the collisionless treatment is given in Table 2. Because of the reduced merger rate, the number of galaxies in the cluster at $z=0$ nearly doubles over the pure SPH case. A modest velocity bias persists, being slightly larger for the more massive galaxies. The spatial distribution depends on the galaxy mass cutoff. The massive subsample is more concentrated than the dark matter while the set of all galaxies above the minimum 32 particle count ($10^{10} M_\odot$ in baryons) is spatially unbiased with respect to the dark matter. The resulting virial mass estimates reflect this trend; the massive subsample underestimates the total mass by a factor of three,

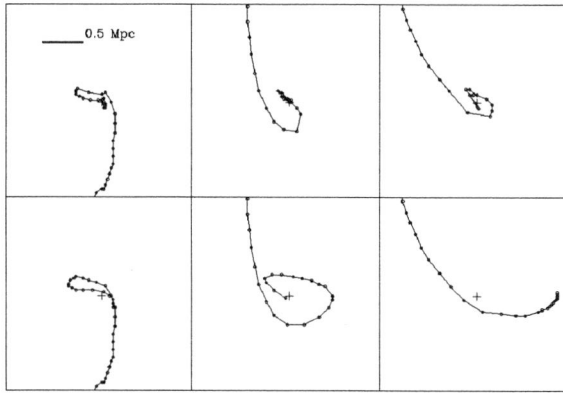

FIGURE 4. Orbits for 'galaxies' in the same cluster when treated as collisional (top row) or collisionless (bottom row) entities. Each row shows the orbit of the same object identified at $z=0.7$, the starting epoch for the collisionless run. From left to right are shown the 2nd, 12th and 14th most massive galaxies identified at that time. The cross marks the center of the cluster at the final epoch. Dots sample the orbits at intervals of roughly 2×10^8 yr, ending at $z=0$. The orbits are clearly sensitive to the dynamical treatment. In the pure SPH treatment, galaxies tend to hit the center and 'stick', whereas their collisionless counterparts sail through relatively unperturbed.

but analysis based on the full sample produces an estimate accurate to within 10%. The different behavior of the two mass groups is not a transient result, since the same trend exists at earlier redshifts. It may be that the result is 'inherited' from the SPH run via the initial conditions; the gravitational clustering being inefficient at erasing the memory of the initial bias.

Table 2 : Cluster Parameters from SPH → N-body Treatment

N_{cut}	N_{gal}	σ_{gal} (km s^{-1})	σ_{gal}/σ_{DM}	R_{gal}/R_{DM}	M_{vir}/M_{true}
32	52	479	0.89	1.08	0.92
128	17	386	0.71	0.56	0.32

To summarize, the issue of biases in the galaxy distribution within clusters remains uncertain. There appears to be a growing trend toward modest amounts of velocity bias, $\sigma_{gal}/\sigma_{DM} \sim 0.7 - 0.9$, but less clear ideas on the relative spatial distributions of galaxies and dark matter. The problem is complicated by the fact that equilibrium models are not appropriate for clusters formed in flat cosmologies, since they will generally have experienced considerable merging on a timescale comparable to their dynamical time.

6. Epilogue

Dynamical modeling of clusters of galaxies has improved significantly in the past few years, with the advent of simulation algorithms capable of handling the coupled evolution of multiple components representing dark matter, intracluster gas and galaxies. The new generation of experiments has yielded insight into the problem of the baryon fraction in clusters, and has provided details on the connection between cosmology and the X-ray morphology of clusters. The latter is a prime example of the type of problem which would be virtually impossible to tackle with a pure N-body approach.

Issues which are more intimately linked to galaxy/star formation (the FOE problem) remain relatively poorly understood. Although definitive answers to the question of dynamical biases for galaxies in clusters remain elusive, results emerging from a variety of independent treatments suggest that galaxies should give a velocity dispersion estimate biased slightly $(10-30\%)$ low with respect to the dark matter. Optical mass estimates are likely to underestimate the total binding mass, but the magnitude of this effect is fairly uncertain.

At this point, clusters appear schizophrenic regarding the value of Ω_o. Their X-ray morphology and abundance prefer $\Omega = 1$ over low Ω_o, but the baryon fraction favors the opposite. Their mass-to-light ratios could point in either direction, depending on the degree of dynamical biasing. This last issue will likely be settled within the next few years by very high resolution simulations incorporating star formation in a self-consistent fashion. In short, the deep waters of the non-linear evolution of clusters remain murky, but the edges of the pool, at least, are beginning to clear.

I would like to thank my collaborators in the above projects — J. Mohr, D. Fabricant, M. Geller, S.D.M. White, J. Navarro, C. Frenk, F. Summers and M. Davis — for allowing me to use our joint results in this article. This work was supported by NASA Theory grant NAGW-2367 and NSF via supercomputer resources.

REFERENCES

[1] Bertschinger, E. 1985, ApJS, 58, 39.
[2] Bertschinger, E. 1987, ApJ, 323, L103.
[3] Bodenheimer, P. 1994, these proceedings.
[4] Bryan, G. & Norman, M.L. 1994, these proceedings.
[5] Carlberg, R.G. 1991, ApJ, 367, 385.
[6] Carlberg, R.G., Couchman, H.M.P. & Thomas,P.A. 1990, ApJ, 352, L29
[7] Cen, R.Y. 1992, ApJS, 78, 341.
[8] Evrard, A.E. 1987, ApJ, 316, 36.
[9] Evrard, A.E. 1988, MNRAS, 235, 911.
[10] Evrard, A.E. 1989, ApJ, 341, L41.
[11] Evrard, A.E., Mohr, J.J., Fabricant, D.G. & Geller, M.J. 1993, ApJ, in press.
[12] Evrard, A.E., Summers, F.J. & Davis, M. 1994, ApJ, in press.
[13] Faber, S.M. & Gallagher, J.S. 1979, ARA&A, 17, 135.
[14] Frenk, C.S. 1994, these proceedings.
[15] Geller, M.J. 1984, Comments on Astr & Sp Sci, 2, 47.
[16] Gunn, J.E. & Gott, J.R. 1972, ApJ, 176, 1.
[17] Gingold, R.A. & Monaghan, J.J. 1977, MNRAS, 181, 375.
[18] Hernquist, L. & Katz, N. 1989, ApJS, 70, 419.
[19] Kang, H., Ostriker, J.P., Cen, R., Ryu, D., Hernquist, L., Evrard, A.E., Bryan, G. & Norman, M.L. 1993, ApJ, submitted.
[20] Kauffmann, G. & White, S.D.M. 1993, MNRAS, 261, 921.
[21] Katz, N. & White, S.D.M. 1993, ApJ, 412, 455.
[22] Lacey, C. & Cole, S. 1993, MNRAS, 262, 627.
[23] Lucy, L.B. 1977, AJ, 82, 1013.
[24] Metzler, C.A. & Evrard, A.E. 1993, ApJ, submitted.
[25] Mohr, J.J., Evrard, A.E., Fabricant, D.G., & Geller, M.J. 1993, in preparation.
[26] Mohr, J.J., Fabricant, D.G., & Geller, M.J. 1993, ApJ, 413, 492.
[27] Navarro, J.N. & White, S.D.M. 1993, MNRAS, in press.
[28] Navarro, J.N. & White, S.D.M. 1993, preprint.
[29] Pearce, F.R., Thomas, P.A. & Couchman, H.M.P. 1993, preprint.
[30] Richstone, D., Loeb, A., & Turner, E. L. 1992, ApJ, 393, 477.

[31] Sarazin, C.L. 1986, Rev Mod Phys, 58, 1.
[32] Shapiro, P. 1994, these proceedings.
[33] Walker, T.P. *et al.* 1991, ApJ, 376, 51.
[34] West, M.J. in *Clusters of Galaxies*, eds. W.R. Oegerle, M.J. Fitchett & L. Danly (Cambridge, Cambridge Univ. Press), p. 65.
[35] White, S.D.M. 1976, MNRAS, 177, 717.
[36] White, S.D.M., Efstathiou, G. & Frenk, C.S. 1993, MNRAS, 262, 1023.
[37] White, S.D.M., Navarro, J.N., Evrard, A.E. & Frenk, C.S. 1993, submitted to Nature.
[38] White, S.D.M. & Rees, M.J. 1978, MNRAS, 183, 341.
[39] Zwicky, F. 1933, Helv. Phys. Acta, 6, 110.
[40] Zurek, W., Quinn, P. & Salmon, J. 1988, ApJ, 330, 519.

Large-scale structure and motions from simulated galaxy clusters

By R. A. C. Croft, G. Efstathiou

Department of Physics, University of Oxford, Keble Road, Oxford, OX1 3RH, UK.

We use high resolution dissipationless N-body simulations to examine the spatial distribution of galaxy clusters on large scales. The Standard Cold Dark Matter (CDM) model and two of its main competitors, Low Density CDM (LCDM) and Mixed Dark Matter (MDM) are compared. The two-point correlation function of simulated clusters is compared with an extended survey of APM clusters, and it is found that the Standard CDM model exhibits a lack of power on *all* scales, whereas the two alternative scenarios are able to match the spatial correlations well. Of the remaining two models, the velocities in the MDM universe have a higher amplitude and their distribution is much broader. We compare these peculiar velocities with observations and find that both models have difficulty in reproducing the observed numbers of very high peculiar velocity clusters. The reliable detection of several more clusters with velocities in excess of 1000 km s^{-1} would render the LCDM scenario very unlikely.

1. Introduction

Clusters of galaxies efficiently trace out the large scale structure of the universe. Measurements of their spatial correlation function in 2 and 3 dimensions [11,3,13] provided some of the first evidence that there is more power in the clustering of matter than can be accounted for by the standard CDM model. More recently, surveys of rich clusters picked from computer generated galaxy catalogues [7,17], and a survey of ROSAT X-ray clusters [18] indicate that their three dimensional two-point correlation function has the following form:

$$\xi_{cc}(r) \approx (r/r_0)^{-1.8} \qquad r_0 \approx 13 - 16 \ h^{-1} \text{Mpc} \qquad (1.1)$$

(the Hubble constant is $H_0 = 100h$ km s^{-1} Mpc^{-1}). We have run a series of N-body simulations to determine whether equation (1) can be reproduced by the standard CDM model, or by variants such as a Low Density CDM or a Mixed Dark Matter model. The simulations, described in Section 2.1, are large enough to resolve individual clusters and sample a large enough volume of space to accurately determine the cluster two-point correlation function.

Measurements of a coherent large-scale velocity field [10] have also been cited as evidence against the standard model. Galaxy clusters play an important role in these measurements too. The observational errors on redshift-independent distances to galaxies (and therefore on their velocities away from Hubble flow, v=z-Hr, where r is the distance and z the redshift) can be large. However the distance errors to rich clusters can be reduced by $1/\sqrt{N}$ where N is the number of cluster galaxies with independent distance measurements. Here we compare cluster peculiar velocities with those measured from N-body simulations.

2. Simulating cluster formation

2.1. N-body simulations

The CDM-like N-body models are described in detail in ref. [5]. A P^{3}M code [9] is used to follow the evolution of the dissipationless component of the mass. We simulate a box of side 300 h^{-1}Mpc, with 10^6 particles per simulation and a spatial resolution of 80 h^{-1}kpc. For each cosmological model, we make 10 runs, using different random phases to generate the initial conditions. We simulate spatially flat universes, with Gaussian initial conditions and the initial power spectrum of mass fluctuations derived from linear theory calculations of the evolution of adiabatic fluctuations [12,19]. The three sets of models are as follows:

(a) Standard CDM, with $\Omega = 1$ and $h = 0.5$.

(b) Low density CDM (hereafter LCDM), with $\Omega = 0.2, h = 1.0$ and a non-zero cosmological constant, Λ, where $\Lambda/(3H_0^2) = (1 - \Omega) = 0.8$, introduced to make the model spatially flat.

(c) Mixed dark matter (hereafter MDM) with 7ev neutrinos: $\Omega_\nu = 0.3, \Omega_{CDM} = 0.6$.

For model (c), the power spectrum of total mass is used [12], but the effect of neutrino thermal velocities is not included, as here we study only the large-scale distribution of clusters and not their detailed internal structure. To find the 'present day', the abundance of rich clusters was fixed to be its observed value [19]. This criterion, together with a best fit to the COBE data requires that the linear amplitude of mass fluctuations in 8 h^{-1}Mpc spheres (σ_8) should be $\simeq 0.57$ for the $\Omega = 1$ models and $\simeq 1.0$ for LCDM. It will be shown in Section 3 that the spatial correlations are insensitive to changes in the value of σ_8.

2.2. Cluster selection

We find clusters by computing the mass contained within non-overlapping spheres of radius r_c centered on mass concentrations found by a percolation algorithm. In [5] we show that the catalogue of clusters picked out is insensitive to variations in r_c (here we use 0.5 h^{-1}Mpc). By a applying a lower mass bound we select clusters with the same space density as the observational sample we are comparing with, in our case the extended APM cluster redshift survey [7] (mean intercluster separation = 30.5 h^{-1}Mpc). As long as there is a roughly monotonic relationship between the mass of a cluster and its luminosity, then we can bypass any other assumptions concerning assignment of galaxies to the clusters.

3. Cluster spatial correlations

The two point correlation function of these clusters is shown in Figure (1) for our three different cosmologies, along with the observational points from the extended APM cluster survey. This survey consists of 364 clusters picked from the APM galaxy survey. The error bars are small, and it can be seen that ξ_{cc} for Standard CDM has too low an amplitude at all separations, unlike the other two models which fit the correlations much better over the full range $2 \leq r \leq 40$ h^{-1}Mpc. Varying the degree of evolution of the mass density field (σ_8) does not significantly alter the results. The thick lines on each plot are ξ_{cc} in redshift space from simulated APM-style catalogues, using the APM survey mask and selection function (see [5] for details). This does not really alter the results, save a slight tilt which brings models (b) and (c) into even better agreement with the data. Spatial modulation of the galaxy distribution by quasars [2] or 'cooperative' effects [4] has been put forward as a potential way of reconciling Standard CDM with observations that show too much large scale power. In the case of these cluster correlations, such effects would also have to work, with a high amplitude, on scales where the mass distribution is strongly non-linear, and it

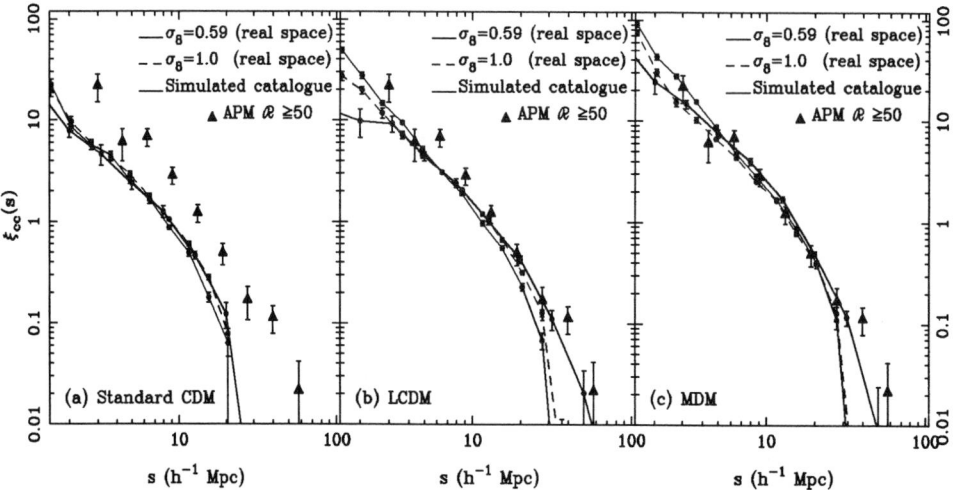

FIGURE 1. The two-point correlation function for simulated clusters with the same space density as APM $\mathcal{R} \geq 50$ clusters (mean intercluster distance $= 30.5\ h^{-1}$Mpc) . ξ_{cc} is plotted in real space for two values of the amplitude of mass fluctuations, σ_8 (thin lines). The error bars are derived from the scatter between 10 simulations for each ensemble. Also plotted (thick lines) is ξ_{cc} in redshift space, using simulated APM-style catalogues (see text). The observed ξ_{cc} calculated from the extended APM cluster redshift survey is shown, with Poissonian error bars.

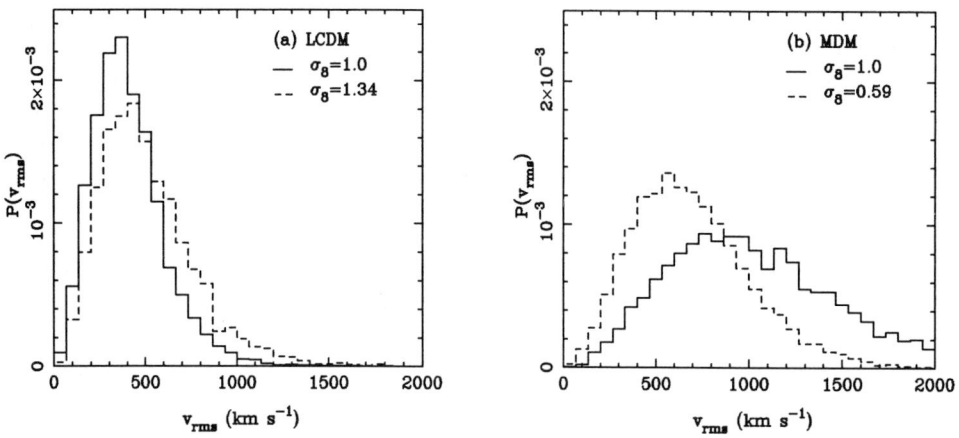

FIGURE 2. The distribution of rms cluster peculiar velocities for two different values of σ_8 in simulations of (a) Low density CDM with a cosmological constant and (b) Mixed Dark Matter.

seems unlikely that this is feasible. In the next section, we therefore choose to concentrate on differentiating between MDM and LCDM (the work is described in more detail in [6])

4. Cluster peculiar velocities

The distribution of three dimensional peculiar velocities for clusters in the LCDM and MDM simulations is shown in Figure (2). The distribution is much flatter, with a higher rms value (700 km s^{-1} as opposed to 400km s^{-1}) for MDM than LCDM. The histograms depend on the value of σ_8, so when comparing with observations, the curve corresponding to σ_8 given in Section (2) should be used.

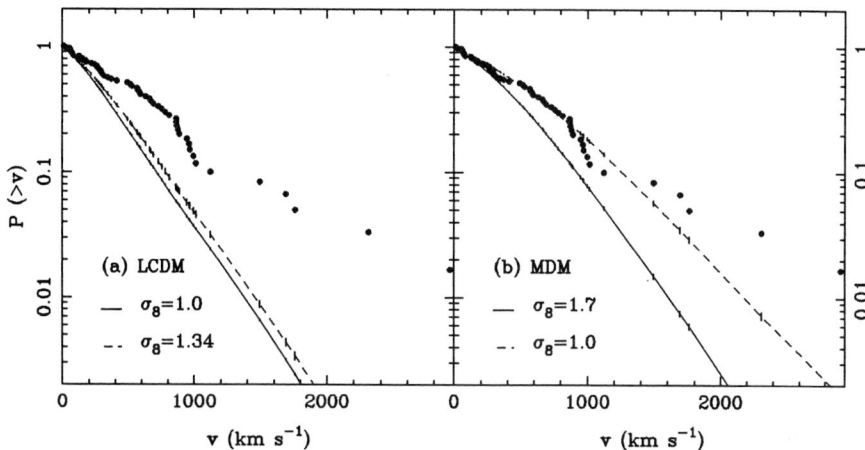

FIGURE 3. The cumulative probability distribution of cluster peculiar velocities. The points in each panel represent the observed sample (which is probably unrepresentative of the universe as a whole). The lines represent the values derived from the 10 numerical simulations for each model, which have been broadened with observational errors. (a) is Low Density CDM and (b) Mixed Dark Matter.

Observationally, the line-of-sight component of velocity is determined using both a redshift (z), and a distance indicator to give $v = H_0 r - cz$. Here we use an observational sample of 65 clusters in total – the distances were calculated using either the Tully-Fisher [1,14,15,16] or $D_n - \sigma$ relations [10]. We choose to exclude the 5 clusters with distance errors $\geq 850 \,\text{km s}^{-1}$. To compare with observations, we broaden our simulated cluster velocity distributions with the observational errors by adding an error to each velocity at random from a Gaussian distribution with dispersion equal to that of the observational error. This process is repeated 100 times for each simulated cluster and the resulting distribution smoothed with a Gaussian filter of width $400 \,\text{km s}^{-1}$.

In Figure (3) the probability of picking a cluster at random with a velocity greater than a certain value is shown. Finding clusters with high velocities is much more likely in the observational sample than in both models. However, the observational errors are certainly extremely large, and there may be important systematic effects - for example the $D_n - \sigma$ velocities are often larger than those determined using other methods. Another problem is the fact that the many of the observed clusters sample the flow to the Great Attractor - our 60 clusters are certainly not a 'fair sample'.

5. Conclusions

The correlation function of rich clusters of galaxies is a controversial subject, but it appears that simulations of the Standard CDM model where clusters are selected in a similar way to observations are unable to match values obtained from real samples. Of the remaining two models that we test, observations of peculiar velocities appear to favour a model with a high proportion of clusters with large velocities, such as MDM. Neither model matches the results that well, though. Whilst the flows detected by surveys are almost certainly real, their magnitude is debatable. For the models to be correct, we must live in a region which more heavily samples the high velocity tail of the distribution. This is particularly true of LCDM, which could be ruled out if the observed values used in this study turn out to be true, and represent a good sample of the universe.

This work was supported by the UK SERC.

REFERENCES

[1] Aaronson, M., Bothun, G., Mould, J., Huchra, J., Schommer, R. & Cornell, M. E., 1986. ApJ, 302, 536.
[2] Babul, A. & White, S. D. M., 1991. MNRAS, 253, 31P.
[3] Bahcall, N. A. & Soneira, R. M., 1983. ApJ, 270, 20.
[4] Bower, R. G., Coles, P., Frenk, C. & White, S., 1993. ApJ, 405, 403.
[5] Croft, R. A. C. & Efstathiou, G., 1994a. MNRAS, 267, 390.
[6] Croft, R. A. C. & Efstathiou, G., 1994b. MNRAS, in press.
[7] Dalton, G. B., 1993. in: *International Workshop on Galaxy Clusters and Large Scale Structures in the Universe*, SISSA, Eds. Giuricin, G., Mardirossian, F., & Mezzetti, M.
[8] Dalton, G. B., Efstathiou, G., Maddox, S. J. & Sutherland, W. J., 1992. ApJL, 390, L1.
[9] Efstathiou, G. & Eastwood, J., 1981. MNRAS, 194, 503.
[10] Faber, S., Wegner, G., Burstein, D., Davies, R. L., Dressler, A., Lynden-Bell, D. & Terlevich, R., 1989. ApJS, 69, 763.
[11] Hauser, M. G. & Peebles, P. J. E., 1973. ApJ, 185, 757.
[12] Klypin, A., Holtzmann, J., Primack, J. & Regos, E., 1993. ApJ, 416, 1.
[13] Klypin, A. A. & Kopylov, A. I., 1983. SovAstLett, 9, 41.
[14] Mathewson, D., Ford, V. & Buchorn, M., 1992. ApJS, 81, 413.
[15] Mould, J., Akeson, R., Schommer, R., Bothun, G., Hall, P., Han, M., Huchra, J. & Roth, J., 1993. ApJ, 409, 14.
[16] Mould, J., Stavely-Smith, L., Schommer, R., Bothun, G., Hall, P., Han, M., Huchra, J., Roth, J., Walsh, W. & Wright, A. E., 1991b. ApJ, 383, 467.
[17] Nichol, R. C., Collins, C. A., Guzzo, L. & Lumsden, S. L., 1992. MNRAS, 255, 21P.
[18] Romer, A. K., Collins, C., MacGillivray, H., Cruddace, R., Ebeling, H., Boringer, H., 1993. in: *International Workshop on Galaxy Clusters and Large Scale Structures in the Universe*, SISSA, Eds. Giuricin, G., Mardirossian, F., & Mezzetti, M.
[19] White, S. D. M., Efstathiou, G. & Frenk, C., 1993. MNRAS, 262, 1023.

Adaptive Smoothed Particle Hydrodynamics and Galaxy Formation

By Paul R. Shapiro[1], Hugo Martel[1], Jens V. Villumsen[2]

[1] Department of Astronomy, University of Texas, Austin, TX 78712, USA

[2] Department of Astronomy, Ohio State University, Columbus, OH 43210, USA

The development of a new Smoothed Particle Hydrodynamics (SPH) method, called Adaptive Smoothed Particle Hydrodynamics (ASPH), generalized for cosmology and coupled to the Particle Mesh (PM) method for solving the Poisson Equation, for the simulation of galaxy and large-scale structure formation, will be described. The accurate numerical simulation of the highly nonlinear phenomena of shocks and caustics which occur generically in the process of structure formation requires enormous dynamic range and resolution. Previously existing numerical methods require substantial modification in order to achieve the required resolution with current computer technology. The ASPH method incorporates new, adaptive, anisotropic smoothing and shock-tracking algorithms, which significantly enhance the resolving power of the SPH method. We describe tests of ASPH versus SPH against the difficult cosmological pancake collapse problem. All cosmological hydro methods should be required to reproduce this test. High resolution 2D simulations of galaxy and large-scale structure formation in the Hot Dark Matter (HDM) model are presented using ASPH, showing that ASPH can resolve pancake shocks with fewer than 40 particles per pancake per dimension.

1. Introduction

What must the capabilities of a successful cosmological hydro method be? Most theories of galaxy formation assume that structure formed when initially small amplitude, primordial density fluctuations grew by gravitational instability to nonlinear amplitude [1]. Such primordial density fluctuations and gravitational instability lead to gravitational collapse, strong shocks, and radiative cooling, occurring over an enormous range of mass and length scales. Shocks, in fact, are the principal mechanism by which gravitational motions are dissipated so as to make formation of bound, star forming objects possible.

Nearly 1D, plane-symmetric collapse to form structures known as cosmological pancakes is a generic feature of the nonlinear growth of a 3D spectrum of density fluctuations. While more complicated structures also arise, the pancakes provide us with an example of a minimally complicated flow which all cosmological hydro methods must be capable of reproducing. 1D pancake calculations [2] show that, in the absence of radiative cooling, a dynamic range of more than 10^3 in density and in length scale is required to resolve even a single pancake. *With* radiative cooling, the dynamic range increases to more than 10^5.

Are existing numerical hydro methods adequate? No. In order to achieve the dynamic range described above, necessary to follow randomly oriented pancake collapse, a 3-D, Eulerian grid (i.e. fixed spatial grid, uniformly spaced, nonadaptive mesh) must have more than 10^3 cells *per dimension*, or 10^9 cells, *per pancake* for simulations *without* radiative cooling and more than 10^5 cells *per dimension*, or 10^{15} cells, *per pancake with radiative cooling!* Current Eulerian methods are limited by existing hardware to $\approx 10^2$ cells per dimension.

Existing Lagrangian hydro methods such as SPH would seem to be a more promising alternative, since they, in principle, adjust their resolution dynamically to follow the flow. Cosmology simulations with SPH have typically used a number of particles $N_{gas} \sim 32^3$ or less, however. (Simulations with $N_{gas} = 64^3$ have recently been reported, but with dynamic range limited by imposing a minimum smoothing length.) As we shall demonstrate, this limitation and the requirement that artificial viscosity be used to treat shocks has kept the capabilities of this method below the level required.

In what follows, we describe the development of a new version of SPH called *Adaptive SPH* (ASPH) which greatly enhances the dynamic range and resolving power of SPH to levels which promise to resolve the shocks which are generic to galaxy and large-scale structure formation flows.

2. Standard Smoothed Particle Hydrodynamics (SPH)

The SPH method [3] is a Lagrangian numerical hydrodynamics method which replaces the continuous baryon-electron fluid by a set of discrete gas "particles" which carry mass and thermal energy and move with the local flow velocity. The particles are essentially moving centers of interpolation for representing the continuous flow variables. As such, they eliminate the need for a numerical grid or the interpolation back and forth between particles and grid. The principal virtue of this method is that, as a Lagrangian method, it has numerical resolution which dynamically adjusts so as to follow the compression, expansion, and distortion of the flow. It has demonstrated great success in a wide variety of astrophysical flow applications.

Isotropic Smoothing Length. The resolving power of the SPH method is related to the "smoothing length" h used for its interpolations, which defines a spherical "zone of influence" centered on each particle. Initially, this $h = h_0$, the mean interparticle spacing (or twice this). In order for the resolution to adjust dynamically to accommodate a changing gas density, the smoothing length must vary in time and space as a function of that density. Ideally, the variations of h should be such as to maintain a constant number of "nearest neighbor" particles, those within a distance of only a few smoothing lengths or less from a given particle. The interpolations which SPH performs in order to evaluate fluid quantities and their gradients anywhere within the computational volume involve sums over the values of those quantities known at the irregularly spaced positions of these "nearest neighbor" particles, weighted by a kernel function W, an isotropic function of the smoothing length, or its gradient. For an isotropic Gaussian kernel W in 3D,

$$W(\mathbf{r}_i - \mathbf{r}_j, h) = \pi^{-3/2} h^{-3} \exp(-r_{ij}^2/h^2), \qquad (2.1)$$

where h is a scalar which varies in space and time according to one of two choices,

$$h = h_p = h_0(\rho_p/\rho_0)^{-1/3} \qquad p = i \text{ or } j, \qquad (2.2)$$

where ρ_i and ρ_j are the density evaluated at the location $\mathbf{r}_i = (x_i, y_i, z_i)$ of particle i and the location $\mathbf{r}_j = (x_j, y_j, z_j)$ of particle j, respectively, $r_{ij} = |\mathbf{r}_i - \mathbf{r}_j|$, and ρ_0 is the initial mean density. The "nearest neighbor" particles are taken to be all those within a distance $r_{ij} \leq 3h$ of particle i.

Artificial Viscosity. Artificial viscosity must be introduced in the standard SPH equations in order to prevent particle trajectory crossing and allow shocks to form.

3. Adaptive Smoothed Particle Hydrodynamics (ASPH)

Standard SPH suffers from the following two limitations which become particularly serious in the generic flows which occur in the formation of galaxies and large-scale structure, involving gravitational collapse through orders of magnitude of compression and strong shock waves:

1) The isotropic variable smoothing length is valid for nearly isotropic compressions or expansions, but breaks down in the presence of strongly anisotropic compression such as in pancake collapse.

2) Shocks require artificial viscosity, but gravitational collapse then results in false "preheating" of supersonically infalling gas, far outside of the actual shock location. This artificially spreads out the shock-heating and can be disastrous for calculations which include radiative cooling, for example.

We solve these problems by (1) replacing the isotropic kernel and smoothing length by a fully adaptive, *anisotropic kernel* and *tensor smoothing length* **H** which tracks the anisotropy of fluid motions; and (2) *tracking the shock* and *restricting the viscous heating to particles at the shock*, as described below.

Anisotropic smoothing tensor **H**. In standard SPH, the zone of influence of a given particle is a sphere centered on that particle, with a radius of order a few smoothing lengths. The spatial resolution in this region of the fluid is given by the smoothing length, and is therefore isotropic. However, in situations involving strong anisotropy, such as 1D or 2D collapses, the resolution required in order to adequately describe the physical system strongly depends on direction. To solve this problem, we introduce the concept of *anisotropic smoothing tensor H*. In this new formalism, the spherical zone of influence of a given particle is replaced by a triaxial ellipsoid. Each ellipsoid is defined by a triad of mutually perpendicular vectors, the semimajor axes of the ellipsoid, which have three components each. We refer to this ensemble of nine components as the *H-tensor* of the particle. The H-tensors are dynamically evolved by the ASPH code using the components of the deformation tensor $\partial v_i/\partial x_j$ to follow the local deformation and vorticity of the flow. In ASPH, the 3D Gaussian kernel is given by

$$W_3(\mathbf{r}_{ij}, \mathbf{H}) = W_1(x', h_1) W_1(y', h_2) W_1(z', h_3) = \left(\frac{e^{-x'^2/h_1^2}}{\pi^{1/2} h_1}\right) \left(\frac{e^{-y'^2/h_2^2}}{\pi^{1/2} h_2}\right) \left(\frac{e^{-z'^2/h_3^2}}{\pi^{1/2} h_3}\right), \quad (3.3)$$

where $\mathbf{r}_{ij} = \mathbf{r}_i - \mathbf{r}_j$, W_1 and W_3 are the 1D and 3D Gaussian kernels, respectively, $h_1 = |\mathbf{h}_1|$, $h_2 = |\mathbf{h}_2|$, and $h_3 = |\mathbf{h}_3|$ are the semimajor axes of the ellipsoid, and x', y', and z' are the projections of $\mathbf{r}_{ij}$ along each of these axes. This technique has three strong advantages over standard SPH:

1) The resolving power of the SPH equations in regions with strong anisotropy can be increased by orders of magnitudes.

2) The H-tensors track the motion of the flow, and consequently each particle keeps roughly the same set of neighbors for many timesteps. This implies that the usually costly nearest neighbor search need be done only occasionally instead of at every time step, resulting in a large speed-up of the algorithm.

3) Caustics and shocks in the flow can be predicted to occur whenever a particle's H-tensor ellipsoid shrinks along one of its axes enough to make the volume go to zero. This forms the basis for our shock tracking algorithm described below.

Shock Tracking and Artificial Viscosity. The evolving H-tensors track the local deformation of the fluid. Without artificial viscosity, particle crossing would occur whenever a particle encounters a shock, and this would result in an "inversion" of the H-tensor for this particle, with the shortest axis shrinking to negative values. The ASPH method makes

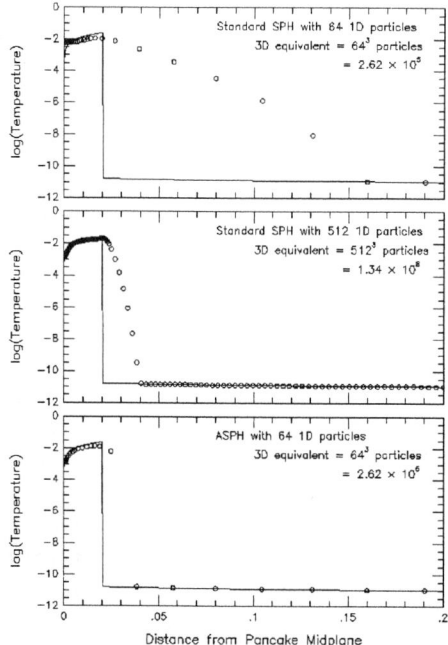

FIGURE 1. Gas temperature (in computational units) for 1D pancake test at $a/a_c = 1.5$. Numerical results (open circles) and exact 1D solution (solid line).

use of this phenomenon by turning artificial viscosity on *only* for particles whose H-tensor is about to undergo this inversion. This automatically restricts artificial viscosity to just those fluid particles which are actually encountering a shock.

4. The Pancake Collapse Test Problem

We focus in what follows on a tough test problem, that of the gravitational collapse of a 1D, plane wave density fluctuation in a universe comprised of baryons and collisionless dark matter. This is the cosmological pancake problem, in which an initially linear amplitude density fluctuation grows to nonlinear amplitude, forms a caustic in the dark matter distribution located in the plane of symmetry of the pancake and strong accretion shocks, one on each side of this central plane, followed by continued infall, phase-mixing of the dark matter, and radiative cooling of the shocked baryon-electron plasma. Detailed, 1D numerical solutions for this problem already exist [2,4,5], as do approximate analytical solutions, for comparison.

The initial condition is just that of the growing mode of a 1-D, sinusoidal, adiabatic, plane wave, cosmological density fluctuation of comoving wavelength λ_p in an Einstein-de Sitter universe. The perturbed comoving position x of the mass element whose unperturbed position is q is given by

$$x = q - \frac{\delta(t_i)}{2\pi}\lambda_p \sin \frac{2\pi q}{\lambda_p} \qquad (4.4)$$

at initial time $t = t_i$ at which the amplitude $\delta(t_i) \ll 1$. The initial density is given by

$$\rho = \frac{\bar{\rho}(t_i)}{1 - \delta(t_i)\cos 2\pi q/\lambda_p}, \qquad (4.5)$$

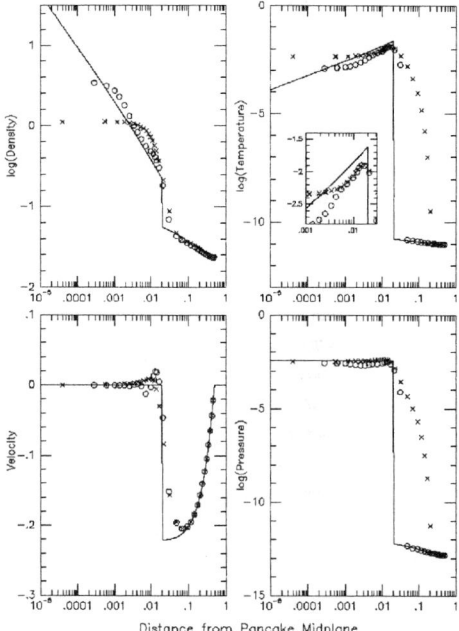

FIGURE 2. 2D tilted, shifted pancake test at $a/a_c = 1.5$. ASPH (open circles), SPH (crosses), and exact solution (solid line). Computational units.

and initial peculiar velocity is given by

$$v_x = -\frac{H_0}{2\pi}\lambda_p \left(\frac{a_i}{a_0}\right)^{-3/2} \delta(t_i) \sin\frac{2\pi q}{\lambda_p}, \qquad (4.6)$$

where a is the cosmic scale factor, $a_i = a(t_i)$, $a_0 = a_{\text{present}} = 1$, H_0 is the Hubble constant, and $\bar{\rho}(t_i)$ is the mean density at t_i. The perturbation is given to both the baryonic and dark matter components. A dark matter caustic forms at $x = q = 0$ and shocks form just outside the center, at time t_c and $a = a_c = a_i[\delta(t_i)]^{-1}$ For the pancake test runs, we have taken $\Omega_{tot} = 1 = \Omega_{DM} + \Omega_b$, with $\Omega_{DM} = \Omega_b = 0.5$, $h = 0.5$, $a_i = 1/28$, $a_c/a_i = 4$, and units in which $\lambda_p = 1$. Radiative cooling is neglected.

4.1. 1D Results: ASPH vs. Standard SPH

We begin with a 1D, plane-symmetric version of both the ASPH and the standard SPH methods, in which the particles may be thought of as plane sheets, and gas and dark matter particles are of equal mass. These 1D calculations are pseudo-3D, in the sense that the smoothing length for standard SPH varies according to $h \propto \rho^{-1/3}$, while the ASPH tensor H component along the collapse direction (evolved using $\partial v/\partial x$) varies as $h_x \propto \rho^{-1}$. We consider a range of gas particle numbers, N_{gas} (i.e. resolutions), distributed across one period λ_p, in a sequence of independent runs.

Figure 1 shows a comparison of the 1D SPH and ASPH results plotted against the analytical solution for a time-slice $a/a_c = 1.5$, well past the time of shock and caustic formation. This figure shows how the two flaws of SPH described above (inability to track a 1-D planar collapse with an isotropic $h \propto \rho^{-1/3}$, and artificial viscosity preheating) force the method to require an enormous value of N_{gas} to match the analytical solution while ASPH does better with very much smaller N_{gas}.

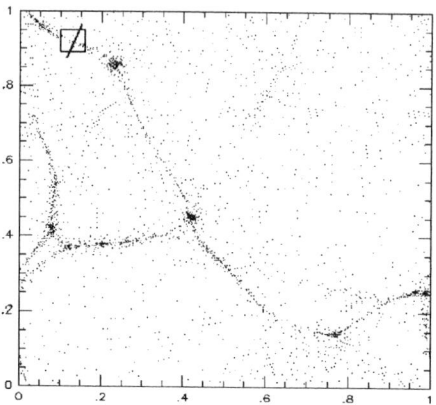

FIGURE 3. 2D HDM results at $z = 0$: ASPH gas particle positions. Small box is enlarged in Fig. 4.

4.2. 2D Results: ASPH vs. Standard SPH

Our ASPH method, in principle, can automatically identify and adjust to accommodate a pancake collapse and shocks along *any* direction, not known *a priori*. Testing this fully requires at least a 2D calculation (i.e. in this case (x,y) Cartesian coordinates). We, therefore, reconsider the pancake problem, except we solve it using a 2D version of the SPH and ASPH methods, respectively, and solve the gravitational force problem by using a standard Particle-Mesh (PM) method [6,7]. For 2D SPH, the smoothing length $h \propto \rho^{-1/2}$.

Tilted, Shifted Pancake We consider a square box with periodic boundary conditions and one edge-on pancake of wavelength $\lambda_p = L_{box}/\sqrt{2}$, oriented with its symmetry plane (*i.e.* density maximum) at 45° with respect to the walls of the computational box, and shifted away from the box center. We take $N_{gas} = N_{DM} = 64 \times 64 = 4096$ particles of each type (equal mass particles) with initial positions displaced by the perturbation away from the same square uniform lattice. This is designed to test that ASPH can adjust to follow the anisotropic collapse of the pancake regardless of its orientation. (*NOTE:* In this case, given the 45° angle of tilt, there are now only 22 gas particles per row perpendicular to the pancake to resolve the flow for each side of the pancake central plane.) The PM calculation uses a square lattice grid with $128 \times 128 = 16,384$ cells per box.

Results are shown at time-slice $a/a_c = 1.5$, along with the exact solution, in Figure 2. The ability of ASPH to match the analytical solution with *no preheating* and with the slope of the postshock profiles in agreement over orders of magnitude of position and density variation is in contrast to the standard SPH results which show preheating, shock spreading, and poor postshock profile fitting.

5. 2D, Cosmological Hydro Simulation Using ASPH/PM: HDM Universe

We apply our ASPH method in 2D to simulate the growth of large-scale structure in a universe dominated by HDM, with density fluctuations given by Gaussian random noise based upon a Harrison-Zel'dovich primordial power spectrum ($n = 1$) with amplitude fixed by the COBE satellite detection of CMB anisotropy. The HDM power spectrum we use is $P(k) = <\delta_k^2> = Ak^{n+1}10^{-1.5(k/k_D)}$, where $k_D = 2\pi/\lambda_D$, λ_D = damping length $\cong 13\,\text{Mpc}(\Omega h^2)^{-1}$ (i.e. 2D power spectrum is same as 3D, but multiplied by $2k$—equal rms $\delta\rho/\rho$ for equal fluctuation scale length).

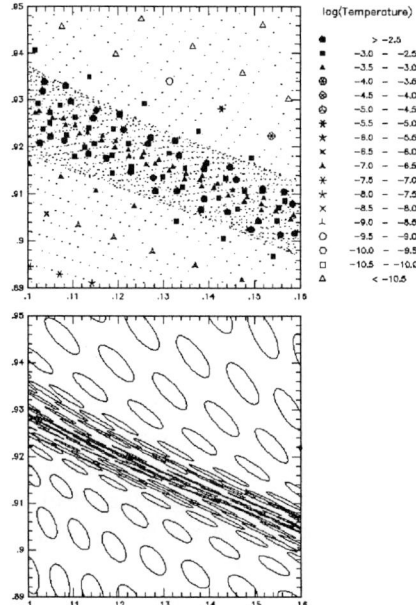

FIGURE 4. Zoom-in enlargement of small box in Fig. 3. (Upper panel) Particle positions: HDM (small dots), gas (other symbols) with temperature bin indicated [computational units= $T(°K)/2.57 \times 10^9$]. (Lower panel) H-tensor ellipses (axes scaled by 0.15)

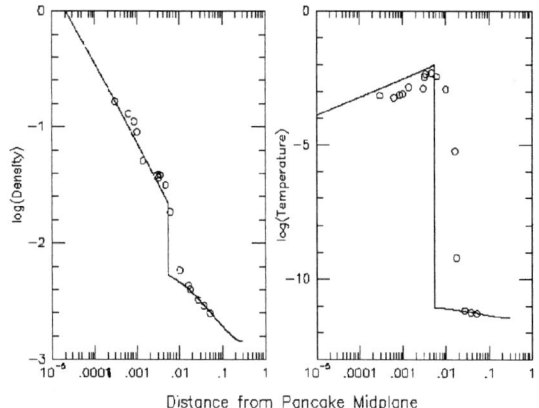

FIGURE 5. Line "cut" thru pancake in Figs. 3 and 4, with gas particles (open circles) and analytical pancake solution (line), for one side of pancake central plane only. Distance units have $L_{box} = 1$.

We take $h = 0.5$, $\Omega = 1$, $\Omega_{DM} = 0.9375$, $\Omega_b = 0.0625$, $N_{particles} = 512 \times 512 = 262,144$ total, $N_{gas} = 128 \times 128 = 16,384$ particles, with initial positions perturbed away from a uniform square lattice in a simulation box of side $L_{box} = 200\,\text{Mpc}$ with 1024×1024 PM cells and periodic boundary conditions, at initial redshift $z_i = 24$. The initial density fluctuations are represented as the superposition of plane wave perturbations with random phases including all wavevectors which satisfy periodic boundary conditions with dimensionless wavenumber $k = L_{box}/\lambda$ in the range $1 \leq k \leq 256$. Radiative cooling is neglected.

5.1. Results: ASPH resolves Pancakes

We show gas particle positions at $z = 0$ in Figure 3, indicating the prevalence of pancake-like structure (in 2D, lines are edge-on pancakes), along with edge-on filaments and clusters at pancake intersections. The small box in the upper left corner, centered on one such edge-on pancake, is enlarged in Figure 4, showing both dark matter and gas particles in the top panel and the anisotropic smoothing tensor ellipses surrounding each gas particle in the lower panel. The temperature symbols for the gas indicate that the two pancake shocks are sharply resolved with a separation of order 10^{-2} times L_{box} with the hottest immediate postshock particles forming two plane layers between which are cooler, previously shocked gas particles, as expected for pancake collapse. To show how well the ASPH results resolve pancake shocks like this, including the postshock profiles, we have plotted in Figure 5 the temperature and density for particles located in a "cut" thru the pancake, in a band centered on the line segment shown inside the small box in Figure 3, drawn perpendicular to the pancake central plane, versus logarithmic distance measured perpendicular to that plane. We use the gas velocity profile in the numerical data to identify the location of the shock, the pancake wavelength, and the bulk peculiar drift velocity of the entire pancake, and plot the exact 1D solution for such a pancake against the numerical results. The numerical results provide a remarkably good match to this idealized 1D pancake solution, indicating that ASPH succeeds in resolving pancake shocks on scales of order 10^{-3} times L_{box}! This required fewer than 40 gas particles per dimension per pancake. Our simulation took approximately 10 hours of Cray Y-MP time running at 153 Megaflops.

We acknowledge NASA NAGW-2399, Robert A. Welch F-1115, the UT System Center for High Performance Computing, the NSF NCSA, and the Ohio Supercomputer Center.

REFERENCES

[1] Peebles, P. J. E. 1993, *Principles of Physical Cosmology* (Princeton: Princeton University Press).
[2] Shapiro, P. R., Struck-Marcell, C. 1985, ApJS, 57, 205.
[3] Monaghan, J. J. 1992, ARAA, 30, 543.
[4] Shapiro, P. R., Struck-Marcell, C., Melott, A. L. 1983, ApJ, 275, 413.
[5] Bond, J. R., Centrella, J., Szalay, A. S., Wilson, J. R. 1984, MNRAS, 210, 505.
[6] Hockney, R. W. & Eastwood, J. W. 1981, *Computer Simulation Using Particles* (New York: McGraw-Hill).
[7] Villumsen, J. V. 1989, ApJS, 71, 407.

Mergers and Galaxy Formation

By Julio F. Navarro

Physics Department, University of Durham, Durham DH1 3LE, UK.

I discuss recent developments regarding the role played by mergers during the formation and transformation of galaxies. I examine briefly the status of the "merger hypothesis"; the proposal that most early-type galaxies are the outcome of major mergers of disk systems. Recent numerical simulations show than an extreme version of this hypothesis, where ellipticals form by the dissipationless merger of two stellar disks similar to those in present-day spirals, may be ruled out. I also report on recent numerical simulations which show that galaxies do *not* merge immediately after their surrounding halos coalesce. The magnitude of this effect may be enough to solve the long-standing "overmerging" problem affecting hierarchical clustering theories of structure formation and to reconcile the thinness of stellar disks with a high density universe.

---

1. Disk mergers and ellipticals

Ever since Alar and Juri Toomre [26] showed that morphological peculiarities such as tails and rings could be understood as the result of collisions between stellar disks, the idea that encounters may be partly or fully responsible for the observed morphology of galaxies has been gradually gaining acceptance. Stellar disks are particularly susceptible to the fluctuations in the gravitational potential that arise during collisions. In fact, the remnant of a major collision between disks leading to a merger is a spheroidal system quite dissimilar from its progenitors and morphologically closer to ellipticals than to spirals. This was recognized by Toomre [27], who proposed that most elliptical galaxies were actually the remnants of ancient collisions between stellar disks.

His argument was mainly statistical; for reasonable, but somewhat arbitrary, choices of orbital parameters and merger frequencies the number of on-going mergers observed today seemed consistent with this "merger hypothesis" for the origin of ellipticals. Actually, this argument is rather weak, as similarly acceptable choices of merger frequencies can produce either too-high or too-low a fraction of ellipticals [12]. Unless the frequency of galaxy mergers can be specified precisely, allowing for the strong dependence of orbital decay times on the orbital parameters and masses of the systems involved, it is difficult to make a case for or against the "merger hypothesis" based only on the number of ellipticals and tidally disturbed galaxies.

Despite this difficulty, the idea of making ellipticals through mergers caught the imagination of many researchers, and a number of numerical studies were soon under way to investigate the dynamical effects of major collisions on model galaxies. The small number of particles that could be used at that time prevented the use of realistic disk galaxy models and, with few exceptions, the premerger galaxies were modelled as spherical systems of particles [10,11,21,22,29,30,32,33]. Despite their simplicity, these experiments provided us with a first insight into the evolution of stellar systems affected by mergers. Some early results seemed to favour the merger hypothesis, and have been confirmed by more sophisticated simulations. For example, i) remnants of mergers tend to develop a density profile that in projection resembles the surface brightness profiles of ellipticals; *i.e.* a de Vaucouleurs law; and ii) energy hierarchies are preserved during a merger; the core and halo of a remnant are predominantly populated by particles in the core and halo of the progenitors, respectively. This inefficient mixing neutralizes arguments against the merger hypothesis based on the presence of metallicity gradients in ellipticals.

The apparent success of the merger hypothesis was, however, marred by other shortcomings. Simple scaling arguments based on the conservation of energy and angular momentum suggest that merger remnants should rotate quite fast and be much less dense than premerger galaxies [34]. This is a major obstacle for the merger hypothesis because ellipticals rotate quite slowly and are much denser than typical spiral disks. It was not until the late 80's that simulations of mergers of multicomponent galaxies proved that the luminous components of galaxies did not conform to the simple scalings mentioned above [1,2]. Disks and bulges are able to shed orbital energy and angular momentum to the surrounding halo, and merge to form a slowly rotating, dense, triaxial spheroid supported by an anisotropic velocity dispersion tensor. The similarity between the dynamical properties of these spheroids and ellipticals thus lifts one important objection to the merger hypothesis [13,14].

The high density of ellipticals has proved more difficult to reconcile with the merger hypothesis. Phase space densities can only *decrease* during a dissipationless merger, and ellipticals have on average higher central phase space densities than disks [6]. However, [19] argued that acceptable disk models may also include a fair fraction of material at high phase-space densities, although this material need not be located at the center of the disk. Recent N-body simulations seem to have tipped the balance against the merger hypothesis; mergers of stellar disks like the ones we see in spirals today fail to produce dense enough spheroids [13]. In view of this latest evidence, opinion seems to be converging towards accepting that an extreme version of the merger hypothesis, where all ellipticals are made by the dissipationless merger of two present-day spiral disks, may be ruled out. If most spheroids have been indeed formed by mergers, their progenitors must have been quite different from present-day spiral disks. For example, it seems likely that these progenitors contained a fair amount of gas, and so dissipative effects not included in these N-body simulations may have been important in determining the structure of the remnants. Dissipative mergers remain an attractive possibility for the origin of spheroidal galaxies, especially if mergers are indeed as common as expected in current hierarchical models of structure formation.

Recent reviews of the status of the merger hypothesis can be found in Barnes & Hernquist [3] and Hernquist [12], where the interested reader may consult further details and references. Let me now turn to the analysis of the role of mergers in the formation of galaxies from a cosmological prespective.

2. Mergers in hierarchical clustering models

Mergers are a natural feature of models where structure grows hierarchically by gravitational instability of small density perturbations present at early times. In these models, mergers are not the exception but the norm during the formation of galaxies, and perhaps all galaxies, irrespective of morphological type, have seen a good fraction of their mass accreted through mergers. Just how large a fraction, as well as when these mergers occur, depend on the details of each specific cosmogonical model and on the coupled interaction between gas, stars, and dark matter. In what follows, I will concentrate on one particular scenario for structure formation, the Cold Dark Matter cosmogony. Although there is mounting evidence that the shape of the CDM power spectrum is inconsistent with observation [25], we may still think of CDM as one example of the more general class of hierarchical clustering theories. Many of the conclusions that apply to the CDM model should also be applicable to various scenarios where structure grows hierarchically.

The growth of structure in hierarchical clustering models has traditionally been studied using N-body simulations, and has recently been supplemented by the development of analytical techniques which describe in detail the evolution of the dark matter component [4,5,16,17]. As a result, it is now possible to predict with reasonable accuracy how the statistical properties of the gravitationally dominant component of the universe evolve in models such as CDM. Once the cosmological model has been specified, it is relatively straightforward to compute the number of mergers a typical galactic halo has suffered during its lifetime, together with the mass spectrum of its merging companions.

Despite the notable success of these models in predicting the evolution of dark halos, we still need to consider how galaxy formation progresses in these dark halos before detailed comparison with observation is possible. The standard lore is that galaxies form when gas, robbed of its pressure support by dissipative effects, is driven to the centers of collapsing dark halos [35]. The gas becomes denser as it collapses and radiates energy even more efficiently. This is a runaway process that can only be halted when the gas forms a centrifugally supported disk or is turned into stars. Following this complicated process with numerical experiments involves solving the equations which govern the evolution of a collisional fluid, including the many additional physical processes which influence its structure: star formation and evolution; energy input from stellar winds, stellar radiation, and supernova explosions; ionization effects from stellar and QSO radiation fields; heavy element enrichment; heat conduction; and magnetic fields. Incorporating these processes presents a serious challenge to any attempt at simulating numerically the evolution of gas and star, not only because of their complexity, but also because many of the physical processes enumerated above are still poorly understood.

Because of these difficulties, many issues regarding the formation of galaxies in hierarchical clustering scenarios remain unresolved. For example, although we know when and how dark halos merge in a specific cosmogony, the same is not true for galaxies. We do know, however, that galaxy mergers should be much less frequent than those of their surrounding halos; it has long been clear that any viable hierarchical clustering scenario for galaxy formation should contemplate mechanisms that delay the merger of galaxies relative to that of their surrounding halos. This is because tidal effects are quite efficient at erasing substructure from halos when they merge; after a few dynamical times the remnant of the merger of two halos regenerates a monolithic core-halo structure which retains little evidence of the collision. On the other hand, in deep potential wells such as galaxy clusters galaxies can retain their identity after the collapse of the cluster, and remain as distinct entities while they orbit in the potential well of their stripped halos.

I will present here preliminary results of a large numerical study aimed at addressing some of these problems. In particular, I would like to address the following questions. Is there any appreciable delay between the merger of galaxies and halos? How well do simple estimates based on dynamical friction fare in predicting orbital decay times for merging systems? What are the typical parameters of the orbit of two merging galaxies? Do these parameters depend on the mass of the galaxies involved? How many satellites has a typical galaxy accreted in recent times? Is this accretion compatible with the present abundance of thin stellar disks? Since these problems require a statistical approach, we have simulated the formation of 30 galactic halos with circular velocities ranging from ~ 100 km s^{-1} to ~ 300 km s^{-1}. This is probably a large enough sample to justify a first examination of the problems mentioned above. This work is part of an ongoing collaboration with Carlos Frenk and Simon White.

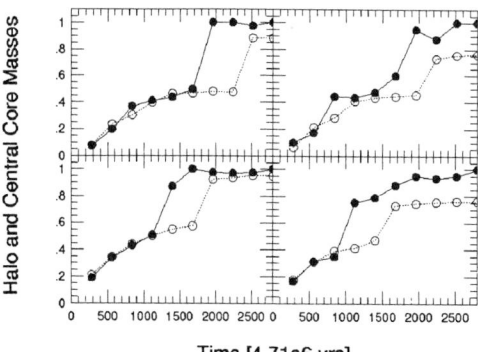

FIGURE 1. The evolution of the mass of the central gaseous core (open circles) and of its surrounding dark halo (filled circles) in four of the 30 experiments. Masses are in units of the total mass of dark material and gas within a sphere of overdensity 200 at $z = 0$.

2.1. The numerical simulations

We have used a gridless, fully Lagrangian code that combines a tree-based N-body algorithm with the Smooth Particle Hydrodynamics (SPH) approach to numerical hydrodynamics. The code is highly adaptive in time and space via the use of individual smoothing lengths and time steps. Several physical processes are modelled in the version of the code used for these runs; including self-gravity, pressure gradients, shock heating, and radiative cooling. We use the cooling curve appropriate for a plasma with primordial composition, and neglect cooling at temperatures below 10^4 K, where radiative cooling becomes ineffective. Our particular implementation of SPH, as well as tests relevant to the issues discussed here, are presented in Navarro and White [23].

Our procedure for generating initial conditions involves several steps. We first carry out a set of N-body simulations of a standard CDM universe with $\Omega = 1$, $H_0 = 50$ km s^{-1} Mpc^{-1}, and fluctuation amplitude, $b = 1/\sigma_8 = 1.53$; these use a P^{3}M code to follow 64^3 particles in a $(30 \text{ Mpc})^3$ box (see Efstathiou et al. 1985). We then identify all clumps at $z = 0$ with circular velocity in the range 100-300 km s^{-1} (measured at the radius where the mean enclosed overdensity is 1000), and pick 10 clumps at random in each of three circular velocity bins of width ~ 20 km s^{-1} and centred on 105, 160, and 240 km s^{-1}.

The particles in each clump are traced back to the initial time, where a box containing all of them is drawn. This box is loaded with 16^3 particles which are perturbed using the *same* waves as before, together with additional waves up to the Nyquist frequency of the new box. The size of this "high-resolution" box is 3.3, 4.8, and 6.6 (comoving) Mpc for systems in the first, second and third circular velocity bins, respectively. Tidal effects due to more distant material are represented using several thousand massive particles as in [15]. A gas particle is placed on top of each dark matter particle in the "high-resolution" region, is given the same initial velocity, and is assigned ten per cent of its mass. Thus we model a universe with baryon density, $\Omega_b = 0.1$. This value is slightly higher than derived from nucleosynthesis arguments, and was chosen in order to promote efficient cooling and the formation of low-mass satellite galaxies, the study of which is another important objective of this study. Initial gas temperatures are chosen to be very low, so that gas and dark matter follow similar trajectories until nonlinear clumps turn around and hydrodynamical effects become important. The gravitational softening in these simulations is 10 kpc, and so we are unable to resolve the internal structure of the gas disks which form at the centres of dark halos. However, our resolution is sufficient to determine when satellites merge into these disks.

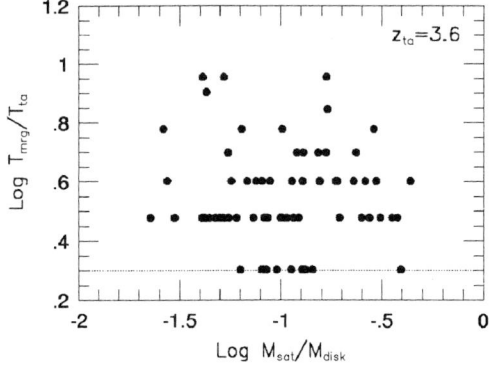

FIGURE 2. Merging times of satellites which turned around at $z \approx 3.6$, as a function of the satellite's gas mass. Merging times are given in units of the satellite's turn-around time T_{ta}. Masses are given in units of the central gaseous core mass after the merger.

2.2. *The evolution of gas and dark matter in a galactic halo*

The overall evolution of these runs is described in detail in [24]. At early times gas radiates efficiently and collects at the centre of virialized dark halos. There it forms cold, tightly bound cores with size determined primarily by the gravitational softening. These cores are surrounded by a hot tenuous atmosphere containing only a small fraction of the gas. As halos collide and merge, dense cores settle towards the centre of each new halo, eventually merging into a single object there. Practically no gas is reheated during this process, and the central core mass increases steadily both through mergers and through additional cooling of hot gas.

These simulations confirm theoretical predictions that the gas that can cool and make galaxies is excessive [7]; in the absence of non-gravitational reheating mechanisms essentially all the baryons in the universe should be locked up at the center of virialized halos, in disagreement with observations. Mergers can also rob the central galaxies of much of their angular momentum. In consequence, gaseous disks formed in hierarchical clustering scenarios end up more concentrated than real spirals. Inclusion of energy feedback from supernovae (or some other physical process preventing efficient cooling at high redshift) is an essential requisite of any satisfactory hierarchical model of galaxy formation.

At late times almost all of the gas is in the form of dense cores. We identify the large central core with the primary galaxy and the cores in orbit around it as satellites or companion galaxies. In order to identify a gaseous clump with a satellite, we require it to be cold ($T \approx 10^4$ K) and dense (it should have more than 30 gas particles within a sphere of radius 20 kpc from its center). Gaseous clumps satisfying this criterion prove to be remarkably resilient to tidal perturbations, and they maintain their identity until they merge with the central core.

2.3. *Galaxy and Halo Mergers*

Galaxy mergers are delayed respect to halo mergers in these simulations. The delays can be easily seen in Figure 1, where we show the time evolution of the mass accreted by 4 systems in our simulations. The mass of the dark halo (filled circles) is always computed within a sphere of mean overdensity 200, while the mass of the central galaxy is computed within 20 kpc from the center of the most massive gaseous core. Careful inspection shows that this includes all the cold, dense gas associated with the central object at all times. If the mass of the halo and the gaseous core were accreted at the same rate, the two curves would overlap. Sudden jumps in mass are associated with merger events, and it is easy

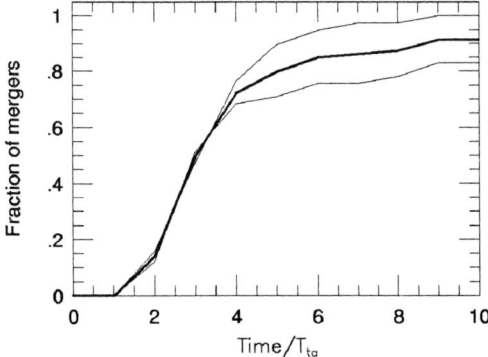

FIGURE 3. Cumulative fraction of mergers as a function of time for an ensemble of satellites that turned around at the same redshift, $z \approx 3.6$. The thick solid line indicates the result for all satellites; thin lines are the result of dividing the sample in two mass bins.

to see that the merger of the central cores may take up to 4 Gyr longer than the dark halos. This is because gaseous cores are denser and harder to disrupt tidally; provided that they are not on a radial orbit, they miss each other at first pericentric approach, and they orbit around each other a few times before finally merging. Although most of the gas is in the central core at $z = 0$ (more than $\sim 80\%$ in all these cases), many systems have small satellites orbiting around the primary.

It is important to ask whether merger delays are statistically important, *i.e.* how many galaxies experience "delayed" mergers and how long on average do they take to merge? Satellites and merging companions first expand away from the central core, reach a maximum radius, turn around at some time T_{ta}, and collapse towards the center. In the case of spherical accretion in an Einstein-de Sitter universe, radial shells reach the center at time $t = 2T_{ta}$. In real systems, the collapse will not be exactly radial, but still the turnaround time T_{ta} is a good unit to express the merger times of a satellite. Figure 2 shows merger times for the 79 satellites that turned around at $z \approx 3.6$, in units of the total mass of the central core (the "disk"). We have chosen to plot only satellites that turned around at about the same time because they constitute a well-defined, homogeneous sample. Only a very small fraction of satellites merge at $2T_{ta}$, and some can even take $\sim 10T_{ta}$ to merge. Satellites which survive longer are still orbiting around the primary at the present time. Only a small fraction of the satellites ($\sim 9\%$) that turned around at $z \approx 3.6$ are still orbiting around the primaries today.

A weak dependence with mass can be seen in Figure 2. This indicates that more massive satellites merge more rapidly than less massive companions, as one would naively expect if the orbital decay of these systems were primarily caused by dynamical friction. This is also seen in Figure 3, where we plot the cumulative fraction of satellites that merged with the central core as a function of time. The thick line corresponds to all satellites, while the thin lines show the result of subdividing the satellite sample in two mass bins. On average, half the satellites which turned around at $z \approx 3.6$ have already merged by $z \approx 1.2$. Massive satellites take less time to merge, so the *mass* of the central core is actually quite close (to within $\sim 30\%$) to the total mass of gas within each halo at essentially all times, even when some satellites may fail to merge [24].

Figure 4 shows a comparison between simple dynamical friction estimates for orbital decay times and the results of the simulations. Dynamical friction times are computed when the satellites first enter the main halo, that is, when they cross for the first time the outer radius of the sphere of mean overdensity 200 surrounding the primary gaseous core.

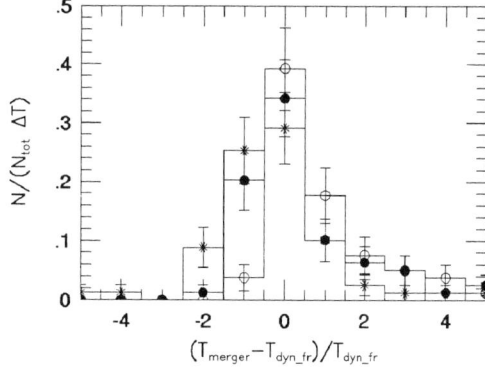

FIGURE 4. Merging times compared with predictions based on the dynamical friction formula of Lacey & Cole [17]. Filled circles take into account the eccentricity and halo of the satellites; open circles assume that all satellites are on rather eccentric orbits, $e = 0.1$; and starred symbols assume that all satellites are on circular orbits. The medians of each distribution are 0.24, 0.95, and -0.32, respectively. A standard KS test shows that all three distributions are significantly different.

The filled circles in Figure 4 show the result of taking into account in the estimates the eccentricity of the satellites' orbits and the masses of their own dark halos. Neglecting the satellites' halos leads to very large overestimates of the merging times. We use the orbit-averaged dynamical friction timescales derived in [17]. Starred and open circles indicate the result of assuming that all satellites are either on circular ($e = 1$) or nearly radial ($e = 0.1$) orbits when they first approach the main core. The eccentricity, e, used in these timescales is defined as the ratio of the angular momentum of the satellite to that of a circular orbit with the same energy, and varies from 0 to 1 for a radial and circular orbit, respectively.

The distribution of eccentricities at the time of first approach is shown in Figure 5, where we also show the result of dividing the sample in two mass bins. The eccentricity distribution is rather flat, with a slight preference towards high angular momentum orbits. Based on the good agreement between the three lines, there does not appear to be any significant correlation between the eccentricity and the mass of a satellite. Figures 4 and 5 indicate that, when proper account is taken of the orbital parameters and dark halos of the satellites, simple dynamical friction estimates give on average a good indication of when these satellites will merge with the central disk.

3. Cosmological implications

Let me now discuss briefly two implications of the results presented in the previous section. The first refers to the amount of mass accreted by galactic disks in recent times. This is important because the presence of thin stellar disks argues against recent mergers for most spirals. Taking as an example the Milky Way, Tóth & Ostriker [28] conclude that not more than $\sim 10\%$ of the disk mass could have been accreted in the form of mergers in the past 5 Gyrs, a result that they interpret as ruling out $\Omega = 1$.

The second implication refers to the shape of the luminosity function in a hierarchical clustering scenario such as standard CDM. Is the delay between galaxy and halo mergers enough to prevent all galaxies in a cluster from merging together to form a supergalaxy at the center? Another related question is whether the frequency of galaxy mergers derived in the previous section has any influence on the faint end slope of the galaxy luminosity function.

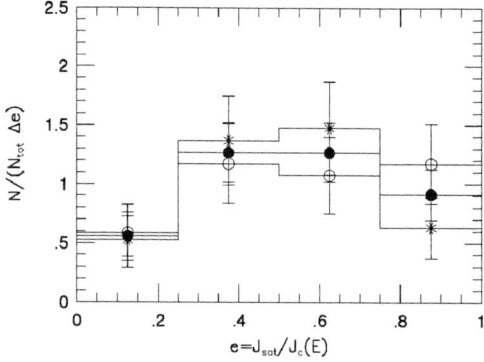

FIGURE 5. The distribution of orbital eccentricities of the satellites' orbits when they first approach the primary (filled circles. Open and starred circles show the result of dividing the sample in two mass bins.

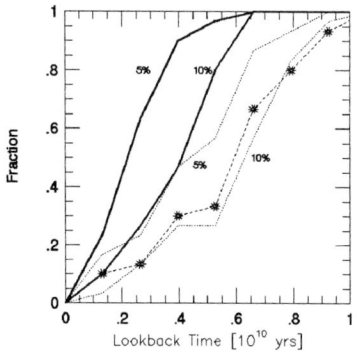

FIGURE 6. The fraction of systems which have accreted more than a certain fraction of their final mass in the time interval indicated (see text).

3.1. *The abundance of thin disks and the value of* Ω

It is straightforward to compute how much material has accreted recently onto halos and onto disks in our 30 simulations. In Figure 6, the solid curves give the fraction of *dark halos* which have increased in mass by more than 5% and 10% respectively over the indicated time interval. Halo masses are again computed within a sphere of mean overdensity 200, centred on the most massive core. Over the past 5 Gyr (*i.e.* since $z = 0.38$), about 70% of the dark halos have accreted at least 10% of their mass and more than 90% have accreted at least 5%.

The dotted lines correspond to the fraction of *disks* which have grown by more than 5% and 10% respectively. We identify the disk mass with the mass of the central gas core. This material always lies within ~ 20 kpc and accounts for 30-50% of the total mass in this region. The difference between the solid and dotted curves indicates that disks are less efficient at accreting mass than are the dark halos in which they are embedded. In fact, fewer than 30% of the disks grow by more than 10% in the last 5 Gyr. This is because most of the recently accreted gas is in the form of small compact satellites which are still in orbit at $z = 0$.

Tóth & Ostriker [28] give several alternative forms of their criterion for the preservation of thin disks. For our purposes, a particularly convenient version is that the Galaxy should have accreted less than 1.5% of the total mass within 25 kpc in the past 5 Gyrs.

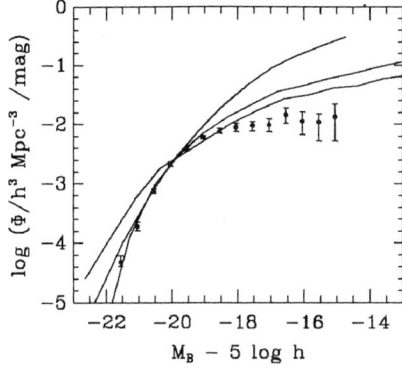

FIGURE 7. The dependence of the predicted B-band luminosity function on the frequency of galaxy mergers in the standard CDM cosmogony.

In our simulations, this radius is over twice the gravitational softening length, allowing us to make reliable mass estimates. The circular velocities, V_c, of our simulated galaxies vary significantly so we rescale this reference radius to $r_{accr} = 25(V_c/V_{MW})$ kpc, where $V_{MW} = 220$ km s^{-1}. Furthermore, our simulations arbitrarily took $\Omega_b = 0.1$ in order to promote efficient cooling and, as a result, they give unrealistically large disk masses [24]. We therefore rescale the particle masses to obtain a value of Ω_b more consistent with primordial nucleosynthesis which, for $H_0 = 50$ km s^{-1} Mpc^{-1}, requires $\Omega_b = 0.05 \pm 0.01$ [31]. We believe this rescaling to be conservative since a real reduction in Ω_b should reduce cooling and affect satellite masses more than primary masses.

The stars in Figure 6 show the fraction of systems which have accreted more than 1.5% of the total mass inside r_{accr} over the time indicated. These results are very similar to the curve corresponding to accretion of 10% of the total *disk* mass. Again, about 70% of the model galaxies have had the relatively undisturbed past required for a thin stellar disk to survive. This conclusion is insensitive to the version of the Tóth and Ostriker criterion which we use.

The fraction of galaxies that remain relatively undisturbed in our models over the last 5 billion years is comparable to the observed spiral fraction among field galaxies. Since the distribution of disk thickness for real galaxies is poorly known, no detailed quantitative comparison is possible at present. Nevertheless it is clear that dismissing $\Omega = 1$ on the basis of the existence of thin disks is premature.

3.2. The shape of the galaxy luminosity function

Examining the effect of "delayed" mergers on the shape of the galaxy luminosity function requires that we incorporate the results of the previous section into a more general model of galaxy formation that takes into account not only the evolution of the dark matter and gas, but also the formation of stars [7,18,36]. We have constructed a semianalytic model which combines the statistical description of the evolution of dark matter halos presented by Cole [7] with specifications for the evolution of the baryonic component derived from the numerical experiments mentioned above. The scheme incorporates the formation and merging of dark matter halos, the shock heating and cooling of baryonic gas gravitationally confined in these halos, star formation regulated by the energy released by evolving stars and supernovae, the merging of galaxies within dark matter halos, and the spectral evolution of stellar populations. Full details can be found in [8].

Figure 7 shows the dependence of this luminosity function on the frequency of galaxy mergers. The top curve on the left (lower on the right) corresponds to assumingthe assumption that *all* galaxies merge inmediately after their halos coalesce. The top curve on the right corresponds to the other extreme; it assumes that galaxies, unlike their halos, *never* merge. The middle curve uses the merger frequencies presented in the last section. All curves are normalized at $M_B - 5\log h = -20$ in order to fit the B-band data of Loveday et al. [20]. This figure shows clearly that the delay between galaxy and halo mergers may be enough to reconcile the shape of the bright end of the luminosity function with the halo mass function in the CDM scenario, avoiding the "overmerging" problem that haunts hierarchical models. It also shows that merging alone is incapable of flattening the faint-end slope of the luminosity function as much as required by observation, as already noted in previous works [7,36]. Strong suppression of star formation in low-mass halos is required in order to produce an acceptable galaxy luminosity function in the CDM scenario.

This work was supported by the SERC.

REFERENCES

[1] Barnes, J. 1988, ApJ 331, 699.
[2] Barnes, J. 1992, ApJ 393, 484.
[3] Barnes, J. & Hernquist, L. 1992, ARAA 30, 705.
[4] Bond, J.R. et al. 1991, ApJ 379, 440.
[5] Bower, R. 1991, MNRAS 248, 332.
[6] Carlberg, R. 1986, ApJ 310, 593.
[7] Cole, S. 1991, ApJ 367, 45.
[8] Cole, S., Aragón, A., Frenk, C., Navarro, J. F. & Zepf, S. 1993, MNRAS, submitted.
[9] Efstathiou, G., Davis, M., Frenk, C. & White, S. D. M. 1985, ApJS 57, 241.
[10] Farouki, R. T. & Shapiro, S. 1982, ApJ 259, 103.
[11] Farouki, R. T., Shapiro, S. & Duncan, M. 1983, ApJ 265, 597.
[12] Hernquist, L. 1992a, in *The Evolution of Galaxies and thier Environment*, ed. J.M.Shull and H.A.Thronson (Kluwer).
[13] Hernquist, L. 1992b, ApJ 400, 460.
[14] Hernquist, L. 1992c, ApJ 409, 548.
[15] Katz, N. & White, S. D. M. 1993, ApJ, in press.
[16] Kauffmann, G. & White, S.D.M 1993, MNRAS 261, 921.
[17] Lacey, C. & Cole, S. 1993, MNRAS 262, 627.
[18] Lacey, C. et al. 1993, ApJ 402, 15.
[19] Lake, G. 1989, AJ 97, 1312.
[20] Loveday, J. et al. 1992, ApJ 390, 338.
[21] Navarro, J. F. 1989, MNRAS, 239, 527.
[22] Navarro, J. F. 1990, MNRAS, 242, 311.
[23] Navarro, J. F. & White, S. D. M. 1993a, MNRAS, in press.
[24] Navarro, J. F. & White, S. D. M. 1993b, MNRAS, in press.
[25] Smoot, G., et al. 1992, ApJ 396, 1.
[26] Toomre, A. & Toomre, J. 1972, ApJ 178, 623.
[27] Toomre, A. 1977, *The Evolution of Galaxies and Stellar Populations*, ed. B.Tinsley & R.Larson, (New Haven:Yale Univ.), 401.
[28] Tóth, G. & Ostriker, J. P. 1992, ApJ 389, 5.
[29] Villumsen, J. V. 1982, MNRAS 199, 493.
[30] Villumsen, J. V. 1983, MNRAS 204, 219.
[31] Walker, T. P. et al. 1991, ApJ 376, 51.
[32] White, S. D. M. 1976, MNRAS 174, 19.
[33] White, S. D. M. 1978, MNRAS 184, 185.
[34] White, S. D. M. 1979, ApJLett 229, L9.
[35] White, S. D. M. & Rees, M.J. 1978, MNRAS 183, 341.
[36] White, S. D. M. & Frenk, C.S. 1991, ApJ 379, 52.

A Model for Galaxy Formation in CDM Cosmology

By M. B. Mosconi, P. B. Tissera, D. G. Lambas

Observatorio Astronómico de Córdoba, Laprida 854, Córdoba, Programa IATE-CONICET, CONICOR, Argentina.

We develop a model to account for the formation of galaxies in a standard biased Cold Dark Matter cosmogony. Galactic halos are identified at different stages of the simulation using an algorithm based on maximal local densities. The model assumes a star formation rate proportional to a power of the local gas density. The evolution of the luminosities in different bands associated with each particle is computed following theoretical evolutionary tracks. For each halo we analyze the dynamical evolution of several properties such as colors, circular velocities, mass to light ratios, and their environmental dependencies. We find that colors in halos consistent with galaxy observations may be obtained by fixing a suitable star formation efficiency parameter. We also obtain a correlation between the absolute magnitude M_{Bj} and the circular velocity of the halos which is in good agreement with the Tully-Fisher relation for spiral galaxies. We find that the halo two-point correlation function depends on both, the circular velocity and the halo color index, with a trend similar to observations. The computed M/L ratios anticorrelate with halo circular velocities and the luminosity function (LF) of the galaxies identified at $z \geq 0.2$ fits the estimated LF derived from deep surveys. The LF obtained at z=0 has an overabundance of faint objects as compared with the observational field LF. However, it provides a good fit to the LF in clusters. We argue that most of the low circular velocity halos expel the bulk of their gas due to the energy input from supernovae at moderate redshifts. Only halos in dense environments were able to retain their gas due to the presence of a high intergalactic pressure confinement. We make a simple model of this situation which provides a reasonable fit to the observed field LF.

---

1. Model

We compute the evolution of the distribution of mass in a CDM universe ($\Omega = 1$, $\Lambda = 0$), assuming that structure grows via gravitational instability of density perturbations present in the early universe. The evolution of the mass distribution is followed by numerical integration of 2×10^5 collisionless particles of mass $m \simeq 7.6 \times 10^9 M_\odot$ in a box of 28 Mpc, $H_o = 50$ km s^{-1} Mpc^{-1}. The simulations were performed using a standard Particle-Mesh code [11] with 128^3 grid mesh and 200 integration time steps $\Delta t = 5.5 \times 10^7$ years. In the adopted model each particle initially contains a fixed fraction of dark and baryonic matter, where $f_b = \Omega_b/\Omega$ is the fraction of baryonic matter. We assume $f_b = f_g + f_r$ where f_g is the initial fraction of gas suitable for being transformed into stars and f_r is the remaining fraction of baryons. We adopt f_g in the range 0.01–0.015 so that critical density in stars is $\Omega_* \leq f_g \leq 0.015$. We compute the gradual transformation of gas into stars for each particle assuming a star formation rate (SFR) induced by gas cloud collisions [12, 15]. This model implies a SFR proportional to the power 3/2 of the particle gas density:

$$d\rho^*/dt = c\rho_{gas}^{3/2} \qquad (1.1)$$

The constant c is a free parameter proportional to the star formation efficiency. It is fixed by requiring the color–color diagram obtained from the simulations to match the observations (see figure 1). In the model, the density of the gas in each particle is proportional to a power of the local mass density, $\rho_g \propto \rho_m^\beta$. Processes of gas cooling and heating during star formation are still difficult to understand. So, this hypothesis may be considered a useful modeling tool [12]. Two values of β were used in the simulations (see Table I) and corresponding values of c to provide a good fit to the color–color diagram.

TABLE I
MODEL CHARACTERISTICS

Model	c	f_g	β
a	0.3	0.015	1
b	0.4	0.010	1
c	0.3	0.015	1
d	0.2-0.4	0.010	2
e	0.2	0.015	2
f	0.3	0.015	1
h	0.3	0.015	1
i	0.7	0.005	1
j	0.1	0.020	1

To compute the star formation history, the simulations are analyzed in equally time-spaced intervals of $\Delta t = 5.5 \times 10^8$ years (10 time steps), starting at $z \simeq 2$ and assuming no stars are present at that time. This is justified since at this epoch the distribution of mass is sufficiently smooth on halo scales where we expect the bulk of star formation to occur. We compute the mass density ρ_m in spheres of comoving radius of 136 Kpc centered on each particle of the simulation. This scale is adopted since the corresponding volume equals the grid resolution volume.

The evolution of the luminosity of each particle in different bands (U, B, V, K) is computed using the evolutionary tracks of Charlot & Bruzual [3] for a single star burst in a star cluster, assuming a Salpeter [16] initial mass function (IMF). We have adopted a mass–to–luminosity ratio of $4M_\odot/L_\odot$ for the stars [9].

2. Statistical Properties of Halos

The statistical properties of 200 hundred halos identified in each simulations are followed in time in order to study the evolution of colors, luminosities and spatial correlation function. This number of halos correspond to a number density of galaxies brighter than $M^* + 2$ adopting a Schechter luminosity function, and satisfies a reasonable contrast criteria, $\delta\rho \geq 64$. The main results are:

a) U–V vs V–B diagram: The distribution of (U–V) and (B–V) colors of the halos in the simulations, shows a remarkable good agreement with observations at redshift $z = 0$. Once the value of β has been fixed, c is the only free parameter of the model. The obtained results are consistent with observations within a rather narrow range of permitted values of c (figure 1).

b) Tully-Fisher relation: We calculated the Tully-Fisher relation finding a good agreement with the observations if the effects of supernova energy injection and the consequent gas removal are taken into account (see figure 2). Protogalaxies with circular velocities V_c less than a critical velocity $V_{crit} \simeq 114$ km/s may have not been able to retain their gas after the first star bursts [6]. V_{crit} is calculated analyzing the balance between gas dynamic binding energy E_{dyn} and the total energy input E_{SN} due to SN explosions in the halos. We find a value of $V_{crit} \simeq 150$ km/s.

c) We have analyzed the halo–halo spatial two–point correlation function at $z = 0$, for

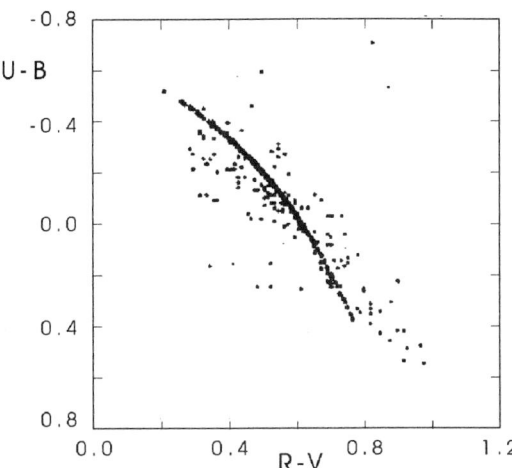

FIGURE 1. The triangles correspond to the distribution of the U–B vs B–V color index obtained for a typical model at $z = 0$. The circles correspond to the observed colors of galaxies taken from the Third Reference Catalog of Bright Galaxies [7].

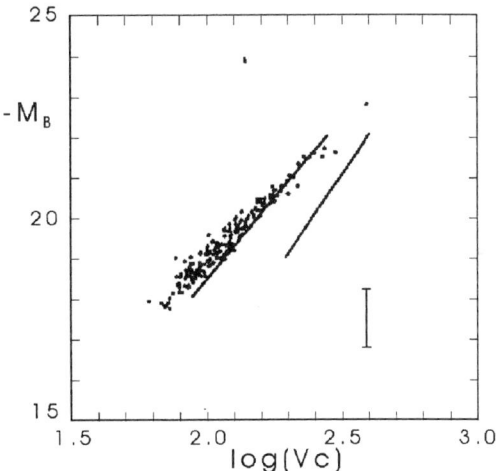

FIGURE 2. Tully-Fisher relation (V_c in km/s). For comparison we show the observed relation for Spirals, with the corresponding error bars, and Ellipticals [1, 10].

halo centers with V_c less than and greater than 150 km/s. We have taken into account the circular velocity V_c and the colors of the center objects. We find that high V_c halos are more strongly correlated than low V_c halos, with a similar power law, $\gamma \simeq 1.8$ (see Figure 3). Due to the small size of the box we do not deal with a fair sample. Nevertheless, the differential effects analyzed are not affected.

3. Luminosity Function

We have computed the luminosity function of the halos in the simulations in the B_j band at different redshifts.

We find at late times, $z \simeq 0$, a large discrepancy at faint magnitudes between our model and observations. The excess of faint objects in the model is in direct relation to the excess of low circular velocity halos according to the Tully-Fisher relation; see for instance [4,10].

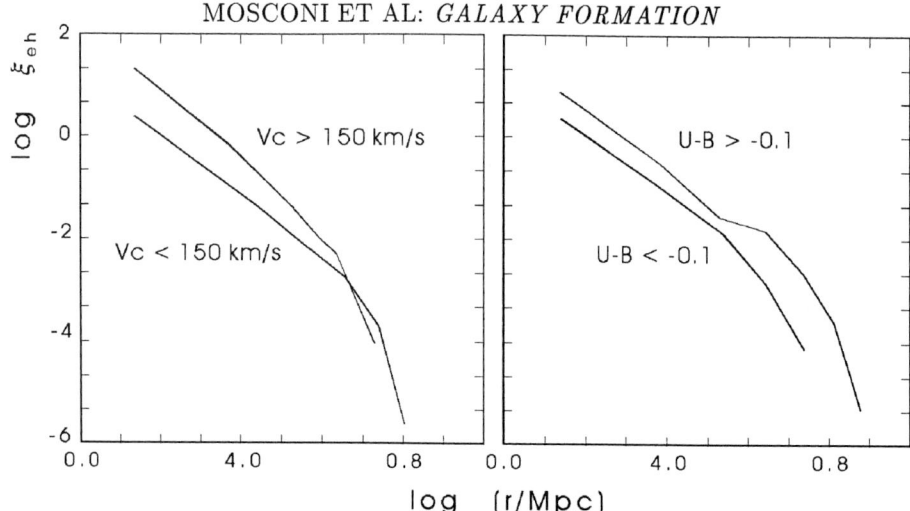

FIGURE 3. Center–halo spatial two–point correlation function dependence on circular velocity and color index. a) for $\xi_{V_c} < 150$ km/s, $\xi_{V_c} > 150$ km/s. b) $\xi_{U-B} < -0.1$, $\xi_{U-B} > -0.1$.

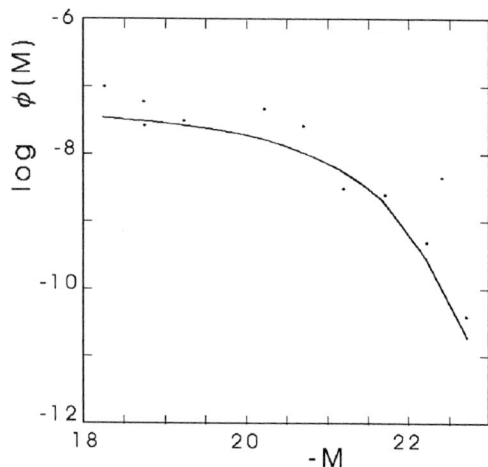

FIGURE 4. Diamonds (model) fit the field luminosity function for $V_c > 150$ km/s and only 1/4 of the halos with circular velocity less than 150 km/s.

A plausible mechanism that might account for these facts could be astrophysical processes related to supernovae energy input. The resulting systems would be at present much fainter than observed dwarf galaxies.

In a simple approximation we may assume that an initial fraction f of galactic halos with $V_c < V_{crit}$ reside in high density regions, and therefore they have retained a substantial amount of gas leaving their luminosities unmodified by SN explosions. The more isolated fraction $1 - f$ of halos with $V_c < V_{crit}$ have lost an important fraction of their gas during the first star bursts, and at $z = 0$ their luminosities are negligible. The resulting LF is shown in the figure 4 for this model with $f = 1/4$. For comparison one can see in this figure the observed LF of field galaxies. We note that there is a good agreement between the results of this particular model and the observed LF. We argue that it may be worth exploring this simple model in more detail.

4. Conclusions

We have developed and analyzed a simple model for the formation of galaxies in hierarchical cosmogonies like the biased CDM model. The models analyzed in this paper may be modified in order to take into account astrophysical processes which are not considered in this simple scheme. We have shown that in spite of the crude approximations the model satisfactorily reproduces several observational properties of galaxies such as colors, the Tully-Fisher relation and spatial correlation functions.

The mass to luminosity ratio M/L is found to be a decreasing function of V_c at $z = 0$ in the range 50–250 indicating that high circular velocity halos are more efficient in forming stars. We have studied possible dependencies of M/L on colors and we have found no significant correlations. The luminosity function of the halos at $z = 0$ fits the observed field LF at high luminosities, and the cluster LF at low luminosities. We propose that star formation in halos with $V_c < V_{crit}$ residing in low density environments is inefficient due to the fact that interstellar gas can be swept out by supernova winds.

At moderate redshifts $z \simeq 0.4$ our model LF provides a good fit to the estimated LF from deep redshift surveys. Both observations and model suggest a strong evolution of the LF at recent epochs.

We find a marked dependence of the spatial center–halo cross correlation function $\xi_{c,h}$ on both V_c and color index of the center objects. These facts may be related to the observed segregation of morphological types. The weak observed dependence of the correlation function on V_c may be successfully accounted for in models where galaxy formation in low V_c halos in the field is strongly suppressed.

We conclude that the models discussed in this paper may provide plausible mechanisms for understanding several observed properties of galaxies and their dependencies on environment within a hierarchical clustering model.

REFERENCES

[1] Aaronson, M. & Mould, J. 1983, ApJ, 265, 1.
[2] Carlberg, R. G. & Couchman, H. M. P. 1989, ApJ, 340, 47.
[3] Charlot, S. & Bruzual, G. 1991, ApJ, 367, 126.
[4] Cole, S. 1991, ApJ, 367, 45.
[6] Dekel, A. & Silk, J. 1986, ApJ, 303, 39.
[7] de Vaucouleurs, G., de Vaucouleurs, A., Corwin, H.Jr., Buta, R., Paturel, G. & Fouque, P. 1991, *Third Reference Catalogue of Bright Galaxies.*
[8] Efstathiou, G., Ellis, R. S. & Peterson, B. A. 1988, MNRAS, 232, 431.
[9] Faber, S. M. & Gallagher, J .S. 1979, ARAA, 17, 135.
[10] Frenk, C., White, S. D. M., Davies, M. & Efstathiou, G. 1988, ApJ, 327, 507.
[11] Hockney, R. W. & Eastwood, J. W. 1988 *Computer Simulation Using Particles*, Adam Hilger, Bristol.
[12] Katz, N. 1992, ApJ, 391, 502.
[13] Lacey, C.G., Guiderdoni, B., Rocca-Volmerange, B. & Silk, J. 1993, ApJ, 402, 15.
[14] Lacey, C.G. & Silk, J. 1991, ApJ, 381, 14.
[15] Larson, R. B. 1969, MNRAS, 145, 405.
[16] Salpeter, E. E., 1955, ApJ, 121, 161.
[17] White, S. D. M. & Frenk, C. 1991, ApJ, 379, 52.

The Formation of a Cluster of Galaxies

By S. N. Dutta, D. N. Spergel

Princeton University Observatory, Peyton Hall, Princeton, NJ 08544-1001, USA.

We investigate the formation of clusters of galaxies in an expanding Universe using a new code that regrids at a region of high density. In particular we investigate two models for the initial conditions, one with the standard CDM power law spectrum with $\Omega = 1$ and with $\Omega = 0.2$. The level of substructure in the final cluster can be used as a discriminant of the cosmic density. We discuss various statistics, that can be measured observationally from clusters of galaxies, that can be used to discriminate between the two models.

---

1. Introduction

The richer clusters of galaxies have for the most part been thought to be relaxed dynamical systems. However, observational evidence has accumulated which seems to go against this picture. Even clusters like Coma that look like perfect examples of relaxed clusters appear to be not so when looked at in detail. Many independent and entirely unrelated observations have pointed in this direction.

The possibility of measuring the redshifts of many galaxies in a clusters have made it possible to use statistics of one dimensional velocity distributions of the clusters. One frequently used technique [10], is to expand the one dimensional velocity distribution in terms of the Gauss-Hermite polynomials. Since the first member of this family of polynomials is a Gaussian, the coefficients of the higher order members are a measure of deviation of the velocity distribution from the Gaussian, which is what is expected in a relaxed dynamical system. Zabludoff [10] used this technique to study several clusters and conclude that many of them are not relaxed.

Another technique that utilizes the measured one dimensional velocity distribution is the Δ-statistic [5]. This involves the definition of a local mean and deviation of a subset of the galaxies and comparing this to the global mean and deviation. Then on plotting this deviation of the local kinematics from the global kinematics in the sky, any pattern is indicative of the deviation of the cluster from Gaussianity. Dressler and Schectman conclude for many of the clusters they investigate considerable substructure exists.

Other techniques rely more on the distribution of mass and light in the sky. Beers and Geller [3] studied the isopleths of number density of galaxies in different clusters and concluded that a large number deviated considerably from circularity. White, Briel & Henry [9] have recently used the ROSAT satellite to observe in considerable detail the morphology of the Coma cluster in X-rays. Even for a cluster that looks as relaxed as the Coma it is apparent from the X-ray plots that it in fact has two components. The dominant part with a smaller component that is falling into it.

All these observations appear to indicate that clusters are not featureless, relaxed objects. In fact almost invariably they have considerable substructure, that appears to stem from infall of bounded objects into the already formed cluster. It is of crucial importance to fix the degree of this phenomenon because this can be a powerful discriminant of the initial conditions. The timescale of the cluster formation was worked out in considerable detail under the assumption of spherical symmetry by Gunn & Gott [6] at almost the inception of the field and analytical knowledge has not progressed much beyond that first attempt. They calculated the time it would take for material *outside* the density peak, where the mean density was still supercritical, to collapse. This is a function of the mean density of the Universe. For the mass scales corresponding to clusters of galaxies, this time in an $\Omega = 1$ Universe, is today [8]. In other words clusters in this Universe are still forming today and can be expected to have substructure. But in an Universe with lower mean density this time is at earlier and substructure will have been substantially erased.

2. Method

The code used is described in detail elsewhere [6]. Here we briefly explain the main features of the approach. We first use a usual particle mesh (PM) code [7, 8] to simulate a 32^3Mpc^3 box of the Universe using a particular set of initial conditions. This region is then visually inspected for a high density region. We choose as our region of interest a 4^3Mpc^3 volume about the highest density peak in the 32^3Mpc^3 box. Next all particles that had at any point of time passed through an imaginary box around this region are flagged. All these particles are broken up and simulated again using a Tree [2] code. The forces on the Tree particles are two fold. One due to other Tree particles, and the other due to all the PM particles that were not broken up into Tree particles. Thus we include both mass infall and tidal forces in our simulations. We achieve the fine resolution of a particle-particle (PP) code without loss of size of the simulation box.

We run the simulation several times for two models. In both models the distribution of the initial density fluctuations is described by a Gaussian probability distribution with uncorrelated phases and a power law power spectrum. The transfer function chosen is for a CDM Universe with three species of massless neutrinos [1]. The only difference in the two Universes is that in one the mean density parameter of the Universe $\Omega = 1$ and in the other $\Omega = 0.2$. We get one realization of each Universe for the same random number generator and look at the same density peak in both. In both cases the simulations are run until the mass fluctuation inside a sphere of radius 8Mpc, σ_8 is unity. We have 10 such realizations for each Universe. Each realization can be treated as three because there are three independent and orthogonal viewing angles for each cluster.

3. ANALYSIS

We have subjected the final results to two different measures of substructure. One is based on the one dimensional velocity distribution of the particles in the cluster, and the other on their distribution in the sky.

3.1. GAUSS HERMITE POLYNOMIALS

The one dimensional velocity distribution of the particles in the high density peak can be expanded in terms of any orthonormal set of functions. In light of our interest in quantifying the deviations of the distribution from a Gaussian, a useful set of functions is the Gauss-Hermite polynomials.

Let the observed velocity distribution function be $\mathcal{L}_0(v)$. This is fitted with the function,

$$\mathcal{L}(v) = \gamma \sum_{i=0}^{N} h_i \mathcal{H}_i, \quad \& \quad \mathcal{H}_i = \frac{e^{w^2/2}}{\sqrt{2\pi}S_s} \frac{H_i(w)}{\sqrt{2^i i!}} \quad (3.1)$$

where h_i are the Gauss Hermite moments, the adjustable parameters. In the above, $H_i(w)$

are the Hermite polynomials and $w = (v - V_s)/S_s$. As mentioned before the fit is most useful in its measurement of the deviation of the velocity distribution from Gaussianity. So we constrain the fit to require, $h_0 = 1, h_1 = h_2 = 0$, since $\mathcal{H}_0$ is the Gaussian function. We try a $\chi^2 = \int [\mathcal{L}_0(v) - \mathcal{L}(v)]^2 dv$ minimization fit by adjusting the parameters, γ, V_s, S_s, h_3, & h_4.

3.2. THE Δ STATISTIC

A statistic that measures substructure in both the projection two dimensional space and velocity space is the Δ-statistic. For this measure we take a "galaxy" (one particle, in our case), and from a sample of a small number ($\simeq 11$) nearby "galaxies" calculate a local velocity mean and dispersion, and compare this to the global velocity mean and dispersion. By nearby galaxies we do not mean the nearest particles. So that the calculated number corresponds better with the observed numbers we use particles in an annular region of radius greater than 1Mpc and less than 2Mpc from the galaxy to calculate the local kinematics. This deviation δ is defined as,

$$\delta^2 = \left(\frac{11}{\sigma^2}\right)[(\bar{v}_{local} - \bar{v})^2 + (\sigma_{local} - \sigma)^2]. \quad (3.2)$$

From this δ we define a cumulative deviation Δ, by summing them over the N_v cluster sample member about whom the local values were calculated.

3.3. DEVIATION FROM CIRCULAR SYMMETRY

Infall of bounded objects into clusters that have already formed will also have an effect that should be visible in the projected surface density. We first convert the projected particle distribution in space into a surface distribution using the cloud in cell assignment of mass to a grid mesh. Since we only have dark matter particles we cannot produce X-ray maps of our clusters, but it is observed that the gas in clusters follow the distribution of dark matter. Then the gas distribution will be given essentially by the potential distribution of the dark matter. So the potential distribution of the cluster projected on the sky should look like the X-ray maps of the cluster. To characterize the deviation from circular symmetry we define a new statistic, the Spergel statistic. We first produce the projected potential distribution. Then define a center by the position of the highest pixel. All positions are now measured relative to this pixel. Then the Spergel statistic is defined as,

$$S = \left| \int \mathbf{x} \frac{I(\mathbf{x})}{I_{max}} d^2\mathbf{x} \right|, \quad (3.3)$$

where, I_{max} is the maximum of the absolute value of the potential distribution.

4. RESULTS

In view of the number of realizations we have of each model we can try to find the distribution functions of the statistics in the different models. To this end, we show the cumulative histograms of the four statistics we have produced from the data. We do not expect all clusters in a $\Omega = 1$ Universe to show more structure that the equivalent cluster in a $\Omega < 1$ Universe because of projection effects.

The value of the h_3 (Eqn. 3.1) is a measure of the infall because it gives the asymmetric deviation from Gaussianity of the velocity distribution. The value of h_4 (Eqn. 3.1) on the other hand is the distance of the single bounded object from relaxation. Fig. 1 shows the cumulative distribution of h_3. As can be seen this statistic cannot distinguish between the two Universes. The K-S test indicates a probability greater 99% that this difference is accidental. Fig. 2 shows the cumulative probability of h_4 in the two Universes. This statistic

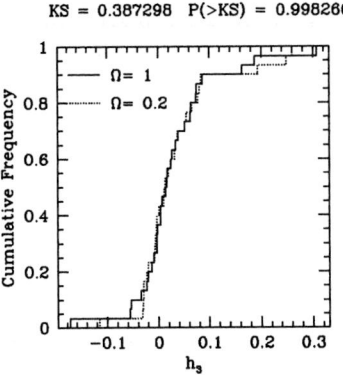

FIGURE 1. The cumulative frequency histogram of the third Gauss-Hermite moments, h_3 for the two Universe models. Numbers at the top are the K–S statistic & the probability of getting a larger K–S statistic by accident.

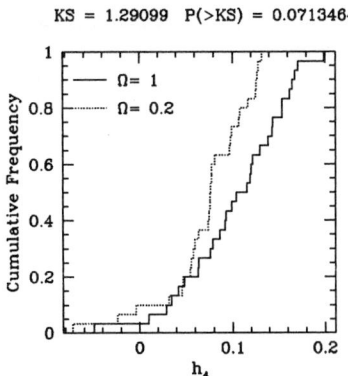

FIGURE 2. The cumulative frequency histogram of the fourth Gauss-Hermite moment, h_4, for the two Universe models. The notation is the same as in Fig. 1.

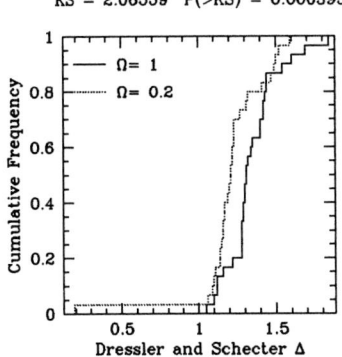

FIGURE 3. The cumulative probablity distribution histogram of the Dressler & Shectman Δ statistic in the two Universes. The notation is the same as in Fig. 1

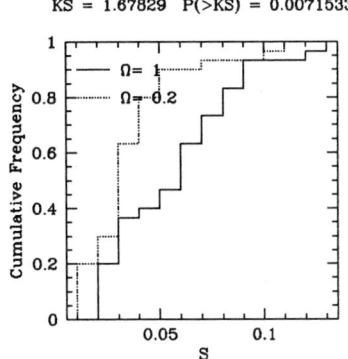

FIGURE 4. The cumulative probability distribution of the Spergel S statistic in the two Universes. The notation is the same as in Fig. 1.

does not appear to distinguish between the two Universes as well and the probability that the two distributions of the variable in the two Universes could be the same is about 7%. The distribution of the Gauss-Hermite polynomials in each Universe is too broad for us to be able to distinguish between the two Universes.

The statistic Δ (Eqn. 3.2) on the other hand appears to be an excellent statistic that distinguishes between the two Universes very well. The cumulative probability density of the statistic is plotted in Fig. 3. As it shows the probability that the distribution in the two Universes are the same is small.

The Spergel statistic S (Eqn. 3.3) also appears to be good discriminant of Ω. Fig. 4 shows the cumulative probability distribution of the statistic for the two Universes. As can be seen from it the K-S test yeilds a very small chance of the two distributions being the same.

5. CONCLUSIONS

This work has attempted to quantify a phenomenon that has long since been believed to have existed. Since in most structure formation scenarios, the time of collapse at a certain mass scale depends on the mean density of the Universe, it should be possible to study the observed level of collapse at a certain mass scale and determine the mean density from there. One problem with this is that it is very difficult to quantify the difference in the degree of collapse in different Universes. This work attempts to quantify this difference with the hope of presenting the observers with a powerful discriminant that will measure mean density of the Universe by means entirely different from the usual painstaking accounting of mass.

We conclude that of the statistics we have chosen to study in detail, the Δ statistic and the S statistic are strong contenders for the measures of the correct model of the Universe.

REFERENCES

[1] Bardeen, J.M., Bond, J.R., Kaiser, N., & Szalay, A.S., 1986, ApJ, 304, 15.
[2] Barnes, J., & Hut, P., 1986, Nature 324 446.
[3] Beers, T.C., & Geller, M.J., 1982, PASP, 1982, 94, 421.
[4] Cen, R., 1990, Ph. D. Thesis, University of Princeton, USA.
[5] Dressler, A., & Schectman, S.A., 1988, AJ, 95, 985.
[6] Dutta, S. N., 1993, Ph. D. Thesis, University of Princeton, USA.
[7] Gunn, J.E., & Gott III, J.R., 1972, ApJ, 176, 1.
[8] Park, C., Ph. D. Thesis, University of Princeton, USA.
[9] Richstone, D., Loeb, A., & Turner, E.L., 1992, ApJ, 393, 477.
[10] White, S.D.M., Briel, U.G., & Henry, J.P., 1993, MPI preprint.
[11] Zabludoff, A.I., 1993, Harvard Smithsonian CFA preprint #3681.

A Hydrodynamic Simulation of Cluster Formation

By Greg L. Bryan, Michael L. Norman

National Center for Supercomputing Applications, 5600 Beckman Institute, D-25, 405 N. Mathews Ave., Urbana, IL 61801, and Department of Astronomy, University of Illinois at Urbana-Champaign, Urbana, IL 61801, USA.

We have developed a hybrid cosmological code to probe the baryonic gas dynamics of large cosmological structures. An adiabatic simulation of large scale structure formation has been performed, using initial conditions specified by the cold dark matter (CDM) power spectrum normalized to the COBE detection. A baryon fraction of 0.06 is adopted and the simulation follows an $85h^{-1}$ Mpc box with 270^3 million zones and 3 million particles. Analysis of the resulting data provides a powerful diagnostic of this particular model. The method shows promise in general by removing another layer of uncertainty between theory and observation. To demonstrate this, we compute an X-ray luminosity function that is straightforward to compare with observations. We also investigate some astrophysics of cluster formation.

---

1. Introduction

Although relatively rare, the high X-ray brightness of large clusters allows them to be seen for great distances and complete samples of relatively nearby clusters have been developed [9]. Previously, others have used the cold dark matter (CDM) prescription to make a quantitative prediction of cluster statistics expected from such a model [8,10]. only the dark, collisionless component, identifying galaxies with the centers of dark halos and comparing the derived clusters against optical compilations, such as Abell's catalog. There are, however, several steps in the analysis, such as identifying cluster characteristics (due to the absence of baryonic matter in the simulations) and comparison against optical observations (projection effects) which increase the uncertainty of the results. Here we model both the hot gas as well as the dark matter and compare the result against X-ray observations which are largely insensitive to projection effects.

The numerical method combines a particle-mesh code which models the dark matter as collisionless particles with an Eulerian grid-based solver for the hydrodynamics. This is based on the third order-accurate piecewise parabolic method [4]. for cosmological systems and have developed a comprehensive test suite [3].

The results described below were computed with initial conditions described by a CDM power spectrum, normalized to COBE [6]. that the variance of the mass overdensity within spheres of radius $8h^{-1}$ Mpc is $\sigma_8^2 = 1.1$. A periodic box with 270^3 zones and 3 million particles of physical size $L = 85h^{-1}$ Mpc with $h = 0.5$ was adopted, where h is the Hubble constant in units of 100 km s^{-1} Mpc^{-1}. A flat universe with $\Omega = 1$ and baryon mass fraction of $\Omega_b = 0.06$ was chosen. The simulation was adiabatic; heating and cooling are due to gas dynamics, not radiative processes.

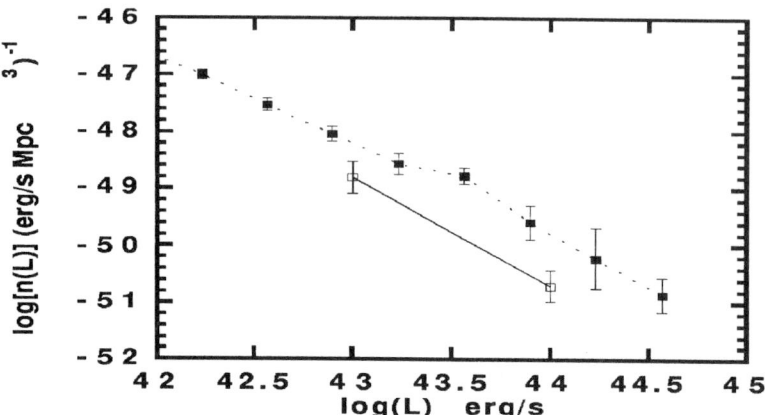

FIGURE 1. The X-ray luminosity function integrated from 2-10 keV; the solid line indicates observations, while the dashed line corresponds to the simulated clusters. Uncertainties are statistical only.

2. The Cluster X-ray Luminosity Function

To identify clusters, we compute the X-ray luminosity from thermal Bremsstrahlung (here integrated from 2-10 keV) for every zone in the simulation. Local maxima above a certain luminosity threshold ($L = 10^{38}$ erg s^{-1}) are found and labeled as clusters. The total luminosity of the cluster is determined by summing the contribution from neighboring zones in a volume corresponding to a sphere of radius $1h^{-1}$ Mpc. We have also done this for $0.5h^{-1}$ Mpc spheres and looked at the results for a variety of redshifts [2]. Since the thermodynamic state of the gas is completely known, we can also analyze other cluster characteristics, but here we restrict ourselves to the better constrained total luminosity.

From this sample of clusters we can compute a luminosity function, shown in Figure 1. The highest luminosity cluster has about $L = 10^{45}$ erg s^{-1} and we expect that this is limited by the finite size of the volume computed. The lower luminosity clusters have larger systematic uncertainties as they are smaller and therefore modeled with fewer cells. The effect of our resolution limit is hard to gauge precisely, although resolution studies indicate that it is unlikely that we have underestimated the luminosity.

Also shown in the figure is the observed X-ray luminosity function from a sample observed with the HEAO-1, Ariel and Einstein satellites [5,9]. approximately correct, the predicted number density of large clusters is about five to ten times larger than observed. The curve in Figure 1 is reasonably well fit by a Schecter function, but with a slope flatter than observations indicate.

3. Cluster Central Temperature Minima

In the process of modeling such a large volume of space we have created a sizable catalog of simulated clusters. Although our resolution is low, we can trace the hydrodynamics of individual clusters. We have observed that some clusters, especially those that are the result of a small number of merging events, exhibit a temperature minimum near or at the central density maximum. Although this behavior is not seen in the majority of clusters, it did occur with sufficient frequency to warrant investigation. It is worth restating that the phenomenon is due to the dynamics of shocks and cosmic expansion, not radiative cooling.

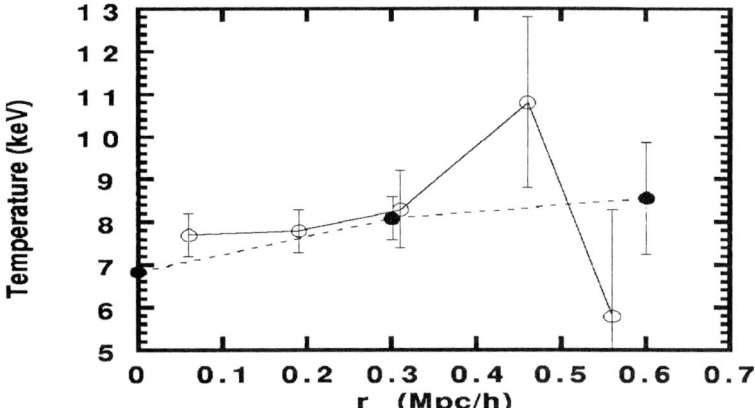

FIGURE 2. Observed temperature minimum in a simulated cluster (dashed line). To create this figure, we have binned neighboring cells and computed a volume weighted average, showing one sigma confidence intervals. We also speculatively plot (solid line) the observed radial temperature distribution of the Coma cluster from Watts, et al. [11].

Figure 2 shows the radial temperature profile of a cluster typical of those showing such minima. We have computed the profile by dividing the cells into several radial bins and computing the average radius and temperature, weighting by volume. If the profile were continued beyond that shown, it would drop due to the presence of other sub-clusters and cold gas.

Although meaningful statistics about the relative frequency of the temperature inversion is difficult to determine because it is almost entirely absent in small clusters, which is almost surely to do with the lack of computational resolution, it is clear that less than ten percent of large clusters exhibit this behavior. However, a visual inspection shows many large temperature inversions which are either distorted or do not lie at the density maxima and hence are not detected by the simple method described above. Inspection of the time evolution of a number of clusters shows that one or more of the components of the cluster form with a central dip but they are effectively erased by merging. This may imply that the presence of a minimum indicates a cluster that either formed from one large nonlinear wave or has not been completely mixed. The principles active here are insensitive to the precise nature of the initial conditions and are more a result of hydrodynamics plus hierarchical clustering (which are observed in a wide range of models). We note that linear analytic calculations [7] have lead to a similar conclusion.

We mention two possible observations of temperature depressions in the centers of real clusters. The ROSAT observations of M87 [1] show a decline from 6 keV to about 2 keV over a similar range of distance as in Figure 2, and a smaller decline is seen in the spacelab observations of the Coma cluster [11]. identifications as isothermal distributions are formally within the uncertainties. Also, the presence of a cooling flow in M87 complicates the interpretation.

4. Conclusions

Although much can be learned from collisionless simulations and self-similarity calculations, the use of hydrodynamics to resolve the non-linear evolution of the gas is not just useful, but necessary as we probe smaller and smaller scales where the gas dynamics become dynamically more and more important. There are also a number of unanswered questions about the evolution of clusters which require more input physics, such as the source of metallicity and the possible presence of magnetic fields.

We thank Renyue Cen, Jeremiah Ostriker and Jim Stone, who have all contributed to this work. Partial support provided by NASA grant NAGW-3152. This simulation was computed on the Convex C3880 supercomputer at the National Center for Supercomputing Applications.

REFERENCES

[1] Böhringer, H., Schwarz, R.A., Briel, U.G., Voges, W., Ebeling, H., Hartner, G., & Cruddace, R.G. 1992, in "Clusters and Superclusters of Galaxies", ed. A.C. Fabian (Kluwer Academic Publishers: Dordrecht), p. 71.
[2] Bryan, G. L., Cen, R., Norman, M. L., Ostriker, P. & Stone, J. M. 1993, Ap.J., submitted.
[3] Bryan, G. L., Norman, M.L., Stone, J. M., Cen, R. & Ostriker, J. P. 1993, Computer Physics Communication, in preparation.
[4] Colella, P. & Woodward, P.R. 1984, J. Comp. Phys. 54, 174.
[5] Edge, A. C., Stewart, G. C., Fabian, A. C., & Arnaud, K. A. 1990, MNRAS, 245, 559.
[6] Efstathiou, G., Bond, J.R., & White, S. D. M., 1992, MNRAS, 258, 1P.
[7] Fang, L.-Z., Bi, H., Xiang, S. & Börner, G. Ap.J. 1993, 477.
[8] Frenk, C.S., White, S.D.M., Efstathiou, G. & Davis, M. 1990, Ap.J. 351, 10.
[9] Henry, J. P. 1992, in "Clusters and Superclusters of Galaxies", ed. A.C. Fabian (Kluwer Academic Publishers: Dordrecht), p. 311.
[10] Kaiser, N. 1992, in "Clusters and Superclusters of Galaxies", ed. A.C. Fabian (Kluwer Academic Publishers:. Dordrecht), p. 323.
[11] Watts, M. P., Ponman, T. J., Bertram, D., Eyles, C. J., Skinner, G. K., Willmore, A. P. 1992, MNRAS, 258, 738.

Simulations of the Formation of Dwarf Ellipticals: Models with no Halo

By E. Athanassoula

Observatoire de Marseille, 2, place le Verrier, 13248 Marseille cedex 04, France.

I briefly present some results of numerical simulations of the formation of dwarf elliptical galaxies, obtained with a multifluid code including self-gravity, star formation, stellar mass loss and heating and cooling of the gas. The resulting models reproduce the observed projected radial light profiles, as well as the relations of several global photometric quantities with luminosity found for samples of dwarf elliptical in the Virgo and Fornax clusters.

---

1. Introduction

The properties of dwarf elliptical galaxies have been revealed by many recent observational studies. Their shape is spheroidal rather than disc-like [6,11,12] and they are poor in gas [4,10,13]. They are mainly found in dense galactic environments and a large number of papers [1,2,3,5,8,13, etc] discuss the photometric properties of dwarf ellipticals in the Virgo and Fornax cluster. Their radial light profiles are found to be largely of two types: exponential and $r^{1/4}$, the latter often being called de Vaucouleurs profiles. The exponential ones can be further subdivided into those that have and those that do not have a nucleus. All these properties are well established and should be reproduced by successful simulations of the formation of these objects. The most important observational results, however, are the trends or correlations found between the luminosity and various global photometric quantities [1,3,5-9,14,15, etc.]. As such we can name the effective radius r_{eff} (radius containing half the measured light), the effective surface brightness μ_{eff} (mean surface brightness within the effective radius), D_{27} (radius of the isophote at 27 mag/arcsec^2), α (the scale length of the exponential radial profile), μ_{exp} (the value of the surface brightness obtained when extrapolating the exponential profile to the center of the galaxy), etc. These trends constitute the most stringent constraints on any simulations of the formation of these galaxies.

In order to do simulations of the formation of dwarf elliptical galaxies, Kevin Prendergast and I wrote a multifluid hydrodynamic code to follow the mass, momentum and energy densities of a number of components. Here I will report only on simulations with two fluids, namely stars and gas. Cases with dark matter will be reported elsewhere. Except for selfgravity our code includes star formation, mass loss from stars, heating of the gas by stars and cooling of the gas. Most of these processes are still not well understood, so we parametrized the star formation and the cooling as power laws of the density and pressure, and the mass loss as an exponentially decreasing function of the mean stellar age. We experimented with two different forms of heating, one modeling heating by supernovae type II, and the other by stellar winds. It is obvious that the parameter space to be explored has many dimensions and that a very large number of runs are necessary for even a cursory coverage. In order to gain simulation time and thus be able to run enough simulations to get an understanding of the main effects, we made an important simplification in restricting our code to one dimension and thus imposing spherical symmetry on the objects. This allowed us to run several hundred simulations in total.

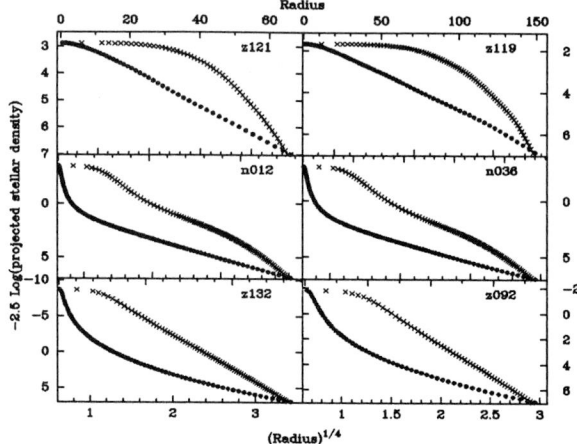

FIGURE 1. Projected stellar density profiles for six simulations as a function of the radius (filled circles - axes at the top of the figures) and of $r^{1/4}$ (crosses - axes at the bottom). The upper row shows two examples of exponentials without a nucleus, the middle one of exponentials with a nucleus, and the lower one of $r^{1/4}$ profiles. The identifier for the simulation is given in the upper right corner of each panel.

2. Results

Most of our simulations start from cold gas at rest. Because of dissipation, however, the memory of the initial conditions is completely lost and one can get nearly identical results from a wide variety of initial conditions. Initially the gas cloud collapses under the influence of its own gravity and this collapse is halted by the formation of a shock. The resulting compression of the gas fuels a burst of star formation. The dynamical evolution after this starburst will depend on the mass of the initial cloud and the parameters for star formation, mass loss and gas heating and cooling. For very low masses and/or high star formation and heating rates the model will become unbound; it will thus first re-expand and then disrupt totally. For intermediate mass cases the gas is blown away but enough stars have formed during the violent star formation phase to make a bound stellar part that survives the outgassing. Finally for the relatively high mass cases nearly all the gas is used up during the violent star formation phase. Contrasting to the initial violent phase the subsequent evolution is quite slow. The stellar part adopts a radial profile whose projected density is mainly either exponential or $r^{1/4}$, in good agreement with the observations, and then hardly evolves further, even though the runs were continued for 30 free-fall times or more. As in the observations, the exponential profiles can be classified in two kinds: those that have nuclei and those that have not. Examples are given in Figure 1, were the log of the projected stellar density is plotted as a function of radius (filled circles) and of $r^{1/4}$ (crosses). If the former are on a straight line for a large radial extent the profile is considered exponential, while if it is the latter it is an $r^{1/4}$. Since the main information on this figure is the form of the radial profiles we do not introduce calibration and give the radii and densities in computer units.

Figure 2 shows three global photometric parameters - the effective radius r_{eff}, the effective surface brightness μ_{eff}, and D_{27} - as a function of absolute magnitude. Observational data have been obtained from photometric studies of the Virgo and Fornax cluster galaxies mentioned in the previous section and are represented by dots, while the results of the simulations are given by asterisks. We see that the agreement is very good, the two sets of data clearly outlining the same trends or correlations. It is worth

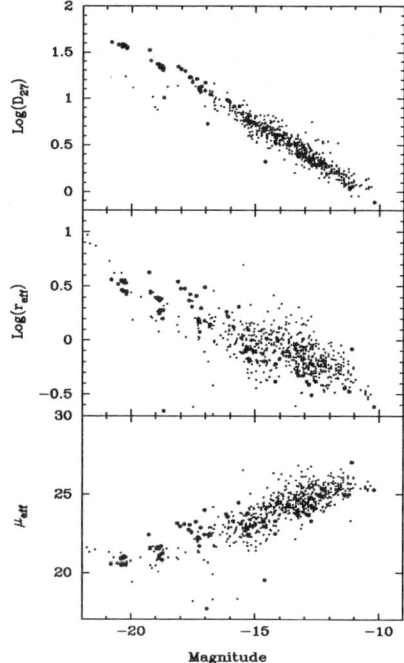

FIGURE 2. $D27$ (upper panel), effective radius (middle panel) and effective surface brightness (lower panel) as a function of magnitude. Observational data are represented by dots and results of simulations by asterisks.

mentioning that we had only two free parameters to help us achieve this, the constant to convert the unit of length in the simulations to kpc and the constant to convert the simulation mass unit into magnitudes. These, however, allow us only to shift the ensemble of the simulation points by corresponding amounts along the x and y axes of the three panels, and in the process give us the correct calibration for the computer units, without affecting the slopes of the regression lines. Thus our models without halo, albeit simplified, can reproduce very well the observed relations. More information on the code, the simulations, as well as on runs with dark matter will be published elsewhere.

REFERENCES

[1] Binggeli, B. & Cameron, L. M. 1991, A&A, 252, 27.
[2] Binggeli, B. & Cameron, L. M. 1993, A&AS, 98, 297.
[3] Binggeli, B. Sandage, A., & Tarenghi, M. 1984, AJ, 89, 64.
[4] Bothun, G., Mould, J., Wirth, A. & Caldwell, N. 1985, AJ, 90, 697.
[5] Bothun, G., Mould, J., Wirth, A. & Gilliwray, H. T. 1986, AJ, 92, 1007.
[6] Caldwell, N. 1983, AJ, 88, 804.
[7] Caldwell, N. 1987, AJ, 94, 1116.
[8] Caldwell, N. & Bothun, G. 1987, AJ, 94, 1126.
[9] Ferguson, H. C. & Sandage, A. 1988, AJ, 96, 1520.
[10] Fisher, J. R. & Tully, R. B. 1975, A&A, 44, 147.
[11] Ichikawa, S. 1989, AJ, 97, 1600.
[12] Ichikawa, S., Wakamatsu, K. & Okamura, S. 1986, ApJS, 60, 475.
[13] Impey, C., Bothun, G. & Malin, D. 1988, ApJ, 330, 634.
[14] Kormendy, J. 1985, ApJ, 295, 73.
[15] Sandage, A. 1983, *Internal Kinematics and Dynamics of Galaxies*, ed. E. Athanassoula, Reidel, 327.

Nonlinear Evolution of Elliptical Perturbations

By Alejandro S. González

Astronomy Centre, University of Sussex, Falmer Brighton, BN1 9QH, U. K

We describe the nonlinear collapse of ellipsoidal perturbations modelled as the superposition of two gaussian distributions. For isolated perturbations the changes in the shape were computed by following a) isodensity contours and b) the motion of the particles. A tidal interaction with a second perturbation was considered to study its effect on the evolution of the first ellipsoid as function of the distance, orientation, shape and mass. When the tidal interaction is strong enough to generate distortions, they tend to be aligned with the major axis of the neighbouring ellipsoid.

1. Introduction

Two main mechanisms to explain the tendency of cosmic structure to asphericity have been proposed. One is that tidal interactions between neighbouring perturbations, in an early stage of the Universe produce important deviations from the initial spherical symmetry of protostructures [2,10,11]. In these models, the growth of anisotropy associated with the changes in the total energy density is calculated in the linear regime, and it is found that the ellipticity grows as $t^{2/3}$. Binney & Silk [2] proposed that alignment of the principal axes of neighbouring rich clusters would be evidence in favour of large-scale tidal interactions.

For primordial perturbations randomly generated in an inflationary epoch, evaluation of the triaxiality distribution function reveals that peaks of the primeval density field are generally triaxial [1,5]. This result, relaxes the usual assumption of spherical initial conditions and provides a natural way of explaining flat structures. A symmetric collapse is highly unlikely. Lin, Mestel & Shu [9] proved that any isolated, uniform ellipsoid remains ellipsoidal as it collapses and has increasing asphericity as a function of time. Later, Zeldovich [12] introduced a powerful method to follow the nonlinear collapse of perturbations considering both, the expansion of the Universe and the presence of neighbouring perturbations.

When the perturbation enters the nonlinear regime, its evolution is more complex because it can otherwise acquire angular momentum, and distortions can be produced for the presence of neighbouring perturbations. Here we use a simple model to explore these aspects.

2. The Model and Equations

Let us take the superposition two symmetric gaussian distributions of matter, close enough to model ellipsoidal structures. Thus, one of our parameters will be the distance

2d between these two distributions. The density of the ellipsoid therefore is

$$\delta(\bar{x}) = \delta_0 \left(\exp\left(-\frac{1}{2}\frac{(\bar{x}-\bar{d})^2}{a^2}\right) + \exp\left(-\frac{1}{2}\frac{(\bar{x}+\bar{d})^2}{a^2}\right) \right), \tag{2.1}$$

where separations and radius are measured in units of a, the central core and $\delta_0 = 1$. In order to apply the Zeldovich formalism, we take the Fourier transform of the density, δ_k, and then calculate the deformation tensor

$$I_{ij} = \frac{1}{(2\pi)^{2/3}} \int \frac{k_i k_j}{k^2} \exp(-k^2/2 + i\bar{k}\bar{x}) d^3k. \tag{2.2}$$

The integration is carried out in spherical coordinates with the radial component aligned with the z-axis, obtaining a diagonal tensor with $Tr I_{ij} = \delta(\bar{x})$ for each of the distributions. By symmetry $I_{11} = I_{22} = (\delta - I_{33})/2$ where

$$I_{33} = (1 + \frac{2}{r^2}) \exp(-r^2/2) - \frac{2}{r^3}\sqrt{\pi} erf(r/\sqrt{2}), \tag{2.3}$$

and erf denotes the error function. The total effect at any point in the ellipsoid is the sum of the effects produced for both distributions, which we must add by referring them to the same reference system by rotating the deformation tensors through an appropriate angle θ around the x-axis. After diagonalization we obtain

$$\lambda_1 = I_{11}^1 + I_{11}^2, \qquad \lambda_{2,3} = \frac{1}{2}(I_{11}^{(1)} + I_{11}^{(2)} + I_{33}^{(1)} + I_{33}^{(2)}) \pm \tag{2.4}$$

$$\frac{1}{2}([I_{11}^1 - I_{33}^1]^2 + [I_{11}^2 - I_{33}^2]^2 + 2[I_{11}^1 - I_{33}^1][I_{11}^2 - I_{33}^2](2\cos(\theta_1 - \theta_2)^2 - 1))^{1/2}.$$

The evaluation of the velocity field is also quite straightforward from

$$\bar{v} = \int \frac{i\bar{k}}{k^2} \delta_k \exp(i\bar{k}\bar{x}) d^3k.$$

Due to the choice of the axes the two components v_1 and v_2 are zero, and the radial component is

$$v_3 = \frac{\sqrt{2}}{r} \exp(-r^2/2) - \frac{\sqrt{\pi}}{r^2} erf(r/\sqrt{2}). \tag{2.5}$$

The sum gives us

$$v_2 = -v_3^{(1)} \sin(\theta_1) + v_3^{(2)} \sin(\theta_2), \qquad v_3 = -v_3^{(1)} \cos(\theta_1) + v_3^{(2)} \cos(\theta_2),$$

3. Results: Isolated Case

3.1. Density Growth Evolution

To study the the agreement between the Zeldovich approximation and linear theory at early times, we fixed a position in the ellipsoid and found how the density changes at that position as a function of time. A good agreement was obtained with a difference of at most 7%.

3.2. Shape Evolution

• We followed the motion of the particles initially placed on an isodensity contour. Since the eigenvalues of the deformation tensor depend on the density and velocity fields in its neighbourhood, the final position of the particles will no longer define isodensity contours for later times (Fig. 1). However, we can describe the changes in the ellipticity

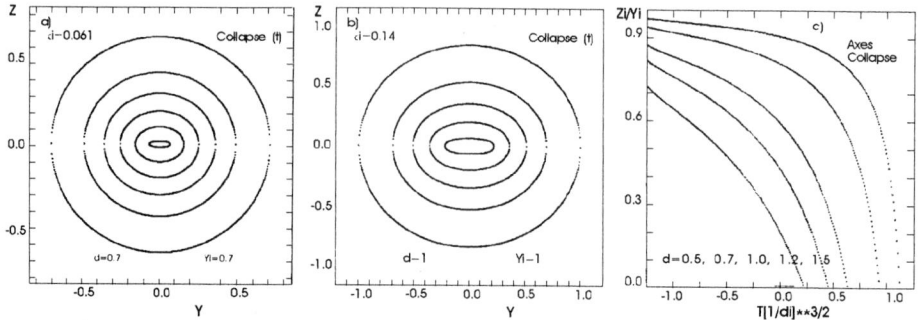

FIGURE 1. a-b) Collapse as function of time: t is scaled as $\delta_0^{-3/2}(y_i, 0)$. c) Instantaneous ratio of minor to major axes

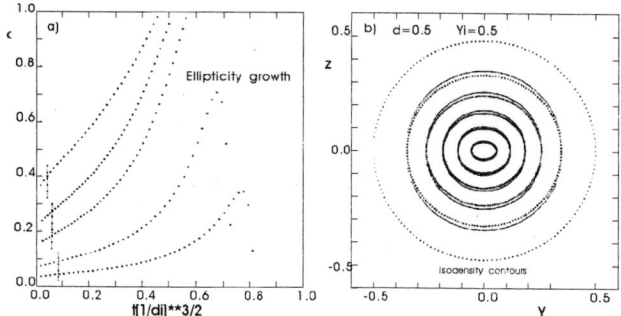

FIGURE 2. a) Ellipticity growth. Lines show the beginning of the nonlinear regime. b) Isocontours from both methods. Dots correspond to the isodensity contours

through the inertia tensor momenta and the relation

$$\frac{3I_{22} - TrI_{ij}}{TrI_{ij}} = \frac{4}{3}\epsilon. \tag{3.6}$$

The inertia tensor was calculated by dividing the ellipsoid into a thin grid. Each of the cubes is considered an element of mass which is taken to be unity. We demanded that the particles do not cross the axes during the collapse to avoid shell-crossing, where the Zeldovich approximation breaks down. Since the path of the particles is given in terms of the peculiar velocity their final position is

$$y = y_0 + v_2 t^{2/3}, \qquad z = z_0 + v_3 t^{2/3}. \tag{3.7}$$

- The second method used was by searching the isodensity contours at a certain time. First, we choose the isocontour density value, δ_i, as that which the particle placed initially at $(y_i, 0)$ will have after a time t_i. Next we evolve all the particles and search for all those which will have the fixed density δ_i. It was found that asphericity grows faster in this case than by following the particles. Figure (2) shows us an example of the ellipticities found by using both methods.

4. Collapse in a Tidal Field

The amplitud and shape of the second ellipsoidal perturbation is determined by the separation of the 3rd and 4th gaussian functions. For this case the deformation tensor

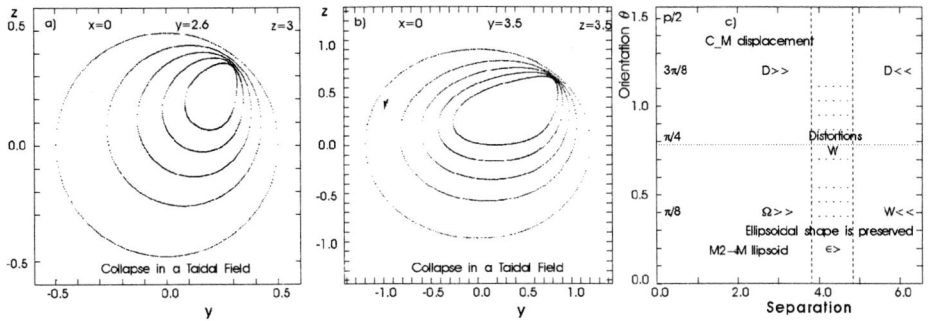

FIGURE 3. a-b) Tidal distortions. The position of the second ellipsoid is indicated and c) Dots show the region in the parametric space where distortions and angular momentum became important

eigenvalues are

$$\lambda_1 = \sum_{i=1}^{4} I_{11}^{(i)}, \qquad \lambda_{2,3} = \frac{1}{2}(\lambda_1 + \sum I_{33}^{(i)}) \pm \qquad (4.8)$$

$$\frac{1}{2}([\sum [I_{11}^{(i)} - I_{33}^{(i)}]^2 + \sum_{i>j} 2[I_{11}^{(i)} - I_{33}^{(i)}][I_{11}^{(j)} - I_{33}^{(j)}](2\cos(\theta_i - \theta_j)^2 - 1))^{1/2},$$

and for the velocity field

$$v_2 = \sum_i v_3^i \sin\theta_i, \qquad v_3 = \sum_i v_3^i \cos\theta_i.$$

These and equation (7) show that the presence of tidal interactions accelerates the ellipticity growth when $|\theta_i - \theta_j| = 0, \pi$, which physically corresponds to the interaction of two parallel ellipsoids. In this case the tidal torque is a minimum. For this to happen, the initial separation has to be about $r > 3.5$. For shorter initial distances there are deviations from the ellipsoidal geometry. For $4 < r < 5$, the most asymmetric ellipsoids are affected in both, the shape and orientation since they experience higher torquing. The strongest effects are reached when the main axes of the ellipsoids form an angle of about 45°. At this angle, important distortions (D) appear which point in the direction of the second perturbation (Fig. 3). The acquisition of angular momentum (Ω) is not surprising. The quadrupole moment associated with each ellipsoid tends to be aligned with the tidal field to minimize the torque. Alignment of structures is expected. On galactic scales, the mean separation is about 4–5 galactic radii, which lies in the region where angular momentum, and distortions are important. Figure 3c summarize some effects for $M_2 = M_{ellipsoid}$. C-M refers to the motion of the center of mass outside the initial ellipsoidal shape. A more detailed description of the results will be presented elsewhere [6].

5. Discussion

If at the epoch of structure formation protostructures spend a long time collapsing quasistatically, having an initial characteristic radius larger than the present radius, tidal interactions could be an effective way to enhance distortions in their shape and establishing non-random orientations with neighbouring structures, besides the angular momentum torquing. The distortions will initially be a function of the mass and distribution of neighbouring perturbations. Due to the short range of this interaction, alignments can be expected at small scales: e.g. between neighbouring galaxies [7], between a galaxy and only its nearest neighbour [4] and even between a galaxy and the radius vector joining it to the cluster centre [8]. In the cosmological scenarios, alignments due to tides might arise in hierarchical models. In cold dark matter models, for example, protostructures collapse almost simultaneously, experiencing important tidal interactions. However, it would be hard to explain with tidal interactions other alignment effects such as those detected at scales of over tens of Mpc [3]. This study supports the idea that small-scale alignments effects can be produced through tidal interactions.

This work was supported by the Universidad Nacional Autónoma de México

REFERENCES

[1] Bardeen, J. M., Bond, J. R., Kaiser, N., & Szalay, A. S., 1986, ApJ, 304, 15
[2] Binney, J. & Silk, J., 1979, MNRAS, 188, 273
[3] Dekel, A., West, M. J. & Aarseth, S. J., ApJ, 279, 1
[4] Djorgovski, S., 1983, ApJ Letters, 274, L7
[5] Doroshkevich, A. G., 1970. Astrophysics, 6, 320
[6] González, S. A. & Thomas, P. A., 1993, in preparation
[7] Gorbachev, B. I., 1971, Soviet Astr, 14, 781
[8] Hawley, D. L. & Peebles, P. J. E., 1975, AJ, 80, 477
[9] Lin, C., Mestel, L. & Shu, F. H., 1965, ApJ, 142, 1431
[10] Palmer, P. L., 1981, MNRAS, 197, 721
[11] Palmer, P. L., 1983, MNRAS, 202, 561
[12] Zel'dovich, Ya. B., 1970, A& A, 5, 84

Particle-mesh Simulations of the Formation of Binary Galaxies and the Spins of Local Group Members

By Sergio Gelato, David F. Chernoff, Ira Wasserman

Department of Astronomy, Cornell University, Ithaca, NY 14853, USA.

We attempt to test the tidal torque theory of the origin of galactic rotation by following the growth of two-peaked density fluctuations in an expanding universe into the nonlinear regime. Our aim is to produce a final state that allows comparison to the Local Group of galaxies, for which reasonably accurate observational data are available. As initial conditions we select isolated pairs of mutually bound peaks from a Gaussian random field. The effect of tides from neighboring groups of galaxies is modeled by imposing an external time-dependent quadrupolar force term. We integrate from a time when the perturbations were still linear to the present epoch with a particle-mesh gravitational code. Hierarchical subgrids are used to achieve adequate resolution in regions of high density such as the central regions of galaxies. We report on the characteristics of the code and on the tests we performed to validate it.

---

1. Introduction

The angular momentum of galaxies is generally thought to arise from the gravitational torques they exert upon one another during their formation [5,8,15]. The magnitude of the induced angular momentum depends on the shape and orientation of the protogalactic cloud, which must be nonspherical and misaligned with the principal axes of the local tidal field for any torquing to occur. Most of the spin-up takes place when the protogalactic material "turns around" from the Hubble flow and starts collapsing; the final angular momentum depends on the time at which this happens. While analytic estimates suggest that the order of magnitude of the induced angular momentum is compatible with observations for reasonable choices of the epoch of galaxy collapse, they cannot provide precise results because of the difficulty of knowing the initial distribution of the matter that will end up in a given galaxy. Detailed numerical simulations are required to overcome this obstacle. Studies have already been made on the spins of halos in large-scale cosmological simulations [1,6], but their mass resolution is rather poor. We attempt to bridge the gap between large-scale and single-galaxy studies by modeling the growth of a pair of galaxies from a two-peaked density perturbation of small initial amplitude. Since the torque on a protogalaxy is not entirely dominated by its nearest neighbor, especially at early times [1,2,7], we allow an adjustable time-dependent external tidal field to act on the pair. In order to obtain realistic initial shapes, relative orientations and peculiar velocities for the peaks, we extract the initial conditions from a Gaussian random field. At this stage, we only consider the dynamics of the collisionless component which presumably dominates the mass distribution. The inclusion of gas dynamical effects is deferred since it would add greatly to the complexity and computational cost of the simulations.

The Local Group is by far the best observed pair of galaxies. To facilitate the eventual comparison of our results to observations, we select initial conditions likely to produce a system similar to the Local Group. The external tidal field can be estimated from present-day data [13], and if necessary improved using approximate solutions for the orbits of nearby galaxies [2,10]. We hope to reproduce the orientations and magnitudes of the spins of the Milky Way and M31 [3], as well as their present-day separation, radial velocity and radial mass profiles (as determined by the orbits of satellite galaxies). The timing argument suggests that the halo of each galaxy extends to large radii ($\gg 50$ kpc); the simulations should provide insight into the mass distribution at these distances. Some choices of cosmological parameters may yield no acceptable solution, in which case they can be ruled out. We hope to improve on Dunn & Laflamme's [2] bound on the epoch of galaxy collapse.

2. Numerical method

The simulation must cover the volume occupied by the two galaxies and by any material they are likely to accrete in a Hubble time. Since the initial comoving separation of the Local Group is ~ 2 Mpc, the minimum simulation size is 4–5 Mpc. Further, since we wish to compare our results to the observed spins of the Milky Way and M31, which are only well known in the central 10–20 kpc, we require a spatial resolution of a few kiloparsecs near the density peaks. The minimum mass resolution is set by the need to have a sufficient number of particles in the spatially resolved galaxy cores, and to keep shot noise at a reasonable level at the beginning of the simulation, when the peak density contrast is $\delta\rho/\rho \leq 0.1$. Both constraints require at least 10^5–10^6 particles. Particle-mesh (PM) schemes are currently the only viable approach for such particle numbers. However, their dynamic range in length falls short of the $\sim 10^3$ we need. Fortunately, we only need high resolution in a few small regions of space, so a hierarchical PM code is suitable.

Our algorithm differs from Villumsen's [14] HPM in various ways. Rather than using a different set of particles at each level of subgridding, we regard the grids merely as a tool to better calculate the forces between particles that exist independently of the subgrids. This allows us to modify the number and placement of the subgrids at will, on every time step. The effects of subgrid forces are propagated back to the main grid via the particles. Since there is no requirement that the particles have equal masses, we can refine the sampling of the initial density distribution in selected regions if we wish. We place a subgrid over any region where the particle number density exceeds a fixed threshold. (High spatial resolution is useless without mass resolution to match.) Since the shapes and sizes of high-density regions may not lend themselves to covering by a single subgrid, we also calculate the forces due to particles in a subgrid in an envelope region a few cells wide that overlaps adjacent subgrids. This extends the benefit of the higher resolution to forces between close pairs of particles that lie across subgrid boundaries and reduces the ringing at the edges of subgrids. Grids may be nested to arbitrary depth, as the need arises.

To allow the imposition of arbitrary external fields, isolating boundary conditions are applied to the top grid. Unlike in codes with periodic boundary conditions, particles may flow into and out of the system. We inject particles according to an extrapolation to later times of the flow into the box in the initial conditions.

3. Tests

Figure 1 (left) compares the interparticle forces calculated with the help of subgrids to the equivalent results with a single-grid version of the code. The particle distribution is white noise, so that small-scale tidal fields dominate the forces. Other mass distributions

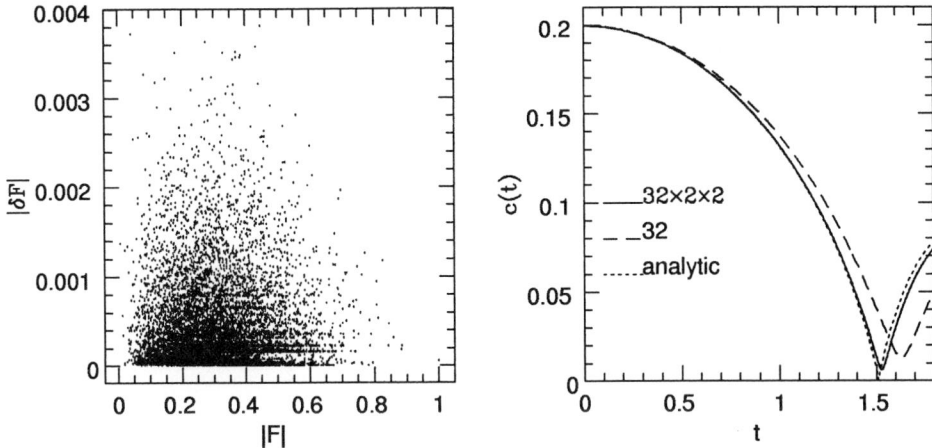

FIGURE 1. Left: scatter plot of the differences between forces computed on a 64^3 grid with one level of subgrids and on an equivalent 128^3 grid without subgrids, versus the forces themselves. Uniform density with white noise. Right: length of the minor axis of a homogeneous 2:1:1 prolate ellipsoid, calculated on a 32^3 main grid with zero and two levels of subgrids, compared to the exact solution.

with more structure yield a comparable result: the subgrid approach produces forces that agree to within 1% with the traditional method.

We subjected the code to a battery of tests of the force law (using two particles on various circular and hyperbolic orbits), of the cosmological expansion (Peebles' hole with compensating ridge [9], evolved for an expansion factor of 100), and of collapse of a homogeneous prolate ellipsoid (with and without an external tidal field). The two-particle tests confirm that deviations from Kepler's laws are only appreciable for impact parameters smaller than three cells. This is normal for PM codes, and effectively defines the smallest length that can be fully resolved. Results for the spherical hole are also in line with other PM codes [12]. For the ellipsoid without external field, figure 1 (right) shows the improvement obtained from subgridding. A similar effect is found when an external quadrupole force is included.

Energy, momentum and angular momentum conservation tests are complicated both by the boundary conditions and by the subgridding. The former require an accounting of particle inflow and outflow and of the work done by external forces, including those that result from the uniform density background outside a nonspherical box. The latter induces spurious changes in the potential energy as subgrids are turned on or off, which are unrelated to the quality of the orbit integration. We have extended the Layzer-Irvine [4,11] energy equation accordingly, and implemented the conservation test to first order accuracy in the timestep. Except for a few cases which we are still investigating, we find energy to be conserved to about the same extent as found by Hockney & Eastwood [4]. Our scheme of surrounding each subgrid with an envelope to ensure a smoother junction between adjacent subgrids introduces some uncompensated forces, which we account for in the momentum conservation test. As expected for a surface effect, we found their contribution to be small. Specifically, the angular momentum they induce is much smaller than the intrinsic angular momentum nonconservation due to the non-radialness of the forces on very small scales in the PM method. The latter must be carefully monitored in production runs. The effect is strongest near density peaks, where short-range contributions dominate the

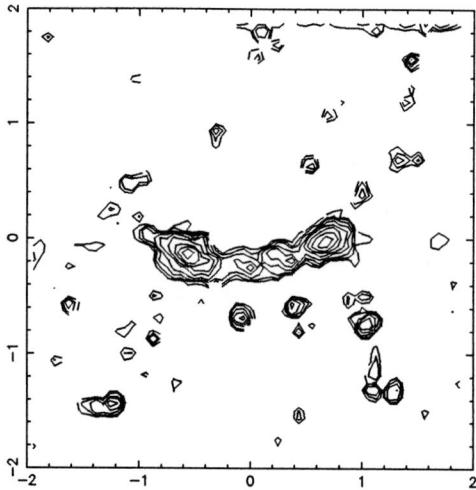

FIGURE 2. Density contours through the midplane of a pair of peaks evolved from a Gaussian field. $\Omega = 1$, $\sigma = 0.9$. Successive contours are spaced by factors of 2. The initial distance between the peaks was 2 length units.

total force. Preliminary results suggest that subgridding does improve angular momentum conservation.

4. Preliminary results

For illustration, we present in figure 2 the final density field in the midplane of a simulation done on a 128^3 grid with 250000 particles in an $\Omega = 1$ universe. The initial conditions were drawn from a Gaussian field with a power spectrum $P(k) \propto k^{-1}$, truncated on scales smaller than $\sim 1/4$ and larger than ~ 8 times the initial peak separation. We chose a pair of peaks of comparable height ($\sim 3.2\sigma$) with no other peak higher than 2σ within a distance of 1.2 times the pair separation, and with a negative relative radial velocity. The system was evolved from an initial r.m.s. field amplitude $\sigma = 0.03$ to $\sigma = 0.9$. During that time, the pair separation decreased by $\sim 30\%$. The final tangential velocity is still only 1/6 of the radial velocity. The angular momenta of the peaks, averaged over regions enclosed by isodensity contours barely high enough to separate the pair, show no particular alignment in the initial conditions, but soon ($\sigma \geq 0.1$) become mostly perpendicular to the line joining the peaks. (No external tidal field was imposed in this run.) Unfortunately, most of the angular momentum is concentrated in a region of radius only two cells within each peak, and is therefore not adequately resolved. We need to rerun this model with our new hierarchical code to remedy this shortcoming.

5. Conclusions

We have developed a new hierarchical particle-mesh code to model the formation of binary galaxies comparable to the Local Group. After extensive testing, this code is now entering production. The tests have convinced us that our subgrid scheme closely reproduces the forces that would be computed by a standard PM code of equivalent resolution, and that where exact solutions are known the code can follow them closely. A preliminary run with realistic initial conditions but only marginal resolution displays the expected qualitative behavior, and confirms the need for subgridding in order to obtain trustworthy quantitative results.

This work was supported by grants NSF-AST-8657467, NSF-AST-9119475 and NAGW-2224 at Cornell University.

REFERENCES

[1] Barnes, J. & Efstathiou, G. 1987, ApJ, 319, 575.
[2] Dunn, A. M. & Laflamme, R. 1993, MNRAS, 264, 865.
[3] Gott, J. R. & Thuan, T. X. 1978, ApJ, 223, 426.
[4] Hockney, R. W. & Eastwood, J. W. 1988, *Computer Simulations Using Particles*, Adam Hilger, Bristol.
[5] Hoyle, F. 1949, in *Problems of Cosmical Aerodynamics*, International Astronomical Union.
[6] Moore, B. & Frenk, C. 1990, in *Dynamics and Interactions of Galaxies*, Springer, Heidelberg.
[7] Oosterloo, T. 1993, AA, 272, 389.
[8] Peebles, P. J. E. 1969, ApJ, 155, 393.
[9] Peebles, P. J. E. 1987, ApJ, 317, 576.
[10] Peebles, P. J. E. 1990, ApJ, 362, 1.
[11] Peebles, P. J. E. 1993, *Principles of physical cosmology*, Princeton University Press, 506.
[12] Peebles, P. J. E., Melott, A. L., Holmes, M. R. & Jiang, L. R. 1989, ApJ, 345, 108.
[13] Raychaudhury, S. & Lynden-Bell, D. 1989, MNRAS, 240, 195.
[14] Villumsen, J. V. 1989, ApJS, 71, 407.
[15] White, S. D. M. 1984, ApJ, 286, 38.

Instabilities of Stellar Discs

By J. A. Sellwood

Department of Physics & Astronomy, Rutgers University, PO Box 849, Piscataway, NJ 08855-0849, USA

I review the wide variety of instabilities of thin, axisymmetric stellar systems, which have so far been discovered in both theoretical work and simulations. These include Jeans-type instabilities which can give rise to rings, bars, spirals and lop-sidedness, as well as bending instabilities which may conceivably be responsible for warps and bulges. In each case, I give a brief discussion of the (suspected) mechanism for the mode, and note the stability boundary, where known. I also re-emphasize the importance of quiet start techniques, both for the extraction of linear growth rates and the suppression of relaxation in N-body experiments.

1. Introduction

Instabilities of stellar systems are studied for three principal reasons: 1) They offer a natural mechanism for the formation of features in galaxies, such as bars, spirals, and possibly bulges and warps. 2) Galaxy formation processes, *e.g.* those discussed by Evrard (this meeting), are likely to create unstable stellar discs, and we need to know how the instabilities will rearrange the distribution of mass and angular momentum after the system has formed. 3) Finally, knowledge of the stability boundaries may be useful in placing constraints on the distribution of velocities or of dark matter within a galaxy. The most widely cited such example was the suggestion by Ostriker & Peebles [41] that the bar instability can be inhibited by a large fraction of dark matter, though such an argument requires dark matter to dominate even at the disc centre [28]. A better example is provided by Merritt's [35] demonstration that the anisotropic model with no black hole proposed by Newton & Binney [40] for the centre of M87 could not remain spherical.

Analytic studies of the stability of stellar systems have so far been largely confined to systems with particularly simple densities or potentials [24,45,67]; see [12] or [5] for reviews. The problem for more realistic cases can be formulated in a general way [23,43,59]. Kalnajs's [26] matrix method has proved the most useful, but the technique has been applied in only a few disc cases [19,27,69,49]. A great deal of what we know of the linear stability, and everything about the non-linear consequences of instabilities, has been learned through N-body simulations.

Here I review some aspects of simulation techniques and issues they raise, before summarising known and suspected instabilities of purely stellar discs, subdividing them into the better-known in-plane disturbances and the newly (re-)discovered bending modes. I do not discuss the stability of gaseous, or of mixed star-gas, discs.

2. Simulation Techniques

The collisionless nature of N-body simulations is limited mainly by the number of particles which can be employed. As the rate of relaxation is always reduced by increasing

N, techniques which permit a large number of particles to be employed are essential. This requirement pretty well dictates that grid or expansion codes be used for the study of the stability of isolated galaxies. The much less efficient tree codes need at least an order of magnitude more cpu time than a grid code for the same number of particles, and the advantages possessed by tree codes – flexible geometry and more efficient treatment of multiple concentrations of particles – are irrelevant in this application. Grid and expansion methods have their limitations also, such as limited spatial resolution (*e.g.* §5) and a fixed geometry. The latter is not a concern for stability work, but renders the methods unsuited to problems in which the mass distribution undergoes wholesale rearrangement (*e.g.* for the encounter problem addressed by Gilbert & Sellwood, this volume). Spatial resolution in density clumps could, in principle, be improved by multi-grid techniques – a refinement which has yet to be widely implemented, however.

2.1. *Noise and Relaxation*

Hernquist & Barnes [14] and Hernquist & Ostriker [16] have emphasized that the rate of relaxation in experiments with a fixed number of particles is almost independent of the N-body technique employed. Weinberg [68] attributed this result to density fluctuations on all scales seeded by a random distribution of particles, and argued that the evolution of even a stable model is driven by the time development of the excited spectrum of large-scale oscillatory or damped modes. He showed that the excited modes can give rise to substantial energy variations for the particles, and that those closest to their stability boundary are the most important, since they oscillate with the largest amplitude.

Thus, Weinberg concluded, it is no surprise that all valid methods seem to produce similar energy variations amongst the particles, since they are seeded with the same initial spectrum and evolve in a manner which is largely independent of the type of short-range force cut-off. However, I am more reluctant than Weinberg to describe this behaviour as a form of relaxation, since I regard the energy variations as resulting from the response of a collisionless system to an imposed spectrum of modes.

Sellwood [50] noted the importance of noise-excited large-scale modes a decade ago, when he found that they masked the growth of the linear mode for which he was searching. He showed that the right way to eliminate this problem is through quiet starts, which were first developed for plasmas [6] and have been standard in cosmological simulations for at least eight years (*e.g.* [10]; see also the paper by Frenk, this meeting). However, few others have adopted them for galactic simulations (an exception is Levine, this meeting). The two essentials of a quiet start are: to create a good equilibrium, carefully selecting a representative set of particles from a distribution function (DF) if at all possible, and to arrange the particles in a regular way to impose some low order symmetry. This crucial second part is extremely easy to achieve and kills the most dangerous "nearly unstable" large-scale modes.

2.2. *Linear Modes*

The clearest indication that the collisionless limit can indeed be approached in N-body simulations is that it is possible to demonstrate impressive agreement between the linear eigenmodes derived analytically in the continuum limit and those found in the simulations. Zang & Hohl [70], Sellwood [50] and Sellwood & Athanassoula [54] showed how this could be achieved with increasing degrees of sophistication. The last of these papers showed that the discrepancy between the complex eigenfrequency predicted by linear theory and that observed in the simulation was typically 5%, and could be still less in simple cases. However, because of the technical difficulties of linear stability analysis for realistic stellar systems, there are few linear theory results which can be used for such checks.

Simulations have historically been used to determine the gross stability of model galaxies and to give an indication of the non-linear evolution. Since it is possible to make quantitative measurements of eigenmodes from the simulations, they are much more powerful, however, and can now be used in place of analytic techniques to determine the form and frequencies of the dominant small-amplitude instabilities of realistic systems. Quantitative estimates of eigenfrequencies allow us to study the effects of varying parameters, such as the degree of pressure support, disc/halo mass ratio, *etc.* , and the simulation also reveals the non-linear consequences of the instability. Of course, analytic methods give the complete mode spectrum without noise or resolution problems, but are usually more laborious than simulations.

3. In-plane Disturbances

3.1. *Axisymmetric Stability*

Toomre's [60] study of axisymmetric Jeans modes led to the familiar criterion that $Q > 1$ for stability, where

$$Q \equiv \frac{\sigma_u \kappa}{3.36 G \Sigma},\qquad(1)$$

σ_u is the radial velocity dispersion, Σ the surface density and κ the epicyclic frequency. The longest wavelength which could be unstable is

$$\lambda_{\text{crit}} = \frac{4\pi^2 G \Sigma}{\kappa^2},\qquad(2)$$

which provides a useful yardstick for Jeans-type instabilities of all kinds.

This locally derived axisymmetric stability boundary has been shown to hold with almost embarrassingly high precision in two global studies [24,69]. There has been some minor subsequent fuss over thickness corrections, which was summarised by Toomre [60], but essentially Toomre's original criterion has withstood the test of time.

3.2. *Bar instability*

The bar instability of stellar discs was discovered in the first large N-body simulations [17,38,18] and confirmed in analytic linear stability studies by Kalnajs [24,27] for two special cases. Because of its virulence, it has subsequently been studied by many authors; recent reviews have been given in [5], (§6.3) and [58], (§9). Zang & Hohl [70] and Sellwood & Athanassoula [54] report excellent agreement between the modes predicted by linear theory and the growth of small-amplitude disturbances in the N-body simulations. Inagaki, Nishida & Sellwood [21] found reasonably good agreement between the non-linear evolution computed in an N-body code and in a direct integration of the collisionless Boltzmann equation; in particular, the limiting amplitude of the bar was very similar in the two types of code.

Toomre [63] argued convincingly that the instability was driven by positive feedback to the swing-amplifier – a beast first identified independently in [13] and [22]. A mode is a standing wave, by definition, but Toomre showed that the bar-instability mechanism is more easily understood in terms of a propagating wave packet. In this viewpoint, a leading spiral disturbance originating near the disc centre propagates outwards towards co-rotation, where the swing-amplifier causes it to shear into a trailing wave of much greater amplitude. The amplified trailing wave then travels back to the centre, where it reflects into a leading disturbance and the cycle repeats, but starting from a higher amplitude. The mode is simply the standing wave which results from a endless wave-train propagating round this cycle, the phase closure condition making the spectrum discrete. Wave action is conserved by an outwardly propagating wave beyond co-rotation, which is presumed to be absorbed at the outer Lindblad resonance. [Mark [34] had already proposed the WASER, a similar feed-back cycle but with a much less powerful amplifier, as a mechanism for possible mild, tightly-wrapped spiral modes.]

Toomre's mechanism suggests three different ways in which the bar mode can be stabilised. The first is to embed the whole disc in a massive, unresponsive halo which shuts off the swing-amplifier because it leads to a mismatch between the azimuthal wavelength of the mode at co-rotation and the local $\lambda_{\rm crit}$ (eq. 2). This solution, which is essentially that advocated in [42], is effective only if sufficient unresponsive mass lies interior to the co-rotation radius [28], and is made still less attractive because it would favour the formation of multi-arm spiral patterns in all non-barred galaxies [55]. The second is to raise the level of random motion in the disc, as advocated in [3]. This method also turns off the swing-amplifier, but again it seems unattractive because the required level of random motion is so high that all collective instabilities are suppressed, inhibiting spiral waves also, and destabilizing the disc to vertical bending modes (see §§4&5) when the surface density is reasonably high. Toomre favoured the third stabilisation method, which is to break the feedback loop while leaving the swing-amplifier active. One way to achieve this is to interpose an inner Lindblad resonance between co-rotation and the disc centre, since in linear theory the Lindblad resonances absorb all incident trailing waves. Sellwood [52], in a direct test of this stabilising mechanism, found that the Lindblad resonance could damp only very mild disturbances, while non-linear trapping occurs for larger-amplitude waves, and the outcome is once again a large amplitude bar!

Some combination of these three mechanisms, *e.g.* a massive, dense bulge and a less responsive warm disc, is presumably responsible for the stability of most galaxies. However, the absence of a bar in a few well studied cases, such as M33, remains a puzzle. This galaxy has a rising rotation curve (which is unfavourable for stabilisation by ILRs), yet it has a well developed bi-symmetric spiral suggesting that it is neither halo dominated nor is there sufficient random motion to inhibit collective effects.

3.3. *Spiral instabilities*

Instabilities which might give rise to the spiral patterns of galaxies have been sought for three decades, with only limited success. A separate and much longer review would be required to do justice to this work, and I will mention briefly only a few aspects here. The essential difficulty faced by all global mode studies is that bi-symmetric instabilities in many plausible models of galaxies do not lead to mild, low-amplitude, grand-design spiral patterns, but to disruptive, large-amplitude bars. On the other hand, a model such as the warm, infinite V=const. disc (which is somewhat inappropriately known as the Mestel disc) was shown by Zang and Toomre [63,69] to have neither bar nor spiral instabilities, essentially because all patterns were damped at their inner Lindblad resonances (ILR). Thus if reasonable pattern speeds possess ILRs they are damped, and if they don't they lead to bars!

C. C. Lin and his co-workers (*e.g.* [4,29]) continue to search for a theory of mild spiral modes based on the hypothesis that spirals are quasi-stationary. Their favoured models have a dense, hot centre which inhibits the bar mode and a "Q barrier" which shields the pattern from damping at its ILR. They argue that rapidly evolving features would have disappeared long ago and that low-growth-rate instabilities in a cool disc, created by gas dissipative processes and star formation, will dominate at later times. However, N-body simulations (Sellwood, in preparation) have shown that a V=const. disc, predicted to be linearly stable in the continuum limit, becomes more active over time and heats up through an instability cycle associated with local changes to the distribution function. This behaviour continues even when the number of particles is increased to two million. Thus particulate galactic discs seem to possess additional instabilities which are not present in the continuum limit, and we therefore suspect that the conditions postulated in the "modal theory" may not occur in galaxies.

The "groove" instability described in [30] and [56] does not require the kind of feed-back cycle on which most other global theories rely. Instead, this instability is driven directly by a local deficiency in the phase-space density of stars, with enthusiastic support from the surrounding disc. It is closely related to the "edge" mode identified by Toomre [63,64], which develops near a steep gradient in the surface density of the disc. Even though groove modes evolve rapidly, they may be able to recur in a self-sustaining cycle, thereby prolonging the period over which strong patterns remain visible [53].

Toomre himself does not favour the idea that spiral patterns are true instabilities, and argues instead that they result from responses of a shearing disc to tidal excitation [63] or to co-orbiting mass clumps, such as GMCs *etc.* [65,66].

3.4. Lop-sided instabilities

Zang [69], in his linear stability analysis of the centrally cut-out V=const. disc, found, somewhat to his surprise and disappointment, that it was dominated by a persistent lop-sided instability. Sellwood's [51] simulation of a model having some resemblance to Zang's, in that it had a dense massive bulge and no extended halo, was also dominated by a lop-sided instability. The idea that these instabilities might be caused by cooperative orbital responses, along the lines discussed for bars by Lynden-Bell [31], was explored without much success by Earn [9].

This instability of directly rotating discs may differ qualitatively from another lop-sided instability that is driven by counter-rotation. This second kind of $m = 1$ instability was first reported in [70] in a series of N-body simulations designed to explore the suppression of the bar instability by reversing the angular momenta of a fraction of the stars; they found that a lop-sided instability was aggravated as more retrograde stars were included to suppress the bar mode. Both Araki's [2] analysis of the Kalnajs discs with retrograde stars and Sawamura's [49] study of "polynomial" disc models found that lop-sided modes became "more unstable" as the fraction of retrograde stars was increased. Merritt & Stiavelli [38] found a lop-sided instability dominated their simulations of 3-D "shell orbit" oblate spheroids with no net rotation; even the most flattened of their models was still too thick to be true discs, however. After this instability was discovered in the simulations, [42] showed that it could have been predicted from a global or local linear stability analysis, but their suggested "mechanism" for the mode amounts to nothing more than restating the well-known fact that the response of orbits to an imposed perturbation is supportive, as long as the forcing frequency in the frame of the mode is below the epicycle frequency; this is a necessary, but by no means sufficient, condition for instability.

4. Bending modes

4.1. Local analysis

Toomre [61], in an important article in an obscure summer-school proceeding, found that a thin sheet of stars will buckle out of the plane when the velocity ellipsoid is sufficiently anisotropic. He called this a Kelvin-Helmholtz or "hose" instability, the latter by analogy with a similar instability in magnetized plasmas [44], while others have variously used the less apt terms of "firehose" or "hose-pipe" instability. Since any relation to water flowing through pipes is very distant, the term bending instability seems more appropriate.

Toomre's local analysis imagined an infinite, uniform, non-rotating stellar sheet of surface density Σ, having a velocity distribution of stars characterised by a flattened spheroid with a Gaussian dispersion σ_u in each direction parallel to the plane and thickened by a typically smaller dispersion σ_w normal to it. When all the stars were confined strongly enough for the thin sheet to bend coherently, he found that a small-scale bend would grow. The reason the system is unstable is that stars moving in a corrugated sheet are forced to oscillate vertically as they pursue their unperturbed horizontal motion; the bend grows if the restoring forces from the perturbation are too weak to provide the vertical accelerations required. This reasoning led Toomre directly to the dispersion relation for a sheet of negligible thickness

$$\omega^2 = 2\pi G \Sigma k - \sigma_u^2 k^2, \qquad (3)$$

which shows that waves on scales greater than $\lambda_J = \sigma_u^2/G\Sigma$ are stabilized by the gravitational restoring force from the perturbation. The thickness cannot be neglected when it is comparable to the wavelength of the bend, and Toomre, in the same article, estimated that the instability was suppressed when the velocity dispersion normal to the plane, σ_w, exceeded some 30% of that within the plane, σ_u. Araki's [1] more detailed analysis of this simple model confirmed that $\sigma_w > 0.293\sigma_u$ for stability. Toomre did not think these modes of much interest, however, because he concluded that the velocity ellipsoid in the Solar neighborhood places the Milky Way "well clear of this stability boundary".

4.2. Global analysis

Hunter & Toomre [20] considered the bending modes of cold stellar discs, *i.e.* those in which all stars move on circular orbits, and found no instabilities when all the stars orbit in the same sense. However, they did briefly consider the consequences of half the stars orbiting in the opposite sense in a Maclaurin disc and noted that in this case "bending instabilities exist for all angular wavenumbers $m \geq 2$." The instabilities were expected to grow vigorously, but the requirement of counter-rotation caused them to note the fact merely as a theoretical curiosity, since their paper preceded the discovery of counter-rotation in NGC 4550 [48] by some 23 years!

Polyachenko [45] generalized this study to bending modes in thin Kalnajs discs – the Maclaurin disc with the family of distribution functions (DFs) given in [24]. The dispersion relation he derived indicates instabilities on all sufficiently short scales, as the Toomre-Araki analysis would lead us to expect for infinitesimally thin discs, and truly global instabilities in discs having a large degree of pressure support. These fierce bending instabilities of uniformly rotating systems, which had previously been all but ignored, have recently been shown to be present in more realistic discs (see §5). Malkov and collaborators [32,33] extended the linear analysis to the bending instabilities of the Freeman [11] family of zero-thickness elliptic discs with arbitrary amounts of pressure support.

None of these studies considered systems of finite thickness. However, both Fridman & Polyachenko [12] and Vandervoort [67] have considered the stability of uniform density spheroids; the highly oblate end of this sequence can be considered as an especially simple kind of thick disc. Both studies find highly oblate stellar systems to be subject to a wide range of bending instabilities, which Fridman & Polyachenko speculate would have eliminated all elliptical galaxies flatter than E7, if the result carries over to more realistic systems.

4.3. N-body work

Bending instabilities have been reported in recent N-body simulations [36,46,15]. Merritt & Hernquist found that non-rotating, prolate stellar spheroids developed large-scale bends when the axis ratio exceeded about 2.5:1 and Raha et al. found that a thin rapidly

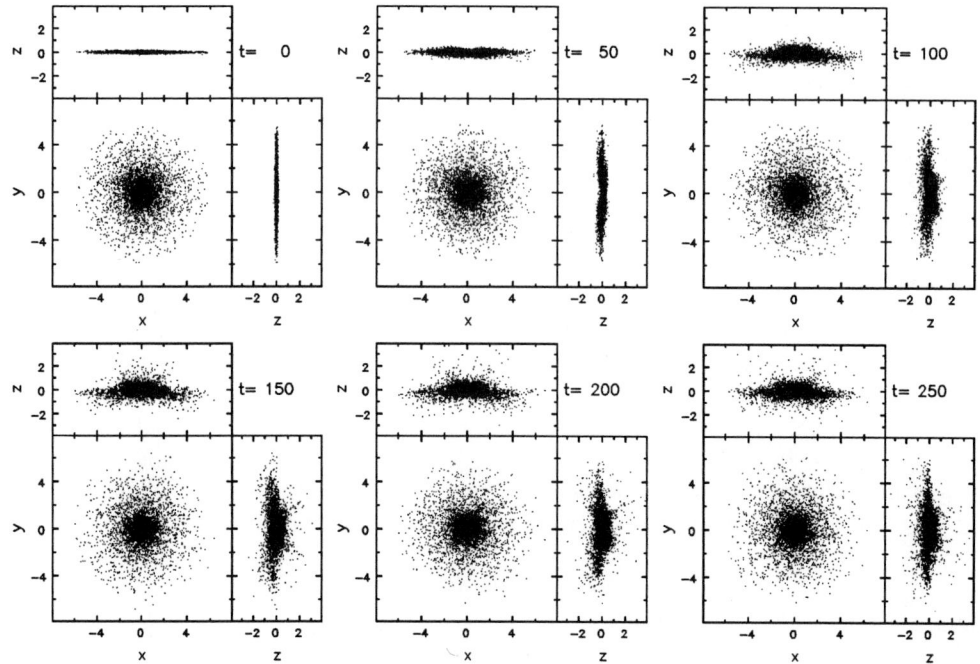

FIGURE 1. The evolution of an N-body realisation of the KT/4 model with two equal counter-rotating components. Only one particle in 40 is plotted, the frames mark the grid boundaries and the time units are dynamical times. Further details are given in Sellwood & Merritt [58].

tumbling bar, which formed in their initially cool disc model, buckled out of the plane. In both cases, the bars buckled even when the velocity dispersion ratio was less extreme than the critical value found by Toomre and Araki, suggesting that bending instabilities in finite systems may be more prevalent than studies of the infinite sheet seemed to indicate. A bending instability in a bar causes it to acquire a box/peanut shape. The same final shape had been reported by Miller & Smith [39] and Combes & Sanders [8] and confirmed in a large number of models by Combes et al. [7]; the authors of this last paper interpreted the formation of the peanut bar shape in terms of scattering at a vertical resonance. While a dynamical instability is a better description of this time-dependent evolution, Combes et al. were undoubtedly on target when they emphasized the 2:2:1 orbit family [58], (§5) as the most important family of orbits in a thick bar, as had already been noted by Miller & Smith.

5. New results

The limited work summarised in the previous section seemed to indicate that bending instabilities should be important in hot, or counter-streaming, stellar discs. Thus when Rubin et al. [48], swiftly confirmed by Rix et al. [47], announced the discovery of two equal counter-streaming components in the disc of NGC 4550, David Merritt and I decided it was about time to examine the stability issue more fully.
We undertook a series of fully 3-D N-body simulations using a standard particle-mesh

code. Setting up the initial thick disc model in equilibrium required considerable care; we really needed a family of 3-integral DFs in order to be able to vary the vertical thickness independently of the velocity dispersion in the plane, but none is known. We therefore adopted Kalnajs's [25] family of 2-integral DFs for motion in the plane, and approximated the velocity distribution normal to the plane using the local Jeans equation [5] (p. 199). We were easily able to employ up to half a million particles, but found we could obtain results of equal quality with less – typically 200K. However, memory limitations restricted us to grids of size 33×129^2 for most runs, though we were able to double the number of cells in each dimension for a few very revealing tests.

As the results and their implications are described in detail in two recent papers [57,37], I will emphasize only a few aspects here. The simulations of these hot discs with no net rotation exhibited six different types of instability: three within the disc plane and three bending modes. The form and vigour of the principal instabilities in any individual model varied with the balance between radial and azimuthal pressure and with disc thickness: an in-plane lop-sided instability was the most disruptive for cool models – in agreement with previous work (§3.4) – while the radially hottest thin discs were disrupted by axisymmetric bending instabilities (bell modes).

The non-linear evolution of the more unstable models created some exotic end states, including lop-sided and warped discs, permanent asymmetries about the initial plane, and pairs of counter-rotating bars. Some of the radially hotter models acquired a thick sub-component in the disc center resembling a bulge in appearance, as shown in Figure 1.

Remarkably, however, the instabilities in a model having intermediate radial pressure caused rather mild changes and led to an apparently stable, moderately thin, and almost axisymmetric disc. The in-plane velocities in this model resemble those reported by Rix et al. [47] for the S0 galaxy NGC 4550, and indicate this galaxy could be stable even without large quantities of dark matter. The stability of this end product also demonstrated that thin axisymmetric systems with modest radial pressure are more stable to bending modes than those having isotropic or radially biased DFs.

We were able to estimate the growth rates in the small-amplitude regime for several dominant modes in each simulation and found that all instabilities are weakened, but most rather slowly, by increasing disc thickness. Unfortunately, we found the growth rates of the bending modes to increase markedly as we improved the resolution of our grid, especially when the disc was very thin. This grid dependence is clearly a cause for concern, but it does imply that our simulations only underestimated the vigour of these modes.

Our semi-analytical work showed that the gravitational restoring force from the perturbation is often unable to stabilize the largest-scale modes, because it is considerably weaker in a realistic inhomogeneous stellar system than in a horizontally infinite and homogeneous slab. We suggested instead that the most important mechanism which stabilizes bending in realistic models is the out-of-phase response of stars that encounter the bend with a frequency greater than their free vertical oscillation frequency κ_z. The strong grid dependence in our numerical experiments led us to identify this mechanism, but we were then able to show that it accounts for the behaviour of the Toomre-Araki infinite slab at short wavelengths and it seems to predict, with some precision, the known stability boundaries to global bending modes in homogeneous spheroids. Moreover, we argued that it predicts instability in any pressure-supported system in which the ratio of orbital oscillation frequencies is less extreme than about 2:1, *i.e.* for which the isodensity contours are rounder than about 1:3, in broad confirmation of Fridman & Polyachenko's [12] (volume 2, p. 159) claim that bending instabilities account for the absence of elliptical galaxies flatter than about E7. (The stable end-product which resembled NGC 4550 is a counter-example and indicates that oblate systems can be yet flatter if the velocities are azimuthally biased).

In our picture, stability to bending is determined primarily by the magnitude of the vertical oscillation frequency, which is determined in turn by the strength of the unperturbed vertical forces to the plane. Somewhat paradoxically, therefore, the more tightly the particles are tied vertically to one another, the more likely the system is to buckle out of the plane. We might expect bending instabilities to thicken stellar systems to the point where the frequency of free vertical oscillation of stars drops below the forcing frequency from all possible bends.

6. Conclusions

Stellar discs can suffer from a wide range of instabilities, depending upon the mass distribution and internal velocity structure. Most instabilities I have reviewed here are present in rotationally supported discs; only one type of lop-sided in-plane instability and bending modes with $m \geq 2$ are provoked by counter-rotation.

I have emphasized the newly re-discovered bending instabilities of axisymmetric discs with some random motion, which had been predicted by Polyachenko [45] but not seen in N-body simulations until recently. They are highly disruptive in hot stellar discs, and have been missed in previous 3-D simulations partly because of poor spatial resolution, but also because most authors investigated cool stellar discs; bending instabilities are much milder in cooler discs and can be suppressed by a moderate disc thickness. In fact, if simulations are to display the most disruptive axisymmetric bending modes, the initial equilibrium must be created from an equilibrium DF.

With some surprise, we concluded that the galaxy NGC 4550 could be stable in the complete absence of dark matter, since both its kinematics and light distribution resemble that of our stable isolated disc. However, how this galaxy arrived in its present state without passing through regions of gross instability is less clear. Unless the dynamics of the inner part of the disc are dominated by a dense dark matter halo (in which case the high velocity dispersion in the plane would be hard to account for) an increase or decrease in its radial velocity dispersion, or a significant reduction in the mass of one of the two counter-rotating components, could easily tip it into an unstable region.

Two further conclusions from our work are that bending instabilities offer promising candidate mechanisms for the formation of bulges, and perhaps also of warps. This last more speculative point was suggested by our finding that warp modes could be the only bending instabilities of moderately hot, directly-rotating thin discs.

Acknowledgments: I would like to thank David Earn, David Merritt and Alar Toomre for comments on the manuscript.

REFERENCES

[1] Araki, S. 1985, PhD thesis, MIT, USA
[2] Araki, S. 1987, AJ, 94, 99
[3] Athanassoula, E. & Sellwood, J. A. 1986, MNRAS, 221, 213 (AS)
[4] Bertin, G., Lin, C. C., Lowe, S. A. & Thurstans, R. P. 1989, ApJ, 338, 78
[5] Binney, J. & Tremaine, S. 1987, *Galactic Dynamics* (Princeton: Princeton University Press)
[6] Byers, J. A. & Grewal, M. 1970, PhysFluids, 13, 1819
[7] Combes, F., Debbasch, F., Friedli, D. & Pfenniger, D. 1990, AA, 233, 82
[8] Combes, F. & Sanders, R. H. 1981, AA, 96, 164
[9] Earn, D. J. D. 1993, PhD thesis, University of Cambridge, UK
[10] Efstathiou, G., Davis, M., Frenk, C. S. & White, S. D. M. 1985, ApJS, 57, 241
[11] Freeman, K. C. 1966, MNRAS, 134, 15
[12] Fridman, A. M. & Polyachenko, V. L. 1984, *Physics of Gravitating Systems*, (New York: Springer-Verlag)

[13] Goldreich, P. & Lynden-Bell, D. 1965, MNRAS, 130, 125
[14] Hernquist, L. & Barnes, J. E. 1990, ApJ, 349, 562
[15] Hernquist, L., Heyl, J. S. & Spergel D. N. 1993, ApJ, 416, L9
[16] Hernquist, L. & Ostriker, J. P. 1992, ApJ, 386, 375
[17] Hockney, R. W. & Hohl, F. 1969, AJ, 74, 1102
[18] Hohl, F. 1971, ApJ, 168, 343
[19] Hunter, C. 1993, *Astrophysical Disks*, ed S. F. Dermott, J. H. Hunter & R. E. Wilson (New York Academy of Sciences)
[20] Hunter, C. & Toomre, A. 1969, ApJ, 155, 747
[21] Inagaki, S., Nishida, M. T. & Sellwood, J. A. 1984, MNRAS, 210, 589
[22] Julian, W. H. & Toomre, A. 1966, ApJ, 146, 810
[23] Kalnajs, A. J. 1971, ApJ, 166, 275
[24] Kalnajs, A. J. 1972, ApJ, 175, 63
[25] Kalnajs, A. J. 1976, ApJ, 205, 751
[26] Kalnajs, A. J. 1977, ApJ, 212, 637
[27] Kalnajs, A. J. 1978, *Structure and Properties of Nearby Galaxies*, ed. E. M. Berkhuisjen & R. Wielebinski, (Dordrecht: Reidel), 113
[28] Kalnajs, A. J. 1987, *Dark Matter in the Universe*, ed. J. Kormendy & G. R. Knapp, (Dordrecht: Reidel), 289
[29] Lin, C. C. & Bertin, G. 1985, *The Milky Way Galaxy*, ed. H. van Woerden, R. J. Allen & W. B. Burton, (Dordrecht: Reidel)
[30] Lovelace, R. V. E. & Hohlfeld, R. G. 1978, ApJ, 221, 51
[31] Lynden-Bell, D. 1979, MNRAS, 187, 101
[32] Malkov, E. A. 1989, *Astron. Zh.*, 66, 1189; Eng. translation: SovAstron, 33, 614
[33] Malkov, E. A., Nuzhnova, T. N. & Sagintaev, B. S. 1991, *Pis'ma Astron. Zh.*, 17, 469; Eng. translation: SovAstronLetters, 17, 200
[34] Mark, J. W-K. 1977, ApJ, 212, 645
[35] Merritt, D. 1987, ApJ, 319 55
[36] Merritt, D. & Hernquist, L. 1991, ApJ, 376, 439
[37] Merritt, D. & Sellwood, J. A. 1994, ApJ, in press
[38] Merritt, D. & Stiavelli, M. 1990, ApJ, 358, 399
[39] Miller, R. H. Prendergast, K. H. & Quirk, W. J. 1970, ApJ, 161, 903
[40] Miller, R. H. & Smith, B. F. 1979, ApJ, 227, 785
[41] Newton, A. J. & Binney, J. J. 1984, MNRAS, 210, 711
[42] Ostriker, J. P. & Peebles, P. J. E. 1973, ApJ, 186, 467
[43] Palmer, P. L. & Papaloizou, J. 1990, MNRAS, 243, 263
[44] Papaloizou, J. C. B. & Lin, D. N. C. 1989, ApJ, 344, 645
[45] Parker, E. N. 1958, PhysRev, 109, 1874
[46] Polyachenko, V. L. 1977, *Pis'ma Astron. Zh.*, 3, 99; English translation: SovAstronLetters, 3, 51
[47] Raha, N., Sellwood, J. A., James, R. A. & Kahn, F. D. 1991, Nature, 352, 411
[48] Rix, H-W., Franx, M., Fisher, D. & Illingworth, G. 1992, ApJ, 400, L5
[49] Rubin, V. C., Graham, J. A. & Kenney, J. D. P. 1992, ApJ, 394, L9
[50] Sawamura, M. 1988, PASJ, 40, 279
[51] Sellwood, J. A. 1983, JComputPhys, 50, 337
[52] Sellwood, J. A. 1985, MNRAS, 217, 127
[53] Sellwood, J. A. 1989, MNRAS, 238, 115
[54] Sellwood, J. A., 1991, *Dynamics of Disk Galaxies*, ed. B. Sundelius, Göteborgs University, Sweden, 123
[55] Sellwood, J. A. & Athanassoula, E. 1986, MNRAS, 221, 195
[56] Sellwood, J. A. & Carlberg, R. G. 1984, ApJ, 282, 61
[57] Sellwood, J. A. & Kahn, F. D. 1991, MNRAS, 250, 278
[58] Sellwood, J. A. & Merritt, D. 1994, ApJ, in press
[59] Sellwood, J. A. & Wilkinson, A. 1993, RepProgPhys, 56, 173
[60] Shu, F. 1970, ApJ, 160, 89
[61] Toomre, A. 1964, ApJ, 139, 1217
[62] Toomre, A. 1966, *Geophysical Fluid Dynamics*, notes on the 1966 Summer Study Program at the Woods Hole Oceanographic Institution, ref. no. 66-46, p111
[63] Toomre, A. 1974, In *Highlights of Astronomy*, 3, ed. G. Contopoulos, (Dordrecht: Reidel)

[64] Toomre, A. 1981, *Structure and Evolution of Normal Galaxies*, ed. S. M. Fall & D. Lynden-Bell, (Cambridge: Cambridge University Press), 111
[65] Toomre, A. 1989, *Dynamics of Astrophysical Discs*, ed. J. A. Sellwood, (Cambridge: Cambridge University Press), 153
[66] Toomre, A. 1990, *Dynamics & Interactions of Galaxies*, ed. R. Wielen, (Berlin: Springer-Verlag), 292
[67] Toomre, A. & Kalnajs, A. J. 1991, *Dynamics of Disc Galaxies*, ed. B. Sundelius, Göteborgs University, Sweden, 341
[68] Vandervoort, P. O. 1991, ApJ, 377, 49
[69] Weinberg, M. D. 1993, ApJ, 410, 543
[70] Zang, T. A. 1976, PhD thesis, MIT
[71] Zang, T. A. & Hohl, F. 1978, ApJ, 226, 521

Star Cluster Dynamics

By Sverre J. Aarseth

Institute of Astronomy, University of Cambridge, Madingley Road, Cambridge CB3 0HA, UK.

The current understanding of star cluster dynamics has been achieved by using a variety of numerical tools, ranging from the brute-force approach through Fokker-Planck, Monte-Carlo and hybrid formulations. In view of likely future developments, this paper concentrates on technical aspects of direct methods which are currently undergoing important changes to match the evolution of computers towards parallel architecture. A brief description of standard integration and special treatments of close encounters is followed by a discussion of Hermite integration based on explicit evaluation of the force and its first derivative. This new scheme has been implemented on the KSR32 parallel computer, as well as the special-purpose HARP machine which aims for teraflop performance. An accurate tree code for collisional N-body problems appears to be more suitable for large-N ($> 10^4$) simulations on conventional supercomputers. The techniques of two-body and multiple regularizations are combined with stellar evolution look-up tables in order to investigate the effect of tidal two body dissipation and physical collisions. Such events are likely to play an important role in a fully consistent globular cluster simulation which should be possible in the near future. Finally, some recent examples of N-body simulations are discussed, with emphasis on open clusters and the effect of primordial binaries in small globular clusters.

1. Introduction

Star clusters form a fascinating class of objects which astronomers are still trying to understand. The main aspects of interest here are concerned with their dynamical behaviour. Although theoretical investigations have led to a general description of the most important processes, it seems hopeless to obtain a self-consistent picture of their evolution by such means. In order to make further progress, the modern astronomer has turned to the numerical approach. In this formidable problem we are faced with a challenge on three fronts: (i) computers, (ii) algorithms, and (iii) astrophysical modelling. By now computers are sufficiently powerful to permit quite extensive calculations to be undertaken; however, to deal with typical globular clusters by direct methods is still beyond the reach of even the biggest supercomputers. Hence to gain increased performance, a new breed of parallel or special-purpose machines will be needed. Another way to study larger systems is to make certain assumptions which permit faster methods to be used. In particular, Fokker-Planck, Monte-Carlo and hybrid formulations have enjoyed some popularity in spite of their approximate nature. Because of likely future hardware developments, such restrictions may soon become less severe, permitting the range of direct methods to be extended. At the same time, the so-called brute-force approach has recently undergone considerable improvements which will facilitate fully self-consistent calculations. A realistic simulation of star clusters also requires appropriate astrophysical processes to be modelled in order to provide diagnostic tools for comparison with observations. In particular, it is desirable to include mass loss from evolving stars, as well as a realistic prescription for finite-size effects.

This paper concentrates on technical aspects of direct N-body methods, including a new collisional tree code. Section 2 reviews different integration algorithms. In Section 3 we describe some important implementations for star cluster modelling. Finally, Section 4 contains a discussion of some recent N-body simulations of open clusters and small globular clusters, whereas Section 5 concludes by outlining future prospects.

2. Direct Methods

The strongly non-linear behaviour of gravity poses many problems for numerical methods. In the following subsections, we describe briefly the main algorithms for direct integration, including special treatments of close encounters. We then outline the new Hermite integration scheme which will be used to simulate globular clusters on special-purpose computers. Finally, the main features of a new collisional tree code are presented.

2.1. Standard Integration

In view of the expensive force calculation in the direct summation method, some kind of difference scheme is desirable to minimize the effort. Since we are dealing with stellar dynamical simulations rather than celestial mechanics orbit determinations, a fourth-order polynomial represents a reasonable trade-off between accuracy and complexity [1]. At the same time, each particle is advanced according to its individual time-step which controls the convergence of the force polynomial. This permits a wide range of time-steps (say $10^6 : 1$), reflecting the large density contrast between the high-density core and low-density halo. Self-consistency requires additional coordinate prediction; however, the extra cost is relatively small since low-order prediction is sufficient.

The introduction of a neighbour scheme leads to further gain in efficiency. Here the idea is to obtain the irregular force due to the $\simeq N^{1/2}$ nearest neighbours on a shorter time-scale than the distant contribution, which is predicted at intermediate times. The so-called Ahmad-Cohen [6]; AC) scheme has proved itself for a variety of problems; the gain with respect to a single summation method is $\simeq (N/3.8)^{1/4}$ [21]. A complete description of such a code will soon be available [3].

2.2. Close Encounters

Unless already present initially, dynamical evolution inevitably leads to the formation of binaries which tend to increase their energy with time. A typical hard binary with semi-major axis $a \simeq r_h/N$ (r_h being the half-mass radius) has about N periods per crossing time, and this ratio may be much larger for a dominant binary. Moreover, the time-step of direct integration scales as $R^{3/2}$, where the separation R may become extremely small at pericentre, $R_p = a(1 - e)$; hence special treatment is called for.

Close two-body encounters are most conveniently studied by the Kustaanheimo-Stiefel [16; hereafter KS] regularization which transforms the singular equation of relative motion into a harmonic oscillator, where the time-step is independent of eccentricity. Since the external force is now of tidal form, i.e. $\propto R^{-3}$, only a few neighbours need be included in the perturbation for large N. Moreover, this description permits unperturbed motion in the absence of significant perturbers, leaving only the corresponding centre-of-mass particle to be advanced in the standard way. The KS treatment may be extended to an arbitrary number of close pairs, cf. [1].

Hard binaries evolve by interactions with other particles, where the most important encounters have small impact parameters. Such critical events are less suitable for study by the KS method which only deals with one dominant two-body term, leading to loss of accuracy and efficiency by frequent switching. Fortunately, strong triple interactions may be treated by a special three-body regularization method [4] which is based on introducing two simultaneous KS transformations, leaving the third interaction as a direct term of smaller magnitude. In view of programming complexities for adding the external perturbation, only the most critical systems are studied in isolation, but even so a number of energetic interactions occur during a typical simulation.

Recent developments of multiple regularization have led to the so-called chain method [29,30]. This method is a generalization of three-body regularization, where a chain of interparticle vectors are selected for treatment by the KS algorithm. An arbitrary number of particles can now be included in the chain, which is updated frequently in response to the changing configuration.

The chain algorithm permits a variety of critical interactions to be studied. We have therefore included the treatment of external perturbations in order to extend the range of applicability, such that the membership can be increased or reduced by the absorption or ejection of a single particle or KS pair. Even so, only relatively compact subsystems ($N \leq 6$) are selected at present because of the rapidly increasing calculation cost of the external perturbation when using the accurate Bulirsch-Stoer [7] integrator. Although chain regularization requires considerable programming efforts, it has already proved itself in cluster simulations and will undoubtedly become an indispensible tool in the future.

2.3. Hermite Scheme

The philosophy behind force polynomials is to save time-consuming summations by using past information together with a corrector. However, a high-order integration scheme can also be constructed more directly by successive differentiation of the acceleration. At first sight this procedure appears rather expensive but a more careful analysis reveals compensating factors. The basic idea of the so-called Hermite scheme is to evaluate the acceleration and its first derivative together without any reference to previous intervals [19,20]. Writing the corresponding Taylor series to third order then yields a corrector in terms of the second and third derivatives which is included at the end of the time-step. This procedure has the advantage of being self-starting; moreover, the corrector is somewhat simpler and therefore faster (factor of 2). On the other hand, both coordinates and velocities must now be predicted. An implementation of the AC scheme [22] indicates a comparable performance even without the benefit of longer time-steps which should be possible due to the increased stability [20].

In order to reduce the prediction cost, it is advantageous to introduce so-called hierarchical time-steps which permit a group of particles to be advanced at the same time. Thus the initial time-steps are truncated to the nearest commensurate value, with subsequent changes up or down by a factor of 2. This requirement leads to further complications in the treatment of close encounters; i.e. when switching solutions from two or more single particles to the combined centre of mass, the new time-step is sometimes reduced to small values. Even so, there is a slight reduction in CPU time with the corresponding AC code for a similar number of time-steps on a standard workstation. This version has been implemented on the parallel computer KSR32 [34].

The Hermite block scheme has also inspired the design of special-purpose hardware for large-N simulations. Thus the original GRAPE machine [15] has now been re-designed to evaluate the acceleration and its derivative [23]. This high-accuracy chip called HARP (Hermite AcceleratoR Pipe) also returns a neighbour list which might be used to implement the AC scheme. The HARP chip can be combined into boards consisting of 16 chips together with one predictor unit to form a powerful parallel computer at relatively modest cost. Thus just one such chip with a peak performance of 1 Gflop out-performs a CRAY-YMP if the force summation contains $\geq 10^4$ particles.

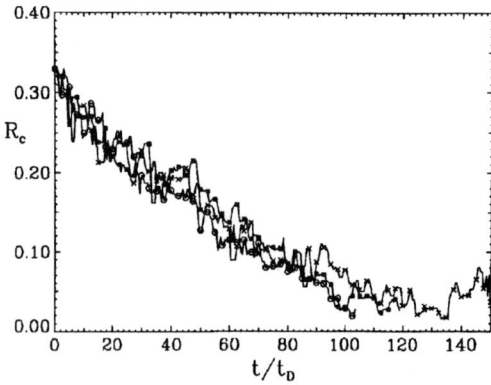

FIGURE 1. Comparison between direct N-body and tree code.

2.4. Collisional Tree Code

Recent tree-based algorithms reduce the cost of force evaluations from $\mathcal{O}(N^2)$ to $\mathcal{O}(N \log N)$. However, these formulations employ a softened potential and are therefore not suitable for star cluster simulations. Here we describe briefly the main features of a new collisional tree code [27] which satisfies more stringent accuracy specifications than can be achieved by traditional means [12].

There are several ways to improve accuracy without reducing the opening angle θ_c unduly, since the latter procedure increases the calculation cost $\propto \theta_c^{-3}$. We adopt a multipole expansion to include octupole terms for increased convergence. Prediction of the quadrupole terms based on a third-order Taylor series permits significant cell deformation without having to re-construct the tree. The integration scheme employs the standard fourth-order polynomial, rather than the usual leap-frog procedure. Note that a high-order integrator is only beneficial if the force errors are acceptably small. Instead of frequent re-determination of the whole tree we have introduced block time-steps for the cells as well as the particle integration, thereby saving time and facilitating vectorization. Finally, close two-body encounters are treated by KS regularization in the usual way. One advantage of this procedure is that only the corresponding centre-of-mass particle appears in the tree structure, limiting further sub-division of cells.

In spite of several refinements, this high-accuracy scheme is still relatively expensive. The approach to the theoretical $N \log N$ relation is therefore rather slow; even by $N \simeq 10^4$ the tree code has not quite reached the asymptotic regime. Comparison with the regularized AC code indicates a break-even value at $N \simeq 10^4$, although further optimization might tilt the advantage. However, it is encouraging that an independent collisional method is now available. Figure 1 shows a comparison of the core collapse phase for an isolated system containing $N = 1024$ equal-mass particles. Here the results for two choices of opening angle (0.5 and 0.7) are essentially indistinguishable from the standard integration. In this case, core collapse occurs on a time-scale of $\simeq 120$ crossing times (t_D). Further examination reveals satisfactory agreement for the general mass distribution as well as the number of escaping particles.

3. Modelling Ingredients

Realistic star cluster simulations require a wide variety of processes to be considered. The corresponding time scales range from hours (even seconds) or days for tidal capture and physical collisions to $\simeq 10^8 - 10^{10}$ yrs for the dynamical evolution itself. In the following subsections, we describe briefly some of the most important features that have already been implemented in the codes.

3.1. External Perturbations

Star clusters are usually located inside a galaxy and hence cannot be considered as isolated systems. We may distinguish between nearly circular orbits with small vertical oscillations for open clusters, whereas globular clusters are characterized by highly eccentric orbits and significant vertical motions.

Open clusters are most readily observed in the solar neighbourhood, where the external perturbation is also better known. Their small sizes permit a linearized expansion of the smooth external potential. This approximation gives rise to simple terms in the equations of motion which are related to the local Oort's constants. The addition of a tidal force lowers the escape barrier and hence speeds up the evolution rate. In addition to the smooth galactic force gradient, nearby interstellar clouds provide an irregular perturbation which may play an important role for some clusters, especially those formed near cloud complexes. This process has been modelled by including a small number of clouds within a spherical region surrounding the cluster [35]. Combining these effects, the disruption rate of open clusters may be compared with observed life times. Although there is qualitative agreement, the initial membership and IMF are rather uncertain.

So far, consistent modelling of tidal effects on globular clusters has not been combined with direct N-body integrations. Most globular clusters describe eccentric orbits and therefore experience a time-varying tidal field. These complications are most conveniently treated by introducing a guiding centre which is integrated separately in a given galactic potential [31].

3.2. Stellar Evolution

A large effort has been devoted to the determination of mass functions in star clusters. In particular, the presence of massive stars plays a crucial role for the dynamical evolution both directly by relaxation effects and indirectly through mass loss. Although there is likely to be a large dispersion in the IMF, it is sufficient for modelling purposes to adopt a simple power-law with prescribed observational limits. As far as open clusters are concerned, stars with maximum masses $\simeq 20\,m_\odot$ experience significant internal evolution on time scales much shorter than the typical ages.

A realistic mass-loss scheme has been adopted using fast fitting algorithms for population I stars [9]. For convenience, mass loss is implemented at discrete intervals when the accumulated contribution exceeds 1%. Because of the high ejection velocity, the mass loss is assumed to occur instantaneously. In this connection, it would seem desirable to add a large kick velocity to neutron stars which are formed by the conventional supernova mechanism, cf. [35], since the small resident population in globular clusters may have another origin. Acceptable energy conservation can be maintained during the calculation by correcting the total potential energy as well as relevant force terms. Here the actual mass loss depends on the location in the HR diagram, since red giant winds, white dwarf formation and supernova events are included. Since these processes occur preferentially in massive stars which tend to be centrally concentrated, the clusters experience significant expansion. Depending on the cluster parameters, i.e. the crossing time in years, there may also be an interesting competition between binary formation and mass loss, both of which act to prevent core collapse.

There is now strong observational evidence for significant binary populations both in open and globular clusters [33,14]. It is therefore necessary to model the process of mass transfer due to Roche lobe overflow in close binaries in order to have a consistent treatment.

3.3. Tidal Capture and Collisions

Nearly all direct cluster simulations to date do not include any finite-size effects. However, the presence of eccentric binaries increases the probability of close encounters within a few stellar radii. One promising channel is connected with strong binary-binary interactions where one component frequently escapes, leaving the third body in a highly eccentric orbit about the inner binary [28]. Such systems may then evolve towards small separations by external or internal perturbations. In view of the considerable technical and astrophysical problems involved, the study of tidal two-body capture and physical collisions offers exciting prospects for the future.

For an idealized treatment, we adopt the classical procedure of Press and Teukolsky [32], together with subsequent modifications. Since this prescription yields the total energy loss during a close passage, it is natural to modify the relative orbit at pericentre where the effect is largest. The close encounters associated with tidal capture are invariably treated by some kind of regularization. This facilitates decision-making as well as the correction procedure itself, since well behaved regularized variables can be used with confidence. By now, the provisional treatment of tidal capture in KS regularization [2] has been extended to three-body and chain regularization, where there are additional complications in determining the pericentre. Modifications of the orbital elements are particularly simple in the absence of mass loss when angular momentum conservation yields an expression for the new pericentre, which actually increases slightly. However, it is still not clear how to treat subsequent close passages if the internal oscillations are not damped.

In view of the steep R-dependence of the energy loss, the tidal dissipation process is only included for reasonably close passages; i.e. about three stellar radii for polytropic models. Here the actual radii are again determined from stellar evolution fitting functions [9]. According to SPH calculations [8], physical collisions of red giant stars take place if $R_p \leq 0.75(R_1 + R_2)$. Hence for two such stars, the expected number of collisions is comparable to the number of tidal capture events because the impact parameter is proportional to R when gravitational focusing is important. This behaviour has been noted in preliminary simulations [2]. In this case, two colliding stars are merged into one object and the relevant two-body energy is monitored for energy conservation.

Simulations of collisions pose many interesting astrophysical problems since there are different channels of outcome for the various interactions. Hopefully, it will soon be possible to employ some of the SPH results in order to have a more consistent treatment of dissipation and mass loss since otherwise the merging of stars could lead to an unrealistic runaway situation in the core.

4. Applications

Dynamical astronomers are fortunate in having a variety of star clusters available for observation. In spite of many difficulties due to membership identification and lack of information about the numerous low-mass stars, there is now a wealth of data which may be used for comparison with simulations. It is becoming increasingly clear that even rich globular clusters show a large dispersion in their internal structure, reflecting differences in initial conditions as well as dynamical history. An attempt to study such diverse objects numerically must therefore be based on rather idealized models. Nevertheless, it would appear that star cluster simulations have already yielded interesting results which add significantly to our understanding of their behaviour.

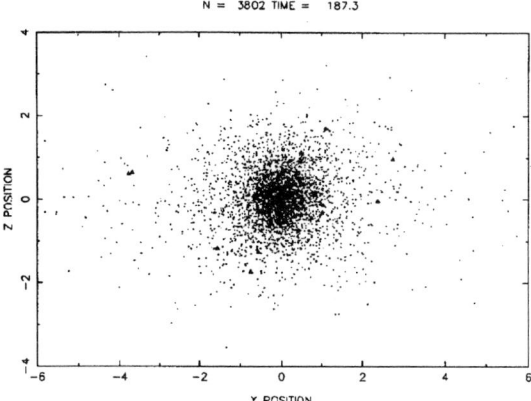

FIGURE 2. Typical open cluster in the xz-plane.

4.1. Open Clusters

Most open clusters only contain less than a few thousand members and can therefore be studied by direct N-body methods without any scaling. An extensive parameter survey by Terlevich [35] obtained typical half-lifes $T_{1/2} \simeq 4 \times 10^8$ yrs for clusters with $N = 1000$ initial members selected from a power-law mass function with exponent 2.75. These simulations included external tidal field and mass loss by supernova events. A characteristic behaviour is that core collapse is halted due to mass loss. Moreover, the absolute escape rate is approximately constant ($\simeq 1$ per 10^6 yrs) during most of the evolution. Somewhat surprisingly, no evidence for smaller values of $T_{1/2}$ was found when the effect of so-called standard interstellar clouds was added. However, such clouds are expected to affect the outer regions of open clusters during the final stages of disruption and may also act to deplete low-mass stars preferentially in the halo.

It is fortuitous that the addition of realistic effects tends to speed up the evolution rate and hence reduces the computational requirements. A more recent simulation (unpublished) with $N = 4000$, which also included tidal field and mass loss, resulted in $T_{1/2} \simeq 9 \times 10^8$ yrs. The xz-distribution is illustrated in Figure 2 at 1.8×10^8 yrs. Here the maximum extent in the x direction corresponds to the classical tidal radius, $\simeq 19$ pc, whereas members inside twice this value are actually retained in the calculation. Note the significant flattening in the vertical direction and that some of the most massive members (denoted by filled triangles) also occupy the outer parts. These simulations provide evolutionary sequences which may be used for comparison with a variety of observed features, such as mass segregation, density and velocity distributions, and overall flattening. A careful examination of the mass function of bound members should also yield vital clues about the IMF of main-sequence stars.

An interesting aspect of open cluster dynamics is connected with the well-known problem of OB runaway stars. Direct N-body simulations of small ($N \leq 300$) isolated systems by Leonard and Duncan [17,18] have attempted to explain these stars in terms of energetic ejections from young clusters. This mechanism appears to be very efficient for clusters containing an initial population of relatively hard binaries which are then involved in strong interactions, although some of these encounters might lead to collisions instead (not modelled here). Detailed comparisons with observations show good agreement on several key features; in particular, it is noteworthy that the simulations can account for

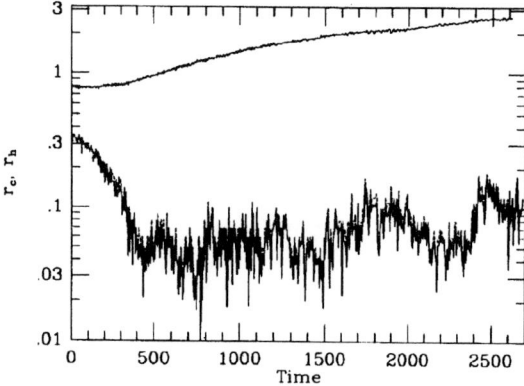

FIGURE 3. Core collapse with primordial binaries.

the actual fraction of runaway binaries and maximum velocities of single stars of $\simeq 200$ km/sec.

4.2. Globular Clusters

A fully consistent N-body simulation of globular cluster dynamics undoubtedly represents the Everest of our subject. Recent efforts have been directed towards understanding core collapse using a relatively modest particle number. In particular, the emphasis has been on studying the consequences of a significant population of primordial binaries which appear to be present in globular clusters [33,14]. On theoretical grounds it is expected that core collapse should be halted at an earlier stage and be less deep than for similar systems of single stars. However, much depends on the actual energy distribution of binaries and their survival in sufficient numbers to be important.

The addition of primordial binaries introduces many interesting complications. This problem has been attacked by two separate collaborations which are both aiming for a systematic exploration [24,25]. Most of these models are still concerned with equal-mass particles, and finite-size effects are neglected. Since the binaries are assumed to have twice the mass of the single particles, the pre-collapse evolution is driven by mass segregation. This leads to strong interactions between binaries and single particles in the core as well as between the binaries themselves, which halt the core collapse. There then follows a stage when binaries are destroyed or ejected from the core with significant velocities, whereas the small remaining binary population becomes harder. The size of the core during post-collapse evolution is in good agreement with simple theoretical considerations [11,26]. Many results of globular cluster modelling have been discussed in the recent review by Hut [13] and will therefore not be covered in this brief summary.

The largest published N-body simulation to date was made for an isolated system with $N = 6000$ single particles and 180 hard initial binaries [5]. A modest power-law mass spectrum with mass ratio 5:1 was employed and the binary masses were selected by splitting every 16th heavy single particle into two equal components; hence in this case the binaries do not experience preferential mas segregation. Figure 3 shows the evolution of the core radius (r_c) and half-mass radius (r_h) during the first $\simeq 1000$ initial crossing times. Here time is given in the recommended standard units [10], with the crossing time $t_D = 2\sqrt{2}$. This part of the simulation required about 3000 CPU hours on the IBM RS6000 workstation ($\simeq 10$ Mflop). The main core collapse terminates after about 150 crossing times, although

TABLE 1. Current status and future prospects.

Machine	N	Code	year
RS6000	6×10^3	ACS	1992
CRAY	1×10^4	ACS	1992/93
CRAY	$\geq 1 \times 10^4$	TREE	1993/94
KSR32	$\simeq 1 \times 10^4$	HACS	1993/94
HARP-1	$\geq 1 \times 10^4$	HITS	1994
HARP-2	$\geq 5 \times 10^4$	HITS	1995

there are later episodes of higher central concentration. Note that the half-mass radius expands significantly due to the energy produced in binary interactions. It is interesting that when primordial binaries are present, the probability of new binary formation is much reduced. Thus newly formed binaries are more easily destroyed by existing harder binaries even if their binding energy is sufficient to survive encounters with single particles.

5. Conclusions

There can be do doubt that star cluster simulations are currently undergoing a stimulating phase. This progress is due to improved performance of computers as well as a more realistic modelling of relevant astrophysical processes. Table 1 summarizes the situation in terms of present and future hardware (Column 1). Typical maximum particle numbers are given in Column 2, whereas Column 3 specifies the corresponding method. Here ACS defines the Ahmad-Cohen scheme, TREE refers to the collisional tree code, ITS represents the individual time-step scheme and the prefix H denotes Hermite integration. A large calculation with $N = 1 \times 10^4$ equal-mass particles has been in progress during the past two years, consuming over 1000 hours on a CRAY-YMP just to reach core collapse [34]. Hopefully this effort can be exceeded next year using the special-purpose HARP-1 chip at a fraction of the cost, whereas the more powerful teraflop machine should allow simulations with $N \geq 5 \times 10^4$ over several thousand crossing times.

On the astrophysical side, we are now in a position to model all the main processes which are known to occur in star clusters. ¿From recent numerical experience, it appears that primordial binaries controls the dynamical evolution. Their interactions also produce hierarchical triple configurations which may subsequently evolve towards physical collisions or tidal capture. These important processes are probably connected with the enigmatic blue stragglers which are now being discovered in increasing numbers, cf. [14].

Since there are excellent prospects for fully self-consistent simulations of small globular clusters, the question of scaling the results to larger N by analytical means should be examined carefully. However, some of the astrophysical parameters break the scale-invariance, restricting this approach to basic considerations [26].

It is appropriate that there has been considerable technical progress on various aspects of the challenging N-body problem. In the first place, the new Hermite integration scheme is ideally suited to the HARP chip design and has already proved itself on standard workstations. The treatment of close encounters by the versatile chain regularization method has also been implemented and promises to play a useful role in future simulations. Finally, the results obtained so far have demonstrated that direct numerical simulations provide a powerful tool for examining the many interesting aspects of star cluster evolution.

REFERENCES

[1] Aarseth, S. J. 1985, *Multiple Time Scales*, ed. J. U. Brackbill & B. I. Cohen, Academic Press, New York, 377.
[2] Aarseth, S. J. 1992, *Binaries as Tracers of Stellar Formation*, ed. A. Duquennoy & M. Mayor, Cambridge University Press, 6.
[3] Aarseth, S. J. 1994, *Computational Astrophysics: Gas Dynamics and Particle Methods*, ed. W. Benz, J. Barnes, E. Müller & M. Norman, Springer-Verlag, New York (in press).
[4] Aarseth, S. J. & Zare, K. 1974, CelestMech, 10, 185.
[5] Aarseth, S. J. & Heggie, D. C. 1993, *The Globular-Cluster Galaxy Connection*, 11th Santa Cruz Summer Workshop, ed. G. H. Smith & J. P. Brodie, ASP Conf. Ser., San Francisco, 701.
[6] Ahmad, A. & Cohen, L. 1973, JCompPhys, 12, 389.
[7] Bulirsch, R. & Stoer, J. 1966, NumMath, 8, 1.
[8] Davies, M. B., Benz, W. & Hills, J. G. 1991, ApJ, 381, 449.
[9] Eggleton, P. P., Fitchett, M. J. & Tout, C. A. 1989, ApJ, 347, 998.
[10] Heggie, D. C. & Mathieu, R. D. 1986, *The Use of Supercomputers in Stellar Dynamics*, ed. P. Hut & S. McMillan, Springer-Verlag, New York, 223.
[11] Heggie, D. C. & Aarseth, S. J. 1992, MNRAS, 257, 513.
[12] Hernquist, L. 1987, ApJS, 64, 715.
[13] Hut, P. 1993, *STScI Mini-Workshop on Blue Stragglers*, ed. R. Saffer, ASP Conf. Ser., San Francisco.
[14] Hut, P., McMillan, S., Goodman, J., Mateo, M., Phinney, E. S., Pryor, C., Richer, H. B., Verbunt, F. & Weinberg, M. 1992, PASP, 104, 981.
[15] Ito, T., Makino, J., Ebisuzaki, T. & Sugimoto, D. 1990. CompPhysComm, 60, 187.
[16] Kustaanheimo, P. & Stiefel, E. 1965, JReineAngewMath, 218, 204.
[17] Leonard, P. J. T. & Duncan, M. J. 1988, AJ, 96, 222.
[18] Leonard, P. J. T. & Duncan, M. J. 1990, AJ, 99, 608.
[19] Makino, J. 1991a, ApJ, 369, 200.
[20] Makino, J. 1991b, PASJ, 43, 859.
[21] Makino, J. & Hut, P. 1988, ApJS, 68, 833.
[22] Makino, J. & Aarseth, S. J. 1992, PASJ, 44, 141.
[23] Makino, J., Kokubo, E. & Taiji, M. 1993, PASJ, 45, 349.
[24] McMillan, S. L. W., Hut, P. & Makino, J. 1990, ApJ, 362, 522.
[25] McMillan, S. L. W., Hut, P. & Makino, J. 1991, ApJ, 372, 111.
[26] McMillan, S. L. W. 1993, *The Globular-Cluster Galaxy Connection*, ed. G. H. Smith & J. P. Brodie, ASP Conf. Ser., San Francisco, 171.
[27] McMillan, S. L. W. & Aarseth, S. J. 1993, ApJ, 414, 200.
[28] Mikkola, S. 1984, MNRAS, 207, 115.
[29] Mikkola, S. & Aarseth, S. J. 1990, CelestMechDynAst, 47, 375.
[30] Mikkola, S. & Aarseth, S. J. 1993, CelestMechDynAst, 57, 439.
[31] Oh, K. S., Lin, D. N. C. & Aarseth, S. J. 1992, ApJ, 386, 506.
[32] Press, W. H. & Teukolsky, S. A. 1977, ApJ, 213, 183.
[33] Pryor, C., McClure, R. D., Hesser, J. E. & Fletcher, J. M. 1989, *Dynamics of Dense Stellar Systems*, ed. D. Merrit, Cambridge University Press, 175.
[34] Spurzem, R. 1993, Personal communication.
[35] Terlevich, E. 1987, MNRAS, 224 193.

Mapping the Galaxy Using Three Different Galactic Potentials

By F. Valera[1], L. Aguilar[2], W. Schuster[2]

[1] Departmento de Astronomía, INAOE, Tonantzintla, Puebla, Apo. 216 and 51, México.

[2] Instituto de Astronomía–UNAM, Ensenada, México.

To understand the structure and dynamics of the Galaxy it is important to have a model of its mass distribution. We study the orbit families in three current models using the method of Surfaces of Section. Although the same orbital families are identified in each model, there are significant differences, mainly for the radial orbits, and where the transition from a bulge to a disk dominated potential occurs. These differences are peculiar to each model and place a limit on our detailed knowledge of the orbital make up of the Galaxy.

1. Introduction

A model of the galactic mass distribution is fundamental to understanding the dynamics of tracers like stars. The resulting dynamics can be used, in turn, to constrain the parameters that went into the construction of a particular mass model. However, our lack of knowledge severely limits the construction of such a model. The usual procedure is to divide the model into an arbitrary number of separate components and to assume a particular functional form for the mass distribution of each (or equivalently their potentials). The parameters introduced by the functional forms are then obtained through a fit to observable quantities like the rotation curve or star counts. There are currently three galactic models which are popular in the literature: Allen & Santillán [1] (AS); Bahcall, Schmidt & Soneira [2,3] (BSS); and Caldwell & Ostriker [5] (CO). Although all three models are consistent with our current knowledge of the galaxy, they differ in the number of assumed components and adopted functional forms. These differences can have a profound effect on the resulting orbital structure because similar potential models can have quite different force fields. This has motivated us to compare the orbital structure found in the three models.

To study these differences we use the method of surfaces of section [4, p. 117]. We have obtained surfaces of section at 4 angular momenta at 6 radial positions for each galactic model.

2. Galactic Models

Mass models of our Galaxy have been assembled using several separate components, traditionally a central bulge, a disk, and a halo responsible for maintaining a flat rotation curve at large galactocentric distances. The BSS model consists of an $R^{1/4}$ bulge, an exponential disk, a spherical halo, and a central oblate nuclear component truncated at 1 kpc from the galactic center. The central component is used to fit the maximum observed in the rotation curve of our galaxy in its central region. The CO model has the three traditional components, but its disk is peculiar and is the difference of two exponential disks. The AS model uses three components also, with the disk and central component modeled using a Miyamoto–Nagai potential [4, p.44]. The BS model is based largely on star counts along different lines of sight. The CO model is based on kinematical quantities like the value of Oort's constants and the rotation curve. The AS model has been proposed as a simple, entirely analytical model, that reproduces the main features of our galaxy.

3. Surfaces of Section

The method of surfaces of section, first proposed by Poincaré and used extensively by Ollongren [6,7] to study an early model of our Galaxy, is a powerful qualitative tool for identifying the presence of a third isolating integral and for classifying the different orbital families that exist within a dynamical system with two effective degrees of freedom. In this study, the two degrees of freedom are given by the cylindrical coordinates R and z (the azimuthal angle being tied to the others by angular momentum conservation). To visualize the structure of orbits in the corresponding four dimensional phase space, the intersection of orbits with the $z = 0$ plane are recorded in the R–$\dot{R}$ plane. To avoid a sign ambiguity in the z component of the velocity intersections are taken with positive $\dot{z}$, only. In other words, we plot the intersections of the stellar orbits with the galactic plane when the star moves toward the northern galactic hemisphere.

4. Families of Orbits

Every star must conserve its energy thus constraining the region its orbit spans in phase space to a three dimensional region. The intersection of this region with the surface of section is a two dimensional region which appears like a sea of points. Orbits of this kind are sometimes called irregular. When the stellar orbit has an additional isolating integral, the region spanned by it is further restricted to a two dimensional region whose intersection with the surface of section is a one dimensional region (line). Orbits of this kind are also called regular. When a regular orbit is further restricted by a resonance between its two main frequencies, the orbit is closed, and the corresponding number of points in the surface of section is finite. These resonant orbits are extremely important because it is the stable resonant orbits that give rise to the families of orbits allowed by a given potential.

5. Results

Surfaces of section have been computed for each of the three studied galactic models at energies that correspond to the potential energy in each model at galactocentric distances of $R_o = 0.5, 1, 2, 4, 8$, and 16 kpc. At each energy we have chosen 4 different values for the angular momenta: $J/J_c = 0.00, 0.50, 0.75, 0.88$, where J_c is the angular momentum of the local circular orbit. At high values of J/J_c the surfaces of section explore regions of the phase space close to the circular orbit at each of the selected radii. As J/J_c decreases, the radial velocity increases and we move toward the region of purely radial orbits, which corresponds to $J/J_c = 0$.

The resulting 24 surfaces of section are shown in Figures 1, 2 and 3 for the AS, BS and CO galactic models, respectively. In each figure, the sections are arranged by increasing R_o from left to right, and by J/J_c increasing from top to bottom. Inner radial orbits thus correspond to the upper left pane, while the outermost near circular orbits correspond to the lower right pane.

All three models are largely dominated by regular orbits, except for the radial orbits

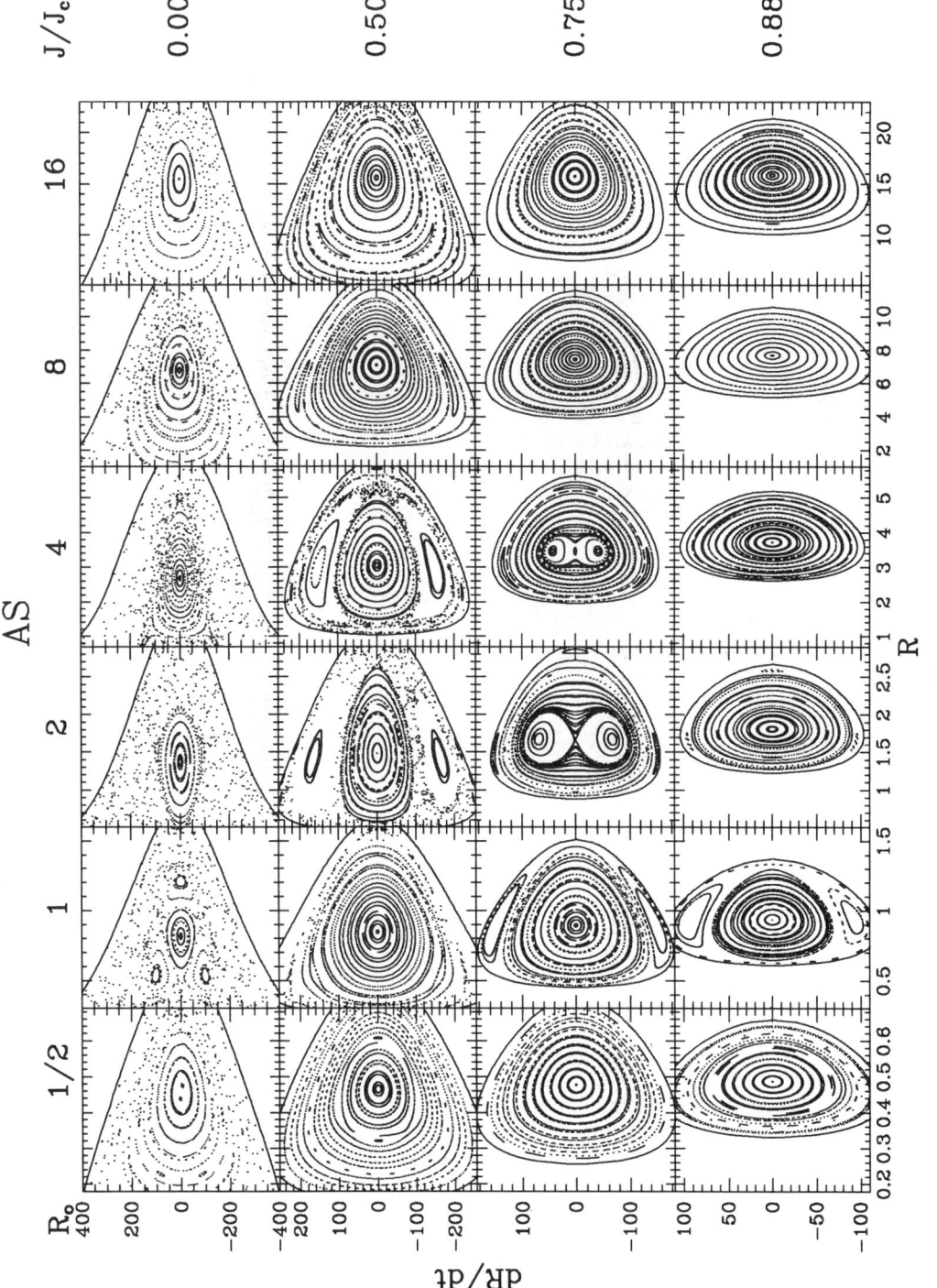

Figure 1

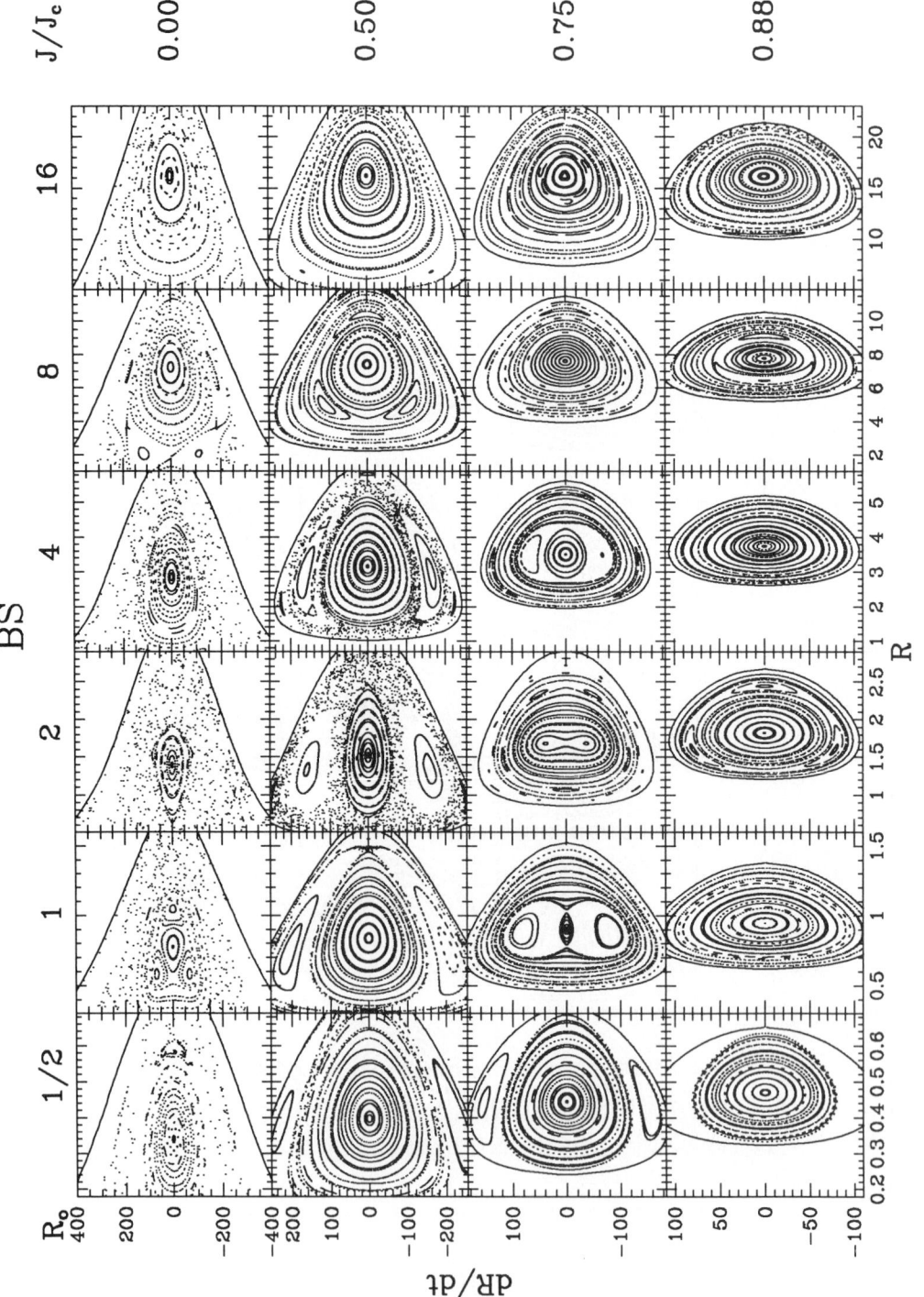

Figure 2

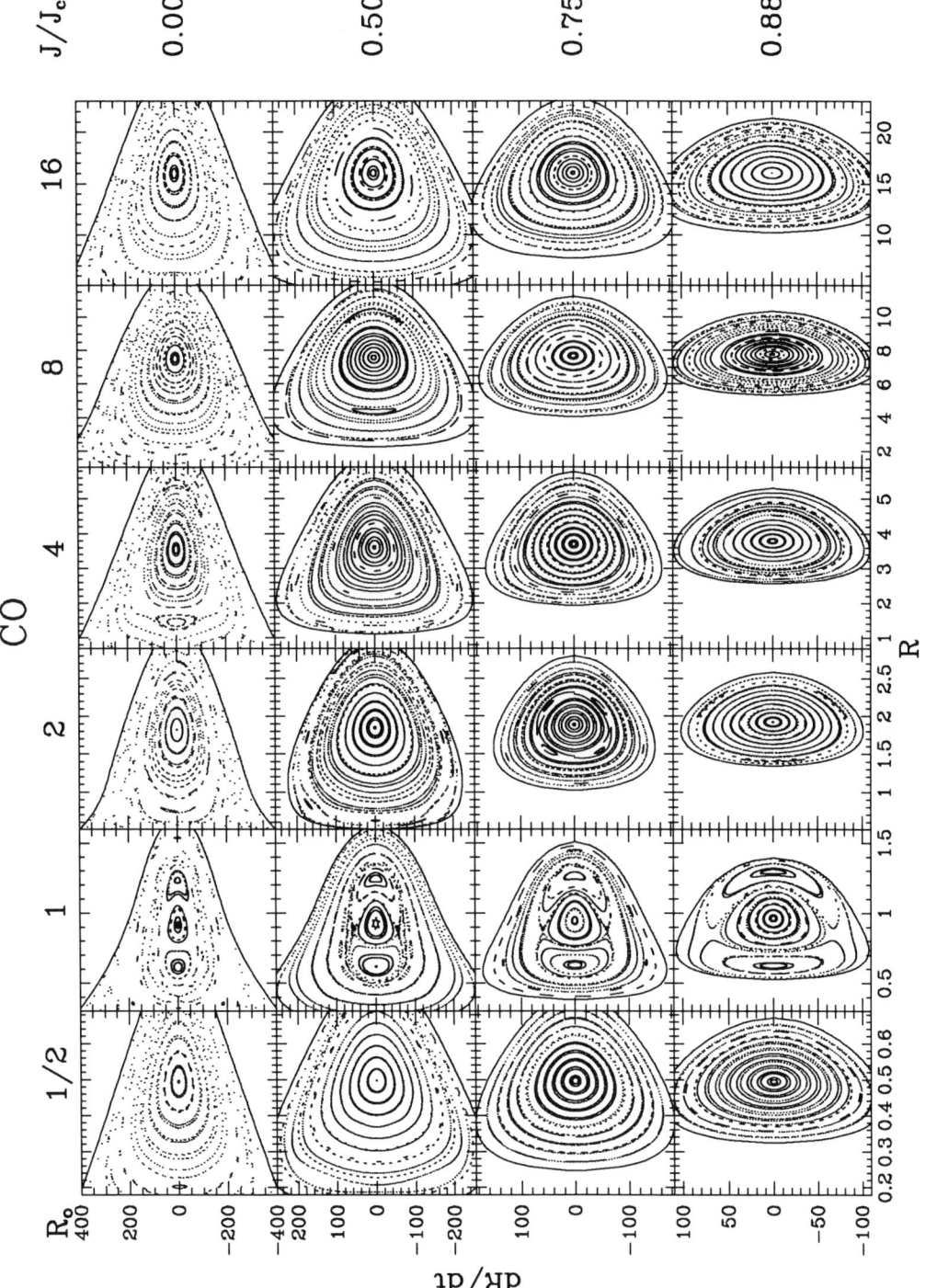

Figure 3

where an irregular orbit "sea" appears. This is not surprising since radial orbits plunge to the rotation axis and have a large chance of passing near the galactic center, where they are scattered in all directions. In fact, the BSS model, which has a truncated nuclear component, is the one that presents, on close inspection, the largest irregular region of the three models.

For high angular momentum orbits the three models present regular orbits only; the family of orbits that oscillate around the local circular orbit are the most dominant. The surface of section nearest the solar orbit ($R_o = 8$ kpc, $J/J_c = 0.88$) is entirely occupied by this family of orbits, except in the BS model where two islands appears symmetrically on both sides of the circular orbit and which correspond to the 1:1 resonance. These islands disappear for larger, and smaller values of R_o.

At intermediate angular momenta the situation is more complicated and differences between the models appear. In all models the region around $R_o =$ 1, 2, 4 kpc, is the most interesting with several families of regular orbits coexisting within an irregular region. This is presumably related to the fact that this is the region where the potential makes the transition from being bulge dominated to disk dominated.

6. Conclusions

The orbital structure generated by three models of the galactic mass distribution has been compared using the method of surfaces of section. All the models are quite regular and similar for high angular momentum orbits but differ for intermediate angular momentum orbits in the region between 1 and 4 kpc. Irregular orbits appear toward the radial orbit region in all three models. Although, as expected, orbits computed for stars with velocities near circular in the three models should give very similar results, real differences will appear for stars with less angular momentum at intermediate galactocentric distances. Conclusions about particular numerical values, derived from orbital parameters of these stars and computed with a particular model, must be viewed with skepticism.

REFERENCES

[1] Allen, C. & Santillán, A. 1991, *Rev. Mex. Astron. Astrof.*, **22**, 255.
[2] Bahcall, J.N., Schmidt, M. & Soneira, R.M. 1982, *ApJ*, **258**, L23.
[3] Bahcall, J.N., Schmidt, M. & Soneira, R.M. 1983, *ApJ*, **265**, 730.
[4] Binney, J. & Tremaine, S. 1987, *Galactic Dynamics*, Princeton Univ. Press.
[5] Caldwell, J. & Ostriker, J.P. 1983, in *Kinematics, Dynamics, and Structure of the Milky Way*, Ed. W.L.H. Shuter, Reidel, Dordrecht.
[6] Ollongren, A. 1962, *Bull. Astron. Ints. Neth.*, **16**, 241.
[7] Ollongren, A. 1965, *Galactic Structure*, University of Chicago Press, Ed. A. Blaauw & M. Schmidt, p.501.

A Quiet Start for an Integrable, Non-Axisymmetric Potential

By Stephen E. Levine

Observatorio Astronómico Nacional, IA-UNAM, Ensenada, B. C., México. US postal address: P. O. Box 439027, San Diego, CA 92143–9027, USA.

A quiet start technique is developed for placing particles on orbits in integrable, non-axisymmetric potentials. The method places particles uniformly on the orbits in action-angle space, making the initial conditions as smooth as possible. This construction is demonstrated by populating a series of self-consistent equilibrium models for the two dimensional perfect elliptic disks. All the methods developed in this investigation can easily be extended to integrable potentials in three dimensions.

---

1. Introduction

In galaxy simulations, and indeed in many numerical simulations, where a system of many ($10^{10} - 10^{12}$) particles is represented by a relatively small number of particles ($10^4 - 10^6$), it is often difficult to measure the growth rates of instabilities because the necessary resolution is lost in the particle discreteness noise. In an effort to combat this problem, *quiet start* techniques have been developed that minimize discreteness noise through uniform distribution of particles [10]. Prior examples of quiet starts in galaxy simulations have all depended upon axisymmetry in the initial conditions. It is possible to generate a quiet start for a non-axisymmetric potential provided the potential is integrable. This paper presents the construction of such a distribution for the two dimensional perfect elliptic disk models.

Perfect elliptic disks are the two dimensional limiting case of the three dimensional perfect ellipsoids [4]; their surface density Σ is given in cartesian coordinates (x, y) by [4]

$$\Sigma(x,y) = \frac{\Sigma_0}{(1+m^2)^{3/2}}, \quad m^2 \equiv \frac{x^2}{a^2} + \frac{y^2}{b^2}. \qquad (1)$$

The elliptic disk potential satisfies Stäckel's criteria, implying that the equations of motion are completely separable [4,7] in confocal elliptic coordinates (λ, μ). Curves of constant λ are confocal ellipses aligned along the minor axis, with $-\alpha \leq \lambda \leq \infty$, while curves of constant μ are hyperbolae, with $-\beta \leq \mu \leq -\alpha$ (see Figure 1) where $\alpha = -a^2$, $\beta = -b^2$. We can write the momenta p_λ and p_μ at any point on an orbit as

$$p_\tau^2 = \frac{(\tau+\alpha)[E+G(\tau)] - i_2}{2(\tau+\alpha)(\tau+\beta)}. \qquad (2)$$

where τ is either λ or μ, and E and i_2 are particular values of the hamiltonian (H) and the second isolating integral (I_2) which specify an orbit and $G(\tau)$ is a monotonically decreasing function [4,13].

Because the potential is not axially symmetric, the orbits in the elliptic disk potential divide into two families, *box* and *loop* orbits. The box orbits resemble combinations of independent oscillations in the x and y directions and have no net angular momentum about the center of the potential. Loop orbits are ellipses or rosettes with a definite sense of rotation about the center of the potential. Because of the separability of the potential, the orbital extremal values of λ and μ, the bounding surfaces, are all coordinate lines, called the *turning points* of the orbit. Loop orbits are bounded by ellipses of constant λ;

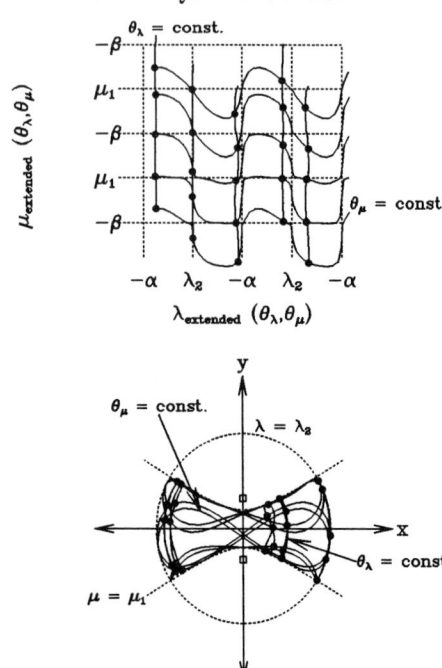

FIGURE 1. In the upper panel, lines of constant $(\theta_\lambda, \theta_\mu)$ on the torus of a box orbit with turning points (μ_1, λ_2) are plotted in the extended elliptic coordinates. Lines of constant θ_λ run approximately vertically; those of constant θ_μ run roughly horizontally. The lines of constant θ_τ are spaced by $\Delta\theta_\tau = 2\pi/5$; black dots mark particles at the intersections of these lines. The lower panel shows the same lines of constant θ_τ plotted in cartesian space, with the coordinate lines bounding the orbit shown as dashed lines. The open squares mark the foci of the elliptic coordinates.

a closed loop orbit is a curve of constant λ. Box orbits are contained within an hyperbola of constant μ_1 and an ellipse of constant λ_2. We have considered maximum streaming models [6], in which all the loop orbits are closed.

A self-consistent model is one such that the distribution function, when integrated over velocities, yields a density which generates the initially specified gravitational potential. We expect this function to be inherently continuous; to set up a discrete representation for N-body testing, we must first select orbits which sample the distribution function, and then choose particles which sample each orbit. Here we assume that a set of orbits has already been chosen, and address the second problem.

2. Populating The Orbits

In a *quiet* distribution of particles [10,11] the particles are initially distributed as uniformly as possible in phase space to minimize random noise in the initial conditions. Sellwood [10] describes a quiet start algorithm for closed loop orbits in axially symmetric potentials. We follow this method for the closed loop orbits and simply place the desired number of particles along the loop, equal fractions of the period apart. For nonclosed loop orbits, we can use the procedure outlined below for the box orbits.

For the box orbits, we have developed a general method for finding the smoothest way to populate a space filling orbit in an integrable potential. This is based on the action–angle formalism [5]: every orbit corresponds to a torus of constant actions (J_λ, J_μ), and the angle coordinates $(\theta_\lambda, \theta_\mu)$ on the torus are cyclic and evolve linearly with time because the Hamiltonian depends only upon the actions. For given (J_λ, J_μ), we will have a quiet particle distribution if we lay down a grid of points uniformly spaced in θ_λ and θ_μ. Because forward evolution in time is area preserving on the torus, they will stay uniformly spaced as long as the potential does not change.

To construct the angle coordinates explicitly, we follow the procedure of Binney & Tremaine [3,§3.5] and begin by assigning a single arbitrary phase space point $(\lambda, \mu, p_\lambda, p_\mu)$ to the angle coordinate $(\theta_\lambda, \theta_\mu) = (0,0)$ on the torus defined by (J_λ, J_μ). We mark out the θ_λ axis by integrating the coupled differential equations

$$\frac{dw_m}{d\theta_\lambda} = [w_m, J_\lambda] \equiv \left(\frac{\partial w_m}{\partial w_1}\frac{\partial J_\lambda}{\partial w_3} + \frac{\partial w_m}{\partial w_2}\frac{\partial J_\lambda}{\partial w_4}\right) - \left(\frac{\partial w_m}{\partial w_3}\frac{\partial J_\lambda}{\partial w_1} + \frac{\partial w_m}{\partial w_4}\frac{\partial J_\lambda}{\partial w_2}\right), \quad (3)$$

where the $w_m (m = 1, \ldots, 4)$ are the phase space coordinates $(\lambda, \mu, p_\lambda, p_\mu)$; only one of the four terms in (3) is ever nonzero, since $\partial w_m / \partial w_k = 1$ only if $m = k$, and is zero otherwise. The θ_μ axis is then marked out similarly. Arnold [2,§49] shows that the coordinate axes defined in this manner remain on the surface of the torus and close upon themselves forming a grid upon the surface. This construction of the angle coordinates gives us a map from the torus onto an orbit in ordinary spatial coordinates.

To find the map explicitly, note that equation (3), and the corresponding equation in μ, yield

$$\left.\frac{d\lambda}{d\theta_\lambda}\right|_{(J_\lambda, J_\mu, \theta_\mu)} = \left.\frac{\partial J_\lambda}{\partial w_3}\right|_{(w_1, w_2, w_4)} \qquad \left.\frac{d\mu}{d\theta_\lambda}\right|_{(J_\lambda, J_\mu, \theta_\mu)} = \left.\frac{\partial J_\lambda}{\partial w_4}\right|_{(w_1, w_2, w_3)}$$

$$\left.\frac{d\lambda}{d\theta_\mu}\right|_{(J_\lambda, J_\mu, \theta_\lambda)} = \left.\frac{\partial J_\mu}{\partial w_3}\right|_{(w_1, w_2, w_4)} \qquad \left.\frac{d\mu}{d\theta_\mu}\right|_{(J_\lambda, J_\mu, \theta_\lambda)} = \left.\frac{\partial J_\mu}{\partial w_4}\right|_{(w_1, w_2, w_3)} \quad (4)$$

The derivatives on the left hand side are taken *on the surface of the torus*, while the right hand derivatives are taken while holding constant the other three phase-space coordinates. There is a similar set for p_λ and p_μ, but because we have two isolating integrals, we need not integrate these as we can compute the momenta as functions of position along the orbit.

Let us work out $\partial J_\tau / \partial w_3$, $\tau = \lambda, \mu$, as an example. By equation (2), the two isolating integrals, E and i_2, associated with the orbit which passes through the phase space point (w_1, w_2, w_3, w_4) can be written explicitly as functions of these coordinates, and so can the momentum p_τ at any point (λ, μ) along the orbit specified by (w_1, w_2, w_3, w_4). That momentum has a derivative at the point (λ, μ) with respect to w_3:

$$p_\tau \frac{\partial p_\tau}{\partial w_3} = \frac{(w_1 + \alpha)(w_1 + \beta) w_3}{w_1 - w_2} \left\{\frac{\tau - w_2}{(\tau + \alpha)(\tau + \beta)}\right\}. \quad (5)$$

Differentiating the action $J_\tau \equiv 1/(2\pi) \oint p_\tau(\tau) d\tau$ and substituting for $\partial p_\tau / \partial w_3$ gives

$$\frac{\partial J_\tau}{\partial w_3} = \frac{1}{2\pi} \oint \frac{\partial p_\tau}{\partial w_3} d\tau = \frac{2\sqrt{2}}{\pi} \frac{(w_1 + \alpha)(w_1 + \beta) w_3}{w_1 - w_2}$$

$$\times \int_{\tau_{\min}}^{\tau_{\max}} \frac{(\tau - w_2) d\tau}{\sqrt{(\tau + \alpha)(\tau + \beta)\{(\tau + \alpha)(E + G(\tau)) - i_2\}}}. \quad (6)$$

The two partial derivatives with respect to w_4 are found similarly. Thus the partial derivatives of equation (4) can all be computed by quadrature, and we can move forward along curves of constant θ_λ and θ_μ by integrating only two coupled equations at any one time.

The torus specified by (J_λ, J_μ) maps into a rectangular region in coordinate space

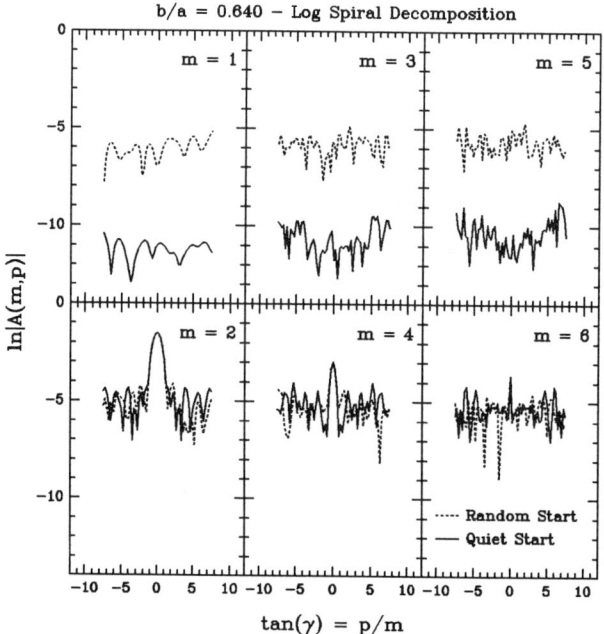

FIGURE 2. Comparison of the power at the initial instant in the first six logarithmic spiral modes for the quiet start (*solid line*) and the random realization (*dashed line*) for the model with axial ratio $b/a = 0.640$.

bounded by the orbit's turning points and the coordinate limits. Every (λ, μ) pair corresponds to four points in cartesian (x, y) space. To remove this four fold degeneracy, we define periodic extensions in both elliptic dimensions by reflections about the rectangle's boundaries, creating the double covering of (λ, μ) space illustrated in figure 1. A point in the above adjoining rectangle will map into our base rectangle by $\tau \to 2\tau_{\max} - \tau$, and a point in the below adjoining rectangle will map into the base rectangle by $\tau \to 2\tau_{\min} - \tau$. This reflection procedure is then continued in both directions, folding the extended plane back into the base rectangle. Along the lines of constant θ_τ, the sign of the x coordinate must change on crossing the y axis ($\mu = -\alpha$ or $\lambda = -\alpha$), and the sign of the y coordinate changes every time the line crosses the x axis ($\mu = -\beta$). The integration of equations (4) is then performed quickly using the Bulirsh-Stoer integration scheme described by Press et al. [8].

At any point (λ, μ) along the line $\theta_\tau = $ constant, the two isolating integrals fix p_λ^2 and p_μ^2. We resolve the sign ambiguity for the velocities v_λ and v_μ by further extending the (λ, μ) space and end up with a well defined mapping from $(J_\lambda, J_\mu, \theta_\lambda, \theta_\mu)$ to the extended space in $(\lambda, \mu, v_\lambda, v_\mu)$ and finally to (x, y, v_x, v_y). Having done that, we are ready to place particles upon each orbit at equal intervals in θ_λ and θ_μ.

3. Discussion

In order to assess the ability of the quiet start method to decrease initial fluctuations in the surface density, we compared our models with models constructed from the same set of orbits which were populated randomly, weighted by the orbital surface density $\Sigma_{\text{orb}}(\lambda, \mu) \propto 1/[v_\lambda v_\mu (\lambda - \mu)]$. The two corresponding representations were decomposed into logarithmic spirals [3,§2.6.4]. These form a complete orthogonal basis set for which the normalized coefficients can be written

$$A(m,p) = \frac{1}{N} \sum_{j=1}^{N} \exp\{i[m\theta_j + p\ln(r_j)]\} \tag{7}$$

where r_j and θ_j are the positions of the N particles in the model [11,12]. As can be seen in Figure 2, the quiet start model shows much less power in the odd log spiral harmonics than the random realization; the substantial power in the even harmonics is due to the elliptical nature of the overall density profile.

Our procedure for generating a quiet start, which minimizes random noise due to particle discreteness, can be carried over to any integrable system including three dimensional perfect ellipsoids and spherical systems. This is potentially most useful in N-body studies which attempt to measure the growth rate of instabilities, because the detection of instabilities which are still in the linear regime is limited by particle noise. For example, Saha [9] found that his linear stability theory was consistent with the results of N-body simulations for highly unstable spherical systems, but predicted slow growing instabilities which could not be seen in the simulations because of particle noise. Allen, Palmer & Papaloizou [1] have constructed an analytic potential–smoothing integration technique which decreases the $\sqrt{N}$ noise associated with binning and softening in N-body codes, and permits better examination of the linear growth regime. Their method would also benefit from a quiet start, because the particle discreteness noise then makes a larger relative contribution.

Thanks to L. Sparke, J. Sellwood, R. French, and T. de Zeeuw. This work was supported by the NASA Theory Program under grant NAGW-2420

REFERENCES

[1] Allen, A. J., Palmer, P. L., & Papaloizou, J. 1990, MNRAS, 242, 576
[2] Arnold, V. I. 1978, *Mathematical Methods of Classical Mechanics*, (New York: Springer-Verlag)
[3] Binney, J., & Tremaine, S. 1987, *Galactic Dynamics*, (Princeton: Princeton University Press)
[4] de Zeeuw, P. T. 1985, MNRAS, 216, 273
[5] Goldstein, H. 1980, *Classical Mechanics*, (2nd ed.) (Reading: Addison-Wesley)
[6] Levine, S. E., & Sparke, L. S. 1994, ApJ, 428, in press
[7] Lynden-Bell, D. 1962, MNRAS, 124, 95
[8] Press, W. H., Flannery, B. P., Teukolsky, S. A., & Vetterling, W. T. 1986, *Numerical Recipes*, (Cambridge: Cambridge University Press)
[9] Saha, P. 1991, MNRAS, 248, 494
[10] Sellwood, J. A. 1983, JCompPhys, 50, 337
[11] Sellwood, J. A. 1989, MNRAS, 238, 115
[12] Sellwood, J. A., & Athanassoula, E. 1986, MNRAS, 221, 195
[13] Teuben, P. J. 1987, MNRAS, 227, 815

Dissipation in Dynamical Models of Galaxies

By J. Palouš[1], B. Jungwiert[2]

[1]Astronomical Institute, Academy of Sciences of the Czech Republic, Boční II 1401, 141 31 Prague 4, Czech Republic.

[2]Center for Theoretical Study, Charles University, Celetná 20, 110 00 Prague 1, Czech Republic.

The problem of collisions of molecular clouds is analyzed using a 3D computer code. The inelastic collisions between the clouds dissipate a fraction of the kinetic energy of their relative motions and a small part of the orbital angular momentum is trapped inside the clouds. The fluctuations of the gravitational potential from close encounters are also taken into account. The output from the simulations is compared with the observed distribution of molecules in nearby galaxies.

---

1. Introduction

In stellar systems with more than 10^6 stars the two-body relaxation time is longer than the dynamical time (the crossing time or the period of rotation). Galaxies with $10^9 - 10^{11}$ stars have a relaxation time, t_{rel}, longer than the age of the galaxy itself, $\sim 10^{10}$ years, or the age of the Universe. Chandrasekhar [2] gives $t_{rel} \sim 10^{17}$ years for stars in our Galaxy.

In N-body simulations, even with the latest progress in hardware and software, the number of particles N cannot be significantly larger than 10^6. The direct integration of orbits allows up to 10^4 particles. The computer GRAPE, devoted to direct N-body computations, will probably increase N up to $\sim 10^5$. In simulations using the particle-mesh scheme (PM) with Fast Fourier Transform, the reasonable N is $10^5 - 10^6$. However, the mesh size restricts the spatial resolution. In simulations of galaxies with these many particles, masses of individual particles are $\sim 10^5 - 10^6$ $M_\odot$. This is considerably higher than masses of individual stars. Heavy computer particles decrease t_{rel} and the two-body particle noise becomes much more important. These spurious fluctuations perturb particle orbits from the underlying mean field and they complicate the interpretation of the output in terms of stellar distribution in observed galaxies. A possible way to overcome this difficulty is proposed by Leeuwin et al. [6].

In galaxies, the stars are coupled with the interstellar medium (ISM) by gravity and star formation. Stars are formed from the ISM and thus the interconnection of the two components is inescapable. Most stars, if not all, form in massive molecular clouds with masses $m_{cl} > 2 \times 10^5 M_\odot$ and diameters $d_{cl} > 30$ pc. A more complete story of galaxies should include both stars and clouds.

For molecular clouds, unlike the collisionless stellar component, two-body encounters play an important role. Close encounters between clouds add gravitational perturbations to the background stellar potential, and also dissipate energy and angular momentum due to the inelasticity of the collisions.

In this contribution, we describe the evolution of a dissipative system of colliding clouds in the background galactic potential. We shall stress the importance of resonances and

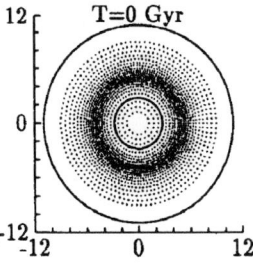 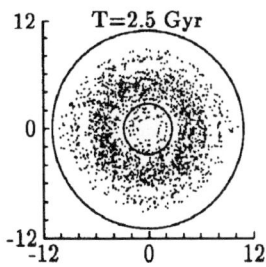 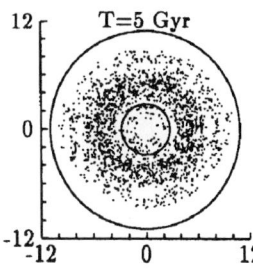

FIGURE 1. Test particles in the axisymmetric potential perturbed with a weak bar. The reference frame corotates with the bar whose major axis is horizontal. Circles represent the ILR and the CR.

show how they may describe the distribution of molecules in galaxies. A more detailed description of the model and of the results is given elsewhere [8].

2. Dissipation and Gravitational Fluctuations

The interstellar molecular clouds are modeled as massive spheres of finite radius. They move in the background gravitational potential of the galaxy; they interact gravitationally and collide when their centers are close enough. A typical cloud has a mass of 3×10^5 $M_\odot$ and diameter of 40 pc.

If the distance between two clouds is smaller than the sum of their radii, they collide inelastically. The inelasticity is introduced through constants β_r and β_t: $V_r^{new} = -\beta_r V_r^{old}$, $V_t^{new} = \beta_t V_t^{old}$, where $V_{r,t}^{new,old}$ are the radial or tangential components of the relative velocity between the two clouds before (old) or after (new) a collision. The parameters β_r and β_t play a role similar to the bulk and shear viscosities in hydrodynamics and they should be understood as statistical quantities. With $\beta_r < 1$ the collisions do not conserve the kinetic energy: a fraction of it is transformed into the internal energy of the colliding clouds and part is radiated away. With $\beta_t < 1$, a certain amount of the angular momentum is lost: it is supposed to be trapped in turbulent motions inside the clouds.

We use the values $\beta_r = \beta_t = 0.2$, which take into account the distribution of relative velocities before collisions: we assume that for velocities smaller than $[G(M_i + M_j)/(R_i + R_j)]^{1/2}$, where $M_{i,j}$ are the masses and $R_{i,j}$ are the radii of clouds i or j respectively, the collisions are completely inelastic and for higher relative velocities we take just arbitrarily $\beta_r = \beta_t = 0.5$. With more colliding partners at lower velocities we get the adopted average values $\beta_r = \beta_t = 0.2$.

The long-range character of gravity, which is particularly important for the stellar component of galaxies, requires large amounts of CPU time, a problem that is sometimes circumvented by sacrificing spatial resolution. In the PM experiments, the small scale fluctuations are smoothed out by the softening and two-body encounters are suppressed. For massive clouds, the two-body encounters are more important than for stars. Therefore, we adopt the opposite approach: we disregard completely the long-range effects for the clouds, and assume that they are included in the background stellar potential. We also consider small-scale fluctuations caused by close cloud-cloud encounters.

The dissipation of the kinetic energy in inelastic collisions produces gravitationally bound groups of clouds. If a group reaches certain critical mass, M_{crit}, we assume that star formation is triggered and an OB association may be formed. This has dramatic consequences. The large amount of energy which is released from massive stars, breaks the gravitationally

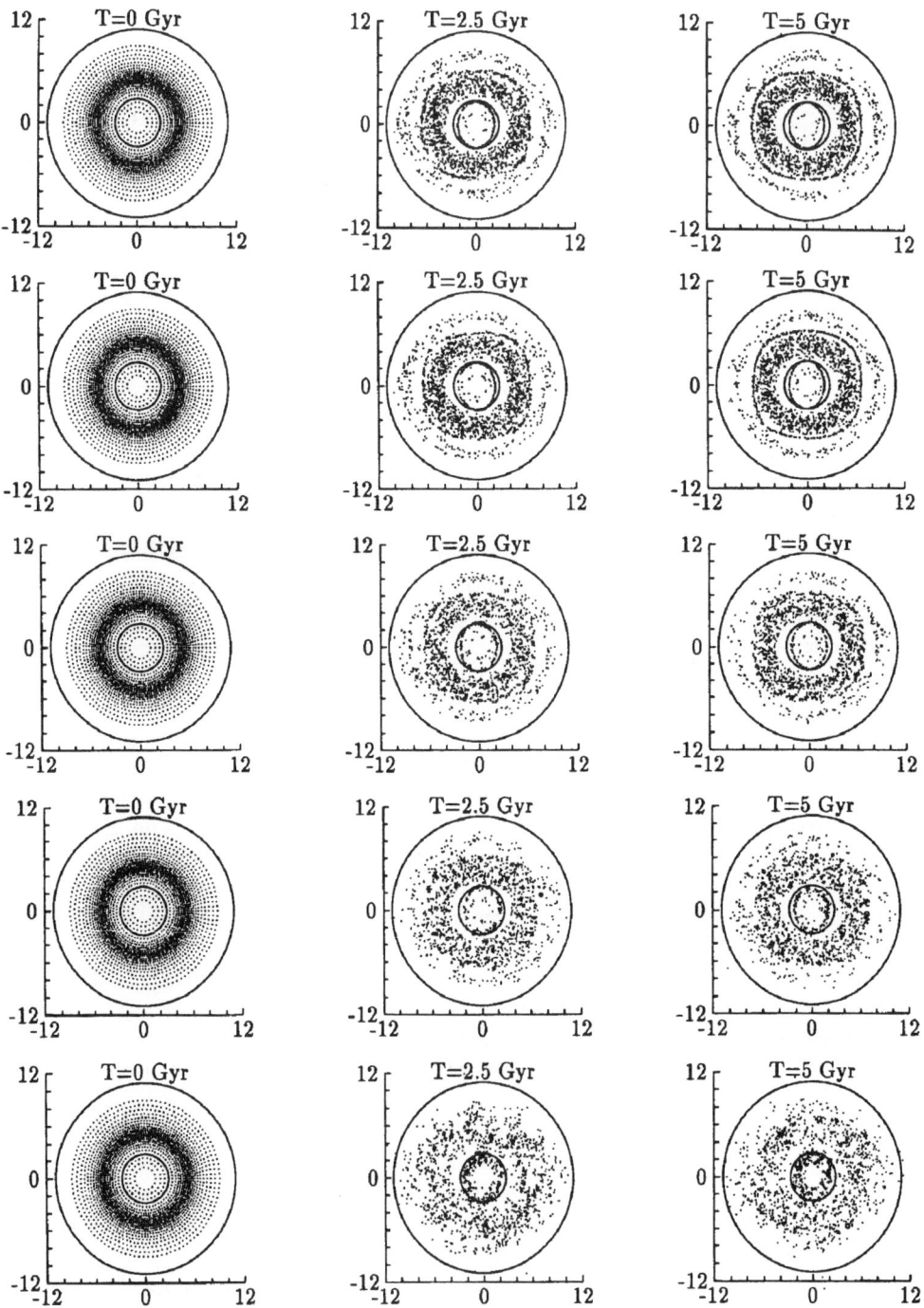

FIGURE 2. Colliding particles. The potential and the reference frame are the same as in Fig. 1. The particle mass increases from the top to the bottom: $0, 10^4, 10^5, 3 \times 10^5, 10^6$ M$_\odot$.

bound group into smaller fragments. In our code, the clouds get an additional velocity push leading to the group's disruption with the final velocity dispersion equal to the initial velocity dispersion given to clouds. This reinjection of trapped clouds into the system counterbalances the kinetic energy losses in inelastic collisions keeping the cloud velocity dispersion nearly constant.

The evolution of the system of colliding clouds is followed in three dimensions. For the stellar background we use a Miyamoto–Nagai potential [7] for the disk and bulge components plus another component representing the halo [1]. This axisymmetric potential is perturbed with a weak triaxial central bar similar to the central asymmetry observed in N-body simulations [10]. A more detailed description of the adopted background potential is given by Palouš et al. [8].

3. The ILR, 4/1 and Other Resonances

First we follow the motion of massless, non-colliding test particles in the background axisymmetric potential with the weak perturbing bar. The distribution of clouds remains roughly axisymmetric throughout the evolution (Fig. 1). This is related to the weakness of the bar. Stronger perturbations would excite more visible non-axisymmetric responses. Next, we include the inelastic collisions. With the same background potential, the evolution is different. After 2 Gyr, a ring near the Inner Lindblad Resonance (ILR) appears. The clouds accumulate in a very thin oval-like arc following the main stable periodic orbit inside the ILR. The orbit is perpendicular to the bar major axis. In the region between the ILR and Corotation (CR), the 4/1 resonance dominates after 2.5 Gyr (Fig. 2).

The combination of the weak bar perturbation, enhancing the orbital crossing, and the dissipative inelastic collisions results in the setting of clouds down to resonances [3]. This is similar to the situation described by Pfenniger [9], when small dissipative forces are introduced into a Hamiltonian system. The trajectories converge to robust stable periodic orbits, which are called attractors, and the occupation of the phase space is reduced to a small subset of the original one.

Subsequently, we introduce the gravitational fluctuations from close cloud–cloud encounters. These fluctuations perturb the orbits and dissolve the narrow ring-like structure formed due to dissipation and gravitational torques. The depth of the local minimum in the potential is competing with the gravity fluctuations (Fig. 2): the 10^4 $M_\odot$ clouds are too light to destroy the rings. On the other hand, the 10^5 $M_\odot$ clouds dissolve the ring connected with the 4/1 resonance. The periodic orbit near the ILR is too robust to be blurred completely with this strength of fluctuations: after 5 Gyr, the ring is still present.

4. The Distribution of Molecules in Galaxies

CO observations of spiral galaxies have discovered molecular ring-type structures [11]. In a large fraction of nearby spirals the gas surface density peaks some distance away from the nucleus, suggesting the presence of a ring. In M101, NGC 3351, and NGC 6951, the strongest CO emission arises from twin peaks located symmetrically about the nucleus and oriented nearly perpendicular to the large-scale stellar bar [5]. The peaks are located where dust lanes curve inwards and intersect the ring. Molecular rings are supposed to be connected with the ILR: clouds settle down to the periodic orbit inside the ILR due to the dissipation in inelastic collisions and the gravitational torque from the bar [4].

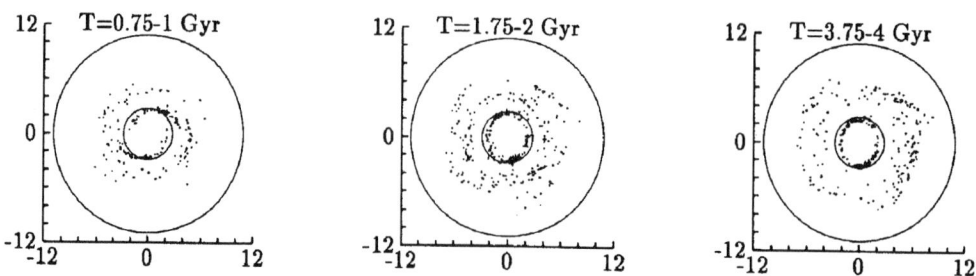

FIGURE 3. Places of fast $V_{rel} > 30$ km s^{-1} collisions for 3×10^5 M$_\odot$ clouds.

5. Conclusions

The presence of rings, which delineate non–intersecting, stable periodic orbits, is derived from the robustness of an orbit and from the amplitude of gravitational fluctuations. In galaxies with lower mass clouds, the rings should be more distinct, since the clouds perturb each other to a lesser extent.

The non-axisymmetric perturbation increases the collision frequency of molecular clouds near the ILR. The locations of fast ($V_{rel} > 30$ km s^{-1}) collisions in our simulation are shown in Fig. 3. These areas display twin peaks nearly perpendicular to the bar major axis near the ILR because of the related high orbital crossing. The dust lanes may be formed there, by the intensive interactions of molecular clouds. We conclude, that the twin CO peaks connected with the dust lanes observed in M101, NGC 3351, NGC 6951, and in some other nearby spiral galaxies, may result from the inelastic collisions of molecular clouds near ILR. The collision frequency is increased in the places perpendicular to the bar major axis, which is the place where the twin CO peaks appear. In the subsequent evolution, the molecular clouds settle down to the periodic orbit inside the ILR forming a molecular ring.

JP is grateful to the Instituto de Astronomía, UNAM for the hospitality during his visit, and thanks Pepe Franco for all his help. The stay was funded as part of an exchange of scientists program between CONACYT–México and the Academy of Sciences of the Czech Republic.

REFERENCES

[1] Allen, Ch. & Santillán, A. 1991, Rev. Mex. Astron. Astrof. 22, 255
[2] Chandrasekhar, S. 1943, Principles of Stellar Dynamics, Chicago University Press
[3] Combes, F. 1993, in Star Formation, Galaxies and the Interstellar Medium, eds. J. Franco, F. Ferrini, and G. Tenorio-Tagle, Cambridge University Press, p. 324
[4] Combes, F. & Elmegreen, B. 1993, A&A, in press
[5] Kenney, J. D. P. 1993, in Star Formation, Galaxies and the Interstellar Medium, eds. J. Franco, F. Ferrini, & G. Tenorio-Tagle, Cambridge University Press, p. 14
[6] Leeuwin, F., Combes, F. & Binney, J. 1993, MNRAS, 262, 1013
[7] Miyamoto, M. & Nagai, R. 1975, PASJ 27, 533
[8] Palouš, J., Jungwiert, B. & Kopecký, J. 1993, A&A, 274, 189
[9] Pfenniger, D. 1992, in Evolution of Interstellar Matter and Dynamics of Galaxies, eds. J. Palouš, W.B. Burton, and P.O. Lindblad, Cambridge University Press, p. 328
[10] Pfenniger, D. & Friedli, D. 1991, A&A, 252, 75
[11] Sofue, Y., 1991, in Dynamics of Galaxies and Their Molecular Cloud Distribution, eds. F. Combes, F. Casioli, Kluwer Academic Publishers, p. 287

Numerical Simulations of Collisions of Spherical Galaxies

By M. M. Vergne, J. C. Muzzio

Facultad de Ciencias Astronómicas y Geofísicas, Universidad de La Plata, and Programa de Fotometría y Estructura Galáctica, Consejo Nacional de Investigaciones Científicas y Técnicas, Argentina.

We used a self-consistent N-body code to simulate encounters of spherical non-rotating galaxies. The galaxies were represented by Plummer spheres ($\rho \propto r^{-5}$) with isotropic velocity dispersions; the mass ratios used for the colliding galaxies were 1:1, 1:2, 1:4 and 1:8. We analyzed the effects on the internal structure of the galaxies caused by collisions that did not result in mergers after a Hubble time (1.5×10^{10} yrs). We estimated quantitatively the changes in mass, energy and size, and discussed possible correlations between these properties and the orbital parameters.

1. Introduction

The study of encounters between galaxies has received much attention since the discovery of peculiar galaxies. Photographs of interacting galaxies [4, 13] show that many pairs of galaxies have peculiar morphological features. It was shown [12], using the restricted three body approach, that tidal forces play an important role in the dynamics of interacting galaxies and can lead to the formation of such features.

The tidal effects change the overall structure of the colliding galaxies. The total internal energy of a galaxy is changed, providing a good measure of the change in its structure [3], as has been estimated by many authors using the impulse approximation. However, for a detailed study of the changes in the size and mass distribution of colliding stellar systems, the use of a self-consistent method, such as N-body simulations, is needed.

Therefore, we present here the results of an investigation of encounters between spherical galaxies of both comparable and unequal masses, where we considered those collisions that did not result in mergers after a Hubble time. We investigate different structural changes in the galaxies (mass, size, energy, etc), specially taking into account the unequal mass collisions, which are less studied than the equal mass collisions.

2. Numerical Simulations

We performed several N-body simulations of encounters of two spherical galaxies. Each galaxy was represented with a system of identical particles interacting via the softened potential

$$\Phi_{ij} = -G \frac{m_i m_j}{[(\vec{r}_i - \vec{r}_j)^2 + \varepsilon^2]^{1/2}} \tag{1}$$

where ϵ, G, and m_i are set equal to 1.

The N-body equations of motion resulting from (1) were integrated using the NBODY2 code developed by S. J. Aarseth.

2.1. Models and Initial Conditions

We considered seven galaxies made up of 50, 90, 150, 225, 300, 360 and 400 particles for our simulations. The galaxies were initially represented by non-rotating Plummer spheres, and truncated at a radius R_l (defined so that 1% of the the Plummer model mass is excluded). Their velocity dispersion was isotropic. Before using them for our experiments, our model galaxies were allowed to evolve in isolation for 3.5 crossing times.

We used four mass ratios for the colliding galaxies: 1:1, 1:2, 1:4 and 1:8, with the combined mass of both galaxies always being 450 particles. The separation of the galaxies at the start of the simulations was large enough for the initial tidal effects to be negligible (larger than ten times the sum of the galaxy's radii). Hyperbolic and parabolic orbits were used for our models. The orbital velocity, in the hyperbolic case, was varied adopting different fractions of the internal energy as orbital energy. We considered different values of pericentric distance ($R_o = 0, 10, 15, 20$ and 30 units). The x–y plane was chosen as the orbital plane, and the initial orbital velocity was in the x-direction. Each experiment was followed for a time interval similar to the Hubble time ($T_H \approx 1.5 \times 10^{10}$ yr).

3. Results

We have only investigated the effects on the internal structure of the galaxies caused by collisions that did not lead to mergers in a T_H.

Each encounter was characterized by two dimensionless parameters which depend on the orbital properties, and are defined as:

$$<E> = \frac{E_{orb}}{E_1 + E_2} \quad (2a)$$

$$<R> = \frac{R_o}{R_{h1} + R_{h2}} \quad (2b)$$

where E_{orb} is the initial orbital energy, $E_i (i = 1, 2)$ is the internal energy of galaxy i, R_o is the pericentric distance and R_{hi} is the half-mass radius of the ith galaxy.

The analysis of the results is limited to the first pericentric passage, because our galaxies do not need more than one passage to become a bound system.

3.1. Mass Loss

When a stellar system suffers an encounter with a second one, its velocity distribution changes, and some stars may acquire enough kinetic energy to escape from the system (tidal stripping). Some authors [5, 11], taking into account that encounters are the most important way to remove matter from galaxies in clusters and groups of galaxies, obtained estimates of this quantity. On the other hand, some of the escapees from one galaxy may be captured by the other one, this effect being maximized for slow encounters. This process have been extensively studied [6, 7, 8, 9] for the particular case of the exchange of globular clusters in clusters of galaxies (cluster swapping).

Our results show:
1. The maximum change in mass is about 18% for mass ratios 1, 1/2 and 1/4 (14% truly lost and 4% swapped). The loss is smaller for mass ratio 1/8 and its maximum value is about 2.6% (1.8% truly lost and 0.8 % swapped). The largest contribution to the mass loss is from the smaller galaxy in the unequal mass collisions, and the larger galaxy only contributes a small percentage of mass to the total loss.
2. The mass loss correlates strongly with the orbital parameters of the collisions, but the exchange of mass exhibits no dependence on them.

3. The minimum exchange of mass corresponds to head on encounters with equal galactic masses. This result was expected, since the relative velocity is high to avoid mergers.
4. The outer parts of both galaxies (in equal-mass encounters), or of the small galaxy (in unequal-mass encounters), lose more mass than their central parts. This effect is more important when encounters are more energetic, interpenetrating and for lower mass ratios.

3.2. Energy Change

Because the collision between two galaxies is an inelastic phenomenon, the system loses and swaps matter and changes its bound mass distribution. Also, the internal energy of the galaxies increases with the loss of orbital kinetic energy. This energy gained by the galaxies during an encounter is a measure of the damage done to them.

From our experiments we found that:

1 - The fractional change of internal energy ($\delta E = (E_f - E_o)/E_o$, E_o and E_f being the initial and final internal energies, respectively) for each galaxy of the pair correlates with the orbital parameters.

2 - The dependence of the internal energy on the mass loss is linear; therefore, we have:

$$\delta E_i = C_m \delta M$$

(taking into account that part of the energy excess acquired by the system is balanced by escapers)

3 - After the encounter, the galaxies are more tightly bound than before.

3.3. Size and Mass-Profiles

In general, the results of others authors [2, 5, 10] show a considerable range of behaviors. In some cases the resulting system is more extended than the original one, and in others it is smaller. This depends on the mass and binding energy changes.

The galaxies change their size as a result of the encounters, and we used the half-mass radius (R_h) to quantify that change.

1 - The final galaxies are more extended than their progenitors.
2 - The expansion is larger for near head-on encounters (strong collisions).
3 - The smaller galaxies of the models with mass ratios 1/4 display a different behavior. In some cases the central parts of the galaxy are concentrated and the outer parts are extended, in others the effect is the opposite.

3.4. Velocity Dispersion Change

When the galaxies do not change their form, the signs of the changes in size (δR_h) and in velocity dispersion ($\delta \sigma$) are opposite (virialized conditions).

The results of our experiments are:

1 - In the models with mass ratios 1 and 1/2, $\delta \sigma < 0$. This is a predictable result because the final galaxies expand ($\delta R_h > 0$).

2 - The mass-profiles show that part of the δR_h is due to deformation of the galaxies for mass ratio 1 and 1/2. The models with mass ratio 1/4 exhibit no correlation between δR_h and $\delta \sigma$. Besides, their mass-profiles show that the galaxies are deformed, so that the signs of δR_h and $\delta \sigma$ cannot be easily predicted.

4. Conclusions

We performed a large number of simulations of collisions of stellar systems represented by non-rotating Plummer spheres, obtaining a good coverage of the space of the parameters (orbits, mass ratios). Our results provide quantitative estimates of the changes suffered by the galaxies due to the collisions and, offer a sample of initial conditions that deserve further, more detailed, studies. Collisions of unequal mass galaxies (mass ratios of 1:8, and even smaller) clearly warrant additional investigations.

REFERENCES

[1] Aarseth, S. J. & Fall, M. 1980, ApJ, 236, 43.
[2] Aguilar, L. A. & White, S. D. M. 1986, ApJ, 307, 97.
[3] Alladin, S. M. 1965, ApJ, 141, 768.
[4] Arp, H. C. 1966, ApJ. Suppl. S., 14, 123.
[5] Dekel, A., Lecar, M. & Shaham, J. 1980, ApJ, 241, 946.
[6] Forte, J. C., Martinez, R. E. & Muzzio, J. C. 1982, AJ, 87, 1465.
[7] Muzzio, J. C. 1986, ApJ, 301, 23.
[8] Muzzio, J. C., Dessaunet, V. H. & Vergne, M. M. 1987, ApJ, 112, 313.
[9] Muzzio, J. C., Martinez, R. E. & Rabolli, M. 1984, ApJ, 285, 7.
[10] Navarro, J. 1988, Ph. D. Thesis, Universidad de Córdoba.
[11] Roos, N. & Norman, C. A. 1979, AA, 76, 75.
[12] Toomre, A. & Toomre, J. 1972, ApJ, 178, 623.
[13] Vorontsov - Velyaminov, B. A. 1977, AASup, 28, 1.

Modeling the Dynamics of the Interacting Galaxy Pair NGC 4676

By S. J. Gilbert, J. A. Sellwood

Department of Physics & Astronomy, Rutgers University, PO Box 849, Piscataway, NJ 08855-0849, U. S. A.

We model the galaxy pair NGC 4676 (the "Mice") as an interacting pair of disk-halo galaxies using a fully self-consistent N-body tree code. By varying the initial parameters of the interaction, we try to match the projected morphological appearance of the system as well as Stockton's kinematic observations. We find the parameters of the encounter to be remarkably tightly constrained. Our best fit model requires a parabolic encounter, which is more energetic than the elliptical orbit recommended by Toomre & Toomre. Our model is still an imperfect fit in a number of minor respects.

1. Motivation

Toomre & Toomre [6] simulated the "Mice" system, NGC 4676 (illustrated in plate # 242 of reference [1]) as a pair of tidally interacting disk galaxies. They derived a set of initial orbital and spin parameters which led to the formation of tails having a morphology resembling those observed in the system. The broad kinematic predictions of their model were confirmed in spectroscopic observations by Stockton [5].

One unpleasant aspect of their model is that they required the galaxies to start from a bound orbit with a comparatively short period which suggests that the pair should have experienced a number of close encounters over the age of the universe. This unlikely feature of their model is, as they were aware, an artifact of their restricted three body simulation method which does not allow for any exchange of energy between orbital and internal motions. Here we attempt to determine the initial orbit of the galaxy pair using more realistic self-consistent N-body simulations, which avoid this artifact.

2. Initial Conditions

Our two identical galaxy models were constructed as follows: We use a non-rotating King model, represented by 5000 particles, for the halo; it has a central potential of $2\sigma^2$, in the usual notation (e.g. reference [3], p 232), which yields a ratio of tidal radius to King radius of 3.2. We adopt units in which Newton's constant, G, the King radius, r_0, and the central density, ρ_0, are all unity. In these units, the halo has a mass of 2.82. We then grow a concentric exponential disk having one tenth the mass of the halo, a scale radius $r_0/4$, and a small initial z thickness, increasing the mass of this component in a smooth (cubic) fashion over a period of five time units ($= 5/\sqrt{G\rho_0}$). After they had reached their full mass, the 1000 particles representing the disk were set in motion on circular orbits and the galaxy model was allowed to evolve in isolation for a further six dynamical times. We then randomly rearranged the azimuthal positions of every particle in the disk in order

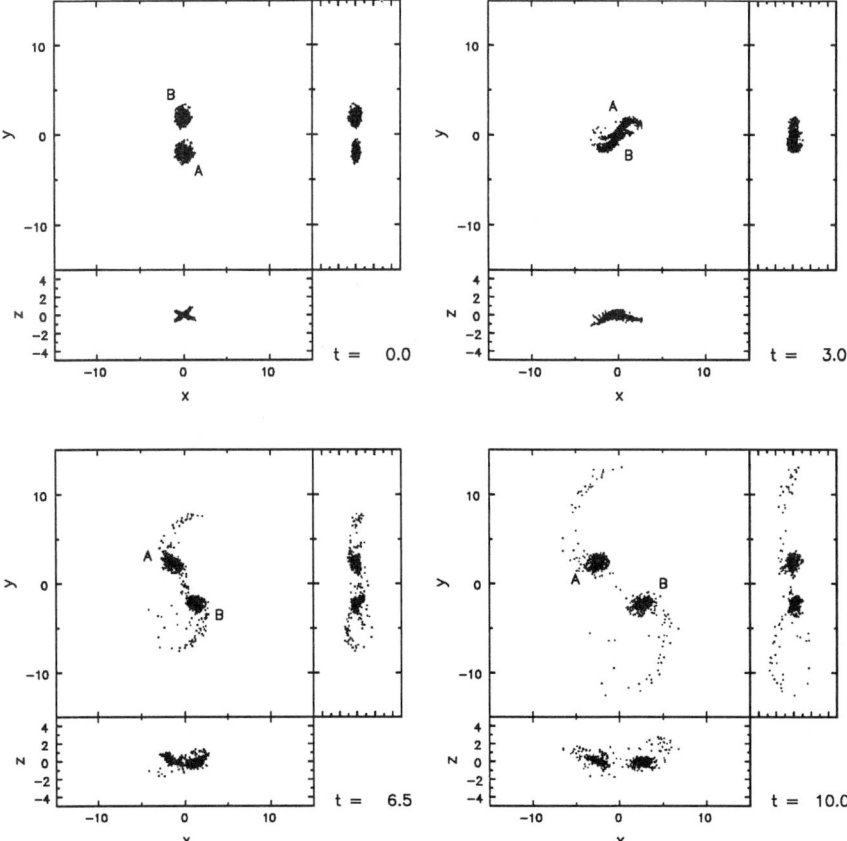

FIGURE 1. The evolution of our "best fit" model seen from three orthogonal projections, with the (x,y)-plane being the orbital plane. Time and distance units are described in the text.

to restore axial symmetry after this relaxation phase. Thereafter, we continued to allow the model to evolve for a further six dynamical times before using two copies of the final model to simulate the interaction.

We computed the evolution of the two disk/halo galaxies using Hernquist's tree code [4], kindly provided by the author. We use a softening parameter of 0.1, an opening angle of unity and include quadrupoles. Our time-step is 0.1 in our adopted units.

We have tried a number of different initial orbital energies and orientation angles of the two disk spin vectors to the orbit plane. We have also varied the time at which we choose to view the system and the viewing angle. We have not experimented with other disk/halo mass ratios, or with unequal galaxies; we always started from the same initial distance between the galaxy centers and have not varied the distance of closest approach.

3. Best Fit Model

The initial orbital parameters for our best result differ slightly from those adopted by Toomre & Toomre for their restricted calculation. Most significantly, we found that the appropriate initial orbital energy was the marginally bound case of a parabolic orbit. We found that when we adopted a bound orbit, the central bodies of the two galaxies had almost merged before the tails were sufficiently developed to resemble those in the "Mice"

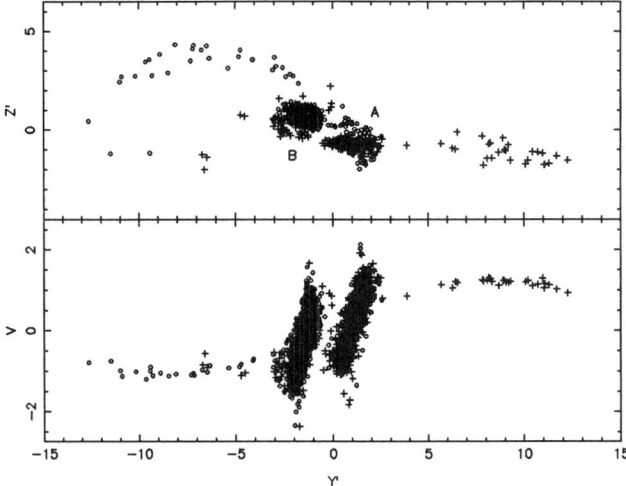

FIGURE 2. Above, the projected distribution of the particles from time 10 shown in Figure 1, viewed from a point slightly above the orbital plane. Below, the velocity components of the particles at this time along the same line-of-sight, with the z' displacements ignored.

system. The rapid braking of both galaxies during an encounter has been noted by many previous authors [2] and is reviewed by Navarro (this meeting).

We placed the two galaxies in an initially parabolic orbit with the pericenter just 1.7 times the disk's outer radius. The disks A & B have inclinations with respect to the orbital plane of $i_A = 20°$ and $i_B = 40°$, with $\omega_A = \omega_B = -90$ in the notation of Toomre & Toomre. Our adopted inclination angles also differ slightly from the 15° and 60° adopted by Toomre and Toomre.

The positions of the disk particles at the start can be seen in the upper left panel of Figure 1. We began our simulation of the encounter at an initial separation of $4r_0$, i.e. with slightly overlapping halos. The other three panels show the evolution of the model during the interaction; only the disk particles are shown throughout. The times are in units of $(G\rho_0)^{-1/2}$.

We have selected this model at time 10, viewed from a direction slightly above the $y-z$ projection of Figure 1 as being the best fit, on the grounds that the shapes and extents of the tails in projection correspond most closely to those observed in the system. The particle distribution in this projection is shown in the upper panel of Figure 2.

The lower panel of Figure 2 shows the line-of-sight velocity components of the particles from this adopted viewpoint. Here we have plotted the velocity of each particle at its projected position along the major axis, ignoring any displacement along the minor axis.

The model reproduces the positions of the central bodies of the two galaxies and the orientations and shapes of the tails as seen in the visual images. The sense of rotation of the two galaxies, and the relative velocity of tail A also agree with data obtained by Stockton [5].

4. Discrepancies

We note here that the model illustrated in Figure 2 is not a perfect fit to the stellar dynamical structure of the "Mice" system in at least three respects. We also have no counterpart to the bright knot of HII regions at the start of tail B – our model is clearly unable to reproduce features associated with gaseous emission.

The tails in our model are broader than in the optical image, suggesting that the particles in the tails have larger random velocities than does the tail material in the "Mice" system. The velocity dispersion amongst the disk stars in our model is representative of the old population and is therefore higher than the extreme population I material which probably makes up the brightest constituents of the tails. Moreover, the velocity dispersion in our model may be higher than that of the real system because of relaxation effects.

The velocities in our model are flat, or even decrease, towards the outer end of tail A, whereas Stockton's data show a steady rise along the entire tail. This discrepancy may be the result of our having begun the simulation when the galaxies were rather close together, which reduces the time over which the tidal forces have acted.

The orientation of the intense region in galaxy B differs in our model from that observed and seems to pose a more serious problem, since the galaxy bodies contain a large portion of the galaxy mass. We have been unable to make a large change to the orientation of this body without also making substantial changes to the tail.

The first two of these discrepancies between our model and the "Mice" system may not be of much significance, whereas we regard the third as more serious. We note, however, that the orientation of galaxy B's low intensity region is roughly in accord with the distribution of our particles.

5. Conclusion

We have found that a self-consistent simulation of a tidally interacting galaxy pair is able to reproduce the observed state of the "Mice" system quite well. Self-consistency has required us to increase the initial orbital energy above that adopted by Toomre and Toomre, since in our model energy is transferred from the orbit to internal motions. We conclude that these galaxies in fact started from close to an unbound orbit and are therefore probably interacting for the first time.

REFERENCES

[1] Arp, H. C. 1966, *Atlas of peculiar galaxies* (Washington: Carnegie Institution); see also *Ap. J Suppl.* **14**, No 123
[2] Barnes, J. E. 1992. *Ap. J.* **393**, 484
[3] Binney, J. & Tremaine, S. 1987. *Galactic Dynamics* (Princeton: Princeton University Press)
[4] Hernquist, L. 1987, *Ap. J. Suppl.* **64**, 715
[5] Stockton, A. 1974, *Ap. J.* **187**, 219
[6] Toomre, A. & Toomre, J. 1972, *Ap. J.* **178**, 623

The Radial Orbit Instability in a Universe with Hubble Expansion

By D. D. Carpintero, J. C. Muzzio

Facultad de Ciencias Astronómicas y Geofísicas, Universidad Nacional de La Plata, Paseo del Bosque S/N, 1900 La Plata, Argentina, and Programa de Fotometría y Estructura Galáctica, Consejo Nacional de Investigaciones Científicas y Técnicas, Argentina.

We investigate the fate of the radial orbit instability in the formation of galactic halos through dissipationless collapse. We present numerical simulations where both the Hubble expansion and the inhomogeneities of the protogalactic material have been taken into account. We find that the radial orbit instability is a transient effect for a wide variety of initial conditions, including those with cosmological–like density fluctuations.

1. Introduction

Since the early works of Hénon [4] and Peebles [6], dissipationless collapse has proved to be very helpful in understanding the underlying dynamics of the formation of galaxies. In particular, Merritt & Aguilar [5] have pointed out the importance of an instability which is present in free fall collapse, namely, the radial orbit instability. Starting from a spherical system, this instability yields a triaxial one (i.e., a barlike system). It occurs when the particles of a system have predominantly radial orbits; its origin remains uncertain, although some insight has been gained from Fridman & Polyachenko's work on gravitating systems [3]. We can see that this instability must be present in a typical dissipationless collapse, for its orbits are almost radial. In fact, Aguilar [2] has found it in a variety of models where the initial velocity dispersion of the system is small. If we now consider an expanding system —like protogalactic material in Hubble expansion— density inhomogeneities can interact gravitationally during a longer period. This would damp the instability, or even prevent it, since the growth of inhomogeneities would destroy radial orbits. To prove this, we have carried out numerical experiments of dissipationless collapses, in which both the Hubble expansion and several degrees of initial density inhomogeneities were included. Throughout we used Aarseth's direct summation code [1] to perform the integrations.

2. Homogeneous Collapses

2.1. *Model 1*

To begin with, we set up a homogeneous Poissonian system by randomly distributing 5000 particles in a sphere of radius unity. These particles were given radial escape velocities (which implies $\Omega = 1$, where Ω is the usual density parameter), and we put a massive particle at the center, having 4% of the total mass, which is set to unity. Figure 1a shows a snapshot of this model, at $t = 0.75 t_{cr}$ (crossing times); we can see that a large number of inhomogeneities were formed. In Figure 1b we see the model at $t = 21 t_{cr}$, when it has settled down to a spherical system. Figure 1c shows the evolution of the ratios a/c (upper

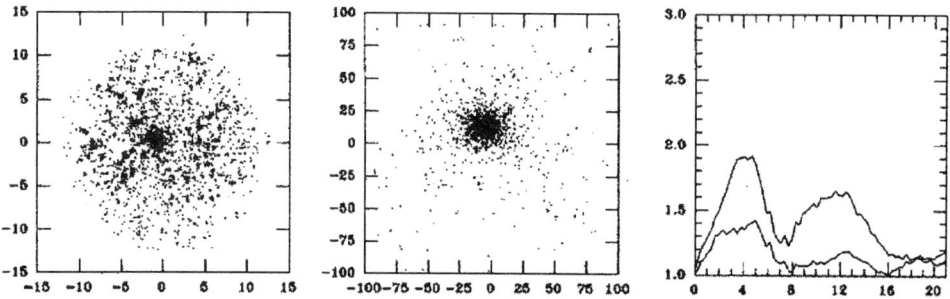

FIGURE 1. Left: snapshot of model 1 at $t = 0.75\,t_{cr}$. Center: same model at $t = 21\,t_{cr}$. Right: ratios of axes of the inertia tensor versus time (see text).

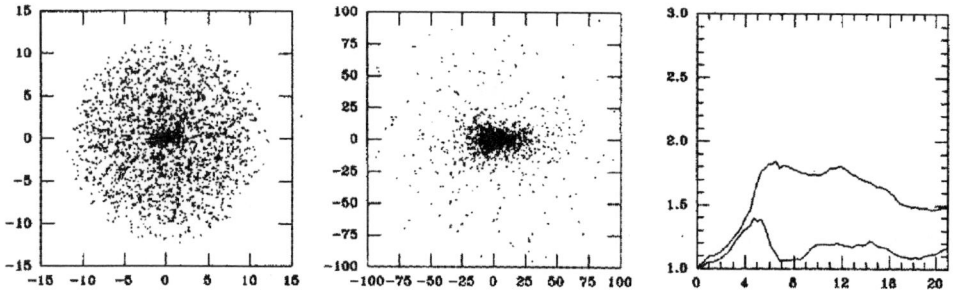

FIGURE 2. Same as Figure 1 but for model 2.

curve) and b/c (lower curve) with time, where $a \geq b \geq c$ are the axes of the moment of inertia tensor of the innermost 70% of the particles, and time is measured in units of crossing times. We can see that the instability appears, but it damps later on.

2.2. Model 2

The second model has the same characteristics as model 1, except for the distribution of particles: the space is divided into identical cubes, and one particle is put randomly in each cube ("quiet initial conditions"). This configuration diminishes the short–range fluctuations in the density. Figure 2 shows the results of this model. As expected, the inhomogeneities are weaker than in model 1, and the bar remains throughout the evolution.

2.3. Model 3

We look for a system in which initial inhomogeneities are absent, to see if this allows the bar to set up and remain. To do so, model 3 was built as model 2, but particles were put in the very central region of each cube, giving the system a crystalline appearance ("crystallized initial conditions"). Figure 3 shows what happened with this model: there is a lack of inhomogeneities throughout the evolution, and the system remains triaxial at $t = 21 t_{cr}$. The axial ratios of the moment of inertia tensor confirm the final triaxiality.

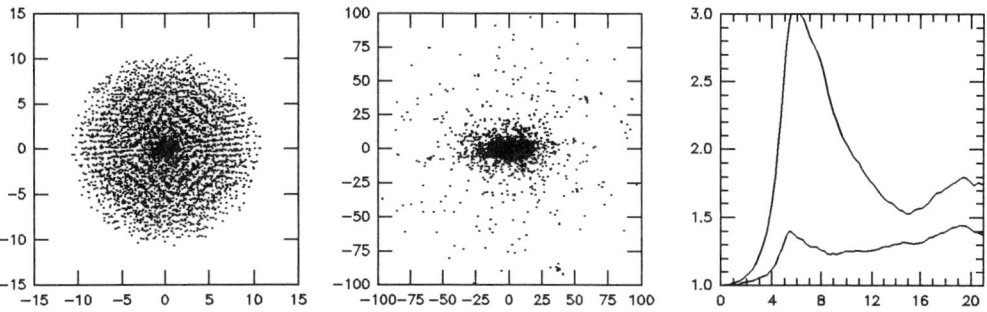

FIGURE 3. Same as Figure 1 but for model 3.

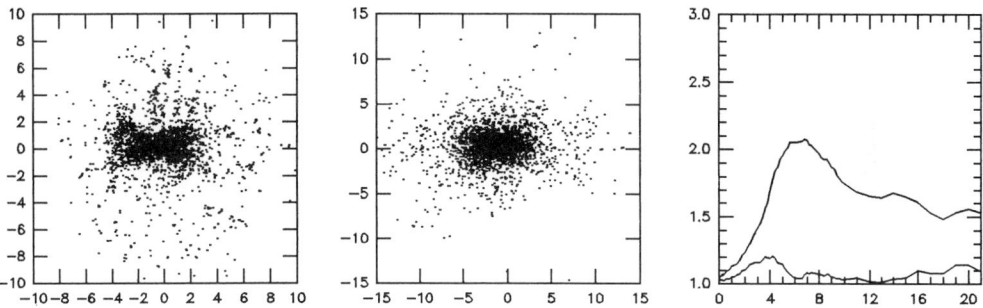

FIGURE 4. Same as Figure 1 but for model 4.

3. Variations

3.1. *Model 4*

We now want to see what happens when we substantially reduce the time of interaction during the expansion phase; if our hypothesis is correct, the system should develop a bar. We set up model 4, similarly to model 1, except that the central particle has 14% of the total mass: this will make the system collapse faster. Figure 4 shows the results. As expected, a bar is still present at the end of the integration.

3.2. *Model 5*

We may wonder if the later result is due to the lack of time, or to the magnitude of the central perturbation. To answer this question, we constructed model 5, in which the density perturbation is again 14%, but it is distributed according to a $\rho(r) = C_1 - C_2 r^2$ law, where C_1 and C_2 are constants. The collapse time will be longer than in model 4, because the inner mass each particle sees is smaller. In Figure 5 we can see that a weak bar appears, but it is finally damped.

3.3. *Model 6*

It is also interesting to analyze whether the large-scale inhomogeneities damp the bar as small-scale ones do. We therefore built a system like model 3, but with a central mass containing 14% of the total mass, and irregular borders. The crystal-like distribution eliminates small-scale fluctuations, whereas large-scale perturbations will be generated by the irregularities of the border surface. We show in Figure 6 the system at $t = 10.02 \, t_{cr}$

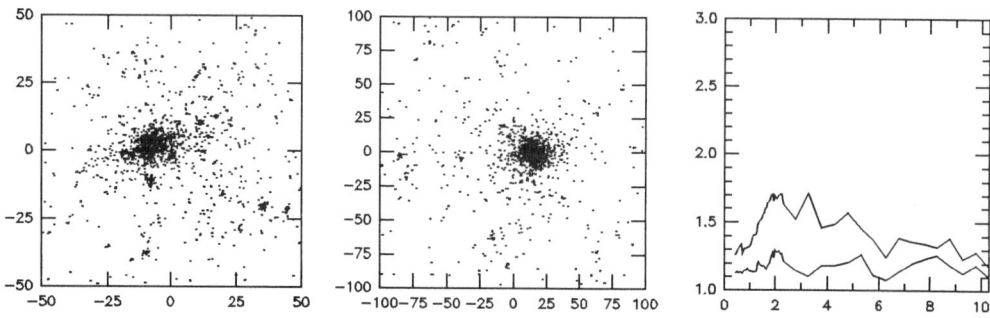

FIGURE 5. Same as Figure 1 but for model 5.

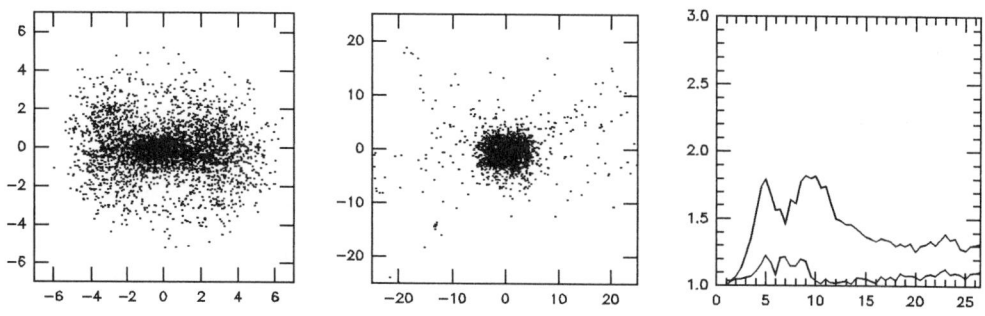

FIGURE 6. Same as Figure 1 but for model 6.

and $t = 26.52\,t_{cr}$. (There are interesting earlier stages in which large condensations of particles develop along the border, but we cannot show them here for the sake of brevity.) We can see that the bar damps again, yielding a nearly spherical system.

4. Cosmological model

4.1. Model 7

If we apply our hypothesis to a system with cosmological inhomogeneities, we can assert in advance that the bar instability will disappear, since these perturbations will be larger than Poissonian, and the latter were able to damp the bar (model 1). We set up model 7 with a density distribution with power spectrum $|\delta_{\mathbf{k}}|^2 \propto k^{-3}$, where k is the wavenumber of the density fluctuations. Figure 7 shows us what we expected, that is, the system is definitely spherical throughout its entire evolution.

We have also run models with other power spectra, obtaining similar results.

5. Conclusions

We showed that the radial orbit instability is damped, or even prevented, when reasonable initial conditions are imposed on a system, namely, to follow the Hubble expansion and to have density fluctuations. This combination allows the inhomogeneities to interact and form groups, and, ultimately, to destroy the radial orbits, damping or preventing the bar.

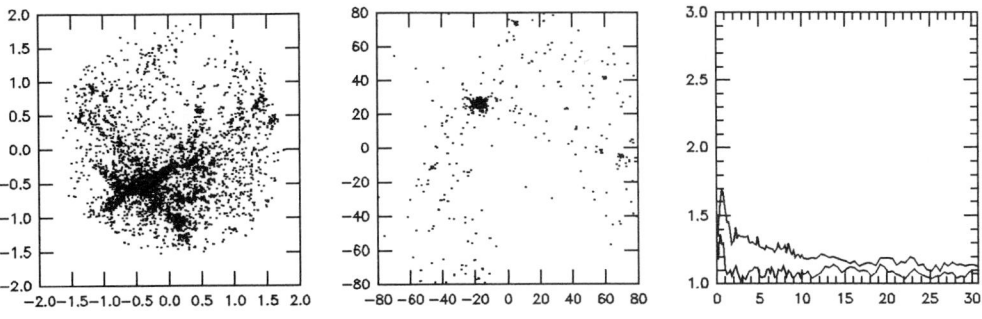

FIGURE 7. Same as Figure 1, for model 7, except that snapshots were taken at $t = 0.18 t_{cr}$ (left) and $t = 30.7 t_{cr}$ (center).

This work was supported in part by a grant from Fundación Antorchas.

REFERENCES

[1] Aarseth, S.J. 1985, in *Multiple Time Scales*, ed. J.U. Brackbill & B.I. Cohen (Orlando: Academic Press), 377.
[2] Aguilar, L. 1988, Cel. Mech., 41, 3.
[3] Fridman, A. M. & Polyachenko, V. L. 1984, *Physics of Gravitating Systems*, Springer Verlag, New York.
[4] Hénon, M. 1964, Ann. Astrophys., 27, 83.
[5] Merritt, D. & Aguilar, L. 1985, MNRAS, 217, 787.
[6] Peebles, P. J. E. 1970, AJ, 75, 13.

Dynamics of Massive Black Holes as a Possible Candidate for Galactic Dark Matter

By Guohong Xu, Jeremiah P. Ostriker

Princeton University Observatory, Princeton, NJ 08544-1001, USA.

If the dark halo of the Galaxy is comprised of massive black holes (MBH), then those within approximately 1kpc will spiral into the center, where they will interact with one another forming binaries which contract due to further dynamical friction and then possibly merge to form more massive objects by emission of gravitational radiation. Several authors have proposed the formation of a very massive galactic nucleus of $10^8 M_\odot$ through mergers. This is excluded on observational grounds, since the observed limit for a Galactic central black hole is approximately $10^{6.5} M_\odot$. We perform detailed N-body calculations using a modified Aarseth code to explore the situation in detail and find that for a "best estimate" model of the Galaxy a runaway does not occur. Instead three body interactions between hard binaries and single MBHs eject massive objects before accumulation of more than a few units, so that typically the center will contain zero, one or two MBHs. We study how the situation depends in detail on the mass per MBH, the rotation of the halo, the mass distribution within the Galaxy and other parameters.

1. Introduction

One possibility for the "dark matter" in the Galactic halo is a distribution of massive black holes ("MBHs"). A possible time of origin is the era of decoupling when the Jeans' mass was approximately $10^6 M_\odot$ [7,13]. Such a population would provide a good mechanism for disk heating [10], and various astrophysical arguments restrict the possible masses to the range $10^2 - 10^7 M_\odot$ [5], with $10^{6.5} M_\odot$ the preferred value according to Lacey & Ostriker [10]. Recently, Hut & Rees [9] argued that the high-mass end of this range is inadmissible since dynamical friction would drag such massive black holes to the Galactic Center, and the outcome would be the accumulation of a central mass of approximately $10^8 M_\odot$. Their arguments are semiqualitative, and ignore several complications, such as the detailed effect of dynamical friction on a binary system, the effect of triaxiality in the Galaxy, the phenomena of the rocket effect from gravitational radiation, the effect of rotation of the whole system, etc. Motivated by these ideas, we decided to perform a detailed direct N-body simulation.

Begelman et al [2] presented essential elements concerning the evolution of massive black hole binaries. The orbital decay of the binaries is dominated by dynamical friction when they are widely separated, and is dominated by gravitational radiation when they are very close to each other. If the decay time is smaller than the time before the next MBH falls to the center, the binary will merge, otherwise three-body interactions between the hard binary and the next incoming MBH could destroy the binary, possibly ejecting all objects from the center via the "gravitational slingshot".

As a result of the three-body interaction, all objects will be kicked out of the central region. If the kick velocity is below the escape velocity, they will return to the center in a spherical system, but in a triaxial system this may not be true, and a disk potential will drag the returned body to an off-center position. Also, when a binary of black holes merges by gravitational radiation, then if the binary is not symmetrical about all axes, there will be a net emission of linear momentum that will impart a recoil velocity to the merged body [14].

The effect of dynamical friction on the orbit of a binary is complicated and not yet understood fully. For soft binaries (with orbital velocity less than the rms stellar velocity), if the orbit is not circular, according to Chandrasekhar's formula [3], the eccentricity of the orbit will increase rapidly towards 1.0. For hard binaries, the result is not clear and some recent analysises find various results [12]. All the above complexities led us to do a detailed direct N-body simulation to try to resolve the problem.

2. Method

We use the Aarseth's [1] direct N-body code implemented with multiple time scales, KS binary regularization, triple and quadrupole regularizations, and gravitational radiation [11]. The MBHs interact with each other and also with a continuous density field representing the non-halo part of the Galaxy. In addition we allow for dynamical friction between the MBH (halo) component and the other components.

2.1. Galaxy Model

We model our Galaxy with a three component model (see [4]): a spheroidal bulge, a thin disk and a pseudo-isothermal halo. We use the Hernquist model [8] for the spheroidal bulge ($r_s = 0.7$kpc, $M_s = 3.533 \times 10^{10} M_\odot$), the Mayamoto-Nagai model for the disk [3 page 44-45] ($a = 6.5$kpc, $b = 0.26$kpc, $M_d = 1.046 \times 10^{11} M_\odot$), and the logarithmic potential for the halo ($R_c = 12.0$kpc, $v_0 = 181.4$km/s). The parameters of this model were fitted by David Spergel (private communication), and we made some further small modifications which will be discussed below. This model has the advantage of a simple force calculation, and a good fit to the observed parameters, such as $\Sigma(R_\odot) = 88.5 M_\odot \text{pc}^{-2}$, $\rho(R_\odot) = 0.012 M_\odot/\text{pc}^3$.

As we assumed, the dark halo is made up of massive black holes with mass of about $10^6 M_\odot$, but since the rotation curve is approximately flat, the total mass of the halo increases as r, so it is impossible to simulate the motion of all the black holes. As we are concerned with black holes in the central region only, and as the dynamical friction can drag a black hole of $10^6 M_\odot$ from about 1 kpc to the center within one Hubble time, we will only simulate the black holes within about 2 kpc of the center region. Thus we break up the halo distribution into two parts. An intermediate distribution of the black holes was introduced:

$$\rho_{bh}(r) \equiv \frac{\rho_{halo}(0)}{1 + (r/r_{bh})^4} \quad (2.1)$$

where r_{bh} is a parameter that varies as the number of total particles simulated changes. Given this density we position our black holes initially with standard Monte Carlo methods. We take as a fixed distribution for the remainder of the halo $\rho_{hf}(r) \equiv \rho_h(r) - \rho_{bh}(r)$.

In order to see the effects of a rotating halo, we constructed models in which we measure the rotation with the λ' parameter which is defined similar to that in Davis et al 1985, but modified for the subsystem of the MBHs,

$$\lambda' \equiv JK^{1/2}/GM^{3/2}M_{tot} \quad (2.2)$$

where J is angular momentum, K is kinetic energy, M is the mass of the active MBH halo, and M_{tot}

$$M_{tot} = \int_0^\infty \frac{\rho_{bh}(r)}{\rho_h(r)} m_{rot}(r) 4\pi r^2 dr \quad (2.3)$$

Model	Halo				Disk	
	m_i $10^6 M_\odot$	ρ_0 $M_\odot \text{pc}^{-3}$	$m(1\text{kpc})$ $10^7 M_\odot$	λ	m_{tot} $10^{10} M_\odot$	$\Sigma(R_\odot)$ $M_\odot \text{pc}^{-2}$
1	1	0.01268	5.279	3.9×10^{-4}	10.46	88.54
2	3	0.01268	5.279	7.2×10^{-5}	10.46	88.54
3	.54	0.01268	5.279	3.4×10^{-4}	10.46	88.54
4	1	0.01268	5.279	5.0×10^{-2}	10.46	88.54
5	1	0.1141	4.502	6.0×10^{-4}	7.448	61.98

TABLE 1. Models with $m_{Bulge,tot} = 3.533 \times 10^{10} M_\odot$ and $m(0.1\text{kpc}) = 5.520 \times 10^8 M_\odot$

is the average mass seen by the active MBH component interior to its position at the beginning of the integration. A reasonable value for λ' is 4.5×10^{-2} [6] for the halo, so we modified our isotropic equilibrium model of the MBH slightly to reach this rotation value. We make the small modification by the following means: let $v_\phi = \beta v_0$, and $\vec{v}_1 = \alpha \vec{v}_0 + v_\phi \vec{e}_\phi$ with $|\vec{v}_1| = |\vec{v}_0|$, where $\vec{v}_0$ is the initial velocity, and $\vec{v}_1$ is the modified velocity.

More details about the models are shown in Table 1. Model 1 is our standard model, and models 2 and 3 are with the same galaxy model parameters, but with different particle mass. Model 4 allows the halo to rotate, and Model 5 is a variant of model 1 but with maximum halo and minimum disk while keeping the rotation curve approximately constant. These models are all compatible with current observations of Galactic structure.

2.2. Dynamical Friction

Dynamical friction in a binary system is very complicated, and no complete theory exists (see [12] and references therein), so we model it in the following way:

(1) For a single body, we use Chandrasekhar's formula directly. But if the body is in the central region, we may replace the mass of the black hole with the minimum of $m(r)$ and M_{bh}, that is,

$$\frac{d\vec{v}}{dt} = -\frac{4\pi \log(\Lambda) G^2 \min(M_{BH}, m(r)) \rho(r)}{v^3} \left[\text{erf}(X) - \frac{2X}{\sqrt{\pi}} e^{-X^2} \right] \vec{v} \quad (2.4)$$

where $m(r)$ is the total mass of the bulge and disk inside the position of the black hole.

(2) For a soft binary, we use the above formula to calculate the dynamical friction of each component separately, then add them together to get the friction on the center of mass body, the same for the perturbation on the binary.

(3) For a hard binary, we use the above formula to calculate the dynamical friction on the center of mass body, but use the formula from Mikkola & Valtonen [12, equations (10,15)], for the perturbation of the orbit of the binary.

(4) For the value of $\log \Lambda$, we use

$$\Lambda = \frac{R}{GM/v_\star^2}, \quad \text{for single body}$$
$$= \min\left(\frac{R}{GM/v_\star^2}, \frac{R}{a}\right), \quad \text{for binary} \quad (2.5)$$

2.3. Tests of the Code

Beside some basic tests, such as producing the correct particle orbit for integrable potentials, energy conservation, close binary decay due to gravitational radiation, we test the dynamical friction part with few particles comparing the results with those of reference

Model	$R < 100$pc		Ejected			Escaped	
	#	m $10^6 M_\odot$	#	m $10^6 M_\odot$	$\bar{R}$ kpc	#	m $10^7 M_\odot$
1	2	2	3	6	2.47	16	1.6
2	0	0	5	15	21.6	45	13.5
3	1	.54	1	5.4	21.1	14	.76
4	4	4	3	6	11.1	17	1.7
5	7	7	5	11	11.9	83	8.3

Model	Bulge		Energy		Merger
	$m_b(100\text{pc})$ $10^8 M_\odot$	%	δE_{df} 10^{56}erg	δE_{gr} 10^{59}erg	#
1	5.455	98.8	2.030	1.18	8
2	5.104	92.5	11.4	13.9	7
3	5.490	99.5	.7560	.127	4
4	5.431	98.4	2.487	.708	5
5	5.223	94.6	7.833	9.79	60

TABLE 2. Results (Time $= 1 \times 10^{10}$ yr)

[15] who used an isothermal model. The results agreed quantitatively without significant differences.

3. Results

The results are collected in Table 2. The second column of the table gives the number of MBHs within a radius of 100 pc and the total mass of these MBHs. We defined an "Ejected" particle to be a particle which sinks to the center once, and is later kicked out beyond 1kpc from the center. And we calculated the gravitational radiation energy by an analytical formula until the separation of the two MBHs is about 6 Schwardschild radius at which point the orbits of the binary will become unstable. The results show that only 0, 1, 2, or 3 particles stay in the central region, which is as predicted in reference [10]. Even the model with maximum halo (model 5) gives results acceptable by observations.

The effect of a disk is significant, making a difference of about a factor of 2 from a simple spherical model. The basic reason is that the disk will prevent the particle which is kicked out by three body interactions from falling back to the exact center if the kick velocity is not big enough for it to escape. Changing the mass of the MBHs gives 0, 1, or 2 MBHs in the Galactic Center, but have different heating effects. We also tried a run which contains different masses for MBHs, that is half of the particles with $10^6 M_\odot$ and half with $2 \times 10^6 M_\odot$, the result is not significantly different from those of models 1 or 2. The rotation of the halo does not affect our results very much. The difference shown in the table can be explained by statistical fluctuations.

It is not very surprising to see that there is not a big mass accumulation in the center of the Galaxy, even though the merger events are not rare. The big masses that form are usually kicked out from the center, either escaping from the galaxy altogether or orbiting at a large radius. We have not included the rocket effect due to gravitational radiation in the results we show in the tables. In fact a small recoil velocity like 300 km/sec does not make a difference according to our calculations. The mass within 100pc of the center results from "core collapse" of the MBH population. Our results show that the essential determinant of accumulation is the density of the dark halo in the central region.

It is a pleasure to thank UNAM for giving G.X. free CPU time on their Cray Y-MP4/432 supercomputer to do these calculations. We thank Man-Hoi Lee and Sverre Aarseth for allowing us to use their N-body code. Discussions with David Spergel, Lars Hernquist, Piet Hut, Martin Rees are grately acknowledged. This work is supported by NASA grant NAGW-2448 and NSF grant AST91-08103.

REFERENCES

[1] Aarseth, S.J., 1985, Multiple Time Scales (Academic Press).
[2] Begelman, M.C., Blandford, R.D., & Rees, M.J. 1980, Nature, 287, 307.
[3] Binney & Tremaine, 1987, Galactic Dynamics (Princeton University Press)
[4] Caldwell, J.A.R & Ostriker, J.P. 1981, ApJ, 251, 61.
[5] Carr, B.J., Bond, J.R., & Arnett, W.D. 1984, ApJ, 277, 445.
[6] Dubinski, J. & Carlberg, R.G. 1991, ApJ, 378, 496-503
[7] Gnedin, N.Y., & Ostriker, J.P. 1992, ApJ, 400, 1.
[8] Hernquist, L. 1990, ApJ, 356, 359
[9] Hut, P. & Rees, M.J. 1992, MNRAS, 259, 27p
[10] Lacey, C.G. & Ostriker, J.P. 1985, ApJ, 299, 633.
[11] Lee, M-H. 1992, PhD Thesis, Princeton Univ., USA.
[12] Mikkola, S. & Valtonen, M.J 1992, MNRAS, 259, 115.
[13] Peebles, P.J.E. & Dicke, R.H. 1968, ApJ, 154, 891.
[14] Redmount, I.H., & Rees, M.J. 1989, Comm. Astrophys. 14, 165.
[15] Tremaine, S.D., Ostriker, J.P., & Spitzer, L. 1975, ApJ, 196, 407.

Galactic Orbits for 280 Halo Stars

By F. Valera[1], W. Schuster[2], L. Aguilar[2]

[1] Departmento de Astronomía, INAOE, Tonantzintla, Puebla, Apo. 216 and 51, México.

[2] Instituto de Astronomía–UNAM, Ensenada, México.

Orbits for a sample of 280 halo stars are computed using three models for the mass distribution of our Galaxy. The orbits are compared and differences are found from model to model. The largest discrepancies are in perigalactica, due to the different mass normalization of the models, and in orbits with low angular momentum, that are scattered by the central galactic bulge. Orbits are catalogued in three broad categories: box, resonant and chaotic orbits. Chaotic orbits represent a large fraction of the computed orbits in all the models ($\sim 40\%$).

1. Introduction

Orbits of a sample of 280 halo stars [10] have been integrated using the potentials of three galactic mass models: Allen & Santillán [2] (AS); Bahcall, Schmidt & Soneira [3, 4] (BSS); and Caldwell & Ostriker [6] (CO). We have used the same 4th order Runge–Kutta integrator in all cases. The potentials of the spherical components are computed analytically while for the axisymmetric components the potentials are interpolated from a grid in the meridional R–Z plane [7]. The orbits are followed for 20 azimuthal periods and the extrema in R and Z are found from these. The orbital eccentricity is computed as $(e = (R_{Apo} - R_{Peri})/((R_{Apo} + R_{Peri})))$. The fractional energy error is 10^{-4} or better in all cases.

The orbits have been classified in three broad categories [8, 9]: box, resonant and chaotic. Box orbits fill in a region in the R–Z plane that is smaller than the one allowed by energy conservation. These are orbits with an extra non–isolating integral of motion. Resonant orbits are further restricted by having a low order resonance between the R and Z motions. Finally, chaotic orbits are the ones that are not restricted by an extra isolating integral of motion other than E and J_z. These orbits fill completely the region allowed to them by the energy conservation. Examples of these orbits are shown in figure 1. It is important to emphasize that this scheme is different from a more recent scheme [5, p.126] that is employed for planar, non–axisymmetric, or triaxial potentials.

2. Results

Figures 2 through 5 show comparisons of the apogalactica (R_{Apo}), perigalactica (R_{Peri}), orbital eccentricities (e), and maximum heights above (and below) the galactic plane (Z_{max}), between the three galactic models. The apogalactica present the larger discrepancies. These discrepancies are systematic and grow with galactocentric distance; They are produced by the different total mass assigned to the Galaxy by each model ($M_{AS} < M_{BSS} < M_{CO}$). Differences in perigalactica do not present a clear trend although they tend to be larger in the center. The AS and BSS models are the more similar. Discrepancies in orbital eccentricities are dominated by radial orbits that come close to the center or travel to large galactocentric distances. Again, the AS and BSS models are the most similar, with the OC eccentricities being systematically smaller. Finally, the maxima in Z present two different trends, a population of orbits with small discrepancies that do not travel more than 10 kpc from the galactic plane, and a second population of points

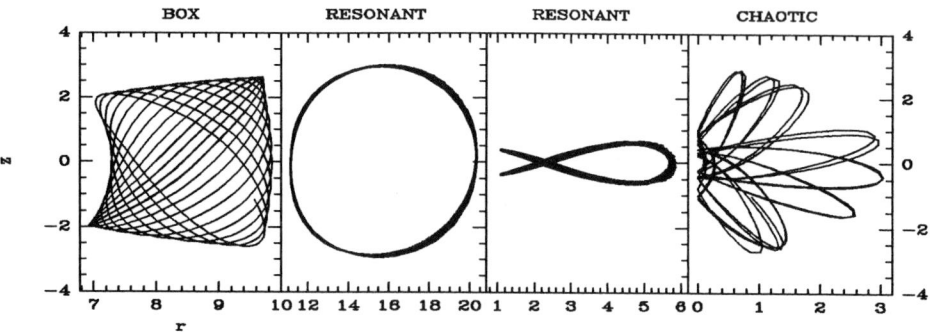

FIGURE 1. Meridional plane projections, from left to right, of a box orbit, a 1:1 resonant orbit, a 2:3 resonant orbit and a chaotic orbit.

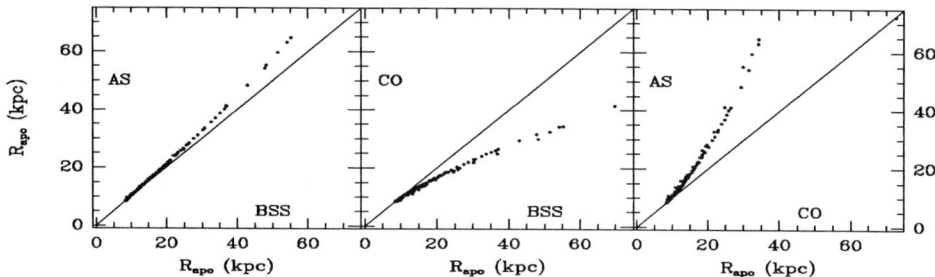

FIGURE 2. Comparison between the apogalactica obtained using the AS and BSS models (left panel), the CO and BSS models (middle panel), and AS and CO models (right panel), for the orbits of the 280 stars in the sample. Units are in kiloparsecs.

with huge discrepancies and which are found at all heights. The second group corresponds largely to chaotic orbits for which Z_{Max} can change substantially from one vertical period to another.

Figure 6 shows histograms of the distribution in R_{Peri} for the three types of orbits that have been classified in each one of the galactic models. In all three galactic models the orbital types behave in much the same way. Box and resonant orbits peak around 2 kiloparsecs while chaotic orbits continue increasing toward the center. This is not surprising, since box and resonant orbits are dominated by the axisymmetric part of the potential and circulate around the galactic axis of symmetry and at some distance from the center. Chaotic orbits, on the other hand, are dominated by the more spherical symmetric components of the potential and come close to the galactic center.

In table 1 the percentage of each type of orbit, found in each galactic model, is presented. The box orbits are the most common, followed closely by chaotic orbits. The resonant orbits are the rarest. The BSS model is the one with the largest percentage of chaotic orbits. This is presumably due to the presence of an extra central component in this model that scatters low angular momentum orbits. Table 2 shows the average R_{Peri} for each type of orbit and galactic model. Notice the low value found for chaotic orbits.

Figure 7 presents the distribution in Z_{Max} according to orbital type and galactic model.

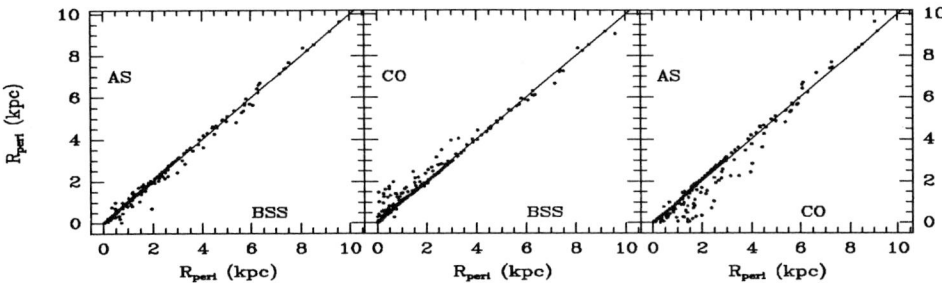

FIGURE 3. Same as figure 2 but for the perigalactica.

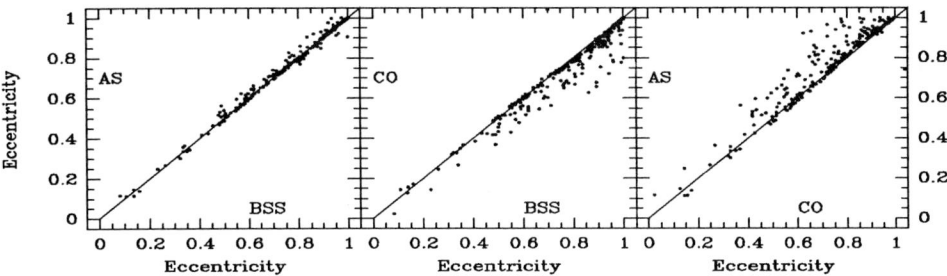

FIGURE 4. Same as figure 2 but for the orbital eccentricities.

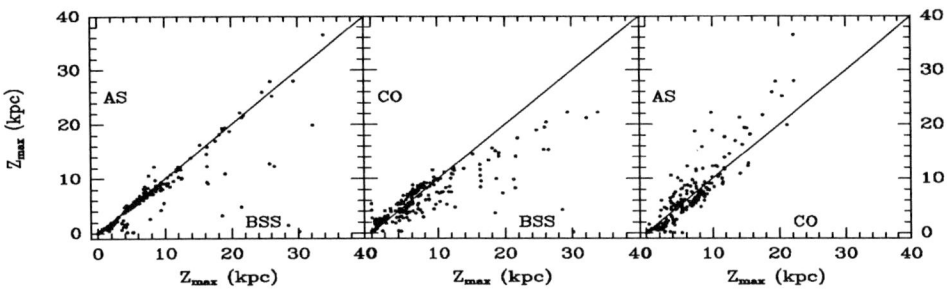

FIGURE 5. Same as figure 2 but for the maxima in distance to the galactic plane.

	box	resonant	chaotic
AS	44.6%	14.3%	41.1%
BSS	42.5%	10.7%	46.8%
CO	50.3%	16.1%	33.6%

TABLE 1. Percentages of the orbital types for each model.

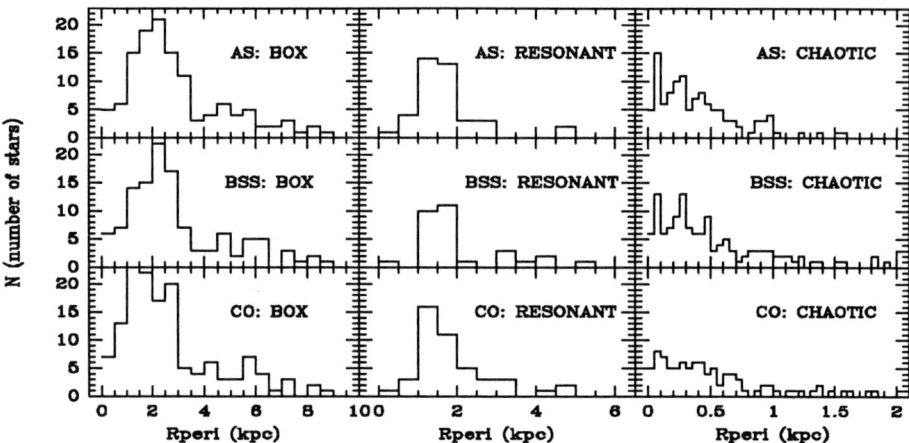

FIGURE 6. Histograms of the distribution in R_{Peri} for the box (left column), resonant (middle column) and chaotic orbits (right column); in the AS (upper row), BSS (middle row) and CO (bottom row) galactic models.

	box	resonant	chaotic
AS	2.1	1.5	0.3
BSS	1.9	1.5	0.3
CO	1.7	1.4	0.2

TABLE 2. Averages of the perigalactic distances (kpc) for the orbital types of each model.

The distribution for all orbits together (upper row) is quite similar for the AS and BSS models, having a peak on the galactic plane and a secondary maximum at $Z \sim 6$ kpc. When the distribution is splitted in orbital types we see that it is the box orbits the ones that contribute to the peak at $Z = 0$ while the chaotic orbits contribute to the second peak. The same trends, although with a larger degree of mixing between box and chaotic orbits, is shown by the CO model. The separation in Z_{Max} between box and chaotic orbits is due to the fact, mentioned earlier, that box orbits are dominated by the axisymmetric part of the potential (disk) while the chaotic orbits are dominated by the more spherical components (bulge and halo), traveling farther away from the disk.

3. Conclusions

This study of 280 halo stars in the solar neighborhood indicates that there is a high average percentage ($\sim 40\%$) of chaotic orbits for the three potentials in the galactic halo. The models of AS and BSS have the higher percentages due to their spherical bulge or central mass components, which dominate the inner part of the potential and scatter those stars penetrating near the galactic center; most chaotic orbits have $R_{Peri} < 1$ kpc.

The existence for a high percentage of chaotic orbits makes it more difficult to study gradients in the Galactic halo. For example, in correlations of $[Fe/H]$ vs Z_{max}, the chaotic

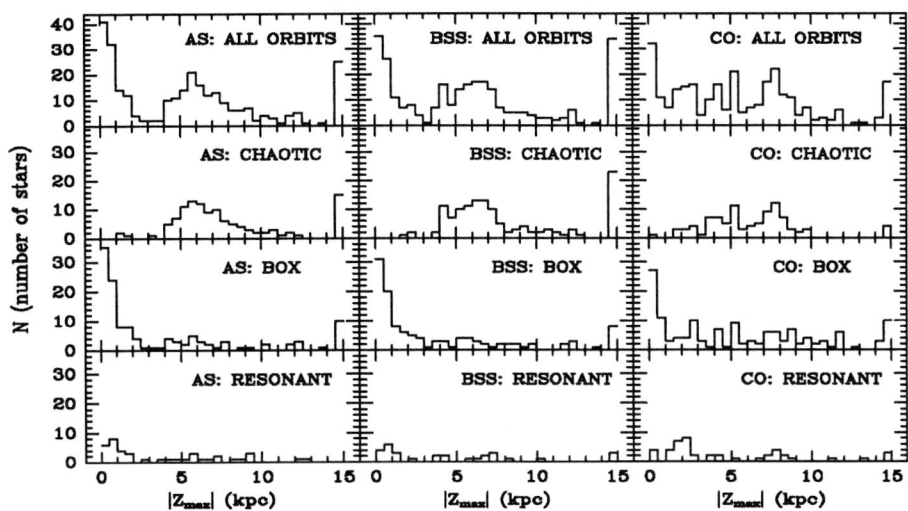

FIGURE 7. Histograms of the distribution in Z_{Max} for all the orbits (upper row), the box orbits (second row), the resonant orbits (third row) and chaotic orbits (last row); in the AS (left column), BSS (middle column) and CO (right column) galactic models.

orbits introduce noise upon the results for stars with more regular orbits. Since for these stars Z_{Max} can vary significatively, it is important to differentiate between the most recent maximum encountered in Z, and the extremum value of the maxima found during a time interval like the age of the Galaxy. The former is proportional to the present value of the Z component of the velocity, while the latter conveys more information about the range in Z explored by the star. A search for vertical gradients using the Z–velocities as an indicator of range in Z, may give misleading results for chaotic orbits.

The models of AS and BSS have results that are more alike, due to presence of central mass components, that the CO model does not have.

Finally, the particular numerical values found for the orbital parameters of an orbit, computed for a specific star or galactic object, and using a unique galactic model, are of limited value. It is the averages obtained from statistical studies with an ensemble of orbits [1], or for a multitude of objects [7], that carries some significance. If, nevertheless, an estimate for a particular galactic object is desired, the best approach is to compute the orbit of the object using at least two galactic models, and use the resulting numbers as the range of possible values introduced by our uncertainty about the detailed form of the mass distribution in the Galaxy.

REFERENCES

[1] Aguilar, L.A., Hut, P. & Ostriker, J.P. 1988, *ApJ*, **335**, 720.
[2] Allen, C. & Santillán, A. 1991, *Rev. Mex. Astron. Astrof.*, **22**, 255.
[3] Bahcall, J.N., Schmidt, M. & Soneira, R.M. 1982, *ApJ*, **258**, L23.
[4] Bahcall, J.N., Schmidt, M. & Soneira, R.M. 1983, *ApJ*, **265**, 730.
[5] Binney, J. & Tremaine, S. 1987, *Galactic Dynamics*, Princeton Univ. Press.

[6] Caldwell, J. & Ostriker, J.P. 1983, in *Kinematics, Dynamics, and Structure of the Milky Way*, Ed. W.L.H. Shuter, Reidel, Dordrecht.
[7] Carney, B.W., Aguilar, L.A., Latham, D.W. & Laird, J.B. 1990, *AJ*, **99**, 201.
[8] Ollongren, A. 1962, *Bull. Astron. Ints. Neth.*, **16**, 241.
[9] Ollongren, A. 1965, *Galactic Structure*, University of Chicago Press, Ed. A. Blaauw & M. Schmidt, p.501.
[10] Schuster, W., Parrao, L. & Contreras, M. E. 1993, AA Suppl., 97, 951.

Galactic Winds from Starburst Galaxies

By A. Habe

Department of Physics, Hokkaido University, Sapporo, Japan. e-mail:habe@phys.hokudai.ac.jp

In starburst galaxies, many supernovae explode in the central kpc region, heating the interstellar gas to more than 10^7 K. Because this temperature is higher than the escape temperature of the gravitational potential of a galaxy, hot gas is ejected. This gas outflow is called a galactic wind. Stratified gas in galactic disks affects the flow of galactic winds, causing their elongation, which has been shown by previous numerical simulations. Galactic winds are usually normal to the galactic plane, and when the gas flow reaches the scale height of the disk gas, the wind is blown out from the disk. These elongated, hot gas flows correlate well with observed X-ray and optical features in starburst galaxies. Recently, interactions of galactic winds with halo gas have been investigated. Halo gas, extending far above the disk, affects the evolution of galactic winds. Halo gas confines the hot gas ejected from these galaxies and sustains X-ray surface brightness. Cool gaseous shell formation is another physical process arising from this interaction and is a possible mechanism for the star formation observed above the disks of starburst galaxies.

1. Introduction

Galactic winds have been proposed to explain the apparent deficiency of cool gas in elliptical galaxies [17]. In elliptical galaxies, the total amount of gas ejected from stars during the age of the galaxy can be as much as $10^{10} M_\odot$. If the gas is cool, we can detect it by radio observations of HI or CO lines. But only small amounts of cool gas have been observed. X-ray halos around massive elliptical galaxies were observed by the Einstein satellite [6]. The X-rays are emitted by hot gas confined by the gravity of massive, dark halos. Because metal lines are detected, this hot gas is thought to be of stellar origin. The total mass of hot gas is as large as that ejected from evolved stars during the Hubble time. Gas is heated by supernovae in elliptical galaxies and the resulting galactic winds are trapped by the dark halo potentials [8,14].

Recent observations suggest that some starburst galaxies have high velocity gas outflows. For a review of these so-called "superwinds" see Heckman et al. [9]. Superwinds in starburst galaxies are energetic and must affect galactic properties; the study of their effects is inherently interesting.

In this paper I review the observational evidence for superwinds, theoretical studies of superwinds, and recent theoretical work on the interaction of superwinds with galactic halos.

2. Observational Evidence for Galactic Winds in Starburst Galaxies

Many active phenomena occur in the central kpc regions of starburst galaxies. Among these are large far-infrared (IR) luminosities, numerous supernova explosions, high velocity gas outflows (on the order of several hundred km s^{-1}) and extended, X-ray emitting, hot gas. Large amounts of molecular gas, which is indirectly related to starbursts, are also observed in these galaxies.

M82 is a nearby, typical starburst galaxy with a superwind. There is ample observational evidence for starbursts in M82. Very powerful IR luminosity — $3 \times 10^{10} L_\odot$ — is observed [23]. Detailed optical studies show that very young star clusters exist in the nucleus of M82 [22]. High resolution radio maps show the presence of more than 40 compact radio sources which are probably young supernova remnants [11,32]. Lo et al. [13] estimate the star formation rate at 3 $M_\odot$ yr^{-1}, within the nuclear region ($r < 1$ kpc). Heckman et al. [9] give convenient formula for the relationship between the star formation rate, $\dot{M}_{SF}$, and the infrared luminosity,

$$\dot{M}_{SF} \sim 25(L_{IR}/10^{11}L_\odot)M_\odot yr^{-1}. \tag{2.1}$$

The effects of supernova explosions on the structure of interstellar matter were studied by McKee & Ostriker [14], Ikeuchi, Habe & Tanaka [10] and Chevalier & Clegg [3]. Because the supernova explosion rate per unit volume is very high in the nuclear regions of starburst galaxies, the volume fraction occupied by supernova remnants is correspondingly high. The volume fraction of hot gas in interstellar space is given by

$$f_{hot} = 1 - exp(-Q_{SNR}) \tag{2.2}$$

where Q_{SNR} is defined by

$$Q_{SNR} = \int_0^{\tau_0} \frac{4}{3}\pi r^3(\tau)(\frac{\gamma_{SN}}{V})d\tau \sim 3.3 E_{51}^{1.28}(\frac{\gamma_{SN}}{V})_{-8} n_{0.1}^{-0.14} P_7^{-1.3}, \tag{2.3}$$

and we consider the SNR before radiative cooling begins. In M82, the supernova explosion rate (γ_{SN}) is estimated as $(\gamma_{SN}/V)_{-8} \sim 1$, and hence $f_{hot} \sim 1$. This means that almost all of interstellar space is occupied by young supernova remnants and hence collision are inevitable. Thus the energy of the supernovae is mainly used to heat the interstellar gas. If gas ejected from supernovae and young stars is heated by energy released by supernovae, its temperature will exceed 10^7 K [3]. This temperature is high enough for the gas to escape the gravitational potential of a galaxy.

Observational evidence suggests the presence of superwinds in starburst galaxies. Biconical flows, which are expected for superwinds, are indicated by optical spectral line observations of starburst galaxies [1,9]. X-ray spectra and large-scale features of the starburst galaxies NGC 253 and M82 are analyzed by Fabbiano [5]. She compared the observed X-ray surface brightness with numerical simulation results for galactic winds of nuclear starburst regions by Tomisaka & Ikeuchi [28] and showed that the X-ray properties are similar to those expected for hot gas ejected in superwinds. In M82, Nakai et al. [20] showed the presence of a CO ring and a CO outflow from the nuclear region with a velocity of 200 km s^{-1}. These molecular gas components surround intense IR emission from the nuclear region.

Because superwinds in star burst galaxies are energetic and must affect galactic properties, it is important to study their possible effects. Many studies are related to this point; I review them in the next section.

3. Theoretical Studies of Galactic Winds in Starburst Galaxies

There have been many studies of galactic winds in starburst galaxies, including galactic bipolar winds [28], active galactic nuclei-like properties due to supernovae in star burst nuclei [24], disk-halo connections [21], and large-scale halo formation [12].

3.1. *Interaction Between Galactic Winds and Disk Gas in Starburst Galaxies*

Tomisaka & Ikeuchi [28] studied bipolar gas outflows from nuclear starburst regions. Since Tomisaka [26] gives a detailed discussion of galactic bipolar outflows, I describe them only briefly here. Tomisaka & Ikeuchi simulated an outflow from a nuclear starburst region in order to explain the X-ray feature of M82. As described in §2, Fabbiano [5] showed that the X-ray feature of M82 is nearly perpendicular to the disk plane. Tomisaka and Ikeuchi assumed rotating gas in hydrostatic equilibrium in a disk potential and a supernova rate appropriate for M82. Their numerical results show that gas ejected from OB stars and supernovae is heated by energy released from supernovae and flows out of the nuclear region. Gas flow is mainly perpendicular to the disk and the shape of shock wave is elongated. If the shock wave reaches the scale height of the disk gas before cooling, the shock wave is blown out. On the other hand, gas expanding in the direction parallel to the disk interacts with the disk gas and stops, since the shocked disk gas is cool. These physical processes are similar to a superbubble in a disk galaxy [15,16,25,27]. They compared the X-ray surface brightness of their numerical results with observations and showed that their numerical results can explain the X-ray surface brightness distribution of M82 seen by the Einstein observatory. Important results are that galactic winds from nuclear starburst galaxies flow perpendicularly to the disks, elongated, hot gas regions are seen as X-ray emitting regions perpendicular to the disks, and the X-ray surface brightness decreases after the blowout of a galactic wind.

3.2. *Interaction Between Galactic Winds and Halo Gas in Starburst Galaxies*

Among the interesting phenomena related to galactic winds in starburst galaxies is the interaction between galactic winds and extended halo gas. Recent X-ray observations indicate that hot gas ejected from M82 is more extended than previously thought [30]. Moreover, large, extended gaseous halos of distant galaxies have been observed though QSO absorption lines. Tomisaka & Bregman [29] recalculated the work of Tomisaka & Ikeuchi [28] in order to consider extended halo gas. They showed that the X-ray luminosity of the hot gas ejected from starburst regions rapidly decreased after blowout of the shock wave, in the case of low-extent halo gas. Tomisaka & Bregman [29] pointed out that if they do not consider more extended halo gas, the model of Tomisaka & Ikeuchi does not explain the large extended X-ray emission region observed in M82. Tomisaka & Bregman assume large extent halo gas and investigate its effect on the dynamics of hot gas ejected from starburst regions. Their numerical results show that the galactic wind is confined by halo gas, an extended X-ray emitting region is formed and X-ray luminosity is maintained at 10^{40-41} erg s^{-1}, until the hot gas expands to about 40 kpc, the size of halo region.

4. Star Formation Triggered by Galactic Winds in Starburst Galaxies

The interaction of a galactic wind with halo gas can induce star formation if a cool, massive gas shell is produced in the process. Observational evidence suggests that star formation induced by such interactions does occur [9].

In starburst galaxy NGC 6240, a nearly edge-on disk galaxy, there are two super-jumbo HII regions. These HII regions are located at $z = -3.5$ kpc and 8.5 kpc and have Hα luminosities of $L(H\alpha) = 4 \times 10^{39}$ erg s^{-1} and 9×10^{39} erg s^{-1}. Because these star formation regions are well away from the disk, a possible mechanism for star formation is the interaction between a galactic wind and the halo gas [9]. Similar evidence of star formation regions distant from galaxies is observed in high redshift radio galaxies where the proposed star formation mechanism is the interaction between radio jets and halo gas [18]. Because more massive gas halos can be expected in high z galaxies [4] and gravitational

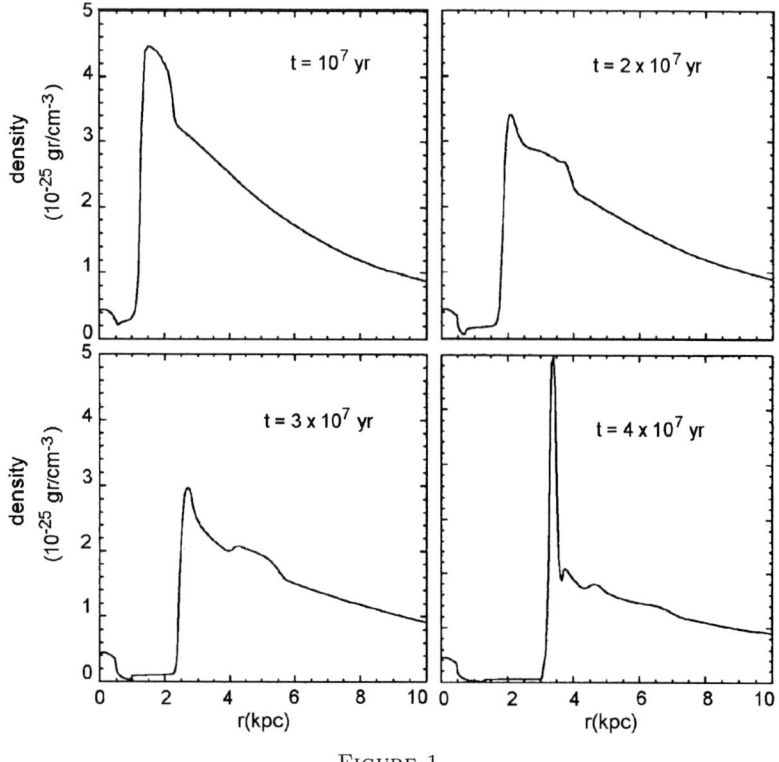

FIGURE 1.

interaction between these galaxies is more frequent, it is expected that star formation triggered by galactic winds of starburst galaxies is important in high z galaxies.

Cooled gas shell formation has been studied by William & Christiansen [35] and Umemura & Ikeuchi [31]. Umemura & Ikeuchi studied interactions of galactic winds and halo gas using one-dimensional hydrodynamic simulations. They examined the structure of the interaction region and the expansion law of shock waves in the halo region. They found that shock waves cool radiatively and cool, dense shells form. A similar process has been suggested for starburst galaxies [33]. It should be pointed out, however, that the shells which form are decelerating, which renders one-dimensional simulations unsuitable.

Decelerating, isothermal shock waves are subject to instabilities and cannot be studied by one-dimensional simulations. In order to understand the star formation process in a cooled, shocked gaseous shell, the instabilities of the shock wave must be considered as well. Yoshida & Habe [36] and Mac Low & Norman [16] have studied the deceleration shock instability, using two-dimensional numerical simulations. Mac Low & Norman showed that in the decelerating shock, linear perturbations can grow until they become nonlinear, after which growth ceases because the lateral velocity of the shocked gas is limited to the sound velocity. Thus, the shell is not destroyed. Yoshida & Habe simulated perturbations with much larger wavelengths, taking self-gravity into account. They showed that perturbations grow faster by coupling the decelerating shock instability and the self-gravitational instability, after enough mass has accumulated in the cooled shell. These studies were done for isothermal, or nearly isothermal, gas.

Evolution of the radiatively cooling shell is interesting, since the shell is subject to thermal instabilities during cooling [2]. Habe & Yoshida [7] further extend Blondin & Cioffi's [2] work to consider thermal instabilities in radiatively cooling gas shells, using two-

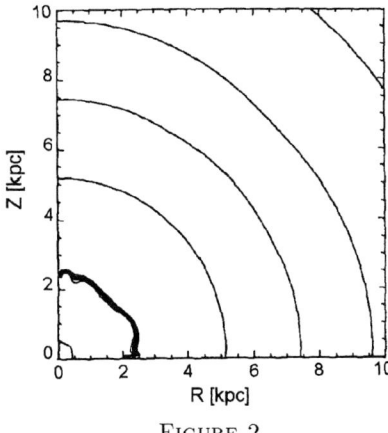

FIGURE 2.

dimensional hydrodynamic simulations. Their second-order, flux-splitting code implicitly solved the energy equation with radiative cooling. We assume Plummer's model for the galactic potential,

$$\Phi(R) = -\frac{GM_d}{\sqrt{R^2 + b^2}}, \quad (4.4)$$

where $R^2 = r^2 + z^2$, $M_d = 10^{11} M_\odot$ and $b = 20$ kpc. The gas density is initially assumed to be

$$n(R) = \frac{n_{h0}}{1 + r^2/r_h^2}, \quad (4.5)$$

where $n_{h0} = 0.2 \sim 0.02$ cm^{-3} and $r_h = 10$ kpc. We adopt this form since Denda & Ikeuchi [4] have shown that the densities of gaseous halos of galaxies observed though QSO absorption lines can be approximated as r^{-2}. The temperature of halo gas is taken to be $T_h = 3 \times 10^6$ K. The energy supply rate, L_{SN}, and mass supply rate, $\dot{M}_{SN}$, are given by

$$L_{SN} = 3.2 \times 10^{42} erg\, s^{-1} \quad (4.6)$$

and

$$\dot{M}_{SN} = 3 M_\odot\, yr^{-1} \quad (4.7)$$

Figure 1 shows the density profile structure of the galactic wind solution. Several separate regions are indicated: the galactic wind region, the inner shock region, the outer shocked region and the halo gas region. When the most inner part of the outer shocked region cools, its pressure decreases and the region is compressed by the higher pressure of the inner shocked region. The cool layer suffers a Rayleigh-Taylor instability in this compression process. Thus the layer accelerates, and density perturbations grow. In Figure 2 we show an example of the radiative cooling stage of a galactic wind. Irregularities appear in the cooled gas layer and they grow in mass. If these enhancements become gravitationally unstable, star formation may occur. This result shows that the thermal cooling stage is important for the evolution of the radiative cooling shell.

5. Summary

We review the observational evidence for galactic winds in starburst galaxies. In starburst galaxies, active star formation and frequent supernovae explosions occur in nuclear regions. We review theoretical studies of galactic winds, particularly the interaction between galactic winds and disk gas, the interaction between galactic winds and halo gas, and the effects of radiative cooling on the evolution of gaseous shells produced by the wind-halo interaction. Stratified gas in the galactic plane affects the flow of galactic winds. Elongation of galactic winds due to this interaction has been shown by previous numerical simulations in disk galaxies. Galactic winds are mainly normal to the galactic plane and when the gas flow reaches the scale height of the disk gas, the wind is blown out from the disk. These elongated hot gas flows correlate well with the X-ray and optical features observed in starburst galaxies. In studies of interactions of galactic winds with halo gas, it is shown that halo gas with large extensions affects the evolution of the galactic wind. Halo gas confines the hot gas ejected from a galaxy and sustains X-ray surface brightness. Another important physical process is cool, gaseous shell formation due to an interaction between the galactic wind and the halo gas. In the cooled gas shell, density enhancements develop by thermal and Rayleigh-Taylor instabilities. This is a possible mechanism for the star formation observed above the disks of starburst galaxies.

REFERENCES

[1] Bland, J. & Tully, R. B. 1988, Nature, 334, 43
[2] Blondin, J. M. & Cioffi, D. F. 1989, ApJ, 345, 853
[3] Chevalier, R. A. & Clegg, A. W. 1985, Nature, 317, 44
[4] Denda, K. & Ikeuchi, S. 1993, PASJ, 45, L1
[5] Fabbiano, G. 1988, ApJ, 330, 672
[6] Forman, W. & Jones, C. 1982, ARAA, 20, 547
[7] Habe, A. & Yoshida, T. 1993, in preparation
[8] Hattori, M., Habe, A. & Ikeuchi, S. 1987, ProgTheorPhys, 78, 1099
[9] Heckman, T. M., Armus, L. & Miley, G. K. 1990, ApJS, 74, 833
[10] Ikeuchi, S., Habe, A. & Tanaka, Y. D. 1984, MNRAS, 207, 909
[11] Kronberg P. P., Beirman P. & Schwab, F. R. 1985, ApJ, 291, 693
[12] Li, F. & Ikeuchi, S. 1992, ApJ, 390, 405
[13] Lo, K. Y., Cheung, K. W., Masson, C. R., Phillips, T. G., Scott, S. L. & Woody, D. P. 1987, ApJ, 312, 574
[14] Loewenstein, M. & Mathews, W. G. 1987, ApJ, 319, 614
[15] Mac Low, M.M., McCray, R. & Norman, M. L. 1989, ApJ, 337, 141
[16] Mac Low, M.-M. & Norman, M. L. 1993, ApJ, 407, 207
[17] Mathews, W. G. & Baker, J. C. 1974, ApJ, 224, 308
[18] McCarthy, P. J., Heckman, T. M. & van Breugel, W. 1987, AJ, 93, 264
[19] McKee, C. & Ostriker, J. 1977, ApJ, 218, 148
[20] Nakai, N., Hayashi, M., Handa, T., Sofue, Y., Hasegawa, T. & Sasaki, M. 1987, PASJ, 39, 685
[21] Norman, C. A. & Ikeuchi, S. 1989, ApJ, 345, , 372
[22] O'Connell, R. W. & Mangano, J. J. 1978, ApJ, 221, 62
[23] Telesco, C. M. & Harper, D. A. 1980, ApJ, 235, 392
[24] Terlevich, R., Tenorio-Tagle, G., Franco, J. & Melnick, J. 1992, MNRAS, 255, 713
[25] Tenorio-Tagle, G., Bodenheimer, P. & Rozyczka, M. 1987, A&A, 182, 120
[26] Tomisaka, K. 1993, this conference
[27] Tomisaka, K. & Ikeuchi, S. 1986, PASJ, 38, 697
[28] Tomisaka, K. & Ikeuchi, S. 1988, ApJ, 330, 695
[29] Tomisaka, K. & Bregman, J. 1993, PASJ, in press
[30] Tsuru, T., Ohashi, T., Makishima, K., Mihara, T. & Kondo, H. 1990, PASJ, 42, L75
[31] Umemura, M. & Ikeuchi, S. 1987, ApJ, 319, 601
[32] Unger, S. W., Pedlar, A., Axon, D. J., Wilkinson, P. N. & Appleton, P. N. 1984, MNRAS, 211, 783
[33] Wada, K., Habe, A. & Sofue, Y. 1993, MNRAS, submitted
[34] White, R. E. III & Chevalier, R. A. 1983, ApJ 275, 69
[35] William & Christiansen 1985, ApJ, 291, 80
[36] Yoshida, T. & Habe, A. 1992, ProgTheorPhys, 88, 251

Shocks and Dust in Active Galactic Nuclei

By S. M. Viegas[1], M. Contini[2]

[1] Instituto Astronômico e Geofísico, Sao Paulo, Brazil
[2] School of Physics and Astronomy, Tel Aviv University, Israel

Numerical simulations of a photoionized emitting gas cloud including shock effects show that the observed infrared continuum of active galactic nuclei (AGN) can be explained by the thermal emission by dust. The shock effect is required to heat the grains to higher temperatures.

1. Introduction

Far-infrared observations of AGN indicate the presence of dust in these nuclei [12]. Until now the models used to account for the observed features have only considered the thermal radiation from grains heated by ultraviolet radiation [2] or X-rays [16]. However, as in supernova remnants, dust could be heated in a shock front [7]. In AGN, shock effects coupled to photoionization have been invoked to explain the observed emission line intensities [13,14] and could provide collisional heating of dust grains. The presence of dust in the emitting clouds of AGN would modify the physical conditions of the gas and, therefore, the calculated line intensities. It will also change the line profiles, and could explain the asymmetric line profiles observed in AGN (see for instance, [17]). Thus, it is necessary to built a self-consistent model including dust and gas excited by shocks and photoionization. In Section 2 we describe the numerical model used to analyse the physical conditions of gas and dust in the emitting clouds of AGN. The results and conclusions appear in Section 3.

2. Numerical Calculations

The computer code SUMA [14] simulates the coupled effects on a gas cloud of shocks due to cloud motions and photoionization due to a radiation source. The code has been improved and updated recently [15] in order to include the processes due to the presence of dust that modify the physical conditions of the cloud. The calculations assume a stationary shock front [3], and its effect on the physical conditions of the cloud depends on the shock velocity and on the pre-shock density. The ionizing radiation is assumed to be a power-law and is characterized by the ionizing flux at 1 Ryd and the power-law index. The cloud is divided in slabs (250 to 300), and in each one, starting at the shocked edge, the ionization and thermal balance equations are solved. The slab widths are variable and depend on the spatial variation of the gas temperature and optical depth at 1 Ryd. In order to account for dust effects, several processes have been included in the computer code: (a) dust opacity assuming spherical dust grains; (b) collisional and radiative heating of the grains [6,7,10]; (c) grain cooling [7]; (d) thermal sputtering of the grains, which changes their sizes [6]; (e) photoelectric heating of the gas [10]; (f) dust emissivity. We have assumed silicate grains with a size $a = 0.2$ μm and a dust-to-gas density ratio in the range $10^{-14} < d/g < 10^{-12}$.

3. Main Results

Our previous analysis of the emission-line intensity ratios indicated that the clouds should be moving out from the active nucleus. Thus, here we consider only ejection cases, where the ionizing radiation reaches the inner side of the cloud and the shock front acts on the opposite edge. A grid of models without magnetic fields have been obtained by varying the input parameters. The main results are the following:

(1) Sputtering destroys the grains near the shock front if the shock velocity is $v_o \geq 300$ km s^{-1} (the presence of a B-field can modify the sputtering rate [9]). Thus, dust may be present in the photoionized region only if it is formed in the cloud before the shock effect (pre-existent dust).

(2) Due to collisional heating, the grain temperature can reach up to 120 K in the postshock zone if $v_o = 500$ km s^{-1}, while it reaches only about 60 K in the photoionized region.

(3) Regarding the continuum spectrum emitted by the cloud, the infrared emission is mainly due to dust, showing a blackbody-like shape (dust emission features should also be present [4], but they are not included here). The IR maximum is proportional to the d/g ratio. If dust is not pre-existent in the cloud, but enters the shock front with the gas, and if sputtering is important, there is no contribution to the thermal IR emission from the photoionized region. Then, the IR spectrum is a blackbody with temperature determined by the shock velocity. If dust is preexistent, the emission from dust at lower temperature (photoionized zone) also contributes and the thermal IR continuum is broader.

Finally, in order to determine the amount of dust in different types of AGN, we used models to fit the observed continuum spectra of QSOs and Seyfert 1 [1], Seyfert 2s [8], and LINERs [11,18]. The fit to the QSO and Sy1 data is not very good, probably because we have only considered the narrow-line clouds, with low velocities (< 600 km s^{-1}) and low densities ($< 10^5$ cm^{-3}), and the contribution of the broad-line clouds could be important. On the other hand, for Sy 2 and LINERs our results give a good fit to the observed spectra. For QSOs, Sy 1 and LINERs the dust-to-gas ratio is low ($10^{-14} <$ d/g $< 10^{-13}$), whereas for Sy 2 it is higher ($10^{-13} <$ d/g $< 10^{-12}$). This indicates that the amount of dust must be higher in Sy 2 than in LINERs, in agreement with results reported previously [12,18]. Only the shock effect can explain the grain temperature necessary to reproduce the observed infrared data.

REFERENCES

[1] Antonucci, R. & Barvainis, R. 1988, ApJ. Letters, 332, L13.
[2] Barvainis, R. 1987, ApJ, 320, 537.
[3] Cox, D. P. 1972, ApJ, 178, 143.
[4] Draine, B. T. 1981, ApJ, 318, 674.
[5] Draine, B. T. & Salpeter, E.E. 1979, ApJ, 231, 77.
[6] Dwek, E. 1981, ApJ, 247, 614.
[7] Dwek, E. 1987, ApJ, 322, 812.
[8] Edelson, R. A., Malkan, M. A. & Rieke, G. H. 1987, ApJ, 321, 233.
[9] McKee, C. F., Hollenbach, D. J., Seab, C. G. & Tielens, A. 1987, ApJ, 318, 674.
[10] Oliveira, S. & Maciel, W. J. 1986, ApSS, 126, 211.
[11] Rowan-Robinson, M. 1992, MNRAS, 125, 787.
[12] Vaceli, M. S., Viegas, S. M., Gruenwald, R. & Benevides-Soares, P. 1993, PASP, 105, 875.
[13] Viegas-Aldrovandi, S. M. & Contini, M. 1989a, ApJ, 373, 405.
[14] Viegas-Aldrovandi, S. M. & Contini, M. 1989b, AA, 215, 253.
[15] Viegas, S. M. & Contini, M. 1993, ApJ, in press.
[16] Voigt, G, M. 1991, ApJ, 379, 122.
[17] Whittle, M. 1985, MNRAS, 213,1.
[18] Willner, S. P., Elvis, M., Fabbiano, G., Lawrence, A. & Ward, M. J. 1985, ApJ, 299, 443.

The Physics of Self-Propagating Star Formation

By G. Tenorio-Tagle

Instituto de Astrofísica de Canarias, 38200-La Laguna, Tenerife, Spain.

1. Introduction

Galaxies constitute a fascinating system of stars, gas, dark matter, radiation, and dynamical forces that interact in ways not well understood to date. Typical spiral galaxies, like our own Milky Way for example, contain sufficient amounts of gas that if turned into stars instantaneously, the Galaxy would be hundreds of times more luminous than at present. In reality, only a small fraction of the available gas seems to be transformed into stars at any given time, as if there exists a self-regulating mechanism that controls not only the star formation rate and process, but also the location where it initiates. The search for this self-regulating mechanism has produced a rapidly growing interest in the physics of propagating star formation and its regulation on galactic scales.

Within the large framework of physical processes that could be relavent to star formation, one attempts to identify the basic elements of the star forming/regulation cycle and carefully weigh their relative importance and the order in which they proceed. Our goal is to clearly define the entire star formation cycle. That is, the sequence of events and fundamental physical processes that eventually lead to conditions suitable for star formation and that is self-consistent in its connection with the various phases of the interstellar medium (ISM). The latter contraint is a major one, often ignored in studies of self-propagating star formation, and must be met if we are to believe the star forming cycle has been completely identified.

2. The Star Forming Cycle and the Physics of Self-Propagation

Stars, particularly massive ones, are catalysts for undeniably spectacular events within galaxies. Here we take the point of view that it is in fact the most massive stars themselves that control the entire star forming cycle, implying total self-propagation and self-regulation, unlike other schemes that lead to molecular cloud formation but that have considered idealized situations only (*i.e.* neglecting the input of energy from massive stars and the structure of the ISM). In our model, we have necessarily adopted many ideas from previous models and regretably cannot give credit to all of them here; however, an attempt was made to do this in several recent papers and reviews [7,8,15,16,21,22].

2.1. *The Star Forming Cycle*

We base our model on the fact that massive stars are born in large groups or associations. Typically, an association in the Galaxy has a mass of $\sim 10^5$ $M_\odot$ and a certain mass distribution (*i.e.* an IMF) among its members. There are few truly massive stars in a given association, but the energy deposition rate into the ISM by these stars is enormous and drives many of the observed features in and around these associations.

Figure 1 is a collage of several hydrodynamical events caused by the energy input from massive stars illustrating what are thought to be some of the steps in the star forming cycle (center panel). Approximate time-scales as well as the physical processes responsible for the phase transitions in the ISM are indicated where they are known.

2.2. The Events Caused by the Appearance of Massive Star Formation

Define $t = 0$ as the time just after star formation has taken place within a giant molecular cloud and the O and B stars in the association are on the main sequence (MS). Photoionization immediately forms an HII region around these stars, heating the surrounding gas to temperatures of $\sim 10^4$ K. We imagine that photoionization is most important for the most massive stars ($M_* \geq 30$ $M_\odot$) which have MS life-times $\leq 10^7$ yr. During this time the HII regions will expand into the surrounding gas at a speed of the order of 10 km s^{-1}, sometimes triggering further star formation [6], until the ionization front reaches the edge of the cloud. At this point, ionizing photons can ionize the intercloud medium creating a large pressure gradient between the ionized cloud and the intercloud gas. The pressure gradient initiates the champagne phase of HII region evolution, and with it the rapid dispersal of the parent molecular cloud [4,19,20]. After a few (4-6) million years, the most massive stars begin to move off the MS causing a drastic decay ($\sim t^{-5}$) in the ionizing photon flux. This naturally leads to the sudden recombination of the ionized matter.

In light of the above cycle, photoionization and recombination are recognized as two of the fundamental physical processes in the star forming cycle. Photoionization heats up the local gas inhibiting further star formation and thus determines the efficiency of the process [2,4,11,13]. In addition, it leads to molecular cloud disruption and provides the dispersed gas with large velocities. Recombination, on the other hand, leads to the reappearance of the HI phase in the galaxy.

2.3. The Mechanical Energy Budget

Massive stars also deposit large amounts of mechanical energy into the ISM during their MS life-time and during their subsequent Wolf-Rayet phase, either in the form of stellar winds or by means of their terminal supernova explosion [1,3,22]. Winds from massive stars result in substantial mass loss (10^{-5} $M_\odot$ yr^{-1}) at large velocities (~ 2000 km s^{-1}), implying that over a life-time of $\sim 5 \times 10^6$ yr a total mechanical energy of 10^{51} erg is deposited into the surroundings. This energy is comparable to both the thermal energy of the HII region and the mechanical energy released during a supernova explosion. Less massive stars ($M_* \leq 30$ $M_\odot$) also have powerful winds; however, their net energy input to the ISM is probably dominated by their terminal supernova explosion. From stellar evolution theory, we know that the lowest mass star expected to terminate as a Type II supernova has an initial mass of ~ 7 $M_\odot$, equivalent to spectral type B3. Furthermore, we know the expected life-time of such stars and given an initial mass function one could work out the total mechanical power of an OB association, as well as the rate at which it is delivered [7,17]. In our Galaxy, an association contains $\sim$20-40 stars with spectral types earlier than B3 in a region less than 100 pc in diameter. This observational result is consistent with a Salpeter IMF, implying that a typical association produces 10 times as many stars in the mass range 7 $M_\odot$ - 30 $M_\odot$ (spectral types B3 - B0) than stars more massive than 30 $M_\odot$.

Several groups in the early 1980s noticed the overwhelming power of OB associations and the impact that they could have on galactic disks [1,10]. This was largely motivated by the discovery of supershells (HI structures expanding with kinetic energies of up to 10^{54} erg) in

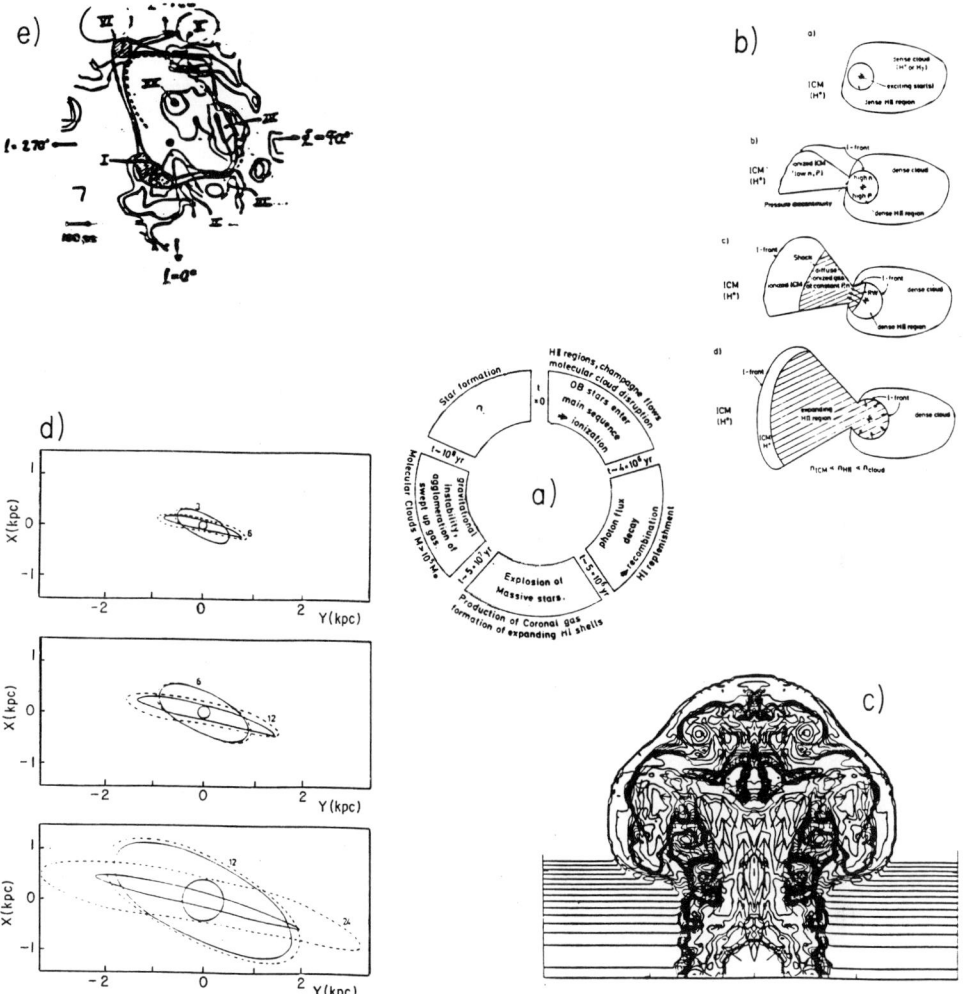

FIGURE 1. The central panel (a) describes the star forming cycle. Time-scales and the fundamental physical processes leading to ISM phase transitions are indicated in the various boxes (from [22]). b) The champagne phase [19,20], showing the dispersal of the parent molecular cloud. c) Blowout of a remnant caused by multiple supernova explosions. The hot gas, representing a small fraction of the total mass involved, goes into the halo while the swept up matter from the disk evolves into a ring-like remnant [22]. d) Calculations of multi-supernova remnants in the plane of the galaxy including the effects of differential galactic rotation at 5, 10 and 20 kpc from the galactic center [23]. e) The Local Superbubble [5]. Indicated are the solar location, the direction towards the galactic center and various well known molecular cloud complexes and stellar associations such as: I.- Lower and Upper Centaurus OB associations. II.- Ophiuchus molecular cloud. III.- Great Dark Rift. IV.- large dust cloud. V.- Per OB2 association. VI.- the Orion OBI association.

the Galaxy [12]. A review of the research on OB associations is given by Tomisaka elsewhere in these proceedings. Here, we simply note that the mechanical energy deposition from evolved OB associations through strong stellar winds and supernova explosions is another fundamental process in the star forming cycle [7,10,22]. This is because it gives rise to the

2.4. The Formation of Giant Molecular Cloud Complexes

From observations we know that molecular clouds only form in certain galactic locations. In the Galaxy, they are located inside of a galactic radius of 10kpc, and it has been found that H_2 dominates over HI when the disk total column density exceeds a threshold value of 10^{21} cm^{-2}. This latter result is also observed in other spiral and irregular galaxies. This is not too surprising given that a minimum column density is required to self-shield the gas from the UV radiation field [9]. With this in mind, we have looked at remnants produced by spatially correlated supernova explosions to see if the giant expanding shells could ever satisfy this opacity criterion.

Our work on multi-supernova remnants suggests many interesting features relevant to the formation of molecular clouds (see also [18]). For instance, there is a sudden fragmentation of the large expanding shells once breakout takes place or when the remnants exceed the dimensions of the gaseous galactic disk. Rayleigh-Taylor instabilities produce rings of matter expanding into the remaining disk whereas the hot gas in the interior of the remnants freely streams into the halo (see lower left panel in Figure 1). Very little disk matter ($\leq$ 10% of the swept up gas) is accelerated away from the disk; thus the remaining structure is one in which the swept up gas is now in a cylindrical remnant in the plane of the galaxy. We followed the evolution of such expanding rings in a differentially rotating disk for different choices of the background density and a range of likely initial energies. The differential rotation causes a strong ellipticity to develop in the remnants while the whole structure rotates, ultimately aligning itself almost perpendicularly to the galactic center direction (see lower left panel in Figure 1). While this deformation of the rings occurs, swept up matter accumulates at the tips of the elliptical remnants, eventually satisfying the opacity criterion for molecular cloud formation ($N \geq 10^{21}$ cm^{-2}). Furthermore, the masses of the newly formed clouds are $\sim 10^{5-6}$ M$_\odot$, similar to the masses of Galactic giant molecular cloud complexes. The upper left panel of Figure 1 shows the solar neighbourhood which displays characteristics of remnants and their associated molecular cloud complexes. Other examples have been found in M33, M31, and HoII. The giant molecular cloud formation criterion is, in our calculations, rapidly fulfilled within the solar circle– in some cases even before the clouds become self-gravitating. This does not occur at larger galactic radii (say 20kpc), however, where despite their size and mass concentration at the tips of the remnants, the resultant clouds remain atomic even if they become self-gravitating.

Our calculations imply that spatially correlated supernova explosions in a differentially rotating galactic disk lead to well-defined remnants causing an organized rather than a chaotic dispersal of matter which promotes the formation of well-spaced giant molecular clouds. If such clouds are the seeds of the next generation of OB associations, this mechanism *determines* the conditions for long range propagating star formation, a phenomenon usually regarded as a stochastic process (see (14]). Furthermore, our results imply that self-propagation and self-regulation are inherent to the star forming process.

3. Evolution in Action and the Physics of Self-Regulation

The basic computational scheme of large-scale propagating stellar formation is based on the parameterization of detailed calculations of the evolution of multi-supernova remnants [16]. This parameterization allows many such complexes to be computed concurrently in a galactic disk. Part of the success of the parameterization relies on the similar evolution of remnants with a large range of initial energies, background densities, galactic disk structures, and rotation curves. Many individual runs were used to fit the logarithmic dependence of each parameter, and in this way a *robot* was developed that mimics the evolution of the remnant without following all of the details. The *robot* keeps track of the location, age, size, shape, inclination, and total swept up mass and has the capability of self-duplication (which occurs when the amount of swept up matter at the tips of the elliptical remnant exceeds that of giant molecular cloud complexes). After a fixed amount of time (the cloud life-time) which can be thought of as the period during which star formation is occurring within the cloud, massive star formation and death eventually lead to the beginning of a new star formation cycle (as discussed in §2.2) and a new *robot* will appear at the location of the cloud. Its initial energy depends on the amount of matter in the parent cloud, as the efficiency of star formation limits the amount of energy expected from such an association (see [17]). *Robots* differ in their cycle times dependent on their location in the galaxy. Thus, some may complete their evolution rapidly and duplicate while others, generated at the same time but at larger galactocentric radii, will still be in their initial cycle. In order to conserve mass, galactic matter is locally depleted after each new cloud is produced, and the background density is correspondingly reduced. Complications develop when the number of *robots* grows such that they begin to interact with each other. To handle these interactions, we broadly define three modes of interaction depending on the location and age of the *robots*– head-on collisions, lateral collisions, and mergers. Each class of interaction has a set of rules that defines how the mass and energy evolve during the encounter [17] in an effort to account for the number of remnants, the number of clouds, and the mass and evolutionary time in each of them.

4. Conclusions

Our models demonstrate that the expansion and ultimate interaction of multi-supernova remnants are key events in large-scale propagating star formation. This follows from the evolution of multi-supernova remnants in a differentially rotating galactic disk, and results in the formation of molecular cloud complexes in a manner consistent with actual cloud complexes in the Galaxy and other spiral and irregular galaxies. The model is self-regulating in that interactions between remnants limit the total population of remnants in the galaxy at any given time. Furthermore, the model is self-propagating given that we initiate the runs with only a few remnants (or one) and observe the formation of many molecular cloud complexes over the duration of the calculation (some 10 Gyr).
The results show excellent qualitative and quantitative with some observed properties of galaxies, such as:
1) The total mass and radial distribution of HI and H_2.
2) The numbers of multi-supernova remnants and massive cloud complexes, and the resultant star formation rate.
3) The surface filling factor of shells.
4) The location of the molecular ring with apparant long spiral arms.
Clearly, such a deterministic, self-propagating star formation scheme has a major impact on the chemical evolution of galaxies and serves as a framework on which to develop a more detailed approach to self-regulation of star formation. The basic element of the scheme we present, the star forming cycle, implies a strong connection between star formation, stellar evolution, the interstellar medium, and galactic dynamics and has a strong influence on the appearance of star forming galaxies.

Acknowledgements: I would like to thank Warren Miller and Pepe Franco for useful comments and suggestions to improve this presentation. Very special thanks to my good frieds and collaborators Jan Palouš and Pepe Franco for allowing me to present some results prior to publication.

REFERENCES

[1] Bruhweiler, F., Gull, T., Kafatos, M., & Sofia, S. 1980, ApJ, 238, L27
[2] Cox, D. P. 1983, ApJ, 265, L61
[3] Cox, D. P. 1993, *Star Formation, Galaxies and the Interstellar Medium*, ed. J. Franco, F. Ferrini & G. Tenorio-Tagle , (Cambridge: Cambridge U. Press)
[4] Elmegreen, B. G. 1983, MNRAS, 203, 1011
[5] Elmegreen, B. G. 1985, Birth and Infancy of Stars, ed. R. Lucas, A. Omont & R. Stora, (Amsterdam: North-Holland).
[6] lmegreen, B. G., and Lada, C. 1977, ApJ, 214, 725
[7] Franco, J. 1992, *Star Formation in Stellar Systems*, ed. G. Tenorio-Tagle , M. Prieto, F. Sanchez, (Cambridge: Cambridge U. Press), 515
[8] Franco, J. 1993, *VIIth IAU Latinamerican Regional Astronomy Meeting, Rev. Mex. Ast. Astrof.*, in press
[9] Franco, J. & Cox, D. 1986, PASP, 98, 1076
[10] Franco, J. & Shore, S. N. 1984, ApJ, 285, 813
[11] Franco, J., Shore, S. N. & Tenorio-Tagle , G. 1994, in preparation
[12] Heiles, C. 1979, ApJ, 229, 533
[13] Larson, R. B. 1988, *Galactic and Extragalactic Star Formation*, eds. R. E. Pudritz & M. Fich, (Dordrecht: Kluwer), 459
[14] Neukirch, T. and Hesse, M.: 1993, *Astrophys. J.*, 411, 840
[15] Palouš , J. 1992, *Star Formation, Galaxies and the Interstellar Medium*, ed. J. Franco, F. Ferrini & G. Tenorio-Tagle , (Cambridge: Cambridge U. Press), 371
[16] Palouš , J., Franco, J. & Tenorio-Tagle , G. 1991, A&A, 227, 175
[17] Palouš , J., Tenorio-Tagle , G. & Franco, J. 1994, MNRAS, in press
[18] Silich, S., Franco, J., Tenorio-Tagle , G. & Palouš , J. 1993, this volume
[19] Tenorio-Tagle , G. 1979, A&A 71, 59
[20] Tenorio-Tagle , G. 1982, *Regions of Recent Star Formation*, ed. R. Roger & P. Dewdney, (Dordrecht: Reidel), 1
[21] Tenorio-Tagle , G. 1991, *Chemical and Dynamical Evolution of Galaxies*, ed. F. Ferrini, J. Franco & F. Matteucci, (Pisa: ETS Editrice), 488
[22] Tenorio-Tagle , G. & Bodenheimer, P. 1988, ARAA 26, 145
[23] Tenorio-Tagle , G. & Palouš , J. 1987, A&A 186, 287

Supernova Explosion Calculations

By J. R. Wilson

Lawrence Livermore National Laboratory

Collapse supernova calculations are discussed which give good agreement with the general features of Type II supernova events. In addition to providing a good model for the energy of explosion, recent calculations give good production of heavy elements via the r-process as well as neutrino output consistent with the observations of 1987A.

1. INTRODUCTION

Observations relevant to the explosion process associated with high mass stars (Type II supernova) are sparse. The precursors, where observed, are high mass ($> 10~M_\odot$). The kinetic energy of the explosion is estimated from the mass and velocity of the dispersing matter to be from 0.5 to 1.5×10^{51} ergs. The late time light emission is considered to arise from Co, Ni, etc, radioactivity so late time luminosity gives estimates of the nuclear burning that occurred during the explosion. How the explosion process itself occurs is weakly constrained by observations.

Results of numerical model calculations are presented. The numerical model is fully general relativistic. The radial positions, temperatures, densities, chemical compositions and neutrino ($\nu_e, \bar\nu_e, \nu_\mu, \bar\nu_\mu, \nu_\tau, \bar\nu_\tau$) energy distributions are followed for mass elements in time. The matter equation of state is assumed to consist of γ, e^-, e^+, π^-, π^0, π^+, K^-, K^0, K^+, n, p, He and a representative heavy element, "iron". All neutrino matter interactions that are thought important are included. Besides solving the equations for hydrodynamics and neutrino transport, thermonuclear burn and mixing length convection are included in the model.

The overall scenario for the supernova is as follows. The central 1.2 to 1.6 $M_\odot$ mass of a high mass ($10 - 30~M_\odot$) star burns to iron. The iron core cools by neutrino emission and rises in density. After the central density rises above about 10^{10} gm/cm^3 the neutrino emission by electron capture on protons and heavy nuclei becomes so fast that the collapse becomes dynamic. The inner 0.5 to 0.7 $M_\odot$ contracts subsonically, but much of the remainder of the iron core implodes supersonically. After the central density rises above nuclear density the pressure increases so rapidly that the infall is halted. The inner 0.7 $M_\odot$ is halted adiabatically, but a shock wave is formed exterior to 0.7 $M_\odot$ in the supersonic matter. The shock wave moves out very rapidly at first but is seriously degraded by neutrino emission and expenditure of appreciable shock energy in decomposing the iron to free nucleons. Typically the shock wave stalls after $\sim$ a tenth second at a radius of a few hundred kilometers. The radius of the proto-neutron star, defined by the radius of the neutrinosphere is about 60 km at this time. Neutrino heating of the matter underneath the shock front slowly forms a hot bubble. The neutrino heating of this hot bubble proceeds for about ten seconds producing the energy of the explosion. The expansion of the hot bubble produces an outward proceeding shock wave which after hours or days reaches the surface of the star and produces the observed light signal. For a detailed analytic analysis of the explosion process see Bethe [2].

2. Particular Details of The Explosion Process

After the central region of the core rises above nuclear density an outward shock wave is formed. But as it moves outward it decomposes the "iron" to free nucleons which requires 8.8 MeV per nucleon. The gravitational energy of the infalling matter is converted to internal energy by the shock so roughly

$$\frac{GMm}{r} \approx \frac{2100}{r} \approx \left(\frac{3}{2} + 3Y_e\right)kT + 8.8 \text{ MeV} \tag{2.1}$$

where r is in km. At a few hundred kilometers the infall energy is consumed mostly as internal energy. The neutrino losses also contribute to the loss of energy behind the shock wave. The net result is that the shock wave stops its outward motion. The conversion of the shock energy to nuclear disintegration energy does have a beneficial effect. The heating at a radius R from the neutrinos emitted from the neutrinosphere at radius R_ν and temperature T_ν can be represented by

$$\dot{E}_+ = K_0 T_\nu^2 \frac{acT_\nu^4}{4}\left(\frac{R_\nu}{R}\right)^2. \tag{2.2}$$

The cooling of matter at temperature T_m is approximated by

$$\dot{E}_- = -K_0 T_m^2 acT_m^4 \tag{2.3}$$

where we have taken the neutrino opacity as $K = K_0 T_\nu^2$. Thus the net heating is

$$\dot{E} = ac\left[T_\nu^6\left(\frac{R_\nu}{2R_m}\right)^2 - T_m^6\right] \tag{2.4}$$

Between 100 and 200 km the matter temperature is sufficiently low, $T_m = 1 - 2$ MeV, that the neutrino heating ($T_\nu = 4 - 5$ MeV) is able to raise the energy of the matter. Below 100 km the matter is still cooling and hence the matter keeps falling onto the proto-neutron star. However, as the heating progresses it slows the subsequent infall. The more the matter heats the more slowly it falls in thus the matter acquires more heat and eventually halts the infall and a hot bubble is formed. Now the heating is limited by the heated material expanding and blowing away. More cool, low entropy matter is heated on the face of the neutron star and blows away.

The above heating formula is derived considering only neutrino capture and emission on nucleons. After the entropy in the hot bubble rises to several tens of MeV neutrino-electron scattering becomes important. At high entropy electron pairs dominate the matter energy. We may thus write

$$E = \frac{11}{3} aT_m^4 \tag{2.5}$$

$$N_{e^-} + N_{e^+} = \frac{7}{4} aT_m^4 / 3kT_m \tag{2.6}$$

$$\sigma = \sigma_0 \frac{3T_\nu T_m}{m_e^2 c^4} \tag{2.7}$$

$$\dot{E} = \frac{\sigma L}{4\pi R^2}(N_{e^-} + N_{e^+}) \tag{2.8}$$

$$\frac{\dot{E}}{E} = \frac{L}{4\pi R^2}\frac{21}{44}\frac{T_\nu T_m}{km_e^2 c^4} = \frac{1}{\tau} \tag{2.9}$$

At a time after collapse of one second τ is about 1/3 sec. Another important component of late time heating is neutrino anti-neutrino annihilation to form electron pairs. Since two thirds of the neutrinos are mu or tau flavors the latter carry two thirds of the neutrino luminosity. The only important ways mu and tau neutrinos can exchange energy with the matter are by electron scattering and neutrino pair annihilation. The heating rate by neutrino pair annihilation is given by

$$\dot{E} \propto \int (\epsilon_\nu + \bar{\epsilon}_\nu) F_\nu(\epsilon_\nu, \Omega_\nu) \bar{F}_\nu(\bar{\epsilon}_\nu, \bar{\Omega})(1 - \Omega_\nu \cdot \bar{\Omega}_\nu)^2 \qquad (2.10)$$

The cross section is proportional to the square of the neutrino energy in the center of mass frame of the neutrino and the anti-neutrino. The angular integral is difficult to evaluate. We have taken snapshots from the computer model output for temperature and density as functions of radius and then solved for the steady state neutrino distribution in angle as well as radius. One to two hundred angular bins are used. From this complete neutrino distribution the neutrino flux and neutrino annihilation rate are evaluated. The neutrino flux is used to check our flux limiter used in the diffusion equation. The neutrino annihilation rate is fitted with a formula in terms of flux limiter parameters.

In order for the neutrino heating to produce the hot bubble, which is necessary for a successful explosion, convection is required. A two dimensional calculation has been made for the convection at and above the neutrinosphere by Miller et al. [7]. While this exterior convection helps the heating process it is not sufficient to produce the explosion process. Convection also occurs at and beneath the neutrinosphere. This convection is complicated to model since it involves neutrino heat flow, neutrino composition flow as well as neutrino viscosity. The development of a computer program is underway, but it requires the model to have as much detail as the spherical model and so its completion is in the far future.

3. Calculations Versus Observations

We will cite results for calculations relevant to supernova 1987a. The initial stellar model was supplied by Woosley et al [11]. The calculated explosion energy is 1.5×10^{51} erg [1]. The estimated energy derived by analysis of the observed light signal ranges from 0.8 to 1.4×10^{51} ergs. The neutrino spectra and time of neutrino emission are shown in figs 1a,1b, 1c. We see as good agreement as is possible with only 18 events [3, 4]. By considering possible degradations of time in the neutrino signal we have arrived at the following limits. The mass of ν_e, the electron neutrino, must be less than $25-30$ eV to not upset the spectra and timing agreement of figure 1. Likewise for a massive Dirac type neutrino the mass must be less than ~ 3 keV in order not to disturb the agreement of figure 1, see Mayle et al [5]. This limit arises from the possible conversion of massive left handed neutrinos into right handed neutrinos which escape the star freely.

Over the last few years people [6, 9, 10] have noted that the hot bubble formed by neutrino heating is an apparent site for the r-process nucleosynthesis of heavy elements. The proper conditions for the r-process arises as a consequence of the following. The anti-electron neutrinos have considerably higher energy than the electron neutrinos. This occurs because near the neutrinosphere the matter is very neutron rich. The density at the neutrinosphere at late times ($5-10$ seconds post bounce) is about 10^{12} gm/cm^3 which drives the electrons into the protons to a high degree. Typically the mean energy of the anti neutrinos is $1\frac{1}{2}$ times larger than the neutrinos at late times. In the heated bubble the density of the matter is relatively low, $10^4 - 10^6$ gm/cm^3. The equilibrium electron

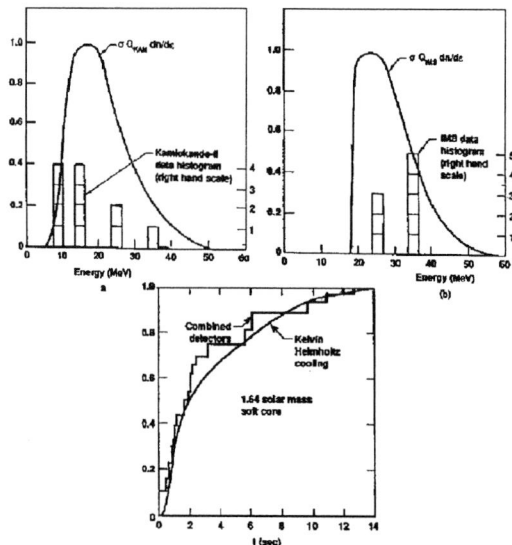

FIGURE 1. (a) The smooth curve is the calculated anti-electron spectra multiplied by the proton capture cross section and the detector efficiency, Q_{KAM}, a function of energy. The histogram is the Kamiokande data. (b) The same as (a) but for the IMB observations. (c) The fractional emitted number of neutrinos versus time. The smooth curve is the model calculation and the stepped curve is a combination of Kamiokande and IMB events weighted by their respective detector sensitivities.

fraction, Y_e is then given by

$$Y_e \approx \frac{\bar{\epsilon}_{\nu_e}}{\bar{\epsilon}_{\nu_e} + \bar{\epsilon}_{\bar{\nu}_e}} \approx 0.4 \qquad (3.11)$$

if the luminosities of the two species are equal and the chemical potential difference of neutron and proton are neglected. From the detailed calculations Y_e is found to lie between 0.40 and 0.46. This neutron excess would not be sufficient except the entropy in the bubble is high, ~ 400. As matter cools in the expanding bubble, first neutrons and protons recombine to form alpha particles at a temperature of about 0.5 MeV. At a little lower temperature the alphas combine to form carbon which subsequently burns to heavy "iron" like nucleii. However, since the alpha burning is a three body collision process, if the density is sufficiently low at the burn temperature the reaction is incomplete. At an entropy of 400 the alpha burn rate is so low that at the completion of this burn 100 to 200 neutrons per heavy "iron" nuclei are present. The neutrons are then absorbed by the various nuclei formed by the thermonuclear burn and by repeated beta decay and further neutron absorption the very heavy elements are formed. Calculations of the heavy elements were made by taking the density, temperature, Y_e time history of matter from the explosion calculation, and processing the matter by computer programs that include hundreds of isotopes and their appropriate reaction networks. The results of a calculation [12] in which no free parameters are involved is shown in figure 2.

If the tau neutrino converts to the electron neutrino via an MSW neutrino flavor mixing effect then, for tau neutrinos of mass greater than 3 eV, the mixing angle, θ, is constrained by $\sin^2 2\theta \leq 10^{-5}$ in order that the r-process occur in supernovae [8]. This arises because the tau neutrinos are more energetic than the anti-electron neutrinos, see equation (3.11).

This work supported by DOE grant #W-7405-ENG-48 and NSF PHY92-08881.

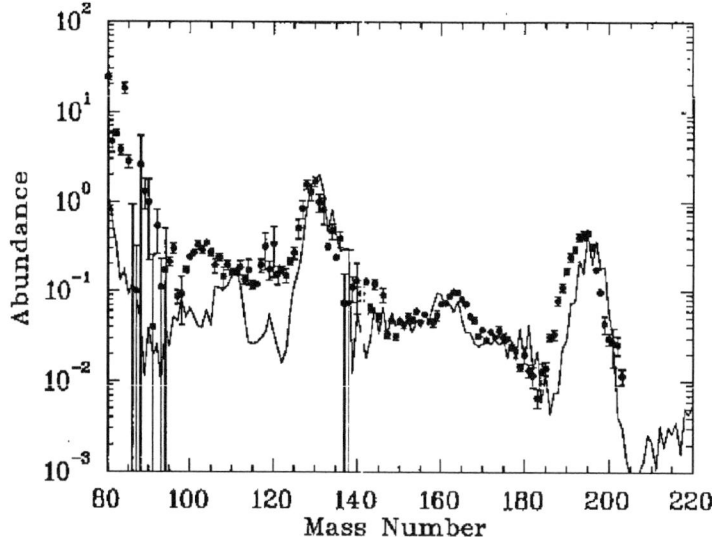

FIGURE 2. Calculated abundances from model compared to the solar abundances [6].

REFERENCES

[1] Arnett, W. D., Bahcall, J. N., Kirshner, R. P. & Woosley, S. E. 1989, ARAA, 27, 629
[2] Bethe, H. A. 1993, ApJ, 412, 192
[3] Bionta, R. M., Blewitt, G., Bratton, C. B. & Casper, D. 1987, PhysRevLett, 58, 1494
[4] Hirata, K., Kajita, T., Koshiba, M. Nakahata, M., Oyama, Y. 1987, PhysRevLett, 58, 1490
[5] Mayle,R., Schramm, D. N., Turner, M. S. & Wilson, J. R., Constraints on Dirac Neutrinos From 1987A, PhysLettB, in press
[6] Meyer, B. S., Mathews, G. J., Howard, W. M., Woosley, S. E. & Hoffman, R. D. 1992, ApJ, 399, 656
[7] Miller, D. S., Wilson, J. R. & Mayle, R. W. 1993, ApJ, 415, 393
[8] Qian, Y., Fuller, G. M., Mathews, G. J., Mayle, R. M., Wilson, J. R. & Woosley, S. E., PhysRevLett, in press
[9] Woosley, S. E. & Baron, E. 1992, ApJ, 391, 228
[10] Woosley, S. E. & Hoffman, R. D. 1992, ApJ, 395, 202
[11] Woosley, S. E., Pinto, P. A. & Weaver, T. A. 1988, Elizabeth and Frederick White Symposium on Supernova 1987 A, ProcAstronSocAust, 7, 355
[12] Woosley, S. E., Wilson, J. R., Mathews, G. J., Hoffman, R. D. & Meyer, B. S., The r-Process and Neutrino-Heated Ejecta, in preparation

Convection in Supernova Cores

By T. Shimizu, S. Yamada, K. Sato

Department of Physics, University of Tokyo, Tokyo 113, Japan.

The most promising theory for the mechanism of type II supernova explosions seems to be the so-called delayed mechanism. In this theory, neutrinos heat matter above the neutrino sphere and cause the stalled shock front to revive. It is well-known that a convective instability occurs behind the shock front; such convection may efficiently carry the energy due to neutrino heating to the location just behind the shock, resulting in a successful explosion. We carried out multi-dimensional hydrodynamic calculations to investigate the effect of convection on the dynamics of type II supernova explosions. We found that convection leads to a successful explosion for a model that otherwise fails to explode without convection, although the explosion energy gained due to convection is still insufficient to explain the observed value.

---

1. Introduction

The explosion mechanism of type II supernovae has for a long time been an unsettled problem. Type II supernovae are believed to be induced by the gravitational collapse of stellar cores. This scenario was confirmed through observations of SN1987A; a supernova with a $\sim 20\, M_\odot$ progenitor that emitted a few 10^{53} ergs of gravitational energy in the form of neutrinos. About 1 % of that energy, a few 10^{51} ergs, went into removing the envelope. The mechanism of collapse-driven supernovae is believed to be the collapsing central core bouncing due to the repulsive action of the nuclear force. In the middle of the core, a shock wave forms and begins to make its way through the core after core bounce. In many realistic calculations, however, the shock wave stalls at a radius of a few hundred kilometers several milliseconds after bounce due to energy losses through neutrino cooling as well as dissociation of nuclei.

Heating by neutrinos that are radiated from the young neutron star is crucial in the delayed explosion mechanism. Wilson [9] found that neutrinos heat matter behind the shock, reviving the stalled shock wave, resulting in a successful explosion. Furthermore, convection was found to play an important role in the delayed explosion in 1-D calculations utilizing mixing length theory [10]. We have carried out 3-D hydrodynamic calculations in order to investigate the effect of convection on the dynamics of type II supernova explosions.

It was suggested by Herant et al. [4] that convective motion carries the thermal energy of the high entropy region to the location just behind the shock front and causes the stalled shock to revive. Although they demonstrated that convective energy transport could lead to a successful explosion, their initial model is rather unrealistic and not satisfactory. Furthermore, if you take as an initial model the results of a 1-D calculation, inconsistencies between 1-D and 2-D codes can cause a false explosion.

Since it is hard to implement a 3-D calculation from the onset of collapse, we started our calculations after the shock was stalled. We used the fact that in the delayed mechanism the shock is stalled at a radius of a few 100 km and lingers for a few 100 msec; that is, we exploited a stationary hydrodynamical solution as an initial model. The initial solution is exactly consistent with the input physics that were assumed in the time evolution calculations.

2. The Stationary Shock Wave

Although we previously completed a series of calculations including convection [7,8,11], here we improve our initial model somewhat. To maintain consistency between the initial model and the hydrodynamical code, we first reproduced the stalled shock wave solving the stationary hydrodynamical equations:

$$4\pi r^2 \rho v = \dot{m}, \tag{2.1}$$

$$v\partial_r v + \frac{1}{\rho}\partial_r p = -\frac{GM}{r^2}, \tag{2.2}$$

$$v(\partial_r e + p\partial_r(\frac{1}{\rho})) = \frac{dq}{dt}, \tag{2.3}$$

as well as the equation of state $p = p(\rho, e)$, where dq/dt is the net neutrino heating rate, M is mass of the young neutron star, and $\dot{m}$ is the infalling mass flux (constant); other variables represent the ordinary ones common to hydrodynamics.

Since we cannot at present include either general relativistic effects or neutrino transport in a 3-D code, our convection calculations neglect these effects. Fortunately, this is a good approximation for calculations of convection above the neutrino sphere where the matter density is not very high and it is not opaque to neutrinos (only 5% of the total neutrinos emitted are absorbed in this region). The effect of neutrinos is instead introduced as an energy source term dq/dt in equation (2.3). We considered neutrino absorption on free nucleons and scattering off electrons and positrons as neutrino heating terms, and electron/positron capture by nucleons and photo, pair, and plasma neutrino emission as cooling terms. The heating and cooling rates are almost identical to those Herant et al. [4] used. The electron chemical potential was included in the electron capture term, and we assumed the neutrino temperature T_ν was 5.0 MeV, the radius of the neutrino sphere was 50 km, and the energy spectrum of neutrinos was a black body distribution.

With regard to the equation of state, we included radiation pressure, the pressure of nuclei and free nucleons, and the pressure of degenerate electrons at zero temperature. The energy density e was assumed to satisfy $p \approx \rho e/3$ when nuclei are completely dissociated into free nucleons, as this assumption facilitates solving the stationary equations (2.1-2.3). We further assumed that cold matter, consisting mainly of Fe nuclei, was infalling above the shock front with a velocity of 80% of the free fall velocity. The shock discontinuity was calculated with the Rankine-Hugoniot relations and placed at a radius of 200 km.

Figure 1 (the initial model) shows that a negative entropy gradient exits, which indicates a site of convective instability. It is also notable that such a negative entropy gradient is a natural consequence of a stationary shock front and that convective instability is thus a general product of the delayed mechanism. This is because in order to support the ram pressure of infalling matter at the shock front, the temperature must rise to ~ 1 MeV behind the shock front. At this temperature, heating dominates cooling; however, with decreasing radius the temperature increases and grows high enough ($\sim T_\nu$) for cooling to dominate around the neutrino sphere since the cooling rate is a steep function of the temperature ($\propto T^6$). There exists, then, a location where the heating rate equals the cooling rate. Now note that the left hand side of equation (2.3) represents an increase of entropy of the infalling matter. That is, the net heating rate vanishes at the peak of the entropy profile [2]. Therefore, the entropy gradient must be negative in the region where the net heating rate is positive, as can be seen in Figure 1.

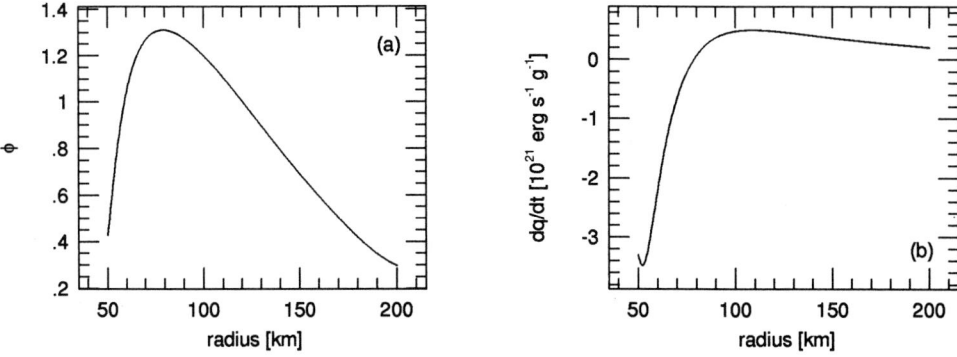

FIGURE 1. The initial model used in our calculations; (a) photon-baryon ratio ϕ (~ 0.1 entropy), (b) net neutrino heating rate dq/dt (heating term minus cooling term); the neutrino sphere and the shock are located at 50 km and 200 km, respectively.

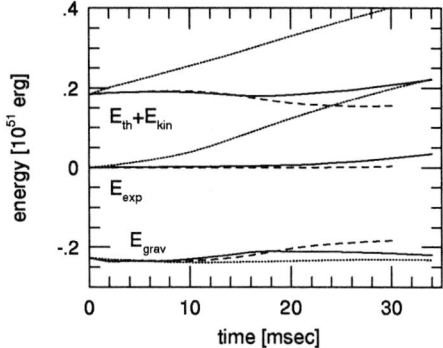

FIGURE 2. The energy time evolution for several cases; solid lines: $\delta\rho/\rho = 50\,\%$; dashed lines: $\delta\rho/\rho = 1\,\%$; dotted lines: 1.5 times the neutrino luminosity.

3. Numerical Simulations of Convection

Hydrodynamical calculations (Eulerian description) were carried out using the Roe method [6] (second-order accurate), modified to include a general equation of state on an uniform Cartesian grid. We checked our hydrodynamical code using standard tests, such as the shock-tube problem and Sedov's similarity solution, and found that the total energy was conserved to within 1% throughout the calculations. The equation of state and the assumptions concerning the neutrinos in the time evolution calculation were exactly the same as in the initial model. In principle, then, it is expected that the shock front will stay at the initial position without any instability. The time evolution was solved only for the region above the neutrino sphere and the hydrodynamical values on both the outer and inner boundaries were fixed to the initial values.

Two calculations with initial density perturbations $\delta\rho/\rho$ of 50 % and 1 % were carried out with the perturbations added to the region between the shock front and the neutrino sphere. Figure 2 shows the time evolution of the sum of both the thermal and kinetic energy ($E_{th} + E_{kin}$) of the system, the gravitational energy E_{grav}, and the explosion energy, E_{exp}, defined by the sum $E_{kin} + E_{th} + E_{grav}$ over mass elements satisfying the condition $E_{kin} + E_{th} + E_{grav} > 0$, for the two cases (the solid and dashed lines correspond

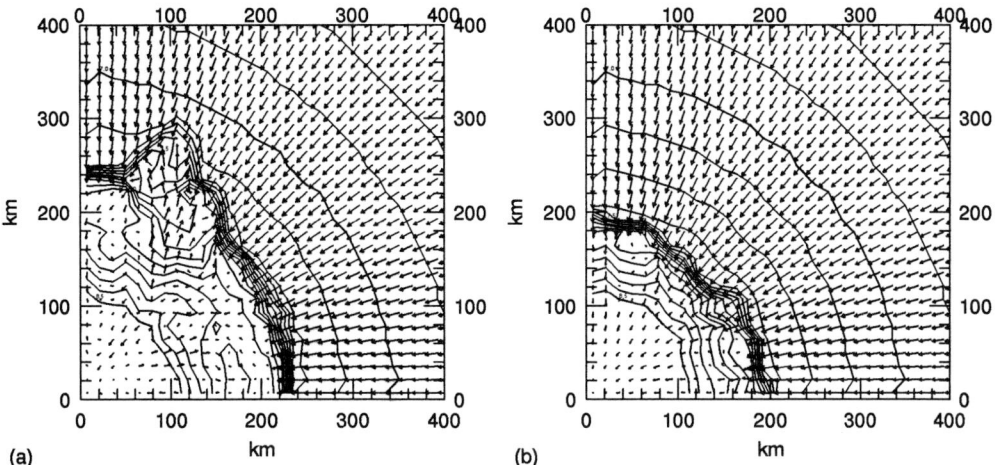

FIGURE 3. Density contours of the section parallel to the equatorial plane 20 msec after the start of the calculation for the two cases, (a) $\delta\rho/\rho = 50\,\%$, (b) $\delta\rho/\rho = 1\,\%$.

to the 50 % and 1 % cases, respectively). It can be seen that in the 1 % case, the shock remained at the initial position for more than 30 msec. In the 50 % case, however, the shock started to move upwards at around 15 msec (i.e. the shock revived). The motion of the shock is confirmed by observing the density contours in Figure 3. In the 50 % case, E_{exp} increased by 0.05×10^{51} erg in 35 msec. Since the shock covered no more than ~ 150 km in 35 msec, an increase of 0.5×10^{51} erg can be expected once the shock escapes from the central core (~ 2000 km). We are in the process of computing longer evolutions in order to estimate the final value of the explosion energy. It is interesting to note, however, that if the neutrino luminosity is increased by a factor of 1.5 (to break the stationary state), then a more energetic explosion is easily obtained as shown by dotted lines in Figure 2.

4. Discussion

We found that convection influences the dynamics of the shock wave. That is, convection leads to a successful explosion for a model that otherwise fails to explode without it. Although the explosion energy gained due to convection is still insufficient to explain observations, it is worth studying the effect of convection further. We expect that stalled shock wave configurations may exist in which convective energy transport is efficient enough to produce explosion energies as high as 10^{51} erg. We did not include the lepton fraction Y_l in these calculations, and it is well-known that the Y_l distribution provides another site of convective instability that is deeper and is expected to transport more energy than the one due to the entropy distribution. We are in the process of including the effect of Y_l in our calculations.

Finally, it was recently shown that a neutrino luminosity boost during a very early phase of the explosion can lead to an energetic explosion [5]. Burrows and Fryxell [1] stressed the importance of convection inside the neutrino sphere towards boosting the neutrino luminosity by performing 2-D hydrodynamical calculations with 1-D neutrino transport. Therefore, it is also important to include neutrino transport in the calculation of convection in order to clarify the effect of the enhancement of neutrinos due to convection.

We acknowledge Cray Research de México, S. A. de C. V., and Universidad Nacional Autónoma de México for Cray allocation time and a warm welcome during my visit. Some parts of the calculations were also performed on a HITAC at KEK (National Laboratory for High Energy Physics, Japan). This work was partially supported by Grant-in-Aids for Scientific Research from the Ministry of Education, Science and Culture of Japan (04234104, 05243103, and 3013).

REFERENCES

[1] Burrows, A. & Fryxell, B. A. 1993, submitted to ApJL
[2] Burrows, A. & Goshy, J. 1993, submitted to ApJL
[3] Epstein, R. I. 1979, MNRAS, 188, 305.
[4] Herant, M., Benz, W., & Colgate, S. 1992, ApJ, 395, 642.
[5] Janka, H. -Th. & Müller, E. 1993, *Frontiers of Neutrino Astrophysics*, Universal Academy Press, Inc., Tokyo, 203.
[6] Roe, P. L. 1981, JCompPhys, 43, 357.
[7] Sato, K., Shimizu, T., & Yamada, S. 1993, *Frontiers of Neutrino Astrophysics*, Universal Academy Press, Inc., Tokyo, 191.
[8] Shimizu, T., Yamada, S., & Sato, K. 1993, PASJ, 45, L53.
[9] Wilson, J.R. 1985, *Numerical Astrophysics*, ed. J. Centrella, J. LeBlanc and R. Bowers, Jones and Bartlett, Boston, 422.
[10] Wilson, J.R. & Mayle, R. 1988, PhysRep, 163, 63.
[11] Yamada, S., Shimizu, T., & Sato, K. 1993, ProgTheorPhys, 89, 1175.

Instabilities in Supernova Explosions

By Bruce Fryxell

Universities Space Research Association, Mail Code 934, NASA/Goddard Space Flight Center, Greenbelt, MD 20720, USA.

It has become clear in recent years that fluid instabilities are a key ingredient of Type II Supernova explosions. As the blast wave propagates outward, it leaves behind a structure which is violently unstable to Rayleigh-Taylor and convective instabilities. Direct evidence that such instabilities exist was seen in the observations of SN1987A. The spectra obtained could only be explained if significant mixing occurred between the light elements in the outer layers of the star and the heavy elements in the core, clearly indicating the presence of non-spherical motion. In addition, the earlier than expected detection of X-rays and γ-rays from SN1987A can most easily be explained if the radioactive elements in the core were mixed outward to regions of lower optical depth or if holes in the density distribution were created by the clumping of matter in the envelope. Both two and three-dimensional calculations of these instabilities, which appear during the first few hours after the explosion, have been performed using the Piecewise-Parabolic Method for hydrodynamics with a wide variety of grid resolutions. The results are able to explain some, but not all, of the observations. In particular, the mixing of heavy elements to high velocity is not reproduced. It now appears that this mixing must be achieved as a result of instabilities which occur within the first few seconds of the explosion. Perhaps more important than the observations of mixing and clumping is the idea that these early time instabilities may play an important role in producing the explosion. Since one-dimensional simulations of Type II supernovae generally produce only weak explosions or do not explode at all, it is possible that multi-dimensional effects are a key ingredient in the explosion mechanism.

---

1. Introduction

It has long been suspected that supernova explosions are not spherically symmetric from observations such as the non-spherical appearance of supernova remnants. In addition, theoretical calculations have shown that the ejecta is subject to Rayleigh-Taylor instabilities [8,13]. Observations of Supernova 1987A provided further convincing evidence that many of the details, and perhaps even the explosion mechanism itself, could only be explained by considering multidimensional effects. These observations include the earlier than expected detection of X-rays and γ-rays, the mixing of hydrogen inward to velocities of a few hundred km s^{-1}, and the mixing of radioactive iron group elements outward to velocities greater than 3500 km s^{-1}. For references to these observations, see eg. Fryxell, Müller & Arnett [15]. A number of groups [1,2,9,16,17,21,22,23,26] have attempted to model these instabilities using different numerical methods and supernova progenitor models. Although these calculations differ in detail, there is general agreement that violent instabilities are excited behind the blast wave during the first few hours after the explosion for all reasonable progenitor models. This paper will present a series of calculations of these instabilities which were performed using a wide variety of grid sizes to show how the results are affected by numerical resolution. The successes and failures of the model in explaining the observations of SN1987A will then be discussed. Finally, preliminary calculations showing the possible importance of instabilities which occur during the first few seconds after the explosion will be described briefly.

2. Numerical Methods

The calculations presented below were performed using the PROMETHEUS computer code [14], which solves Euler's equations for compressible gas dynamics using the Piecewise-Parabolic Method (PPM) [11,25]. The original PPM method has been extended to include a general equation of state [10] (ideal gas plus radiation pressure for these simulations) and the ability to advect an arbitrary number of separate fluids so that the effects of mixing can be calculated. Simulations have been carried out in both two and three dimensions using either cylindrical or spherical coordinates. For the two-dimensional calculations, only one quadrant of the star was modeled. Rotational symmetry about the vertical axis and equatorial symmetry across the horizontal axis were assumed. For the highest resolution calculations, the grid was allowed to expand homologously so that the unstable layers would contain as many zones as possible at early times when small wavelength modes dominate the solution. In the three-dimensional calculations, only a narrow wedge of the star was simulated in order to maintain high resolution.

The initial model was chosen to give a good representation of the properties of Sanduleak -69° 202, the progenitor star of SN1987A. In addition, one-dimensional calculations of the explosion and light curve agree very well with the observations. The innermost 10% of the star in radius, taken at a point in time just before core collapse, was then interpolated onto a one-dimensional uniformly zoned grid containing 6000 zones. The explosion was initiated by instantaneously depositing 10^{51} ergs of energy (half kinetic and half internal) into the inner few zones. The propagation of the shock was then followed for 300 s using a one-dimensional version of PROMETHEUS. Following this early portion of the explosion with a multidimensional code would be very expensive as a result of the high temperature and correspondingly small time step in the inner zones. The growth rate of the instability is sufficiently small during this initial period that the overall results are not affected by using this approach. Significant growth of the instability is not observed until approximately 2000 s after the explosion. However, other instabilities which may develop on a much earlier time scale are of course eliminated by this procedure. The results of this calculation were then mapped onto a multidimensional grid. The outer 90% of the star in radius was interpolated onto the multidimensional grid using the original initial model. In order to break the initial spherical symmetry of the problem, a random perturbation of 10% amplitude was applied to the radial velocity at each grid point. The flow was then followed for 13,000 s, at which time the instabilities had extended outward to the pre-explosion radius of the progenitor star (3×10^{12} cm).

3. Results

3.1. Effects of Numerical Resolution

Calculations were performed using a wide variety of grid sizes to check the dependence of the results on numerical resolution. Figure 1 shows the density contours obtained using four low-resolution cylindrical grids ranging in size from 50×50 to 200×200. On the 50×50 grid, the flow is very poorly resolved. There is only a hint of the instability in the flow. A single finger forms along the vertical axis, but at this resolution, it is hard to tell if it is a real physical feature or a numerical artifact associated with the symmetry axis. It is probably a combination of the two. There is also an indication of a second finger forming near the

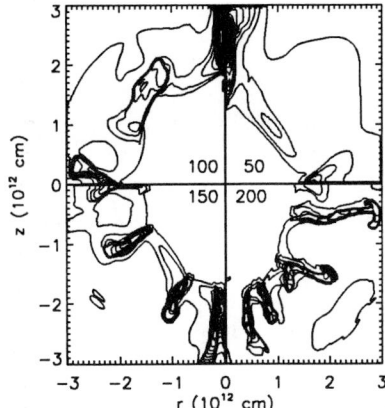

FIGURE 1. Equally spaced contours of constant density for calculations performed using four different sized cylindrical grids. Each quadrant is labeled by the number of grid points per spatial dimension.

horizontal axis. At 100 × 100, The Rayleigh-Taylor finger just above the horizontal axis is much more clearly defined, but the feature along the vertical axis remains questionable at this resolution. At a resolution of 150 × 150, there are four distinct fingers extending radially outward. This plot begins to illustrate the expected behavior of a Rayleigh-Taylor instability with narrow dense fingers separated by wider diffuse bubbles, although the grid is still too coarse to reproduce the detailed structure of these features. Finally, at 200 × 200, the calculation is beginning to attain sufficient accuracy to resolve the "mushroom cap" structure which is characteristic of Rayleigh-Taylor instabilities. However, only one of the fingers is sufficiently resolved to see this effect.

Figure 2 shows the results obtained on four higher resolution cylindrical grids. At 250 × 250, most of the fingers are starting to develop mushroom caps. In fact, the four calculations appear quite similar, indicating that a "converged" solution may have been achieved. All four show the same basic behavior. The width of the unstable layer changes very little with resolution, and the wavelength of the instability remains approximately the same, even though the effective wavelength of the initial perturbation differed by a factor of two between the 250 × 250 and 500 × 500 calculations. This indicates that there is a preferred mode for the instability which is obtained for grid sizes greater than about 400 × 400. The primary difference between the calculations in this set is that the amount of small scale structure increases significantly with resolution. This, however, has no effect on the major results of the calculation. The fingers are composed primarily of the elements from carbon to silicon. These elements are accelerated to higher velocities than predicted by spherically symmetric calculations and are also mixed outward in radius. The bubbles between the fingers are composed primarily of hydrogen which is being mixed inward to low velocities.

It is possible to test for convergence by plotting, for example, the width of the mixing layer as a function of the number of grid points in each dimension, as shown in Figure 3. Note that the horizontal axis covers a range of a factor of 10 in linear resolution. In other words, the total number of zones varies by a factor of 100. This is a much wider range of grid sizes than is normally used for convergence studies. For grid sizes less than about 150 × 150, the flow is so poorly resolved that it is difficult to even measure the width of the mixing layer, so that portion of the graph should not be taken too seriously. However, for higher resolutions there is a clear trend for the mixing width to increase with resolution

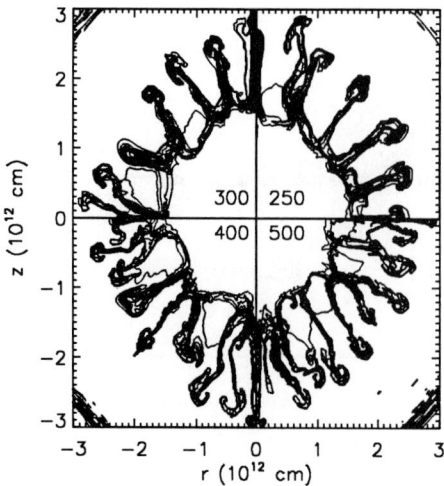

FIGURE 2. Equally spaced contours of constant density for calculations performed using four different sized cylindrical grids. Each quadrant is labeled by the number of grid points per spatial dimension.

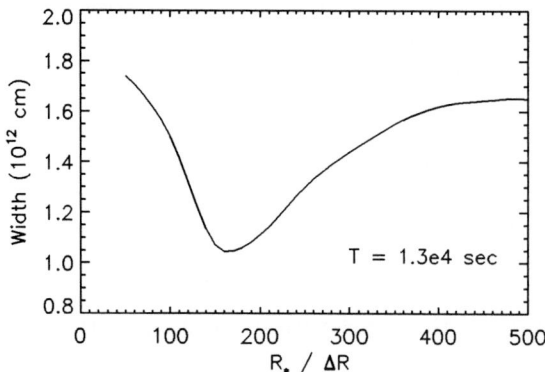

FIGURE 3. Width of the mixing layer as a function of the number of grid points per spatial dimension. The plot covers a factor of 10 in linear resolution and shows the approach to a converged solution.

until reaching a constant value at a grid size of about 400 × 400. This graph shows clear indication of a converged solution.

However, the appearance of a converged solution in Figure 3 is misleading. Although it is clear that the calculations are converging, they are actually converging to an incorrect solution. A simple stability analysis of the initial model [22] shows that there are actually two separate regions of the star which are unstable to Rayleigh-Taylor modes. The outer fingers are composed primarily of helium. The inner set of fingers contain primarily the elements from carbon to silicon. These two regions are fairly close to each other in radius, and unless very high resolution is used, the two regions merge, so that only a single instability is seen. The second instability is clearly visible on a grid of 1000 × 1000, as shown in Figure 4. One might imagine that the presence of a second instability would increase the amount of mixing in the calculation, but in fact, the opposite is true. Because

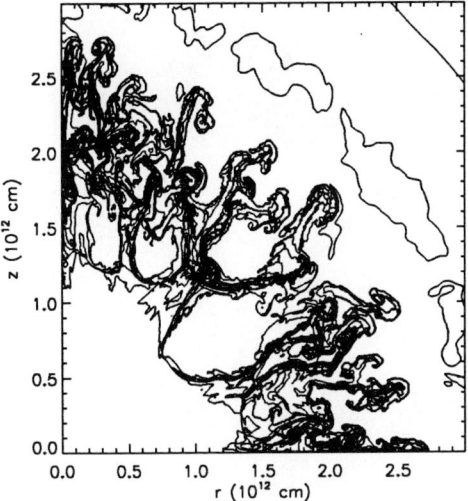

FIGURE 4. Density contours obtained using a 1000 × 1000 cylindrical grid. At this resolution, two separate instabilities appear in the flow.

of the interaction of the two instabilities, the fingers are bent at an angle and no longer extend out radially. This has the effect of decreasing the mixing width.

It is possible to obtain even greater resolution than show in Figure 4 by using a moving grid. For these calculations, the initial grid contained only the inner 20% of the star in radius. As the explosion progressed, the grid was allowed to expand homologously, so that the outer boundary of the grid remained slightly beyond the unstable layer. By doing this, the number of zones in the unstable region is maximized during the critical time when the instability is forming. This is very important, since at early times, short wavelength modes dominate the solution. As the instability evolves, it undergoes an inverse cascade to longer scale modes. Because of this, the resolution required to resolve all the important features decreases with time.

The highest resolution calculation performed to date used a spherical grid with 800 zones in radius and 400 zones in angle. Note that, since the initial grid contained only the inner 20% of the star in radius, 4000 radial zones would have been required in order to obtain the same resolution with a fixed Eulerian grid covering the entire star. Thus, this calculation has approximately 4 times the radial resolution of the calculation shown in Figure 3. A gray scale plot of the density is shown in Figure 5. The results are qualitatively different from the lower resolution calculations described above and those obtained by other groups. As a result of the more accurate representation of the interaction of the two instabilities and the ability to resolve shorter wavelength modes at early times, the unstable region appears more like a turbulent layer and less like a collection of independent radial fingers. This has the effect of further decreasing the width of the unstable layer and reducing the amount of mixing which occurs. This new type of behavior is not a result of changing from cylindrical to spherical coordinates. The calculation shown in Figure 4, which was performed using a fixed cylindrical grid, already shows the beginning of the trend. In addition, calculations performed on a moving cylindrical grid appear very similar to the spherical results. The fact that it is possible to obtain the same result using two different coordinate systems gives one additional confidence about the accuracy of the results.

The width of the mixing layer is plotted as a function of resolution for the entire set of calculations in Figure 6. The portion of the graph shown in Figure 3 is the leftmost

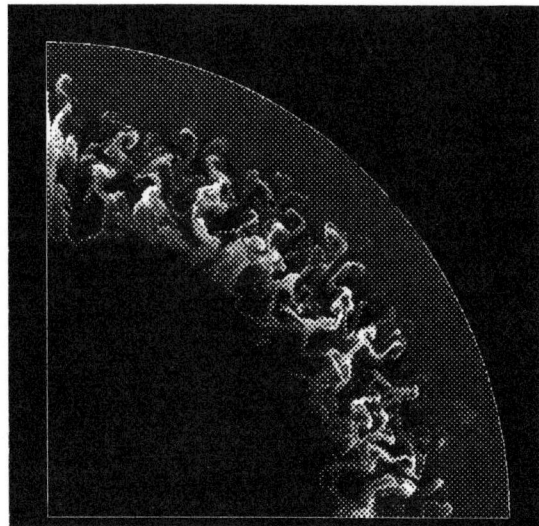

FIGURE 5. Gray scale plot of density obtained using an expanding 800 × 400 spherical grid. At this resolution, the unstable region appears more like a turbulent layer.

eighth of this plot. The entire plot covers a factor of 80 in linear resolution and shows convergence to a second solution with a considerably smaller mixing width than indicated in Figure 3. Plots of other relevant quantities, such as the maximum velocity for each nuclear species, show similar behavior. It should be emphasized that the fact that two separate "converged" solutions are obtained is not at all a pathological case. Many problems in astrophysics contain multiple length and time scale phenomena, and a separate "converged" solution corresponding to each of these scales can be expected. In some ways, the case illustrated here is unusually simple, since there is no physics in the problem except adiabatic hydrodynamics. For more complex situations in which other physical processes become important, a vast multitude of "converged" solutions can be expected. It only makes sense to talk about a truly converged solution to a given problem after all of the important length and time scales have been resolved. In general, it is not adequate to investigate convergence by simply doubling the number of zones or particles per spatial dimension and comparing the results. It is first necessary to understand the physics of the problem being solved sufficiently to know that all of the important scales in the problem are being resolved.

For calculations such as those described above, more and more fine detail structure will appear as the grid resolution is increased. In this sense, it will never be possible to obtain a completely converged solution to this problem, unless a physical viscosity is added to limit the development of short wavelength modes. However, the calculation plotted in Figure 5 has resolved all of the important length scales in the problem. In particular, the two unstable regions are well separated and both instabilities are clearly visible. It is unlikely that higher resolution of the small scale structure will produce any significant change in the behavior of observable quantities such as the mixing width. In this sense, it is reasonable to describe the result shown in Figure 5 as a converged solution. However, in nonlinear problems such as this one, it is not always easy to determine ahead of time what all of the important length scales will be. If there is another unforeseen physical phenomena in the problem which has not been resolved by these calculations, the conclusion that Figure 5 represents a converged solution may not be valid.

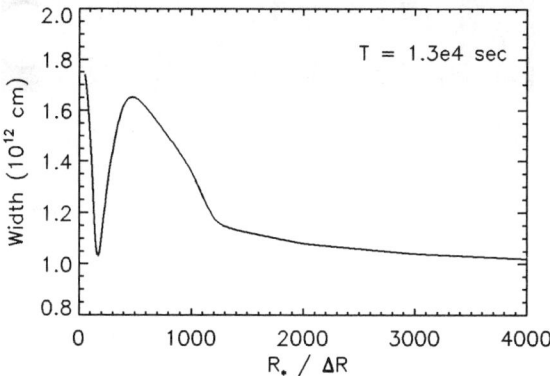

FIGURE 6. Width of the mixing layer as a function of the number of grid points per spatial dimension. The plot covers a factor of 80 in linear resolution and shows a second converged solution at very high resolution.

3.2. Comparison With Observations

The results described above can explain many, but not all, of the observations obtained from SN1987A. In particular, the mixing of hydrogen inward to velocities as low as a few hundred km s^{-1} is reproduced by the calculations. In order for the X-rays and γ-rays to be detected as early as they were, significant clumping in the ejecta must occur. This is clearly present in the figures described above. Although calculations have not yet been performed to determine exactly when the hard radiation would be observed in these models, the column density shows a factor of two variation with angle. Thus, it seems reasonable to expect that the hard radiation would be detectable about a factor of two earlier in time that expected from a spherical model. This is exactly what was observed.

The primary failure of this model is in explaining how the radioactive iron group elements formed during the explosion are accelerated to velocities greater than 3500 km s^{-1}. The heavy elements in the above calculations reach velocities no higher than about 1800 km s^{-1} and there appears to be no easy way to create the required mixing as a result of these instabilities. Of course higher velocities could be obtained if a larger explosion energy had been used. However, if an explosion energy large enough to explain the high velocities had been used, the calculated light curve would no longer agree with the observations. It was originally thought that the radioactive decay of the nickel would create sufficient energy to accelerate the material [1], but simulations showed that this effect was negligible. It was also thought that the instability might be more violent in three dimensions. However, full three-dimensional simulations showed that the mixing width was almost exactly the same as predicted by the two-dimensional calculations [21]. It is now generally believed that mixing of the heavy elements must occur very soon after the explosion. The possible importance of instabilities which occur during the first few seconds will be discussed in the next section.

4. Early-time Instabilities

As described above, instabilities which occur during the first few seconds of the explosion may be required to produce sufficient outward mixing of the radioactive iron group elements. However, what may be even more important is that these instabilities may be an integral part of the explosion mechanism. Most realistic one-dimensional models of the explosion mechanism either fail to explode or produce only weak explosions [3,24]. A number of possible instabilities have been investigated. These include overturn of the entire core as a result of unstable lepton gradients [4,12], convective instabilities caused by neutrino heating of the matter outside the neutrinosphere [18], and a doubly diffusive "neutron finger" instability [24]. However, none of the models invoking these instabilities has yet resulted in strong explosions.

This section briefly describes preliminary calculations of another instability resulting from unstable entropy and composition gradients which develop within the protoneutron star behind the stalled shock [5,6]. Similar calculations have been carried out by Janka & Müller [19]. The code was modified for this calculation to include a nuclear equation of state [7,20], gravitational forces, and a simplified neutrino transport scheme for both electron-type and muon-type neutrinos. The transport was calculated only in the radial direction. Transport in the angular direction was assumed to be negligible for these preliminary experiments. The initial model was taken from a one-dimensional collapse calculation of a 20 $M_\odot$ star [7]. The effects of general relativity were not included in either the one-dimensional or two-dimensional models. A spherical grid was used for the calculation with 300 radial zones and 200 zones in the angular direction. The region from $r = 25$ km ($M = 0.7$ $M_\odot$) to $r = 175$ km ($M = 1.1$ $M_\odot$) contained 200 uniformly space zones in radius. An additional 100 non-uniformly space radial zones covered the region out to $r = 2000$ km. The inner boundary condition was treated as a solid wall. A constant neutrino luminosity through this boundary was assumed.

The result of the calculation was the development of a violent convective instability within the core which advected the neutrinos across the neutrinosphere to regions of lower optical depth. The neutrino luminosity from the core was increased by about a factor of two. This model, does not produce an explosion in one-dimensional calculations. Furthermore, in two-dimensional adiabatic calculations, no explosion occurs. However, when both the effects of the convective instability and neutrino transport are included, an explosion does result on a time scale of 20-30 msec, although the strength of the explosion has not yet been determined.

It is too early to say if this instability is the missing ingredient in the Type II supernova mechanism. Considerable improvements in the physics, particularly in the treatment of the inner boundary and the neutrino transport algorithm, are required before the results can be considered reliable. In addition, the calculations will have to be extended to much later times to determine the strength of the explosion. It is unlikely that this particular instability will produce the additional mixing of the heavy elements required to explain the observations of SN1987A, since it occurs well inside the region where these elements are formed. However, this instability may provide a perturbation in the flow which will seed later instabilities. Before a reliable model can be constructed to explain the mixing, a calculation will have to be performed which is capable of self-consistently treating the interaction of all of the instabilities which occur during the explosion, from time scales of tens of milliseconds to a few hours.

REFERENCES

[1] Arnett, D., Fryxell, B. & and Müller, E. 1989, ApJ (Letters), 341, L63.
[2] Arnett, D., Fryxell, B. & and Müller, E. 1991, in: Proceedings of The ESO/EIPC Workshop: SN1987A and Other Supernovae, eds. I.J. Danziger, L.B. Lucy, & W. Hillebrandt, ESO, Garching, in press.
[3] Bruenn, S. W. 1992, First Symposium on Nuclear Physics in the Universe, held in Oak Ridge, TN, in press.
[4] Bruenn, S. W., Buchler, J. R., & Livio, M. 1979, ApJ (Letters), 234, L183.
[5] Burrows, A. & Fryxell, B. 1992, Science, 258, 430.
[6] Burrows, A. & Fryxell, B. 1993, ApJ (Letters), in press.

[7] Burrows, A. & Lattimer, J. 1985, ApJ (Letters), 299, L15.
[8] Chevalier, R. 1976, ApJ, 207, 872.
[9] Chevalier, R. & Klein, R. 1978, ApJ, 219, 994.
[10] Colella, P. & Glaz, H. M. 1985, J. Comput. Phys., 59, 264.
[11] Colella, P. & Woodward, P. R. 1984, J. Comput. Phys., 54, 174.
[12] Epstein, R. I. 1979, MNRAS, 188, 305.
[13] Falk, S. W. & Arnett, W. D. 1973, ApJ (Letters), 180, L65.
[14] Fryxell, B., Müller, E. & Arnett, D. 1989, Max-Planck-Institut für Astrophysik Preprint 449.
[15] Fryxell, B., Müller, E. & Arnett, D. 1991, ApJ, 367, 619.
[16] Hachisu, I., Matsuda, T., Nomoto, K., & Shigeyama, T. 1991, ApJ, 368, L27.
[17] Herant, M. & Benz, W. 1991, ApJ (Letters), 370, L81.
[18] Herant, M. Benz, W. & Colgate, S. A. 1992, ApJ, 395, 642.
[19] Janka, H.-T. & Müller, E. 1993 in: Frontiers of Neutrino Astrophysics, Universal Academy Press Inc., Tokyo, Japan, in press.
[20] Lattimer, J. M. 1981, Ann. Rev. Nucl. Part. Sci., 246, 995.
[21] Müller, Fryxell, B. & Arnett, D. 1991, Chemical and Dynamical Evolution of Galaxies, eds. F. Ferrini, F. Matteucci & J. Franco, ETS Editrice, Pisa, 394.
[22] Müller, E., Fryxell, B. & Arnett, D. 1991, Astron Astrophys, 251, 505.
[23] Müller, E., Fryxell, B. & Arnett, D. 1989, Proceedings of The ESO/EIPC Workshop: SN1987A and Other Supernovae, eds. I.J. Danziger, L.B. Lucy, & W. Hillebrandt, ESO, Garching, in press.
[24] Wilson, J. R. & Mayle, R. 1992, Phys Repts, in press.
[25] Woodward, P. R. & Colella, P. 1984, J. Comput. Phys., 54, 115.
[26] Yamada, Y., Nakamura, T., and Oohara, K. 1990, Prog Theor Phys, 84, 436.

Superbubbles and Supernova Remnants in Magnetized Interstellar Media

By Kohji Tomisaka

Faculty of Education, Niigata University, 8050 Ikarashi-2, Niigata 950-21, Japan.
(tomisaka@ed.niigata-u.ac.jp.)

The importance of the interstellar magnetic field is studied in relation to the evolution of superbubbles and supernova remnants (SNRs), using two- and three-dimensional numerical magnetohydrodynamical (MHD) simulations. The superbubble is a large multi-supernova remnant driven by sequential supernova explosions in an OB association. The evolution is affected by density stratification in the galactic disk: after the size is 2–3 times greater than the density scale-height, the superbubble expands preferentially in the z-direction until it finally breaks out of the disk (blow-out). The magnetic field component parallel to the galactic disk, however, acts to prevent perpendicular expansion in the z-direction. Thus, density stratification and the magnetic field have exactly opposite effects on the evolution of the superbubble. We present the results of 3D MHD simulations which include both effects. We find that: (1) for models where the bubble would blow out from the disk in the absence of a magnetic field, (e.g., mechanical luminosity $\sim 10^{37}$erg/s, scale height ~ 200 pc, density ~ 0.3cm^{-3}), the addition of a 5 μG field can confine the bubble to $|z| \lesssim 200$ pc for $\simeq 20$ Myr. (2) if the field strength in the halo decreases as $B \propto \rho^{1/2}$ (and hence has a larger scale height than the density distribution), then the superbubble eventually blows out, even if the magnetic field in the mid-plane is as strong as $B = 5\mu$G. The dynamics of the SNR are also affected by the magnetic field, especially in the later stages when the ram pressure in the post-shock region becomes comparable to the magnetic pressure. By 2D MHD simulations we show that the expansion of the hot cavity is blocked in the direction perpendicular to the field but facilitated parallel to the field. In the very late stages of the SNR, the volume of the hot cavity decreases due to magnetic tension. The effects of magnetic fields are significant, and models of interstellar matter driven by supernova explosions must account for their presence.

1. Introduction — What is a Superbubble?

A superbubble is a complex of an OB association, the surrounding, X-ray emitting, hot gas, and a corresponding HI hole/shell. Three examples are known in our Galaxy: Cygnus [4], Orion-Eridanus [6,25], and the Gum nebula [24]. In external galaxies, superbubbles are observed as HI shells and holes: the LMC [9,21], M31 [2], M33 [7], and M101 [14]. Superbubble sizes range from 100 pc to 1 kpc and they cannot be explained by a single supernova explosion (the size of an ordinary supernova remnant (SNR) is $\lesssim 50$pc). The energy required to produce a superbubble is 5×10^{51}erg s^{-1} – 10^{54}erg s^{-1} [32]. Two models for superbubbles have been proposed: (1) a large SNR driven by sequential supernova explosions in an OB association, and (2) a complex formed by the collision of a high-velocity cloud and the galactic disk [30]. We confine ourselves to the first model and discuss its evolution. For review papers in this field, see [28,32,36,38].

2. Evolution of a Non-Magnetic Superbubble

As a first step, we examine the expansion of a superbubble in the simple approximation of spherical symmetry [3,40]. A self-similar solution by [42] gives the expansion law for a shock wave driven by a steady energy release as

$$R_s = 271 \text{pc} (L_{\text{SN}}/3 \times 10^{37} \text{erg s}^{-1})^{1/5} (\rho_0/6 \times 10^{-25} \text{g cm}^{-3})^{-1/5} (t/10 \text{Myr})^{3/5}, \quad (2.1)$$

where L_{SN} is the mean mechanical luminosity ejected by supernovae, i.e., $L_{\text{SN}} = E_0/\Delta\tau$ using the total energy ejected by a SN (E_0) and the mean time interval between two supernovae ($\Delta\tau$).

Because the typical size of a superbubble is much greater than the perpendicular density scale-height of the galactic disk, superbubble evolution should be affected by density stratification within the disk. This makes the problem 2-dimensional. Pioneering work on this problem was done by Chevalier & Gardner [5], in which the evolution of a SNR in the galactic halo was studied, albeit with a small 20×39 mesh. This class of problem came under serious attack in the mid-1980's using large, 2D hydro-codes [13,18,32,33,38,39] and semi-analytic approximations [1,15,16,17].

The evolution is characterized by two phases. At first the expansion is spherical because the density gradient has little effect before the bubble expands. Then, if L_{SN} is sufficiently large, the bubble enters a new phase where the expansion accelerates in the direction of the density gradient. Although dependent upon the density distribution, after the size is larger than $2 - 3H$, the shock accelerates upward. In an exponential atmosphere $\rho = \rho_0 \exp(-|z|/H)$, the expansion of the shock front accelerates after passing $z \simeq 2.9H$ (Mac Low et al. 1989). Finally, the superbubble punches out of the disk and the hot gas contained in the bubble flows into the galactic halo.

What about less-energetic bubbles? If the thermal pressure of the ambient ISM is dominant before the blowout, i.e. $\rho_0 v_s^2 \lesssim p_0$, the bubble will remain confined within the disk. Using equation (1), the ram pressure becomes comparable to the thermal pressure at an age $t_P \simeq 79 L_{38}^{1/2} n_0^{3/4} p_{-12}^{-5/4}$ Myr, where $L_{38} = L_{\text{SN}}/10^{38} \text{erg s}^{-1}$ and $p_{-12} = p_0/10^{-12} \text{dyn cm}^{-2}$. Assuming the condition for blow-out to be $R_s(t_P) > \alpha H$ [16], the critical luminosity becomes

$$L_{\text{crit}} = 0.59 \times 10^{37} \text{erg s}^{-1} (\alpha^2/5) H_2^2 n_0^{-1/2} p_{-12}^{3/2}, \quad (2.2)$$

with $H_2 = H/100\text{pc}$. The value of $\alpha \simeq \sqrt{5}$ is an estimate from the numerical results of [17]. Thus, if L_{SN} is much smaller than L_{crit}, the bubble remains confined to the galactic disk.

2.1. Numerical Simulation

Here we examine the evolution of a non-magnetized superbubble for the case of a realistic model atmosphere. The distribution proposed in [8] is adopted as

$$n(z) = \frac{n_0}{0.566} \left[0.395 \exp\left[-\frac{1}{2}\left(\frac{z}{90\text{pc}}\right)^2\right] + 0.107 \exp\left[-\frac{1}{2}\left(\frac{z}{225\text{pc}}\right)^2\right] + 0.064 \exp\left[-\frac{|z|}{403\text{pc}}\right] \right], \quad (2.3)$$

which yields an effective scale-height of $H_{\text{eff}} \equiv \int_0^\infty \rho dz/\rho_0 \simeq 180\text{pc}$. Model parameters are chosen to be $n_0 = 0.3 \text{cm}^{-3}$ and $T_0 = 8000\text{K}$, giving $L_{\text{crit}} \sim 2 \times 10^{37} \text{erg s}^{-1}$ ($H_{\text{eff}}/180\text{pc}$)^2 $(n_0/0.3\text{cm}^{-3})^{-1/2}$ $(p_0/7 \times 10^{-13} \text{dyn cm}^{-2})^{3/2}$. In Figure 1 we show the density distribution and velocity field. Numerically, we employed van Leer's [41] monotonic scheme. In this model $L_{\text{SN}} = 3 \times 10^{37} \text{erg s}^{-1} > L_{\text{crit}}$. As expected, the expansion of the bubble accelerates in the z-direction after $z_s \gtrsim 200$ pc. In contrast, radial expansion near the mid-plane stops at an age of $t \simeq 20$Myr.

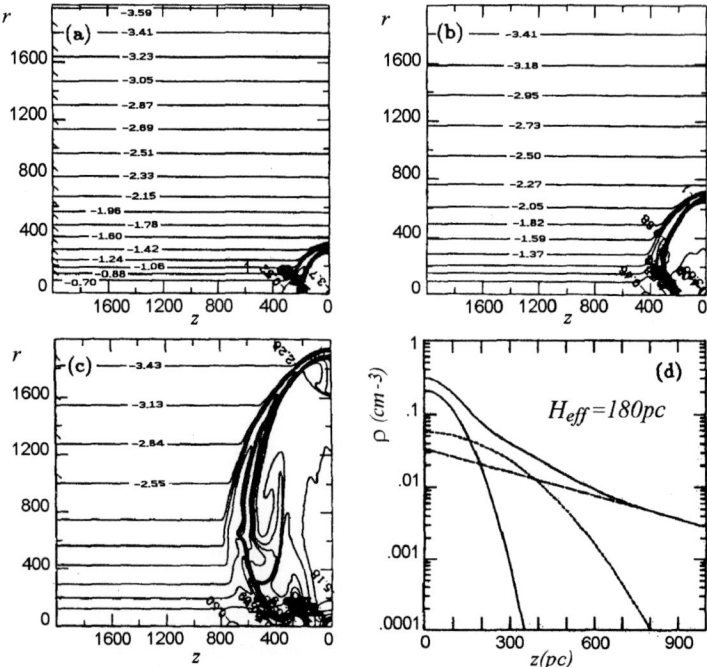

FIGURE 1. The evolution of a non-magnetized superbubble in the Dickey & Lockman [8] atmosphere (d). Three snap-shots at 10Myr(a), 20Myr(b), and 38Myr(c) are shown.

3. Magnetic Superbubbles

The magnetic pressure in the ISM is $p_{\mathrm{mag}} = B_0^2/8\pi \simeq 10^{-12} \mathrm{dyn\ cm}^{-2}(B_0/5\mu\mathrm{G})^2$. Including both magnetic and thermal pressures, the critical luminosity becomes $L_{\mathrm{crit}} \simeq 3 \times 10^{37} \mathrm{erg\ s}^{-1}\ (H_{\mathrm{eff}}/180\mathrm{pc})^2\ (n_0/0.3\mathrm{cm}^{-3})^{-1/2}\ (p_0/1.7 \times 10^{-12} \mathrm{dyn\ cm}^{-2})^{3/2}$. If $L_{\mathrm{SN}} \simeq 3 \times 10^{37} \mathrm{erg\ s}^{-1}$ the superbubble may be confined to the disk by magnetic pressure.

3.1. Two-dimensional Evolution

Although the problem is inherently 3-dimensional, the structure of the magnetic superbubble can be studied more precisely using 2-dimensional symmetry. This problem has been studied with a semi-analytic method [12] and with 2D MHD codes [20,26,37]. The characteristic time scale, t_P, was defined in the previous section. If the magnetic field controls the evolution, the time-scale is found by replacing the thermal pressure with the magnetic pressure, e.g., $t_P \simeq 79 L_{38}^{1/2} n_0^{3/4} (B_0/5\mu\mathrm{G})^{-5/2}$ Myr. Early on ($t \ll t_P$) the flow is spherically symmetric, while in the later phases ($t \gtrsim t_P$) the flow is collimated parallel to the magnetic field.

Figure 2 shows the evolution of a superbubble with cylindrical symmetry. We employed van Leer's Monotonic Scheme [22,41] and the Constrained Transport method [10] for the magnetic induction. See [37] for a detailed description of this numerical scheme.

Characteristic features of magnetic superbubbles are: (1) The outermost shock is a fast-mode MHD shock wave, since the magnetic fields break upon reaching the front. (2) The shell propagating parallel to the magnetic field is thin ($\Delta z \simeq 12$ pc; see Fig. 2a), while that propagating perpendicular is thick ($\Delta r \simeq 104$ pc). This is because the magnetic pressure prevents the shell from contracting in the radial direction. In contrast, the magnetic field has no effect near the polar cap (the z-axis). (3) The expansion of the hot cavity, occupied by tenuous, field-free gas, is anisotropic. The MHD shock front expansion is also

Table 1. Model Parameters

Model	n_0 (cm^{-3})	B_0 (μG)	T_0 (K)	$L_{\rm SN}$ (erg s^{-1})	$B_x(z)$
A	0.3	3	8000	3×10^{38}	const
B	0.3	5	8000	3×10^{37}	const
C	0.3	5	8000	3×10^{37}	$\propto \rho^{1/2}$

anisotropic, with expansion perpendicular to the magnetic field (the r-direction) faster than expansion parallel to it (the z-direction); that is, $\dot{Z}_s < \dot{R}_s$. Expansion of the contact surface between the shell and the cavity is greatly discouraged in the r-direction because the magnetic tension prevents the cavity from expanding ($\dot{R}_c < \dot{Z}_c$).

From numerical results it is shown that the volume of the hot cavity is reduced to $\sim 60\%$ of the volume in the nonmagnetic case. Furthermore, the expansion of R_c is strongly decelerated and, in the $B_0 = 5\mu$G model, it stops completely before 30 Myr. This implies that expansion perpendicular to the galactic disk is blocked by the magnetic field, since the magnetic field is parallel to the galactic disk on a global scale.

3.2. Three-dimensional Evolution

Taking both the magnetic field and the density gradient into account, the problem becomes three-dimensional. Three-dimensional MHD simulation of superbubbles was done in [35], although using the adiabatic assumption. Here we report new results of a full simulation including radiative cooling. We assume a magnetic field configuration of $\mathbf{B}= (B_x,0,0)$. Two models of initial magnetic configuration are studied: (a) B_x constant for all z, and (b) constant Alfvén speed, i.e., $B_x(z) = B_0(\rho(z)/\rho_0)^{1/2}$. The density distribution is again chosen to be similar to that of Dickey & Lockman [8]. The parameters are summarized in Table 1. We adopt the same numerical scheme as in the previous section, but using 3-dimensional Cartesian coordinates. The space is divided into cubic grids 10 pc on a side; we initially use $51 \times 51 \times 81$ zones, then change to $81 \times 81 \times 131$ zones.

Figure 3a shows the structure of model A, which corresponds to the parameters for blow-out (Table 1). Nine Myr after the SN explosions, the shock front reaches the upper boundary at $z = 800$ pc. Because $\ddot{Z}_c > 0$ indicates positive acceleration, the effect of the density stratification seems more important than that of the magnetic field. This is expected, because $L_{\rm SN} \gg L_{\rm crit}$.

In contrast, model B corresponds to the parameters for $L_{\rm SN} \simeq L_{\rm crit}$. Figure 3b shows a snapshot at the age of 18 Myr. Due to the low mechanical luminosity and strong magnetic field, the hot cavity of the superbubble is confined below $z \lesssim 300$pc, although the MHD wave front has propagated away from the OB association located at the origin (this can be seen more clearly in the video animation). In the halo region the Alfvén speed is high and the MHD wave propagates much faster than in the disk region. In this stage the hot cavity expands to $Z_c \simeq 350$pc, $X_c \simeq 300$pc, and $Y_c \simeq 150$pc. In the direction perpendicular to the magnetic field the expansion has stopped ($\dot{Y}_c = 0$) while in other directions the bubble still expands as $\dot{Z}_c \simeq \dot{X}_c \simeq 15$km s^{-1}. The expansion speed in the z-direction, however, does not indicate acceleration. We conclude that if $L_{\rm SN} \simeq L_{\rm crit}$ the superbubble can be confined by the magnetic tension. Is this still true if the magnetic field strength decreases with $|z|$?

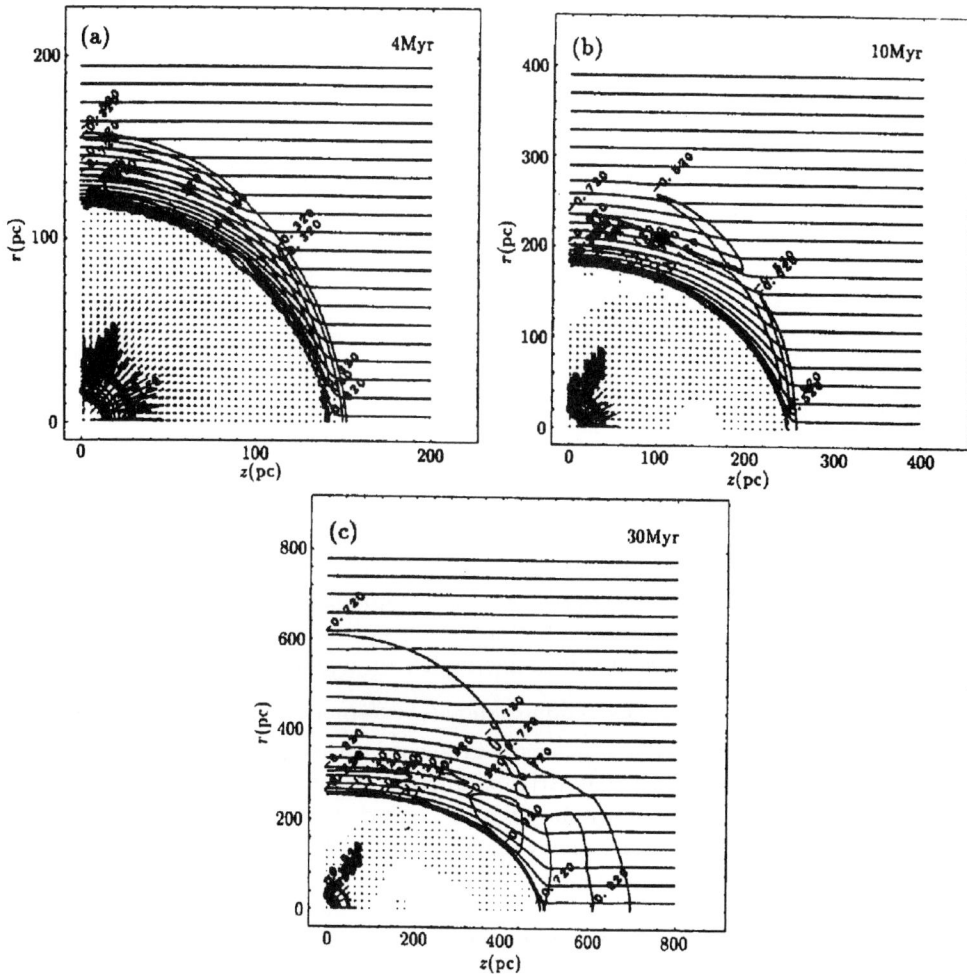

FIGURE 2. Structure of the superbubble with $n_0 = 0.3 \text{cm}^{-3}$, $T_0 = 8000\text{K}$, $B_0 = 3\mu\text{G}$, and $L_{\text{SN}} = 3 \times 10^{37} \text{erg s}^{-1}$. We employ a square grid of size $\Delta z = \Delta r = 2\text{pc}$. Initially the magnetic field runs parallel to the z-axis (horizontal). Three snapshots are shown: $t = 4\text{Myr}$ (a), 10Myr (b), and 30Myr (c). Pressure contour lines and magnetic field lines are plotted.

In model C, the magnetic field strength decreases with height as $B_x(z) \propto \rho^{1/2}$. Figure 4 shows the structure of the corresponding superbubble. This is a cross-sectional view of the yz-plane ($x = 0$) at an age of $t = 21.5$ Myr. It shows clearly that while the bubble expands toward the halo, the shell near the galactic mid-plane is contracting due to the magnetic tension. Comparing Figures 3b and 4, the effect of the distribution of the magnetic field is clearly seen: When the magnetic field strength decreases at large z, the superbubble easily breaks through the disk. The thermal pressure drops after the rapid expansion of the bubble. Thus, the magnetic tension assumes a relatively important role in this phase. In model C the shell contracts in the y-direction (perpendicular to the magnetic field) near the mid-plane of the galactic disk. In contrast, if the magnetic field has uniform strength at all z, the superbubble is confined to the galactic disk. In the latter case there is no rapid decrease in pressure so the shell does not contract as long as the OB association undergoes active SN explosions.

In conclusion, superbubbles can be confined by magnetic fields if (1) $L_{\text{SN}} \lesssim L_{\text{crit}}$, where

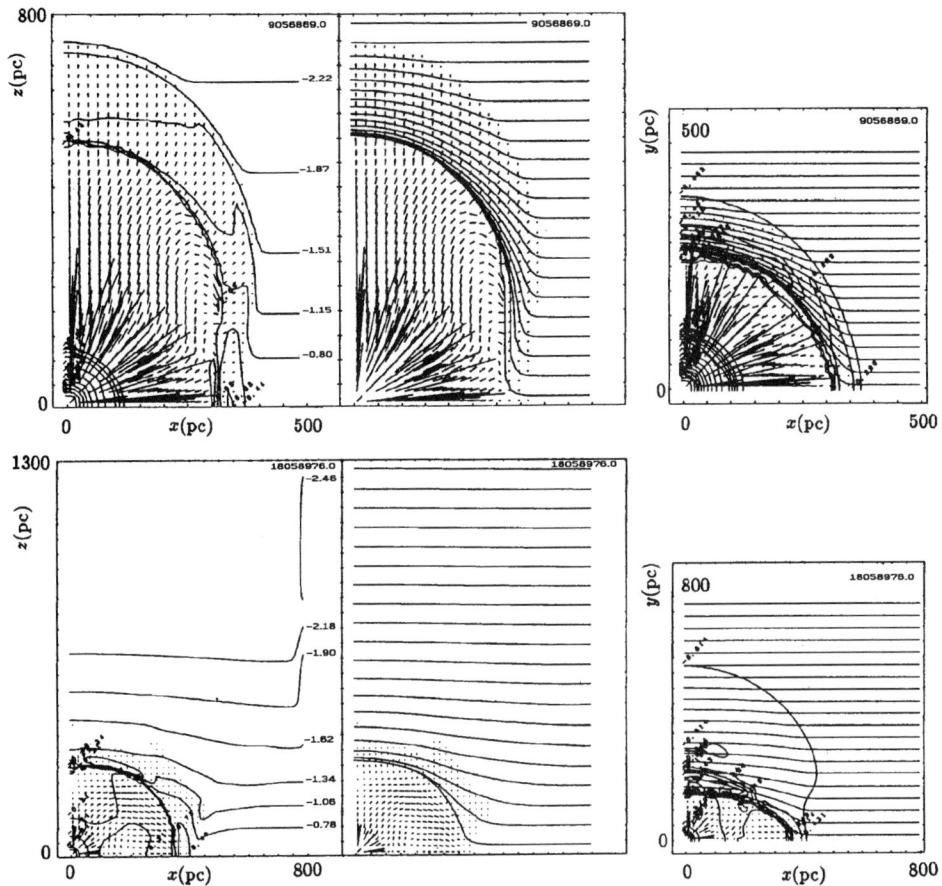

FIGURE 3. The structure of a 3-dimensional superbubble in the magnetized ISM. (a) in model A we assume $L_{\rm SN} = 3 \times 10^{38}$ erg s^{-1} and $B_0 = 3\mu$G. This corresponds to the case of $L_{\rm SN} > L_{\rm crit}$. Pressure contour lines and magnetic field lines at $t = 9.06$ Myr are plotted for the cross-section of $y = 0$ (left) and for $z = 0$ (right). (b) in model B we assume $L_{\rm SN} = 3 \times 10^{37}$ erg s^{-1} and $B_0 = 5\mu$G. This corresponds to the case of $L_{\rm SN} \simeq L_{\rm crit}$. Shown is a snapshot at $t = 18$Myr.

$L_{\rm crit}$ includes the effect of the magnetic fields, and (2) the strength of the magnetic field does not decrease rapidly with height.

3.3. Other Effects

Other factors that should be taken into account are galactic rotation and the shear of this motion [34,23,27]. The characteristic time-scales of rotation and the shear are estimated respectively as $\tau_{\rm R} \sim 1/\Omega_0 \sim 40{\rm Myr}(\Omega_0/26{\rm km\ s^{-1}\ kpc^{-1}})^{-1}$ and $\tau_{\rm S} \sim (ld\Omega/dR)^{-1} \sim 320{\rm Myr}\ (l/1{\rm kpc})^{-1}(\Omega_0/26{\rm km\ s^{-1}\ kpc^{-1}})^{-1}(R_0/8.5{\rm kpc})$, where Ω_0, R_0, and l are the angular speed of galactic rotation, distance from the galactic center, and the typical size of a superbubble, respectively. Since the active SN-explosion phase continues for ~ 50 Myr for an OB association [19], in the late phase ($t \gtrsim \tau_{\rm R}$) the effect of the Coriolis force appears as a deformation force on the shell. The $\alpha\omega$-dynamo mechanism driven by a superbubble was studied recently in [11].

In this section, the evolution of the superbubble in a plane-stratified, magnetized medium was examined.

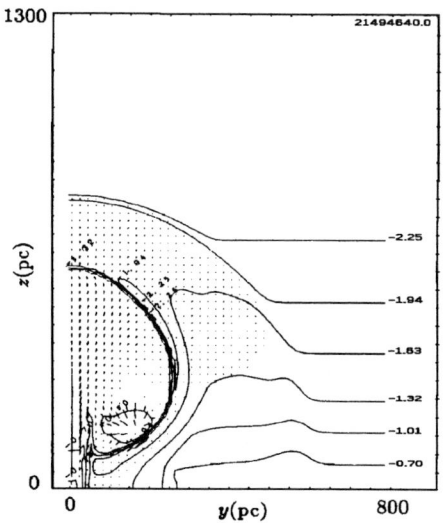

FIGURE 4. The cross-sectional view of the superbubble in the yz-plane ($x = 0$). The magnetic field runs perpendicular to this plane. In model C, the magnetic field strength decreases with height.

4. SNRs in the Magnetized Interstellar Medium

In a late phase of SNRs, the magnetic pressure becomes relatively important compared to the ram pressure exerted on the shock front. Thus, the SNR dynamics should be affected by magnetic fields in a way analogous to superbubbles. Using a one-dimensional, spherically symmetric MHD code, Slavin & Cox [29] showed that (1) the volume occupied by the (MHD-)shocked material is larger than that of a non-magnetic SNR; (2) in contrast, the volume of the hot tenuous cavity is much smaller than that of a non-magnetic SNR. This volume first increases as the SNR expands, but then begins to decrease because the cavity contracts as a result of the magnetic field tension. Their MHD simulation, however, is one-dimensional.

Here, we present results of a two-dimensional simulation of a magnetized SNR. The parameters are the same as those in [29]: $E_0 = 5 \times 10^{50}$erg is the SN explosion energy, and $n_0 = 0.3 \text{cm}^{-3}$, $T_0 = 8000$K, and $B_z = B_0 = 5\mu$G are used for the initially uniform ISM. The code is identical with that of the previous section, in which the meshes are evenly spaced as $\Delta z = \Delta r = 10$pc for 200×200 meshes.

For $t \lesssim 0.2$Myr, the SNR is spherical and the magnetic field seems to have no importance in the dynamics. The outer shock front and hot cavity expansion are nearly spherically symmetric. After $t \sim 0.2$Myr, the magnetic field becomes relatively important and the expansion of the SNR begins to depart from spherical symmetry. At $t \simeq 1$Myr the expansion of the hot cavity stops in the r-direction. For $t \gtrsim 1$Myr, the cavity contracts radially — although it expands in the z-direction. Two Myr after the explosion, the hot gas in the SNR is confined to a cigar-shaped region $(R, Z) \simeq (40, 100)$pc, while the hot gas in the non-magnetic SNR occupies a spherical region of $r \lesssim 80$pc.

Figure 5 shows how the hot cavity volume evolves for various magnetic field strengths. It can be seen that a magnetic field of $B_0 \simeq 5\mu$G decreases the volume of the hot cavity to $\lesssim 20$ % of that of a non-magnetic SNR. We conclude that in the very late stages of a SNR the volume of the hot cavity decreases because of magnetic tension.

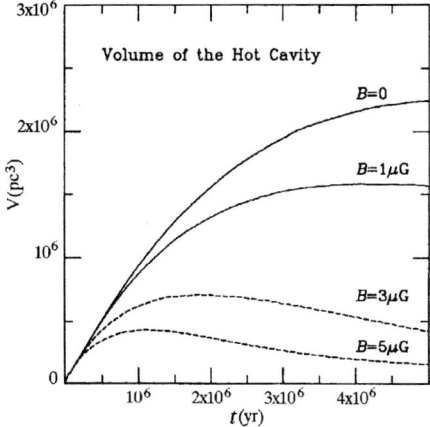

FIGURE 5. The time evolution of the volume occupied by the hot cavity of a SNR. Each curve corresponds to a different magnetic field strength.

This work was supported in part by a Grant-in-Aid from the Ministry of Education, Science, & Culture (05640306).

REFERENCES

[1] Bisnovatyi-Kogan G. S., Blinnikov, S. I., & Silich S. A. 1989, 154, 229.
[2] Brink, E. & Bajaja, E. 1986, AA, 169, 14.
[3] Bruhweile, F. C., Gull, T. R., Kafatos. M., & Sofia, S. 1980, ApJL, 238, L27
[4] Cash, W., Charles, P., Bowyer, S., Walter, F., Garmire, G., & Riegler, G. 1980, ApJL, 238, L71.
[5] Chevalier, R. A. & Gardner, J. 1974, ApJ, 192, 457.
[6] Cowie, L. L., Songaila, A., & York, D. G. 1979, ApJ, 230, 469.
[7] Deul, E. R., & den Hartog, R. H. 1990, AA, 229, 362.
[8] Dickey, J. M., & Lockman, F. J. 1990, ARAA, 28, 215.
[9] Dopita, M. A., Mathewson, D. S., & Ford, V. L. 1985, ApJ, 297, 599.
[10] Evans, C. R., & Hawley, J. F. 1988, ApJ, 332, 659.
[11] Ferrière, K. M. 1992, ApJ, 391, 188.
[12] Ferrière, K. M., Mac Low, M. M., & Zweibel, E. G. 1991, ApJ, 375, 239.
[13] Igumentshchev, I. V., Shustov, B. M., & Tutukov, A. V. 1990, AA, 234, 396.
[14] Kamphuis, J., Sancisi, R., & van der Hulst, T. 1991, AA, 244, L29.
[15] Koo, B.-C., & McKee, C. F. 1990, ApJ, 354, 513.
[16] Koo, B.-C., & McKee, C. F. 1992, ApJ, 388, 93.
[17] Mac Low, M. M., & McCray, R. 1988, ApJ, 324, 776.
[18] Mac Low, M. M., McCray, R., & Norman, M. L. 1989, ApJ, 337, 141.
[19] McCray, R. & Kafatos, M. 1987, ApJ, 317, 190.
[20] Mineshige, S., Shibata, K. & Shapiro, P. R. 1993, ApJ, 409, 663.
[21] Meaburn, J. 1980, MNRAS, 192, 365.
[22] Norman, M. L., & Winkler, K.-H. A. 1986, in *Astrophysical Radiation Hydrodynamics*, ed. K.-H. A. Winkler, & M. L. Norman (Reidel, Dordrecht), p.187.
[23] Palouš, J., Franco, J., & Tenorio-Tagle, G. 1990, AA, 227, 175.
[24] Reynolds, R. J. 1976, ApJ, 206, 679.
[25] Reynolds, R. J., & Ogden, P. M. 1979, ApJ, 229, 942.
[26] Shapiro, P. R., Mineshige, S., & Shibata, K. 1991, in *The Interstellar Disk-Halo Connection in Galaxies*, ed. H. Bloemen (Kluwer, Dordrecht), p.417.
[27] Silich, S. A. 1993, ApSpSc, 195, 317.
[28] Spitzer, L. Jr. 1990, ARAA, 28, 71.
[29] Slavin, J. D., & Cox, D. P. 1992, ApJ, 392, 131.
[30] Tenorio-Tagle 1981, AA, 94, 338.

[31] Tenorio-Tagle & Bohdenheimer, P. 1988, ARAA, 26, 145.
[32] Tenorio-Tagle, G., Bohdenheimer, P., & Różyczka, M. 1987, AAp, 182, 120.
[33] Tenorio-Tagle, G. Różyczka, M., & Bohdenheimer, P. 1990, AAp, 237, 207.
[34] Tenorio-Tagle, G., & Palouš 1987, AAp, 186, 287.
[35] Tomisaka, K. 1990, ApJL, 361, L5.
[36] Tomisaka, K. 1991, in *The Interstellar Disk-Halo Connection in Galaxies*, ed. H. Bloemen (Kluwer, Dordrecht), p.407.
[37] Tomisaka, K. 1992, PASJ, 44, 177.
[38] Tomisaka, K., & Ikeuchi, S. 1986, PASJ, 38, 697.
[39] Tomisaka, K., & Ikeuchi, S. 1988, ApJ, 330, 695.
[40] Tomisaka, K., Ikeuchi, S., & Habe, A. 1981, ApSS, 78, 273.
[41] van Leer, B. 1977, JCompPhys, 23, 276.
[42] Weaver, R., McCray, R., Castor, J., Shapiro, P., & Moore, R. 1977, ApJ, 218, 377 (errata 220, 742).

3-D Models for Supershells in a Cloudy Medium

By S. A. Silich[1], J. Franco[2], J. Palouš[3], G. Tenorio-Tagle[4]

[1] Main Astronomical Observatory of the Ukrainian Academy of Sciences, 252127, Kiev-127, Goloseevo, Ukraina.

[2] Instituto de Astronomía UNAM, Apartado Postal 70-264, 04510 México D.F., México.

[3] Astronomical Institute, Academy of Sciences of the Czech Republic, Boční II 1401, 141 31 Prague 4, Czech Republic.

[4] Instituto de Astrofísica de Canarias, 38200-La Laguna, Tenerife, Spain.

We present 3-D numerical simulations of expanding bubbles in a plane-stratified cloudy medium, with cloud filling factors in the galactic plane ranging between 0.1 and 0.3. The effects due to galactic differential rotation, gravity, cloud evaporation, and interstellar drag have been included. The radiative cooling of the gas inside cavity is also considered, and the X-ray emission is estimated.

1. Introduction

Gaseous galaxies, including the Milky Way, display a variety of large expanding bubbles or supershells [9,19]. They have a complex morphology and their sizes often exceed the gas disk thickness. The hydrodynamical evolution of such large-scale structures has been studied with 2-D and 3-D numerical codes by a number of authors [2,10,13,19-22], see reviews in [8,20].

The evolution of supernova remnants depends on the structure of the ambient interstellar medium [5,7,23,24] and, with one exception [11], previous calculations of superbubble evolution have not considered the inhomogeneities of the interstellar medium (*i.e.*, the effects of the density contrast between the cloud component and the intercloud medium). Here we discuss the evolution of multi-supernova remnants in a cloudy medium and describe some of the observable properties of evaporation-dominated superbubbles.

2. The ISM Model and Equations

Two interstellar components have been included in our model: the intercloud medium and spherical HI clouds, with a mean internal density $n = 10$ cm^{-3}, a radius $R_{cl} = 5$ pc, and mass $M_{cl} = 165$ M$_\odot$. We do not consider the molecular cloud component, which is concentrated in a relatively small number of giant molecular clouds, because the problem in a three-component fluid is far beyond the scope of the present study. Following Dickey & Lockman [6], the large-scale HI distribution is represented by three plane parallel components with different scale-heights. The vertical structure of the cloud number density is assumed to follow the main HI disk component, and the cloud filling factor $f(z)$ (*i.e.*, the fraction of the disk volume occupied by clouds), decreases with increasing z as

$$f(z) = \frac{\rho_{cl}(z)}{\rho_{tot}(z)}, \tag{2.1}$$

where ρ_{cl} is the average mass density in clouds, and ρ_{tot} is the sum of the three HI components.

We assume a continuous energy input rate from the OB-association and, in units of 10^{38} erg s^{-1}, here we describe the results of models with $L_{38} = 0.315$. The hydrodynamical scheme is based on the thin layer approximation, as described by [2,3,13,15,18,21], and we used a Cartesian galactocentric coordinate system with the origin at the Galactic centre. The 3-D shell is approximated by a number (typically N=1522) of Lagrangian elements. The motion of each element is described by the equations of mass and momentum conservation, and the equation of total energy balance. Following Cowie et al. [5] and Ostriker & McKee [15] we introduce the frictional drag force

$$\mathbf{F}_{dr} = \frac{\mu}{\lambda_{cl}} (\mathbf{u} - \mathbf{v}_{cl}) |(\mathbf{u} - \mathbf{v}_{cl}) \cdot \mathbf{n}|, \tag{2.2}$$

and the thermal evaporative mass loss rate (in g s^{-1})

$$\dot{M}_{ev} = \begin{cases} 3.75 \times 10^4 T^{5/2} a_{cl} D \sigma^{-5/8}, & \sigma > 1 \\ 2.75 \times 10^4 T^{5/2} a_{cl} D, & 0.03 < \sigma < 1 \end{cases} \tag{2.3}$$

where μ and $\mathbf{u}$ are the mass and velocity of the Lagrangian element, $\mathbf{n}$ is the unit vector normal to the shell surface, $\mathbf{v}_{cl}$ is the cloud velocity, a_{cl} is the radius of the cloud in parsecs, λ_{cl} is the cloud mean free path, T is the temperature of the hot gas inside the cavity, D is a parameter ($0 \leq D \leq 1$) which depends on the cloud properties (geometry, B-field strength, etc.), and σ is the saturation parameter defined by the formula

$$\sigma = \frac{T}{1.54 \times 10^7 \mathrm{K}} \frac{1}{n_{int} \, a_{cl} \, D}, \tag{2.4}$$

where n_{int} is the intercloud gas number density.

The expansion of the shell is followed in a differentially rotating galactic disk, and we use the nearly flat rotation curve derived for our Galaxy by Wouterloot et al. [26].

3. The Assumptions and Results

The model results for a galactocentric distance of $R = 8.5$ kpc, with two positions of the OB-association relative to the Galactic plane ($z_0 = 0$ and 100 pc) and two values for the filling factor, $f = 0.1$ and 0.3, are shown in Figures 1-3. The following main assumptions have been made:

1) The cloud size distribution was not taken into account and spherical clouds of only one given size, and mass, were considered.

2) The equilibrium cooling function of Raymond et al. [17] was used. As a rough correction to include the radial metallicity gradient in the Galaxy, it was multiplied by $\xi = Z/Z_\odot$ (where Z is the metallicity) in the temperature range $3 \times 10^5 - 3 \times 10^7$ K.

3) Neither the distortion of clouds during their interaction with the expanding shell, nor the decrease of the cloud radius due to mass evaporation were considered. This last process was partially included by assuming that the number of clouds inside the cavity, $N_{cl,tot}$, is reduced in direct proportion to the mass evaporation rate dM_{ev}/dt

$$\frac{dN_{cl,tot}}{dt} = \sum_{i=1}^{N} n_{cl}(z)(\mathbf{u} - \mathbf{v}_{cl}) \mathbf{n}_i d\Sigma_i - \frac{1}{m_{cl}} \frac{dM_{ev}}{dt}, \tag{3.5}$$

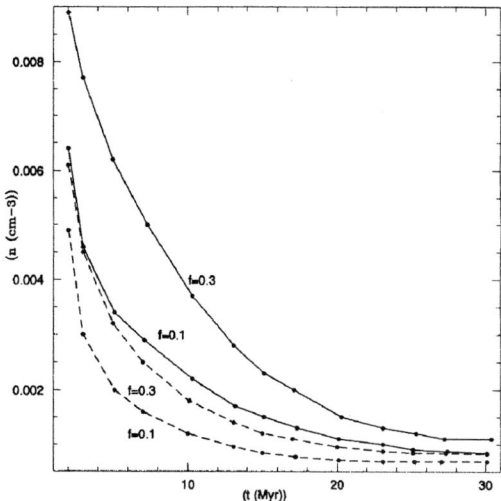

FIGURE 1. The gas number density inside an expanding superbubble in a clody medium as a function of time. The simulations were done with $R = 8.5$ kpc and $L_{38} = 0.315$, for OB-associations located at the galactic symmetry plane (solid lines) and at 100 pc above the galactic plane (dashed lines).

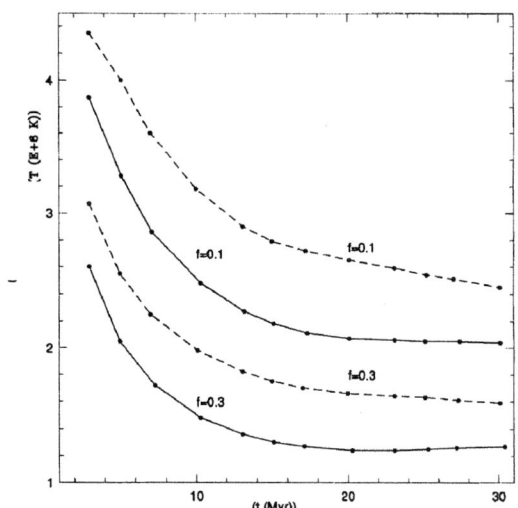

FIGURE 2. The gas temperature inside the superbubble as a function of time for the cases described in Fig. 1.

where n_{cl} is the number of clouds per unit volume, $d\Sigma_i$ is the surface area of a Lagrangian element, and m_{cl} is the cloud mass.

4) We use the approximation to the X-ray emission between 0.2-4 KeV described by Chu & Mac Low [4]. This approximation, which is probably good to 25% for equilibrium cooling with solar abundances [25], is independent of temperature over the range 2.5×10^6 to 10^8 K.

5) We do not consider possible variations of the thermodynamical variables inside the

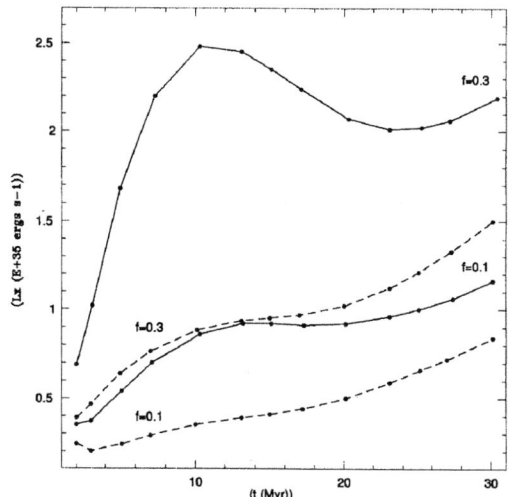

FIGURE 3. X-ray luminosity of the superbubble as a function of time for the cases described in Fig. 1.

remnant and only calculate the average gas number density n, the average temperature T, and average pressure P. Of course, without knowledge of the distribution of these quantities, our calculations of the superbubble luminosity are only rough estimates. Thus, the results are a first approach in the study of the general properties of the dynamics and radiative cooling of bubbles evolving in a cloudy medium.

4. Conclusions

The calculations shown here were performed for a galactocentric radius of $R = 8.5$ kpc, and two locations from midplane (at $z = 0$ and 100 pc). Two different values of the cloud filling factor in the plane of Galaxy were explored: $f(z = 0) = 0.1$ and 0.3. The results of these calculations, shown in Figures 1-3, indicate that the cloud component does not strongly influence the final sizes of the expanding supershells. Other properties, however, are affected by the presence of the clouds in the cavity. The average gas number density inside the evaporation-dominated bubble decreases to about 10^{-3} cm^{-3} and remains nearly constant after 15-20 Myrs. The temperature inside the cavity declines to $1 - 3 \times 10^6$ K at about 10 Myr, and maintains this value for up to an age of 30 Myr. The total X-ray luminosity from the superbubble depends on the structure of the interstellar medium and on the position of the energy source above the galactic plane. For instance, for a cloud filling factor of $f_0 = 0.3$ and an energy input rate $L_{38} = 0.315$, the X-ray luminosity reaches a maximum, of about 2.5×10^{35} erg s^{-1}, at about 12 Myr. In the case of $f_0 = 0.1$, there is a first peak of about 10^{35} erg s^{-1} and later grows slowly above that value due to the increase of the total superbubble volume. A more detailed discussion, and additional different cases, will be presented elsewhere [19].

JF was partially supported by DGAPA-UNAM through the grant IN103991, and a CRAY R&D grant. JF and GTT were partially supported by the EEC grant CI1*-CT91-0935. JP was funded as part of the scientific exchange program between CONACyT-México and the Academy of Sciences of the Czech Republic. The work of SAS was partially supported by a grant from CONACyT-México. JP, SAS, and GTT thank the hospitality of the Instituto de Astronomía, UNAM. Some of the simulations for this work were performed with the CRAY/YMP of the Supercomputing Center–UNAM.

REFERENCES

[1] Bisnovatyi-Kogan, G. S. & Blinnikov, S. I. 1982, Astron. Zh, 59, 876
[2] Bisnovatyi-Kogan, G. S., Blinnikov, S. I. & Silich, S. A. 1989, Ap. Space. Sci, 154, 229
[3] Bisnovatyi-Kogan, G. S. & Silich, S. A. 1991, Astron. Zh, 68, 749
[4] Chu, Y.-H. & Mac Low, M.-M. 1990, ApJ, 365, 510
[5] Cowie, L. L., McKee, C. F. & Ostriker, J. P. 1981, ApJ, 247, 908
[6] Dickey, J. & Lockman, F. 1990, ARAA, 28, 215
[7] Franco, J., Tenorio-Tagle, G., Bodenheimer, P. & Różyczka, M. N. 1991, PASP, 103, 803
[8] Franco, J., Bodenheimer, P., Tenorio-Tagle, G. & Różyczka, M. 1992, *Evolution of Interstellar Matter and Dynamics of Galaxies*, ed J. Palouš et al., (Cambridge: Cambridge U. Press), 83
[9] Heiles, C. 1979, ApJ, 229, 533
[10] Igumentshchev, I. V., Shustov, B. M. & Tutukov, A. V. 1990, A&A, 234, 396
[11] Kunze, R., Yorke, H.W. & Spurzem, R. 1992, *Evolution of Interstellar Matter and Dynamics of Galaxies*, ed J. Palouš et al., (Cambridge: Cambridge U. Press), 77
[12] Kuijken, K. & Gilmore, G. 1989, MNRAS, 239, 571
[13] Mac Low M.-M. & McCray, R. 1988, ApJ, 324, 776
[14] Mac Low M.-M., McCray, R. & Norman, N. L. 1989, ApJ, 337, 141
[15] Ostriker, J. P. & McKee, C. F. 1988, Rev. Mod. Phys., 60, 1
[16] Palouš, J., Franco, J. & Tenorio-Tagle, G. 1990, A&A, 227, 175
[17] Raymond, J. C., Cox, D. P. & Smith, B.W. 1976, ApJ, 204, 290
[18] Silich, S. 1992, Ap. Space. Sci, 195, 317
[19] Silich, S., Franco, J., Palouš, J. & Tenorio-Tagle, G. 1994, in preparation
[20] Tenorio-Tagle, G. & Bodenheimer, P. 1988, ARAA, 26, 145
[21] Tenorio-Tagle, G. & Palouš, J. 1987, A&A, 186, 287
[22] Tomisaka, K. & Ikeuchi, S. 1986, PASJ, 38, 697
[23] Tomisaka, K. 1992, PASJ, 44, 177
[24] Tomisaka, K. 1993, this volume
[25] White, R. & Long, K. 1991, Apj, 373, 543
[26] Wouterloot, J. G. A., Brand, J., Burton, W. B. & Kwee, K. K. 1990, A&A, 230, 21

Producing the Soft X-ray Background with Multiple Supernova Remnants

By Randall K. Smith

Department of Physics, University of Wisconsin-Madison, Madison, WI 53706, USA.

The genesis of the soft X-ray background could be a supernova remnant heating the gas to 10^6 K. However, keeping the gas at this temperature would be difficult without some external pressure confining the gas. The suggestion is that this could be supplied by a wall of matter swept up and left over from a previous explosion, so-called "inertial confinement." Initial exploratory models, in 1 dimension, show that this idea has merit. Two single supernovae, exploding from the same position but separated by 3 million years, leads to a hot bubble that seems to radiate in much the same fashion as observations suggest for the local soft X-rays.

1. Introduction

The origin of the soft X-ray photons that fill our sky has been a puzzle since they were first mapped [7]. These X-rays, between 70 eV and 278 eV, could not be completely extragalactic since their intensity did not drop to zero at the plane, when galactic hydrogren would certainly absorb any extragalactic emission. However, the intensity of the X-rays below 0.25 keV does show an anticorrelation with HI column density, but the variation does not correspond to that expected by a constant source being absorbed by intervening matter [1]. Experimental work then concentrated on trying to find shadows in the soft X-rays from molecular clouds, and thus get some idea of the distance to the X-ray source(s). Despite strenuous efforts, it was only after ROSAT was launched that the first shadow was found, in the Draco region [2,11].

Due to the difficulty of finding shadows in the soft X-ray background, it is thought that the X-rays seen originate from within a few hundred parsecs of the sun. Various physical methods of creating the observed spectrum were examined in Williamson et al. [12], with the conclusion that the only physical picture that could not be ruled out by observations was nearby hot gas emitting X-rays via spectral lines (the dominant process for gas with cosmic abundances at temperatures between $0.5 - 4 \times 10^6$ K). Recent observations of S VIII and Si VIII emission lines by the Diffuse X-ray Spectrometer, made almost 20 years after Williamson's paper, have provided the first direct evidence for this picture [5,10].

Currently, the only spectral information available is from the Wisconsin survey, and is in the form of count rates from different spectral bands. These bands, the Be, B, and C, are defined by the K-edge absorption of their respective elements, with the Be band responding mainly to the softest X-rays and C band the hardest. With this set of observations, it is possible to define a set of criteria for a successful model of the situation. Any explanation of the source of the soft X-rays must have the X-rays produced by spectral lines, with the correct luminosity and the correct spectrum. McCammon's measurements for the soft X-rays vary across the sky of course, but on average there was a C band count rate of between 100 - 200 counts sec^{-1}, and a B/C ratio of about 0.3.

2. Modelling the Response of the Local ISM to a SNR

Perhaps the simplest model is that we are inside a supernova remnant (SNR), and the X-rays are coming from the shock-heated gas around and in the shell of the remnant. This idea was first explored by Cox & Anderson [3]. More recently Edgar & Cox [4] (hereafter EC) have searched the parameter space of single explosions in a one-dimensional hydrocode with non-equilibrium cooling and a magnetic field, and could not find any solutions meeting the above criteria. Their closest approaches used super-energetic supernovae and magnetic fields much larger than the currently accepted range of 3 to 5 μG. Less energetic supernovae were not luminous enough, and without a strong magnetic field, the remnant put too much of its energy into expansion, and thus the luminosity was too low. Despite these extreme conditions, in almost all cases, the B band to C band ratio was too high, averaging between 0.5 and 0.8; thus, the X-rays were too soft even with these conditions. Edgar & Cox then suggested that using multiple supernovae, instead of a single oversize explosion, might solve some of these problems.

Why would multiple supernovae of lesser total energy lead to brighter, harder X-rays? The answer is confinement. Since the above models were done in one dimension, the magnetic field provided only a pressure term to the hydrodynamics. The need for a high magnetic field in EC, then, suggests that the real problem with the models was the lack of sufficient exterior pressure confining the shock-heated material behind the blast wave of the supernova. We also know that SNR tend to collect material into a dense shell that slowly dissipates into the ISM. This shell's inertia, then, might provide the pressure necessary to confine an explosion, and thus increase the luminosity of X-rays coming from a second supernova inside the remnant of the first. A quick calculation, assuming one supernova per 50 years per galaxy, a galactic radius of 15 kpc and thickness of 200 pc, says that there should be 1 supernova per 100 kpc radius sphere every 3 million years. So it is certainly justifiable to consider what might be the effects on the soft X-ray spectrum of a supernova inside an active supernova remnant.

2.1. Numerical Method

The hydrocode used for these calculations was a simple one dimensional Lagrangian code written by the author. It included the following terms:
- Magnetic field used only as a pressure term, with $p_B \propto n^2$
- Thermal conduction and saturation of same.
- Radiative cooling using the Kahn approximation.

In addition, the B and C band count rates were approximated by running a Raymond & Smith model [8,9] of gas with solar abundances in equilibrium at various temperatures and densities and then calculating the response due to the resulting spectra in each band.

The code was tested on a selection of analytically soluble problems and on the same model explosion used in EC. Despite performing well on the analytic problems (a Sedov blast wave and a constant-gravity exponential atmosphere), the match to the EC model problem was only approximate, good only to a factor of 2 overall in the B and C bands. I attribute this to using equilibrium approximations for the cooling and spectral response, as opposed to the detailed non-equilibrium calculations in EC. However, examining the detailed evolution of the model and comparing with EC shows that the qualitative match is quite good. In light of this, the code was deemed adequate for an examination of the multiple-supernova case, since it does have the important quality of being fast.

The speed of the code is so important because there is a large space of possible parameters and it is difficult to constrain this space beforehand with either physical intuition or observational results. The complete code is expected to contain both non-equilibrium cooling and dust heating and sputtering. In this case, both the elemental abundances and

TABLE 1. Model Parameters

Model	n_0	$<B>$	$E_{51}(1)$	$E_{51}(2)$	Delay
1	0.2	11.0	5.0	0.0	0.0
2	0.1	5.0	0.5	0.5	0.1
3	0.1	5.0	0.5	0.5	1.0
4	0.1	5.0	0.5	0.5	3.0

TABLE 2. Results of Model Calculations

Model	B/C ratio					C Band Count Rate				
Time (Myr)	0.5	1.5	2.5	3.5	4.5	0.5	1.5	2.5	3.5	4.5
EC	.634	.524	.480	.482	.511	194	134	279	562	951
1	.477	.736	.783	.710	.580	1100	191	175	345	965
2	.383	.912	.890	.482		288	19.2	17.8	10.5	
3	.978	.366	.627	.687	.702	47.5	80.7	35.3	31.1	33.5
4	.978	1.96	2.20	.464	.704	47.5	2.18	1.41	119	35.8

the ionization state of each element must be tracked, which effectively means the code has one extra dimensionto calculate, with the resulting slowdown. Even a one dimensional code can consume a significant amount of computer time in this circumstance!

2.2. Results of Single and Multiple Explosions

The following parameters were used to define each model:
- Ambient density (nuclei cm^{-3})
- Ambient magnetic field (μG)
- Explosion energy of first supernova (10^{51} erg)
- Explosion energy of second supernova (10^{51} erg)
- Delay between first and second explosions (10^6 years)

4 models were calculated, with the parameters shown in Table 1. The results (along with the results from EC) are shown in Table 2, showing the luminosity in the C band and B/C band ratio at 1 million year increments starting at 0.5 million years.

Model 1, the EC standard model, was done primarily to test the code. As previously stated, despite the good match in pressure, density, and temperature evolutes, the B and C count rates, while qualitatively the same, differ by as much as a factor of 5 (though usually less). Note that the C band count rate of 1100 counts s^{-1} at 0.5 Myr is a result of using the Kahn approximation to the cooling curve, which cools slightly slower than non-equilibrium cooling. The EC results also start out with C band count rates over 1000 counts s^{-1} but cools before 0.5 Myr. In any event, the overall discrepency is not unexpected, since the count rates are a sensitive function of temperature and density. While equilibrium methods cannot find the "right" model of the local ISM, they can direct us to it.

The other 3 models examined simple cases of 2 supernovae exploding. They used realistic supernova energies of 0.5×10^{51} erg, differing from each other only in the delay between explosions. Each was run for 4.5 million years, with the exception of model 2, which was halted at 3.5 million years since it had almost completely cooled by that time.

3. Conclusions

The results give hope, but show that many of the same problems occur as in EC. Model 2, where the second explosion comes only 0.1 Myr after the first, is almost indistinguishable from a single explosion of equal total energy. It rapidly fades, rapidly getting far less luminous than observations. Model 3, with a 1 Myr separation but otherwise identical to model 2, manages to keep a harder overall spectrum and stay brighter. The only difference is that in model 2, after 0.1 Myr, the first supernova has not begun to form a dense, cold shell. Part of the material has cooled, but not nearly enough to hold the second supernova. When the second supernova hits the partially formed shell, it merges with it and reheats it. Soon, the merged shocks cool and form a shell that rapidly dims. In model 3, however, the first supernova has formed a shell about 20 pc thick with an average density of nearly double ambient and much more dense than the material inside the shell. The second supernova then bounces, and in the end puts more energy into radiation than model 2.

Model 4, two normal supernovae separated by 3 Myr, shows the most promise. It also happens to be the average separation time between supernova events in a given area. As can be seen from the table, the first supernova never reaches even 100 C band counts s^{-1}, and rapidly fades to invisibility. However, the second supernova exploding in its wake does reach above 100 C band counts s^{-1} 0.5 Myr after the explosion. Then it rapidly settles down to look very similar to model 3's late time evolution. While the luminosity and hardness of the spectrum are not quite what is needed, they are within striking distance. A fuller examination of the parameter space may well find a model with only a few reasonably-sized supernovae that could explain the soft X-ray observations.

The author would like to thank Don Cox and Dick Edgar for many helpful discussions and useful advice. This work was supported by a National Science Foundation Graduate Fellowship and by NASA Grant NAGW-2532 to the University of Wisconsin–Madison.

REFERENCES

[1] Burrows, D. N., McCammon, D., Sanders, W. T., & Kraushaar, W. L. 1984, ApJ, 287, 208.
[2] Burrows, D. N. & Mendenhall, J. A. 1991, Nature, 351, 629.
[3] Cox, D. P., & Anderson, P. 1982, ApJ, 253, 268.
[4] Edgar, R. J., & Cox, D. P. 1993, ApJ, in press.
[5] Edgar, R. J. 1993, private communication.
[6] Kahn, F. D. 1975, Proc. 15th Int. CR Conf. (Munich), 11, 3566.
[7] McCammon D., Burrows, D. N., Sanders, W. T., & Kraushaar, W. L. 1983, ApJ, 269, 107.
[8] Raymond, J. C., & Smith, B. W. 1977, ApJSupp, 35, 419.
[9] — 1987, Informally distributed update to 1977 paper.
[10] Sanders, W. T., Edgar, R. J., Juda, M., Kraushaar, W. L., McCammon, D., Snowden, S., Zhang, J., Skinner, M. A., Jahoda, K., Kelley, R., Smale, A. Stahle, C., & Szymkowiak, A. 1993, *EUV, X-Ray, and Gamma-Ray Instrumentation for Astronomy III*, ed. Oswald H. W. Siegmund, Proc. SPIE 2006, in press.
[11] Snowden, S. L., Mebold, U., Hirth, W., Herbstmeier, U, & Schmitt, J. H. M. M. 1991, Science, 252, 1529.
[12] Williamson F. O., Sanders, W. T., Kraushaar, W. L., McCammon, D., Borken, R., & Bunner, A. N. 1974, ApJ, 193, L133.

Formation of Molecular Clouds in Expanding Supershells: 3-D Models

By S. Ya. Mashchenko, S. A. Silich

Main Astronomical Observatory of the Ukrainian Academy of Sciences, 252127, Kiev-127, Goloseevo, Ukraina.

The evolution of the shells produced by a collection of supernova explosions from OB-associations is investigated with a 2.5-dimensional hydrodynamical scheme. For the case of our Galaxy, the conditions for molecular cloud formation are fulfilled only at galactocentric distances smaller than 15 kpc and when the explosion centers are located at heights below 100 pc from the galactic plane. The masses of the molecular clouds can reach 10^6 $M_\odot$ and the half-width of the slab containing the newly born clouds is smaller than 100 pc.

---

1. Introduction

During the last two decades measurements of the soft X-ray background and O VI absorption lines, along with the discovery of huge expanding neutral and ionized hydrogen supershells, have drastically modified our picture of the interstellar medium. It now seems obvious that powerful energy sources may control the large-scale structure and energy balance of the galactic gas.

It has been proposed that huge expanding supershells sweep the ambient gas and provide a feedback control to the star formation activity in gaseous galaxies [4-8,10-14]. The possibility of molecular cloud formation in these expanding supershells is one of the main aspects of this picture. Here we examine this possibility with several numerical simulations of superbubble evolution in our Galaxy.

2. Numerical Technics

We have used a 2.5-dimensional numerical hydrodynamical scheme, based on the thin layer approximation, that was discribed in Bisnovatyi-Kogan & Silich [2] and Silich [12]. We work with the Cartesian galactocentric coordinate system connected with the Galactic centre. At the beginning of the calculations, the spherical shell is split into N=1522 Lagrangian elements distributed parallel and perpendicular to the Galactic plane.

The motion of every Lagrangian element is described by the equations of mass and momentum conservation, and by the equation of the total energy balance. The motion of the whole remnant is described by the set of 7N+1 equations and this set of equations is solved numerically with the Adams scheme.

3. The Model

The calculations were performed at three different galactocentric radii: R =5 kpc, 8.5 kpc and 15 kpc. We used the rotation curve given by Wouterloot et al. [15]

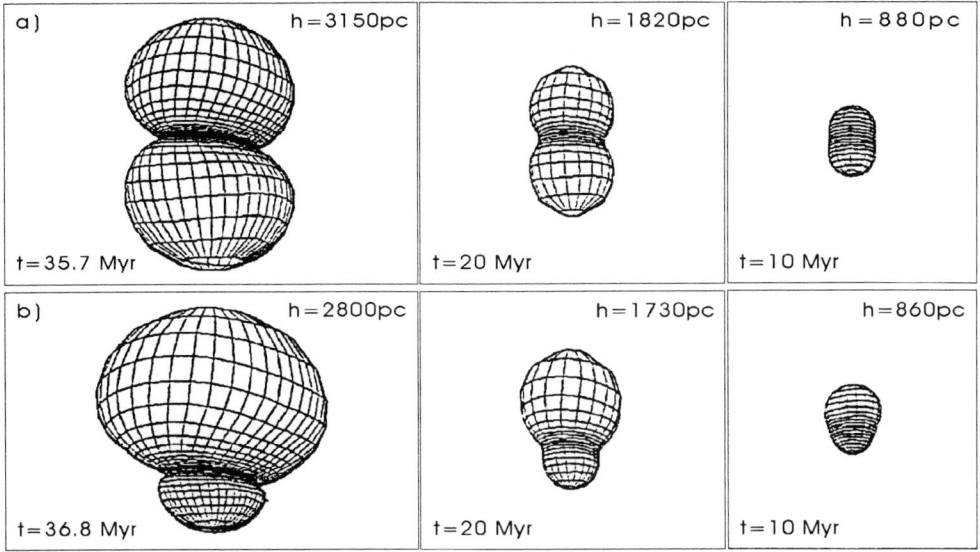

FIGURE 1. Superbubble evolution with $L_{38} = 1.05$ at R=8.5 kpc for two different positions of the OB-association above the Galactic plane. a) $z_{OB} = 0$; b) $z_{OB} = 50$ pc (h is the distance between the top and the bottom of the shell).

$$V_R(R) = 220(R/8.5\text{kpc})^{0.382} \text{ km s}^{-1} \quad (1)$$

and a three component model for vertical structure of the gaseous disk [3]:

$$n(z) = \frac{n_1}{\exp(z/H_1)^2} + \frac{n_2}{\exp(z/H_2)^2} + \frac{n_3}{\exp(|z|/H_3)}. \quad (2)$$

In the solar vicinity, these parameters are: $n_1 = 0.395$ cm^{-3}, $H_1 = 127$ pc; $n_2 = 0.107$ cm^{-3}, $H_2 = 318$ pc; $n_3 = 0.064$ cm^{-3}, $H_3 = 403$ pc.

It was assumed that the value of the scale heights, H_i, change with galactocentric distance in the same way as the half-width of the HI layer, and that the relation between the maximum values of all three different gas components, $n_i(z = 0)$, remain constant along the Galactic disk. The z-component of the gravitational field was described with the analytical formula in Kuijken & Gilmore [9]. We have also assumed that the energy sources (i.e. the OB-associations) could be located at three different heights: 0, 50, and 100 pc from the Galactic plane. The energy input rate L(t) was assumed constant with three different values (0.315, 1.05 or 3.15×10^{38} erg s^{-1}) and stops after 30 Myr, when all the massive stars in the OB-association explode as supernovae.

It was assumed that the molecular gas appears in those parts of the shell, where the column number density exceeds the critical value discussed in Franco & Cox [6] and Arshutkin & Kolesnik [1]:

$$N_c = (Z_\odot/Z) \times 10^{21} \text{cm}^{-2}, \quad (3)$$

where Z is the metallicity. The results of the calculations are presented in the Figures 1-3.

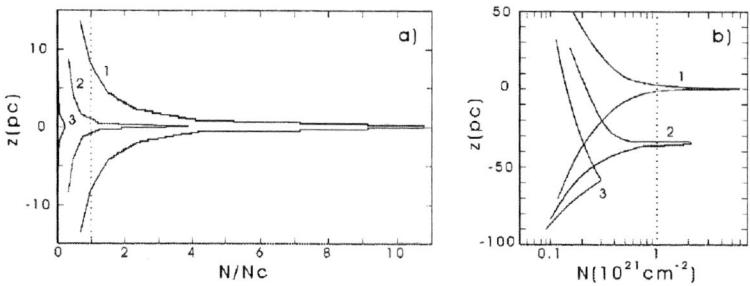

FIGURE 2. Distribution of the relative column density N/N_c along the shell. a) OB-association with $L_{38} = 0.315$ in the Galactic plane. Curves 1, 2, and 3 correspond to the galactocentric radii R=5 kpc, 8.5 kpc, and 15 kpc. b) OB-association at 8.5 kpc, with $L_{38} = 1.05$ above the Galactic plane. Curves 1, 2, and 3 correspond to the z_{OB}=0 pc, 50 pc and 100 pc.

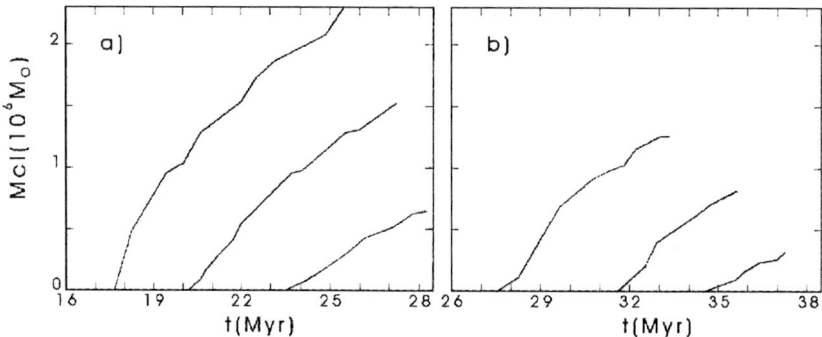

FIGURE 3. The mass of the molecular gas as a function of time. Curves 1, 2, and 3 correspond to different energy input rates: L_{38}= 0.315, 1.05, and 3.15. a) R=5 kpc b) R=8.5 kpc.

4. Results

1. The column densities across some sections of the expanding shells can exceed the critical value $N_c = (Z_\odot/Z) \times 10^{21}$ cm^{-2} at the late stages of the bubble evolution. The largest supershells could therefore effectively transform atomic hydrogen into molecular clouds where star formation will be achieved.

2. The formation of molecular cloud complexes is possible at distances smaller than 15 kpc from the Galactic centre and when the parent OB-association is located below 100 pc from the Galactic plane.

3. The newly formed molecular gas is confined to a slab with a half-width smaller than 100 pc, in agreement with the observed distribution of molecular clouds in our Galaxy.

We thank Jan Palouš for useful discussions and a description of the galactic model that was used in the calculations. We also thank Pepe Franco and Jane Arthur for useful suggestions to improve this report. This work was partially supported by a grant from the American Astronomical Society for the former Soviet Union astronomers.

REFERENCES

[1] Arshutkin, L. N. & Kolesnik, I. G. 1984, Astrofizika, 21, 147.
[2] Bisnovatyi-Kogan, G. S. & Silich, S. A. 1991, Astron. Zh, 68, 749.
[3] Dickey, J. & Lockman, F. 1990, ARAA, 28, 215.
[4] Franco, J. 1991, *Chemical and Dynamical Evolution of Galaxies.*, ed. F. Ferrini, J. Franco and F. Matteucci, (ETS Editrice; Pisa), 506.
[5] Franco, J. 1992, *Star Formation in Stellar Systems*, ed. G. Tenorio-Tagle, M. Prieto and F. Sanchez, (Cambridge U. Press; Cambridge), 515.
[6] Franco, J. & Cox, D., P. 1986, PASP, 98, 1076.
[7] Franco, J. & Shore, S. N. 1984, ApJ, 285, 813.
[8] Kolesnik, I. G. & Silich, S. A. 1989, Astrofizika, 30, 296.
[9] Kuijken, K. & Gilmore, G. 1989, MNRAS, 239, 571.
[10] McCray, R. & Kafatos, M. 1987, ApJ, 317, 190.
[11] Palouš, J., Franco, J.& Tenorio-Tagle, G. 1990, A&A, 227, 175.
[12] Silich, S. 1992, Astrophys. Space. Sci, 195, 317.
[13] Tenorio-Tagle, G & Bodenheimer, P. 1988, Ann, Rev. A. Ap., 26, 145.
[14] Tenorio-Tagle, G & Palouš, J. 1987, A&A, 186, 287.
[15] Wouterloot, J. G. A., Brand, J., Burton, W. B. & Kwee, K. K. 1990, A&A, 230, 21.

2D Simulations of SN Remnants with a Moving Precursor

By F. Brighenti[1], A. D'Ercole[2]

[1] Dipartimento di Astronomia, Università di Bologna, Via Zamboni 33, I-40126 Bologna, Italy.

[2] Osservatorio Astronomico di Bologna, Via Zamboni 33, I-40126 Bologna, Italy.

Numerical models for the time-dependent evolution of supernova remnants evolving through a moving medium are presented. A proper description of the gas close to the progenitor is obtained through a nested grid. Realistic models are constrained by several time scales connected to the presupernova evolution and its velocity through the interstellar medium. Arch-like remnants may be explained by these models. Barrel remnants remain problematic to be interpreted, and we favour the possibility that they are generated by asymmetric explosions rather than by asymmetries of the circumstellar medium.

1. Introduction

Before to explode as supernovae, massive stars modify the status of the circumstellar matter (CSM) creating large cavities mainly through the action of the stellar wind during the main sequence (MS) phase. Supernova ejecta are therefore expected to expand within such wind-driven bubbles. The effects of the processed environment on the evolution of supernova remnants (SNRs) have been investigated by a number of authors both through analytical approaches [2] and numerical simulations [3,7]. The remnant moves almost freely across the low density cavity and becomes suddenly observable in X-ray when reaches the edge of the bubble. 2D simulations [8] have shown an efficient turbulent mixing between the ejecta and the shocked wind. In all the papers mentioned so far the star is assumed to be at the rest with respect to the interstellar medium, and the SN explosion occurs at the center of the bubble. Actually the star moves and the explosion is off center. Numerical simulations have been developed in order to explain the jet observed in the Crab nebula [4]. Off-centered SN explosions in pre-existing bubbles have been simulated [6] without attempting to compare them to any particular SNR, just focusing on general morphological properties. The parameter choice in this latter paper is dictated by numerical rather than astrophysical reasons. We present here two models of SNR expansion in which parameters are constrained essentially by the evolution of the progenitor star.

2. Models

2.1. Slowly moving progenitor

In this model we consider a typical O6 star which, during the main sequence phase, blows a wind with velocity $V = 2000$ km s^{-1} suffering a mass loss rate $\dot{M} = 10^{-6}$ $M_\odot$ yr^{-1}. The wind luminosity at this stage is then $L_w = 0.5\dot{M}V^2 = 1.27 \times 10^{36} = 1.27 L_{36}$ erg s^{-1}. Before to explode, the star enters the red supergiant (RSG) phase, which is assumed to last 7×10^5 yr. During this time the wind velocity is $v = 20$ km s^{-1} and the mass loss rate $\dot{m} = 10^{-5}$ $M_\odot$ yr^{-1}. Finally, we assume that the star moves with velocity $v_\star = 17$ km s^{-1} ($M = 1.5$) across an external medium with density $n_0 = 1$ cm^{-3}. Given the spatial extension of the meshes (0.5 pc), we adopted a refined nested grid around the star in order to simulate properly the isotropic stellar wind inside a region smaller than the length scale given by the stand-off distance of the bow shock around the star, of the order of 4 pc.

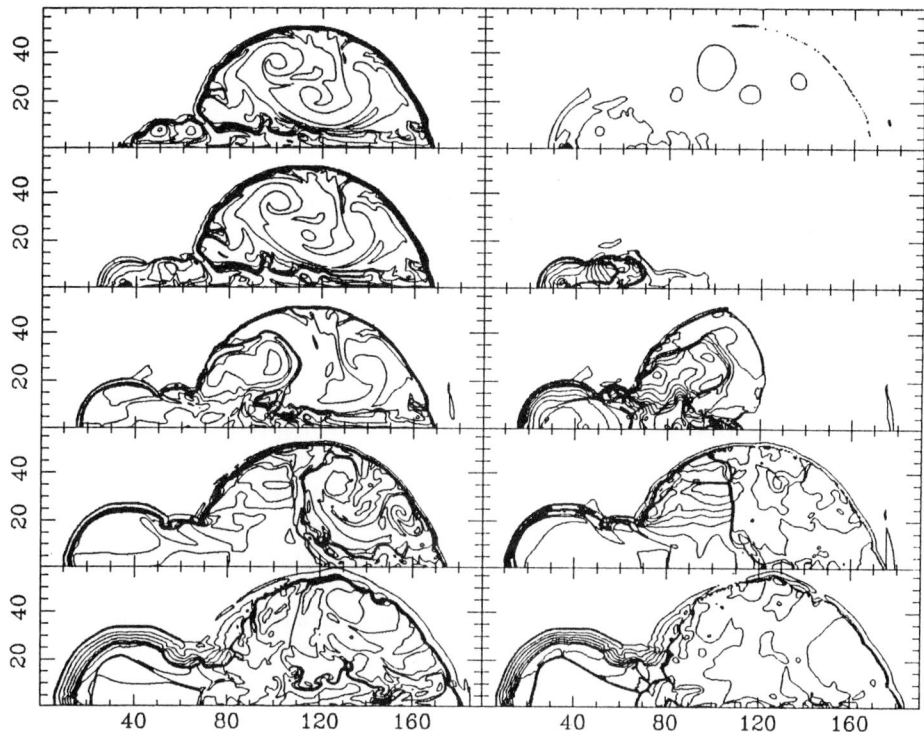

FIGURE 1. Density (left column) and pressure contours of the SNR generated by a star moving with a velocity $v_\star = 17$ km s^{-1} through a medium of density $n_0=1$ cm^{-3} The top panels illustrate the CSM just before the SN explosion, after the RSG phase. The other panels show the SNR expansion at several times, namely at $t = 13 \times 10^3, 68 \times 10^3, 193 \times 10^3, 398 \times 10^3$ yr after the explosion. Labels indicate distances in parsec. The star is located on the horizontal axis at nearly 35 pc.

Provided $v_\star < c_b$, where $c_b \sim 300$ km s^{-1} is the sound speed in the hot bubble, the bubble remains isobaric and spherical irrespective of the stellar location, as far as the star remains inside the original cavity. When the star crosses the edge of this cavity the bubble becomes distorted. Thus, the relevant parameters here are the duration of the main sequence phase $t_{\rm MS} = 4.4 \times L_{36}^{-1/6}$ Myr [5], the time at which the wind blown bubble stalls $t_s = 2.2 \times (L_{36}/n_0)^{1/2}$ Myr, and the time at which the star reaches the edge of the spherical bubble $t_r = 2 \times (L_{36}/n_0)^{1/2}(v_\star/20{\rm kms}^{-1})^{-5/2}$ Myr. For our set of parameters we get $t_{\rm MS} = 4.2$ Myr, $t_s = 2.5$ Myr and $t_r = 3.3$ Myr. This means that the bubble stalls when the star is still in the main sequence stage and still inside the stalled cavity. Given the stalling radius by $R_s = 43.3(L_{36}/n_0)^{1/2}$ pc, the star starts to exit from the bubble at $t = R_s/v_\star = 3.2$ Myr. Just before the explosion, the star has traveled ~ 30 pc outside the bubble. The edge of the tunnel carved by the star through the unperturbed ISM develops mild Kelvin-Helmholtz instabilities (Fig. 1). At the early beginning the remnant expands leftward and upward showing a nearly Sedov profile. The remnant assumes therefore an edge on arch-like aspect, the ring being interrupted in the direction of the tunnel, where the density of the bubble gas is very low. The portion of the remnant moving through the bubble quickly expands across the cavity interacting with an increasing fraction of its boundary. The SN blast, however, is reflected inward without producing any significative compression at the bubble edge (see the front at ~ 110 pc in the fourth panel of Fig. 1).

This means that the only observable region of the remnant remains the shell close to the explosion place.

2.2. Fast moving progenitor

Usually isolated massive stars move with high velocity through the ISM [1]. We therefore computed a model in which $v_\star = 80$ km s^{-1}. Also in this case we assume $n_0 = 1$ cm^{-3}. For this model several considerations are in order. Given the high star velocity, it is $t_r \ll t_s$ and a bow shock geometry develops rather soon instead of that of the classical bubble. The bow shock becomes steady (apart for the presence of Kelvin-Helmholtz instabilities) much before the end of the main sequence. During the RSG phase this structure is left behind and will tend to dissipate, but a similar configuration is recreated, although on a smaller scale because the wind luminosity is now 10^{-3} lower (the wind parameters during this phase are the same as in the previous model). Such a structure can not be properly described on our numerical grid because of its poor spatial resolution (0.17 pc). In any case, the SN explosion of a moving RSG star has been already studied [4]. If however the star goes through a Wolf-Rayet stage, the stellar wind is revitalized approaching mechanical luminosities which are greater or comparable with those reached during the main sequence. Our simulation actually describe this phase (for this reason we assumed here an higher value of the wind velocity, $V = 3000$ km s^{-1}, compatible with a Wolf-Rayet wind). The first panel of Fig. 2 thus shows the structure of the CSM around a moving Wolf-Rayet star before it explodes as a SN. The other panels show the expansion of the SNR at several times. The remnant moves rather freely through the tunnel carved by the star. The shock driven in the tunnel does not propagate parallel to the symmetry axis, but bounces several times between this axis and the contact surface. Because of the increasing pressure, the tunnel tends to expand compressing the gas at its edge and giving rise to a "cylindrical" shell. Instead, the SN blast propagating across the unperturbed ISM closely resembles the spherical Sedov solution, although the "sphere" is slightly bended toward the trailing part of the motion. This nonradiative phases ends at $t \sim 30000$ yr, when cooling becomes substantial and the shell becomes thin and very dense. Although a well definite "cylindrical" shell is formed along the tunnel wall, the combination of density and temperature is such that the surface brightness is up to ten times lower than the surface brightness of the shell expanding across the unperturbed medium. Thus, also in this case the appearance of the remnant is of the arch-like type, without the presence of any appreciable tail.

3. Conclusions

The results shown in this paper indicate that arch-like SNRs may be due, at least in part, to moving precursors rather than to ISM inhomogeneities. Barrel remnants can not be reproduced by our models, unless the star explodes *inside* the bubble, before crossing its edge. In this case the SN blast reaches the bubble's edge progressively, starting from the closest side. Initially the SNR will be arc-like. Later on the arch will extend while the brightness of its central part will fade out. At this stage the remnant could appear barrel-like. However, given the high number of barrel-shaped SNRs, we feel that a more general mechanism, as an asymmetric explosion, must be invoked to explain them.

REFERENCES

[1] Blaauw, A. 1992, in "Massive Stars: Their Lives in the Interstellar Medium", eds. J.P. Casinelli and E.B. Churchwell, Astr. Soc. of the Pac. Conf. Series, in press.

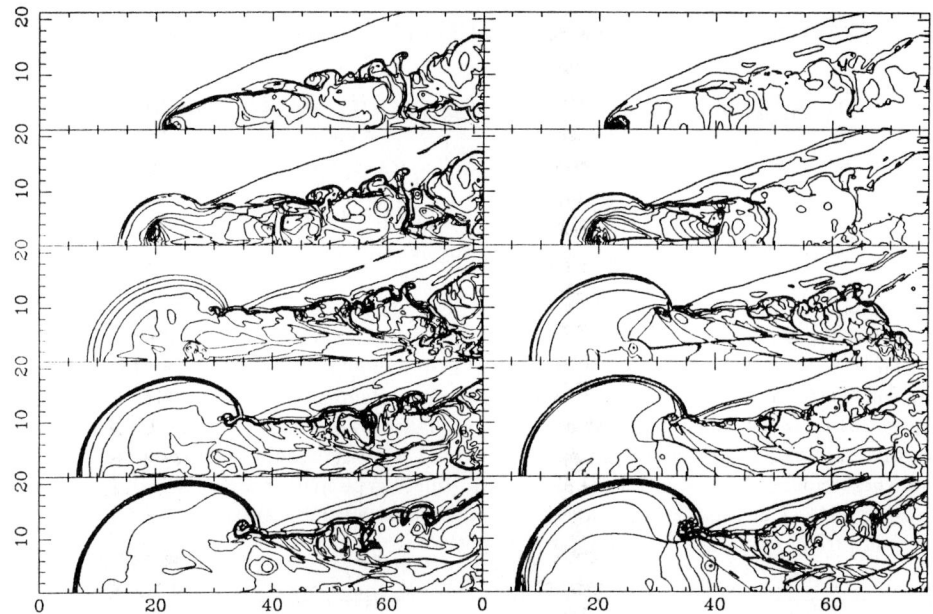

FIGURE 2. The same as Fig. 1, but for $v_\star = 80$ km s^{-1}. The top panels show the CSM before the explosion, after the Wolf-Rayet phase. The other panels show the SNR expansion at $t = 5 \times 10^3, 21 \times 10^3, 35 \times 10^3, 53 \times 10^3$ yr.

[2] Chevalier, R. A. & Liang, E. P. 1989, ApJ, 344, 332.
[3] Ciotti, L. & D'Ercole, A. 1989, AA, 215, 347.
[4] Cox, C. I., Gull, S. F. & Green, D. A. 1991, MNRAS, 250, 750.
[5] McKee, C.F., Van Buren, D., Lazareff, B. 1984 ApJ, 278, L115
[6] Różyczka, M., Tenorio-Tagle, G., Franco, J. & Bodenheimer, P. 1993, MNRAS, 261, 674.
[7] Tenorio-Tagle, G., Bodenheimer, P., Franco, J. & Różyczka, M. 1990, MNRAS, 244, 563.
[8] Tenorio-Tagle, G., Różyczka, M., Franco, J. & Bodenheimer, P. 1991, MNRAS, 251, 318.

Mass Loaded Astrophysical Flows

By S. J. Arthur[1], J. E. Dyson[2], T. W. Hartquist[3]

[1] Instituto de Astronomía, UNAM, México
[2] Astronomy Department, University of Manchester, U. K.
[3] Max Planck Institut für Extraterrestrischephysik, Garching bei München, Germany.

The interstellar medium is far from uniform: massive stars preprocess their environment with strong, fast stellar winds before finally, in some cases, exploding as supernovae. The properties of these stellar winds change as the stars evolve, and it is most unlikely that an isotropic circumstellar medium will be maintained. The effects of a clumpy circumstellar medium on stellar winds and supernova remnants include ablation of clump mass by the flow. This ablated mass can significantly alter the properties of a planetary nebula wind or a supernova remnant. We have modelled these modified flows with an Eulerian, Godunov-type hydrocode. The results show that observations such as unusually high electron temperatures and acceleration in planetary nebula haloes are naturally explained by this mass loading model.

1. Introduction

By the term *mass loading* we simply mean addition of mass to a flow. The flows can be stellar winds such as those found in planetary nebulae and Wolf-Rayet nebulae, or the flow of hot gas inside a supernova remnant. The mass can come from dense clumps ejected during the red giant phase of stellar evolution, fragmentation of shells due to Rayleigh-Taylor instabilities, or simply from clouds present in the interstellar medium. Mass can find its way into the flow by a variety of processes. In this paper we will discuss conductively driven evaporation and hydrodynamic ablation as possible mechanisms, though we show that the former is not important in astrophysical flows.

2. Mechanisms for Mass Loading of a Flow

2.1. *Conductively Driven Evaporation*

Theories of the evaporation of isolated, spherical, cold clumps in a hot tenuous plasma have assumed the conduction to be due solely to electrons and to occur parallel to the magnetic field direction [8]. Depending on the ratio of the electron mean free path to the temperature scale height, conduction is defined to be either *classical* or *saturated*, with saturated conduction being applicable in young supernova remnants where the electron mean free path can be greater than the temperature scale height. Evaporation rates can be determined for each case as follows: $\dot{m}_{\rm CLASS} = 2.75 \times 10^4 T^{5/2} R \phi_s$ and $\dot{m}_{\rm SAT} = 3.75 \times 10^4 T^{5/2} R \phi_s \sigma_o^{-5/8}$, where T is the plasma temperature, R is the cloud radius in parsecs, ϕ_s is a factor around unity, that reflects the uncertainty in the calculation of the heat flux, and σ_o is a saturation parameter.

This theory is all very well, but does not take into account the flow of the hot plasma around the cold clump. The hydrodynamic effects of such an interaction will severely limit the efficiency of the conduction process. For example, Kelvin-Helmholtz and other fluid instabilities create vorticity, winding up the magnetic field and suppressing conduction. Also, the flow of the gas around the clump will cause the magnetic field in the evaporating gas to be bent back around the clump, again suppressing conduction. Plasma instabilities will lead to turbulence from which electrons are scattered, thereby reducing their mean free path. Further, neutral gas in the clumps will increase the cooling rate by one to two orders of magnitude, thus inhibiting evaporation of clump gas.

2.2. Hydrodynamic Ablation

The differences between subsonic and supersonic flow past an embedded clump result in different ablation rates [11]. In the subsonic case, the Bernoulli effect leads to pressure variations along the clump boundary. The clump then expands perpendicular to the upstream direction. Mass loading occurs when the mass flow through the mixing region is comparable to the clump mass loss rate. This ablation rate is found to depend on the upstream Mach number according to $\dot{m} = M_u^{4/3}$, where $\dot{m}$ is the ablation rate and M_u is the upstream Mach number.

In the supersonic case, the flow past the clump is a lot more complicated. In the absence of either experimental or numerical results, a comparison is made with flow past a solid body, which is applicable when the clump is much denser than the tenuous medium flowing past it. A bow shock forms upstream of the clump, and a turbulent wake behind it. It is in this region that mixing occurs. In practice, for supersonic flow past a compressible clump, there would also be shocks transmitted into the clump, which would cause compression parallel to the flow direction, and expansion perpendicular to it. The expansion of the clump is essentially independent of the upstream Mach number of the flow, and the mass loading rate is taken to be constant in this instance.

A third hydrodynamic way of adding mass to a flow is due to crushing of clouds by shock waves. Numerical studies by a variety of authors [12,17,19] have shown that transmitted shocks propagate into the cloud, which is first compressed and then reexpands. Fluid instabilities caused by the flow of the intercloud gas past the cloud then lead to its fragmentation and final destruction. A detailed study of the shock-cloud interaction [12] has shown that the cloud is completely destroyed in about 3 cloud crushing times, where this timescale is defined by $t_{cc} = R_c/V_{cs}$, where R_c is the cloud radius, and V_{cs} is the velocity of the transmitted cloud shock, given by $V_{cs} \sim (\rho_T/\rho_{cl})^{1/2} V_s$, where ρ_T and ρ_{cl} are the densities in the tenuous medium and cloud respectively, and V_s is the velocity of the incident shock wave. Furthermore, it is found that for small cloud-intercloud density contrasts the cloud fragments hierarchically into progressively smaller fragments, whereas for large density contrasts destruction is by continuous erosion of small fragments that rapidly become comoving with the flow.

Mixing of clump mass with the flow will generally occur in a turbulent mixing layer. Astrophysical flows possess large Reynolds numbers, equivalent to very small viscosities, hence in the bulk of the fluid they can by treated as ideal. However, near an obstacle, the small viscosity means that all components of velocity must go to zero at the body surface, and so a thin boundary layer forms next to the surface of the obstacle. If the obstacle itself is a fluid, the boundary layer will extend into that, too. The thickness of the boundary layer, and hence its Reynolds number, grows with distance from the leading edge of the body, and at some point the flow in the boundary layer passes its critical Reynolds number and becomes turbulent. An analytic treatment of turbulent boundary layers [3] shows that the layer extends comparable depths into both the hot flowing gas and the cool clump.

3. Observations and Models

3.1. Clumpy Wolf-Rayet Ring Nebula, RCW 58

The Wolf-Rayet ring nebula RCW 58 consists of an expanding shell enveloping clumps of stellar ejecta [5,19], where the shell is driven by a wind blown bubble. Absorption spectra show that the ionization potential (*i.e.* temperature) of a variety of ions is correlated with their velocities [18], with the ions with the highest ionization potential possessing the highest blue shifted velocities. For instance, C IV (representing a temperature of 1.5×10^5K) has a velocity of -150 km s^{-1}, while Fe II (10^4K) is seen at -102 km s^{-1}, with the proper motion of the central source being measured at -14 km s^{-1}. It is not possible to model this temperature and velocity range with a standard stellar wind bubble model [11,18]. If it is assumed that the emission comes from behind the outer, radiative shock, then the slope of the velocity-temperature correlation would go the wrong way. On the other hand, if the emission was produced behind an inner stellar wind shock, then the velocity spread obtained for the required temperature range would be far too small for the observed stellar wind terminal velocity that has been measured.

A simple mass loaded stellar wind model [1], that assumes mass is added to the flow in the clumpy region outside the nebula core (radius > 0.45 parsec), including the effects of radiative cooling, successfully reproduces the observed expansion speed and radius of the nebula for an age of 13,000 yrs, consistent with that derived from observations [19]. Moreover, the temperature range 1.5×10^5K to 10^4K corresponds to a velocity range of -135 km s^{-1} to -64 km s^{-1}, though the correlation is not linear. This is probably because collisional ionization does not hold in the shocked wind plus ablated clump material [1].

3.2. Core-Halo Planetary Nebulae

Observations of some planetary nebulae with clumpy cores and less clumpy haloes [14,16] show electron temperatures of 15000K, which is much higher than the 10^4K that can be maintained by photoionization. More detailed observations of an ionized clump in the halo of NGC 6543 [14] show that this high temperature gas has very small velocity widths, of the order 5 km s^{-1}, suggesting that it is the shocking of gas originally moving at 20 – 25 km s^{-1} (*i.e.* Mach 2) that produces this emission. The ionized dusty tails of clumps observed in the planetary nebula NGC 7293 [15] have shapes that suggest that the flow is transonic rather than supersonic. Observations of multi-shell planetary nebulae show that some have accelerating haloes [6,7]. None of the above observations can be explained by a standard two wind model for planetary nebulae.

We propose a model for NGC 6543 in which isothermal mass loading of the fast stellar wind occurs only in the central core (radius < 0.13 parsecs). We assume an interclump density of 10cm^{-3} both in the core and the halo, though in practice this ambient density does not affect the results. The stellar wind terminal velocity is taken to be 2000 km s^{-1}, and the stellar wind mass loss rate is 10^{-7} M$_\odot$ yr^{-1}, both values consistent with observations [13]. With these parameters, the mass loading rate required to give a particular Mach number on exit from the mass loading core region can then be estimated analytically [2]. It is found that for a mass loading rate, Q, of $3.0 \times 10^{-33} M^\alpha$ g cm^{-3}s^{-1}, where M is the Mach number of the flow, and α is 4/3 or zero, depending on whether the flow is subsonic or supersonic, the mass loaded wind leaving the core has a Mach number of around unity and a velocity of about 15 km s^{-1}, and slowly accelerates to 30 km s^{-1} at the inner edge of the dense outer shell. The shocking of such a mass loaded wind around a clump at the edge of the core region would produce the observed high temperature emission. For a nebula radius of 1 parsec, and age of 58,000 yrs, the ratio of ablated mass to wind mass is found to be 124. In practice, as long as the mass loading rate is above a certain value, the Mach number of the flow leaving the core will be around unity [2]. This result is also insensitive to changes in interclump density. In Figure 1 we show the distributions of density, velocity and Mach number in the nebula.

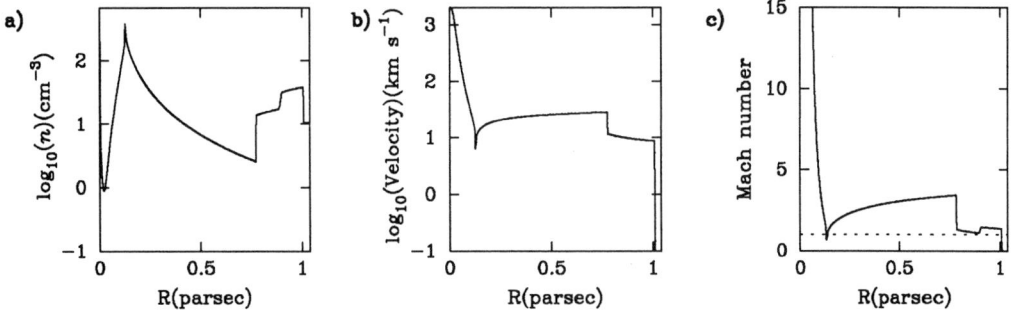

FIGURE 1. Model of a mass loaded planetary nebula showing a) $\log_{10}$(density), b) $\log_{10}$(velocity) and c) Mach number.

3.3. Supernova Remnants

Several models have been put forward for mass loaded supernova remnants in an attempt to explain such diverse observations as X-ray haloes, centrally peaked X-ray emission, and the local soft X-ray background [4,9,21]. These models are all self-similar, do not include radiative cooling, and in some instances require contrived ambient density distributions to satisfy conditions for self-similarity [4,9]. Models assuming conductively driven evaporation are bound to give centrally peaked X-ray emission, since the mass loading rate depends strongly on the temperature, which is highest in the centre of a supernova remnant. However, as we have seen earlier, the case against conductively driven evaporation is strong, and so we present models of hydrodynamically mass loaded supernova remnants, including the effects of radiative cooling. Cooling is likely to become important, even if the ambient density is low, when the mass loading rate is high enough. Since hydrodynamic ablation, in the subsonic case, depends on the Mach number, the amount of ablated mass in the centre of such a mass loaded SNR is likely to be low, since the velocity goes to zero at the centre while the pressure remains finite. We are therefore unlikely to produce centrally peaked X-ray emission from such remnants.

With these caveats we have embarked on a study of the effects of hydrodynamic ablation plus radiative cooling on the evolution of SNR, although we stress that these models are qualitative rather than quantitative, and serve only to show how different levels of mass loading could affect the expansion and X-ray appearance of a remnant. To demonstrate this most effectively, we have evolved all our models (which have an explosion energy of 10^{51} ergs) to a radius of 8 parsecs, and have adopted 2 values for the ambient density (0.1 cm^{-3} and 10 cm^{-3}), and a range of mass loading values, from a factor of zero to 100 times the characteristic mass loading rate of 2.6×10^{-36} g cm^{-3} s^{-1}. Mass loading is assumed to occur everywhere in the remnant with the same dependence on flow Mach number as was described earlier. A real remnant, however, is likely to evolve in a medium containing regions with different degrees of clumpiness, caused by the action of winds from the progenitor star and neighbouring stars and, of course, the ejecta themselves could be clumpy. In Figure 2 we plot the expected X-ray surface brightness profiles (in the range 0.1

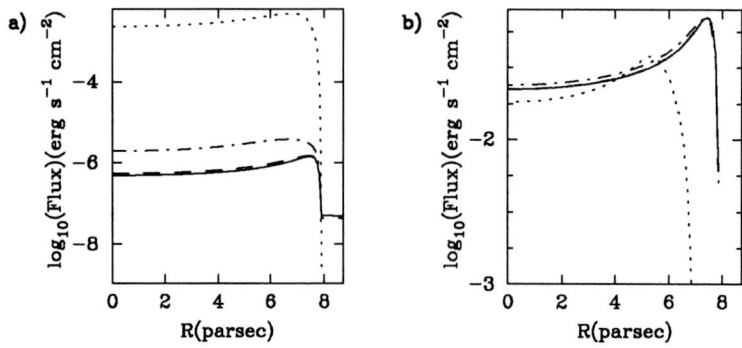

FIGURE 2. X-ray surface brightness profiles for mass loaded supernova remnants with **a)** ambient density = 0.1 cm^{-3}, and **b)** ambient density = 10.0 cm^{-3}

n_0 (cm^{-3})	Line Style	Q	Age (yr)	Total Luminosity (erg s^{-1})	Ablated Mass (M$_\odot$)
0.1	solid	0	770	1.53×10^{34}	0.0
	dashed	1	780	1.70×10^{34}	0.43
	dot-dashed	10	920	5.10×10^{34}	5.36
	dotted	100	3610	1.86×10^{37}	232.0
10.0	solid	0	7800	1.51×10^{38}	0.0
	dashed	1	7820	1.52×10^{38}	4.4
	dot-dashed	10	7950	1.57×10^{38}	45.1
	dotted	100	10640	7.14×10^{37}	735.5

TABLE 1. Supernova remnant model parameters

to 10 keV) for an absorbing column density of 3.0×10^{21} cm^{-2}. In Table 1 we summarise the models. As can be seen from the figures, a combination of high mass loading rate and low ambient density can produce a very bright X-ray remnant, while for the higher ambient density cooling becomes an important factor when the mass loading rate is high. However, since non-equilibrium effects will be important behind the shock and, due to the mixing of cold clump gas with the hot remnant gas, will almost certainly be important in the interior of the remnant as well, these results must be taken simply as an indication of the ways in which mass loading can lead to a departure from the "standard" Sedov remnant.

4. Conclusions

There is no doubt that clumps and clouds exist in the interstellar medium in locations where they could, and almost certainly do, affect the properties of winds and supernova remnants. The simple mass loaded models described above show that previously "anomalous" observations such as the velocity-temperature correlation in RCW 58 and the high electron temperatures in some planetary nebula haloes, can be naturally explained as a result of mass loading. However, the actual mass loading process needs to be studied further, giving full attention to the rôle of turbulent mixing layers. For a more detailed discussion of mass loading, see Hartquist and Dyson [10].

REFERENCES

[1] Arthur, S. J., Dyson J. E. & Hartquist T. W. 1993, MNRAS, 261, 425.
[2] Arthur, S. J., Dyson J. E. & Hartquist T. W. 1994, *submitted*.
[3] Cantó, J. & Raga A. 1991, ApJ, 372, 646.
[4] Chièze, J. P. & Lazareff B. 1981, AA, 95, 194.
[5] Chu, Y-H. 1982, ApJ, 254, 578.
[6] Chu, Y-H. 1989, in IAU Symp. 131 *Planetary Nebulae*, ed. S. Torres-Peimbert (Dordrecht: Kluwer), 105.
[7] Chu, Y-H., Jacoby G. H. & Arendt R. 1987, ApJS, 64, 529.
[8] Cowie, L. L. & McKee C. F. 1977, ApJ, 211, 123.
[9] Dyson, J. E. & Hartquist T. W. 1987, MNRAS, 228, 353.
[10] Hartquist, T. W.. & Dyson J. E. 1993, QJRAS, 34, 57.
[11] Hartquist, T. W., Dyson J. E., Pettini M. & Smith L. J. 1986, MNRAS, 221, 715.
[12] Klein, R. I., McKee C. F. & Colella P. 1990, in *Supernovae*, ed. S. E. Woosley (Springer-Verlag), 696.
[13] Lucy, L. B. & Perinotto, M. 1987, AA, 188, 125.
[14] Meaburn, J., Nicholson R., Bryce M., Dyson J. E. & Walsh J. R. 1991, MNRAS, 252, 535.
[15] Meaburn, J., Walsh J. R., Clegg R. E. S., Walton N. A., Taylor D. & Berry D. S. 1992, MNRAS, 255, 177.
[16] Middlemass, D., Clegg R. E. S., Walsh J. R. & Harrington J. P. 1991, MNRAS, 251, 284.
[17] Nittman, J., Falle S. A. E. G. & Gaskell P. H. 1982, MNRAS, 201, 833.
[18] Smith, L. J., Pettini M., Dyson J. E. & Hartquist T. W. 1984, MNRAS, 211, 679.
[19] Smith, L. J., Pettini M., Dyson J. E. & Hartquist T. W. 1988, MNRAS, 234, 625.
[20] Stone, J. M. & Norman M. L. 1992, ApJ, 390, L17.
[21] White, R. L. & Long K. S. 1991, ApJ, 373, 543.

The effects of a magnetic field in the evolution of cosmic ray mediated shocks

By Byung-Il Jun[1], David. A. Clarke[2], Michael. L. Norman[1]

[1] National Center for Supercomputing Applications, Beckman Institute, 405 N. Mathews Ave., Urbana, IL 61801 USA

[2] Harvard-Smithsonian Center for Astrophysics, 60 Garden Street, Cambridge, MA 02138, USA.

We study the acceleration efficiency of MHD cosmic-ray shocks by using a two-fluid model. Our model includes the dynamical effect of magnetic fields and cosmic rays on a background thermal fluid. We explore the time evolution of plane- perpendicular, piston-driven shocks numerically. We find that the magnetic field plays an important role in the acceleration efficiency of the MHD cosmic-ray shocks. The acceleration of cosmic-ray particles becomes less efficient in the presence of strong magnetic pressure because the field makes the shock less compressive. This effect is more pronounced at low Mach numbers than at high Mach numbers.

1. Introduction

The diffusive acceleration of cosmic rays (hereafter CR) in shock waves is now widely accepted as the source of the galactic CR and the synchrotron emitting electrons in radio sources such as supernova remnants and extragalactic radio jets [1,2]. The back-reaction of the accelerated particles on the background fluid can modify the structure of the shock. Drury and Völk [5] studied this effect using a two-fluid model of CR propagation. This two-fluid model is justified if one assumes that the CR are almost isotropic with respect to the thermal background plasma and the interaction between the CR and the gas can be represented through a scalar pressure. Since particles gain energy both by scattering back and forth across the shock by magnetic irregularities and via particle drifts in the magnetic field at the shock [11], it is natural to expect the role of a magnetic field is also important in cosmic-ray mediated MHD (hereafter CRMHD) shocks. Nevertheless, most of previous work has neglected the dynamical effects of mean magnetic fields. Here we investigate the dynamical effects of a magnetic field on the acceleration efficiency considering plane-perpendicular shocks.

2. Equations and Numerical Method

By neglecting the motion of the scatterers relative to the fluid, the dynamics of the system can be described by the two-fluid equations including a diffusion term which describes the scattering of energetic particles [3,4].

$$\frac{\partial \rho}{\partial t} + \nabla \cdot (\rho \vec{u}) = 0, \quad (2.1)$$

$$\rho \frac{\partial \vec{u}}{\partial t} + \rho (\vec{u} \cdot \nabla) \vec{u} = -\nabla (P_g + P_c) + \frac{1}{4\pi} \left(\nabla \times \vec{B} \right) \times \vec{B}, \quad (2.2)$$

$$\frac{\partial e_g}{\partial t} + \nabla \cdot (e_g \vec{u}) = -P_g (\nabla \cdot \vec{u}), \quad (2.3)$$

$$\frac{\partial e_c}{\partial t} + \nabla \cdot (e_c \vec{u}) = -P_c (\nabla \cdot \vec{u}) + \nabla \cdot (\kappa \nabla e_c), \quad (2.4)$$

$$\frac{\partial \vec{B}}{\partial t} = \nabla \times (\vec{u} \times \vec{B}), \quad (2.5)$$

$$P_c = (\gamma_c - 1) e_c, \quad (2.6)$$

$$P_g = (\gamma_g - 1) e_g, \quad (2.7)$$

where ρ is the gas density, $\vec{u}$ is the gas velocity, P is the pressure, B is the magnetic field, e is the internal energy density, γ is the specific heat ratio, and κ is the spatial diffusion coefficient. The thermal gas is denoted by a subscript g and the CR gas by a subscript c.

Equations (2.1)-(2.7) are solved numerically using ZEUS-3D which is a fully three-dimensional CRMHD code. Algorithms used to solve the MHD equations are described in Stone and Norman [8,9] and includes a von-Neumann & Richtmyer artificial viscosity to stabilize shocks and a second order upwinded, monotonic interpolation scheme taken from van Leer [10]. The code is finite differenced on an Eulerian mesh and fully explicit in time. Magnetic fields are transported using Constrained Transport [6] modified with the Method of Characteristics [9]. In addition to the equation of MHD, ZEUS-3D has been extended to solve a two-fluid model for the CR component. The major difference between equations 2.3 and 2.4 is the diffusion term in the latter.

We solve the diffusion term explicitly and explicit diffusion schemes are limited by the diffusion time step, namely;

$$\Delta t \leq 0.25 \frac{\Delta x^2}{\kappa}. \quad (2.8)$$

One can see that this makes severe restrictions on the time step as the resolution increases. To avoid this problem, we "sub-cycle" the diffusion term. Generally, the dynamical time step is much larger than the diffusion time step and it becomes enormously expensive to update the hydrodynamics on the diffusion time scale. To avoid this problem, one may update the CR energy on the diffusion time scale until the accumulated time reaches the dynamical time scale. Only then do we perform a general MHD update.

We have applied the artificial viscosity only to the gas energy equation to stabilize the subshock. This is justified on both physical and empirical grounds since CR particles do not experience a viscous interaction. The diffusion makes the gas shock smooth and a foot will appear in front of the shock. Therefore, any viscous heating in the foot is amplified by the shock and swept downstream generating higher entropy than expected analytically. We have observed this extra heating in the thermal gas pressure as did Drury and Falle [4]. Because the diffusion is also playing a role in smoothing the shock, we only need to apply the artificial viscosity to the subshock region itself. This is accomplished by comparing the two slopes of gas pressure and CR pressure. Thus, if

$$\frac{\Delta P_{g,i,j,k}}{P_{g,i,j,k}} > n \times \left(\frac{\Delta P_{c,i,j,k}}{P_{c,i,j,k}} \right) \quad (2.9)$$

where n is a constant of order unity and $\Delta P_{g,i,j,k}$ is the van Leer slope of the gas pressure, then we apply the artificial viscosity. n is a free parameter and can be adjusted as convenient. The value for n used in this work was 6. In the calculation of a magnetosonic Mach 10 shock resolved with 300 zones, the artificial viscosity generated 40 % too much heating of the thermal gas if applied to the entire shock region. However, restricting the artificial viscosity to the subshock region only (via eq. 2.9), the post shock gas pressure

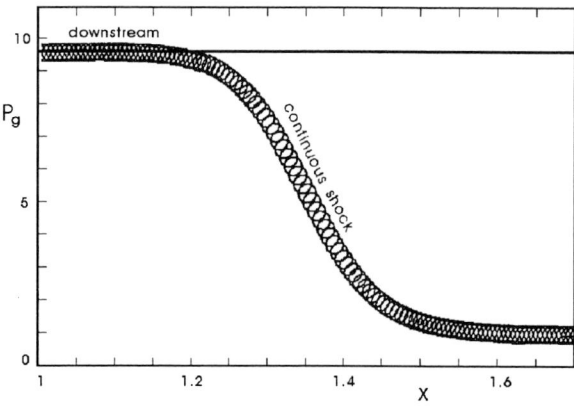

FIGURE 1. The comparison of analytic solution and numerical solution of Mach 10 CRMHD shock. Open circles represent numerical solution and a solid line shows the analytic solution of downstream state.

was correct to within 1 % (Figure 1). This error was reduced further by increasing the numerical resolution.

3. Results

The time-dependent evolution of energy components of the plane-perpendicular, piston-driven shock for different diffusion coefficients and magnetic pressures, P_b, is shown in Figure 2a. The magnetosonic Mach number of the shock is 10. This shock wave is propagating through a uniform background of $\rho = 1$, $P_g = 1$, $P_c = 1$, $P_b = 0.1$, and $\kappa = 1$ resolved by 300 uniform grid zones. The ratio of specific heats, γ_c is assumed to be 5/3. The calculation is carried up to $t/t_d = 105.9$ where t_d is the diffusion time scale defined as κ/u_s^2 where u_s is the velocity of a shock wave. The steady state is reached at this time. The numerical results are compared to the exact downstream state obtained by Jun, Clarke & Norman [7] and are found to agree to within less than 1%. Figure 2b has same parameters as Figure 2a except that $P_b = 1.0$. At early times, kinetic energy is transferred to internal energy by shock heating. As the system reaches steady state, the CR energy density becomes the dominating factor. From the comparison of Figs. 2a and 2b, one can see that the presence of a magnetic field makes conversion of kinetic energy to CR energy less efficient. This effect is seen prominently in Figs 2c and 2d for which $M = 3$ and $\gamma_c = 4/3$. These calculations are carried out to the same evolution time, $t = 144 t_d$. Thus the role of magnetic field is important for low Mach numbers as expected from the analysis of steady state solution [7].

In Figures 2, the "problem time" corresponds to "t" in the text. Problem times 3.0 (Fig. 2a), 2.0 (Fig. 2b), 5.0 (Fig. 2c), and 3.2 (Fig. 2d) correspond to evolution times $105.9 t_d$, $105.9 t_d$, $144 t_d$, and $144 t_d$, respectively. Note that the diffusion time scale is different in each case because it dependenta on the shock velocity.

4. Conclusion

We have studied the effect of the magnetic field in the acceleration efficiency of CRMHD shocks and obtained the following conclusions: i) The conversion efficiency of kinetic energy to CR energy decreases as the magnetic field pressure increases, and ii) the dependency of

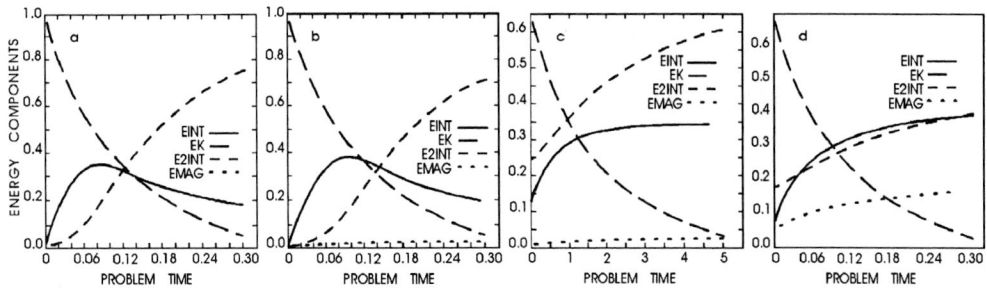

FIGURE 2. Energy components normalized by the total energy. a) Mach 10 and $P_b = 0.1$, b) Mach 10 and $P_b = 1.0$, c) Mach 3 and $P_b = 0.1$, d) Mach 3 and $P_b = 1.0$. EINT, EK, E2INT, and EMAG represent internal gas energy, kinetic energy, internal CR energy, and magnetic energy, respectively.

the CR-mediated shocks on the magnetic field is higher at low Mach number than at high Mach number.

REFERENCES

[1] Bell, A. R. 1978, MNRAS, 182, 147.
[2] Blanford, R.D. & Ostriker, J. P. 1978, ApJ, 221, 129.
[3] Drury, L. O'C. 1985, in Cosmical Gas Dynamics, ed. F. D. Kahn (Utrecht:VNU Science Press), p.131.
[4] Drury, L. O'C. & Falle, S. A. E. G. 1986, MNRAS, 223, 353.
[5] Drury, L. O'C. & Völk, H. J. 1981, ApJ, 248, 344.
[6] Evans, C. & Hawley, J. F. 1988, ApJ, 33, 659.
[7] Jun, B., Clarke, D. A. & Norman, M. L. 1993, ApJ submitted.
[8] Stone, J. M. & Norman, M. L. 1992a, ApJS, 80, 753.
[9] Stone, J. M. & Norman, M. L. 1992b, ApJS, 80, 791.
[10] van Leer, B. 1977, J. Comput. Phys., 23, 276.
[11] Webb, G. M., Axford, W. I. & Terasawa, T. 1983, ApJ, 270, 537.

Particle Acceleration with Spontaneous Excitation of Alfvén Waves in the Magnetospheres of Neutron Stars

By Hitoshi Hanami

Physics Section, College of Humanities and Social Sciences, Iwate University, Morioka 020, Japan.

We studied a particle acceleration process in the magnetosphere of a neutron star due to a current circuit as a means of producing gamma-ray bursts. Planetary objects periodically falling in the magnetosphere work as the battery in the circuit system. The physical conditions on the surface of the neutron star then form a closed current circuit providing good conversion of the kinetic energy of rotating objects to that of magnetosphere oscillation. The magnetosphere system is unstable to a feedback instability which excites Alfvén waves and induces strong particle acceleration.

1. Introduction

High-energy particle acceleration is one of the many unsolved physical processes in astrophysics as are gamma-ray bursts and pulsar non-thermal emission. Gamma-ray bursts have defied explanation for fifteen years since their discovery [6]. Recent observations imply that the sources have strong magnetic fields ($\sim 10^{12}$G) which is consistent with the model of cyclotron absorption for the observed spectral feature [9]. This suggests that the burst process is related to magnetospheric activity around neutron stars as they are the only objects whose magnetic field is strong enough. Furthermore, the discovery of planets around pulsars has been reported [15] leading us to make an interesting speculation. That is, that planets and comets, rotating in the magnetosphere of neutron stars, spontaneously excite Alfvén waves and induce high-energy particle acceleration driving gamma-ray bursts. This paper deals with an electromagnetic coupling process using a current circuit analogy in the magnetosphere of neutron stars and the possibility of strong particle acceleration.

2. Spontaneous Excitation of Alfvén Waves and Particle Acceleration in Magnetospheres

Magnetospheres can be a self-exciting system when energy is supplied from outside of the system through falling planets or rotation of the central star. Hanami [3] has shown that rotation of an object such as a planet produces an electric potential gap in the circuit which is formed by the magnetic field column between the object and the surface of the neutron star, shown here schematically in Figure 1. The circuit is closed by currents on the surface of the neutron star which should have Pedersen conductance sufficient to carry current across the magnetic field. A closed circuit with Pedersen current could result in particle acceleration through a coupling between the magnetosphere and the surface that induces spontaneous excitation of Alfvén waves by alternating current in the magnetosphere. We consider the energy relaxation process from the wave to the particles below.

3. High Energy Particle Acceleration with Alfvén Waves

Coherency of the excited wave is advantageous for particle acceleration and the accelerated particles can be converted to high energy photons such as the observed gamma-rays.

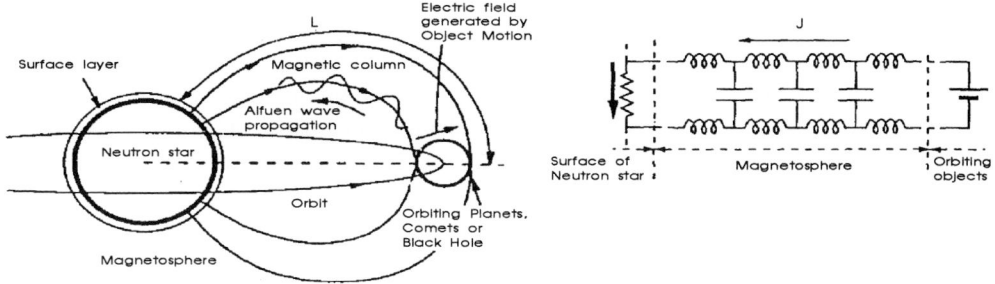

Using the numerical simulation code described in [5] with Alfvén waves highly excited as mentioned above, we attempt to learn more about the acceleration mechanism using numerical simulations. Acceleration processes with compressional Alfvén waves propagating perpendicular to the magnetic field have been studied previously. In this paper, we examine an acceleration process where the Alfvén waves propagate parallel to the magnetic field. We used a relativistic, electromagnetic particle-in-cell code in one spatial dimension for the simulations. The code solves the full set of Maxwell's equations for the electromagnetic fields, and the particles are advanced in time using the relativistic Lorentz force equation. The geometry of the model is one-dimensional real space oriented along the mean magnetic field, and the field and density distributions are considered only along this direction. However, each particle has three phase space coordinates for the velocity. Furthermore, periodic boundary conditions were used.

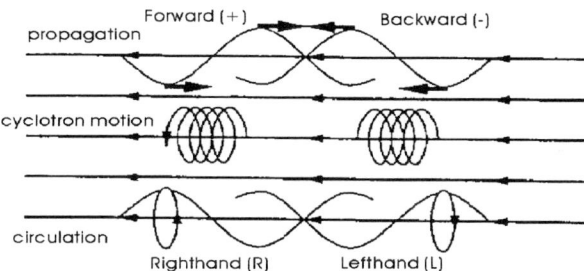

In our simulations, we start from the state in which the Alfvén waves are excited at long wave lengths and the particle motion is made consistent with the input wave. The Alfvén wave has four degrees of the freedom: waves along the magnetic field line are circularly polarized. For right-handed (R) waves, the electric vector rotates clockwise as we look along the field direction; left-handed (L) wave electric vectors rotate counter clockwise. There are two oppositely propagating (forward and backward) waves for each circularized mode. In Figure 2, we show the simulation geometry and the mixing variety of input waves. In Figure 3 (a), we show the Alfvén wave input as the initial data. The right (left) hand side of Figure 3 represents the motion of ions (electrons). The top (bottom) panel of the figure shows the velocity phase space (real space vs. v_x, v_y, v_z velocity). For several thousand time steps, the Alfvén wave propagates monotonically in one direction or oscillates as a standing wave. We take this to imply that our numerical scheme and initial conditions are well representative of the plasma properties in the linear region. We also verified the accuracy of the numerical scheme with linear Alfvén wave theory. After a few thousand time steps, the wave modulates and the sinusoidal shape collapses. This feature is shown in Figure 3 (b). After several thousand more time steps, the wave shapes have collapsed and the particles for the case with a mass ratio of plasma particles equal to 10 is shown in Figure 4. The two graphs show the difference between the mixing ratio of right

(R) and left (L) circulating waves. The top and bottom graphs in Figure 4 represent the case of mixing R and L waves and L waves only, respectively. We see that for the case of mixed R and L waves, particle acceleration is more efficient than that of only L waves, even if the total input wave energy is the same for both cases. We also found that the mixing of oppositely-propagating waves is efficient for particle acceleration. This point suggests to us that coherent wave excitation in our model is important.

4. Discussion

Before a general discussion, we derive the basic properties of this system. The scale L of the magnetic column between the neutron star and the object determines a characteristic frequency $\simeq \frac{1}{\tau_1}$ along with the typical Alfvén speed in the magnetosphere. Since the density is very low and the Alfvén speed may be nearly the light speed c, the frequency $\frac{1}{\tau_1}$ of this system is given by $\omega \sim \frac{\pi c}{2L} \sim 45 \times 10^3 \left(\frac{L}{100\,\mathrm{km}}\right)^{-1}$. The integrated Pedersen conductibility of a surface layer with thickness h is

$$\Sigma_{P,0} = en_e h/m_e \nu_{ei}(1 + (\omega/\nu_{ei})^2)^{-1} \simeq 3 \times 10^{17}(h/0.1\,\mathrm{km}) \tag{4.1}$$

when the surface temperature is $T \simeq 10\mathrm{eV}$. Since $\Sigma_{P,0} Z_0 = 1.2 \times 10^8 \gg 1$ if $V_A \simeq c$, the current decay time scale is $\tau_2 \simeq 10^8 \tau_1$.

Some gamma-ray bursts exhibit structure on time scales as short as the detector resolution (a few ms). The duration of gamma-ray bursts span from a few hundred ms to 1000's of seconds. If the typical value for the shortest time scale τ_1 is related to the variability of gamma-ray bursts, then the circuit has a characteristic scale of $L \simeq 100$ km. On the other hand, using the smallest time scale for τ_1, the current decay time τ_2 would be larger than the burst duration time when the surface temperature is 10eV. This suggests that the duration time may be determined from physics outside of this circuit analysis.

Given the circuit analysis above for the time scale of the variability and its relation to the length scale L, we find that an object must rotate within the 100 km radius in our model. Additionally, the energy of the gamma-ray burst comes from the kinetic energy of the object,

$$E_{p,kinetic} = \frac{GMm_p}{R} = 2.7 \times 10^{38}(M/M_\odot)(m_p/10^{-14}M_\odot)(R/100\mathrm{km})^{-1}, \tag{4.2}$$

where M is the mass of the neutron star and m_P is that of the object, in our model. Comparing this energy to the typical energy observed in gamma-ray bursts,

$$E_\gamma = 10^{42}(F/10^{-6}\mathrm{ergs\ cm}^{-2})(D/100\mathrm{kpc})^2, \tag{4.3}$$

we estimate the mass of the object to be

$$m_p = 3.8 \times 10^{-11}\,\eta^{-1}(R/10^{-6}\mathrm{ergs\ cm}^{-2})(D/100\mathrm{kpc})^2 M_\odot, \tag{4.4}$$

where F is the observed fluence and D is the distance of the burst source with an energy conversion efficiency η. Recent BATSE observations [7] report that the burst sources are distributed isotropicly but not homogeneously. This suggests a disk-like distribution is unacceptable, and any galactic halo distribution must be at least 50 kpc distant. Although the existence of a halo population of neutron stars is not verified, we consider such a population as the source in our models. Given the distance restriction imposed by the BATSE observations, the mass m_P needed to explain the burst energy is not unreasonable for a comet or an asteroid.

We should also point out that our model is more flexible for explaining the gamma-ray bursts than episodic accretion onto a neutron star which needs more fine tuning. As is

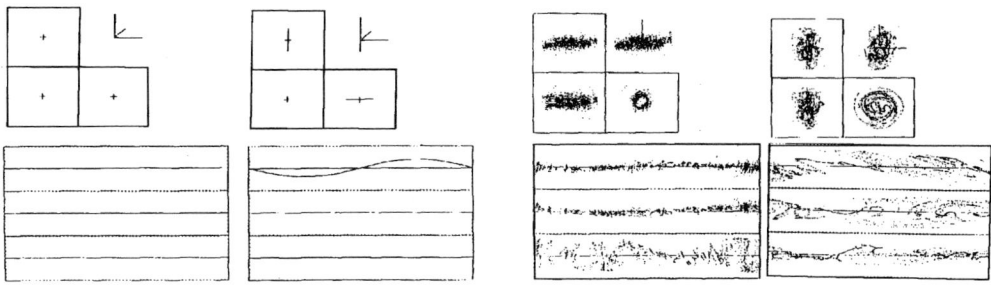

well known for our solar system, comets have a variety of orbital parameters and most of them have very eccentric orbits. This fact suggests that the time interval during which a comet could interact with the magnetosphere in our model spans a wide range, possibly explaining the variety of observed burst duration times. There is, however, a problem with the interaction of the magnetosphere with a comet or planet if the current flows as DC. In this case, the object may begin to melt due to Ohmic heating given typical planet compositions of metals like iron. If the orbiting object is black hole (which is also a conductor [14]), however, this is not a problem.

A key point in our model is the Pedersen conductivity which reaches a maximum when

$$(n_e/3 \times 10^{26} \text{cm}^{-3}) \ln\Lambda (T/10eV)^{-3/2} (B/10^{10}G)^{-1} \simeq 1. \tag{4.5}$$

Even if we consider the high density region just below neutron dropping, we can consider the Pedersen conductivity as significant in the current circuit. ¿From this point of view, a cooled neutron star can make Pedersen current most efficiently. Furthermore, if the surface becomes as hot as $T \simeq 1 MeV$, the duration time, related to the current decay time, should be comparable with τ_1 which is related to the time variability. Thus, it is not appropriate for explaining long duration bursts, but supports the idea that gamma-ray bursts occur in older neutron stars whose surface temperature is lower than 10^5 K, with ages greater than 10^8 yr, given cooling processes [11,12,13].

This work has been partly supported by a Grant-Aid for Scientific Research on Priority Areas by the Ministry of Education, Science and Culture (04233101).

REFERENCES

[1] Blaes, O., Blandford, R., Madau, P. & Koonin, S. 1990, ApJ, 363, 612.
[2] Goldreich, P. & Lynden-Bell, D. 1969, ApJ, 156, 59.
[3] Hanami, H. 1993, preprint.
[4] Harwit, M. & Salpeter, E.E. 1973, ApJL, 186, L37.
[5] Hoshino, M., Arons, J., Gallant, Y. A. & Langdon, B. 1992, ApJ, 390, 454.
[6] Klebesadel, R.W., Strong, I. & Olson, R.A. 1973, ApJL, 182, L185.
[7] Meegan, C.A., Fishman, G.J., Wilson, R.B., Paciesas, W.S., Pendleton, G.N., Horack, J.M., Brock, M.N. & Kouveliotou, C. 1992, Nature, in press.
[8] Melia, F. 1990, ApJ, 351, 601.
[9] Murakami, T. et al. 1988, Nature, 335, 234.
[10] Nakamura, T. & Piran, T. 1991, in preprint.
[11] Nomoto, K. & Tsuruta, S. 1987, ApJ, 312, 711.
[12] Ruderman, M.A. 1991, ApJ, 366, 261.
[13] Shibazaki, N. & Lamb, F.K. 1989, ApJ, 346, 808.
[14] Thorne, K.S., Price, R.H. & Macdonald, D.A. (ed.) 1986, "Black Holes The Membrane Paradigm".
[15] Wolszczan, A. & Frail, D.A. 1992, Nature, 355, 145.

Instability of C-shocks in the ISM

By G. Tóth

Department of Astrophysical Sciences, Princeton University, Princeton, NJ 08544, USA

C-type MHD shocks in the partially ionized ISM are studied in the two-fluid approximation. The ionized fluid can move along the magnetic field lines, while it interacts with the neutral fluid via ion-neutral elastic scattering. I use an explicitly flux conserving 2-dimensional Eulerian FCT (Flux Corrected Transport) code to study the dynamics of two-fluid shocks. A numerical instability intrinsic to two-fluid problems was discovered, and a fix to the problem is proposed. The code can successfully simulate C-type shocks. The results of the linear stability analysis by Wardle are confirmed, and the non-linear behavior of the instability is explored.

1. Introduction

There are several mechanisms driving shocks into the interstellar medium: supernova explosions, strong stellar winds, and collisions of molecular clouds. The partially ionized ISM can be treated as two distinct fluids coupled by the ion-neutral friction. When the fractional ionization is small, the neutral fluid carries most of the density and inertia, but only the ion fluid interacts with the dynamically important magnetic field. If the shock speed v_s is lower than the ion Alfvén speed $v_A^{(i)} \equiv B/\sqrt{4\pi\rho^{(i)}}$ the ions can build up a magnetic precursor ahead of the shock, and the ion flow remains continuous. There is an ion-neutral slip (ambipolar diffusion) throughout the shock front, thus – in the "standard shock frame", where the shock front is at rest – the ions are decelerating the neutral fluid running into the shock at a speed of v_s. If the neutral fluid remains cold due to the weakness of the shock, or because of effective cooling, the neutral flow will be supersonic everywhere and the flow variables will vary continuously, hence the shock is called C-type. Shocks with $v_s \lesssim 25\,\mathrm{km\,s^{-1}}$ in a gas with $10^2\,\mathrm{cm^{-3}}$ ambient density and 10^{-4} fractional ionization are C-type [4]. The maximum velocity rises to around $50\,\mathrm{km\,s^{-1}}$ in denser gas with lower ionization [5].

It was recognized by Wardle [7] that, while the models of the chemistry and physics of C-type shocks assumed a steady state planar shock front, in reality all but the weakest shocks are subject to an instability analogous to the Parker instability. The magnetic field lines can buckle across the shock front and the ions flow along the field lines due to the force of ion-neutral friction. The ion density increases in the troughs (i.e. in the bends closer to the downstream flow), thus the neutral drag will be stronger at these points than at the crests of the magnetic field, and the field lines will bend further (see Figure 1). The linear analysis [7,8,9] showed that there will be growing modes with wavelength on the order of the shock thickness L_{flow} and with an e-folding time that can be much shorter than the flow time t_{flow} through the shock. Therefore the steady state models are of dubious validity, and a fully dynamical simulation is called for to model the physics of C-type shocks which fulfill the instability criterion. Another consequence of the instability is the formation of high density lumps of the neutral fluid, a possible place for low mass

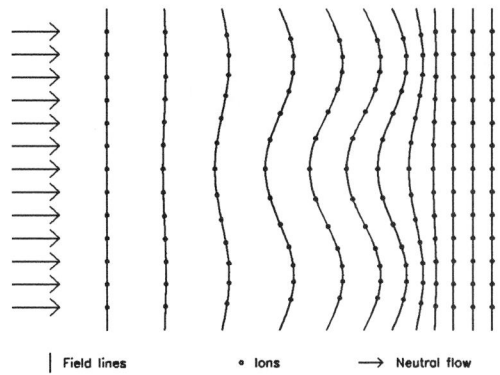

FIGURE 1. Schematic picture of the Wardle instability

$\sim 0.1 M_\odot$ star formation. One needs, however, to know the non-linear behavior of the instability to check if these speculations are correct. There is, of course, a possibility that the instability saturates at a low amplitude, or that the necessary approximations in the linear analysis make the analytic results very different from the solutions of the exact equations; these questions give further motivation for numerical simulations.

2. Modeling C-type Shocks

2.1. *Equations*

The fluid equations are written in terms of the conservative variables: $\rho^{(f)}$, $m_j^{(f)}$, $e^{(f)}$ and B_j, the mass, momentum, energy densities of the fluid f, and the magnetic field strength, respectively. The superscript f is either i for ions, or n for neutrals, while the lower index j denotes any of the three spatial variables, x, y, and z. The momenta are simply $m_j^{(f)} \equiv \rho^{(f)} v_j^{(f)}$. The energy densities contain kinetic, thermal and magnetic contributions. The friction between the two fluids introduce source terms in the momentum and energy conservation equations with opposite signs for the ions and the neutrals. The drag force is $\mathcal{F}_j^{(i)} = -\mathcal{F}_j^{(n)} = \alpha \rho^{(i)} \rho^{(n)} [v_j^{(n)} - v_j^{(i)}]$, where $\alpha \equiv \langle \sigma v \rangle / (M^{(i)} + M^{(n)}) \approx 3.7 \times 10^{13}$ cm^{3}s^{-1}g^{-1} is the coupling constant and $M^{(i)} \approx 30 M_H$ and $M^{(n)} = 7/3 M_H$ are the mean ion and neutral particle masses, assuming $n_{He}/n_H = 10\%$, and the hydrogen is fully molecular.

It proved to be useful for the numerical stability of the calculations to replace the differential equation for the ion energy density by an approximate algebraic equation. As Chernoff [2] pointed out, the ion heat capacity is low thus the heating rate must be ≈ 0, i.e. the ions are heated to an equilibrium temperature $T^{(i)} = T^{(n)} + M^{(n)}(v^{(i)} - v^{(n)})^2/(3 k_B)$ which determines the ion energy density. Numerical tests confirmed that the approximation hardly changes the other flow variables, while the temperature remains smoother than if it was integrated from the differential equation. The energy transfer due to elastic scattering, needed for the neutral energy density equation, simplifies to $\mathcal{E}^{(n)} = -\mathcal{F}_j^{(i)} v_j^{(i)}$.

Though the code can handle any form of cooling function, in this paper $\gamma^{(n)} = 1.001$ will be used for the neutral adiabatic index, which effectively implies an isothermal neutral gas, i.e. a very efficient cooling. This simplification allows easier comparison with the linear analysis.

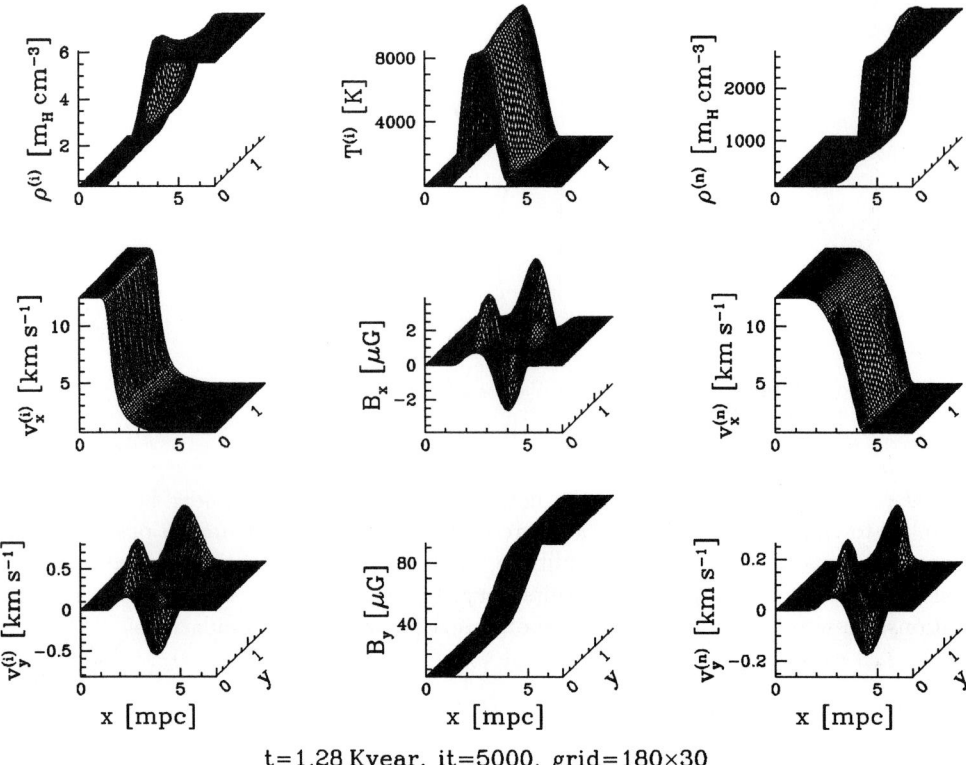

FIGURE 2. The perturbation in the linear stage for a perpendiculer shock

2.2. Numerical Method and Tests

The method of choice was the Flux Corrected Transport (FCT) scheme [1], which offers an explicitly two dimensional Eulerian difference scheme on an optionally nonuniform rectangular grid. It is known to work well for magnetohydrodynamic shock waves, and has a version that can conserve the divergence of the magnetic field to the accuracy of numerical truncation errors [3]. It is relatively easy to code, but sufficiently powerful for our purposes. The interaction between the two fluids introduce a new time scale which limits the time step to $\Delta t < D/\max(\alpha\rho^{(n)} + \alpha\rho^{(i)})$, where the maximum is taken over the grid, and D is a constant of order unity. To relax this restriction I suggest a minor modification to the FCT algorithm: the source terms should be calculated as an appropriately weighted average over neighbouring cells. This way the numerically most unstable staggered mode is eliminated, without an essential sacrifice in the accuracy.

As a specifically two-fluid test, I modeled how a C-type shock is built up when a piston is driven into a uniform medium. The result agrees very well with the high accuracy numerical solution of the steady state equations. Self-consistency and convergence are checked by grid refinement, and by rotation of the initial conditions relative to the grid. Comparisons to the linear theory also confirm that the code can accurately model the dynamics of C-type shock waves.

The initial conditions for studying the Wardle instability are set up by solving the steady state equations first and then perturbing the velocities of the fluids by a small amplitude perturbation ($\approx 10^{-3}$ km s^{-1}) upstream of the shock front. The boundary conditions are continuous upstream and downstream, and periodic in the orthogonal direction. The width of the computational box therefore determines the possible wavelengths for the growing modes. Usually the velocity perturbation is a single sine wave with a wavelength equal to the width of the grid. Figure 2 shows the result of a typical simulation in the linear regime.

2.3. Results

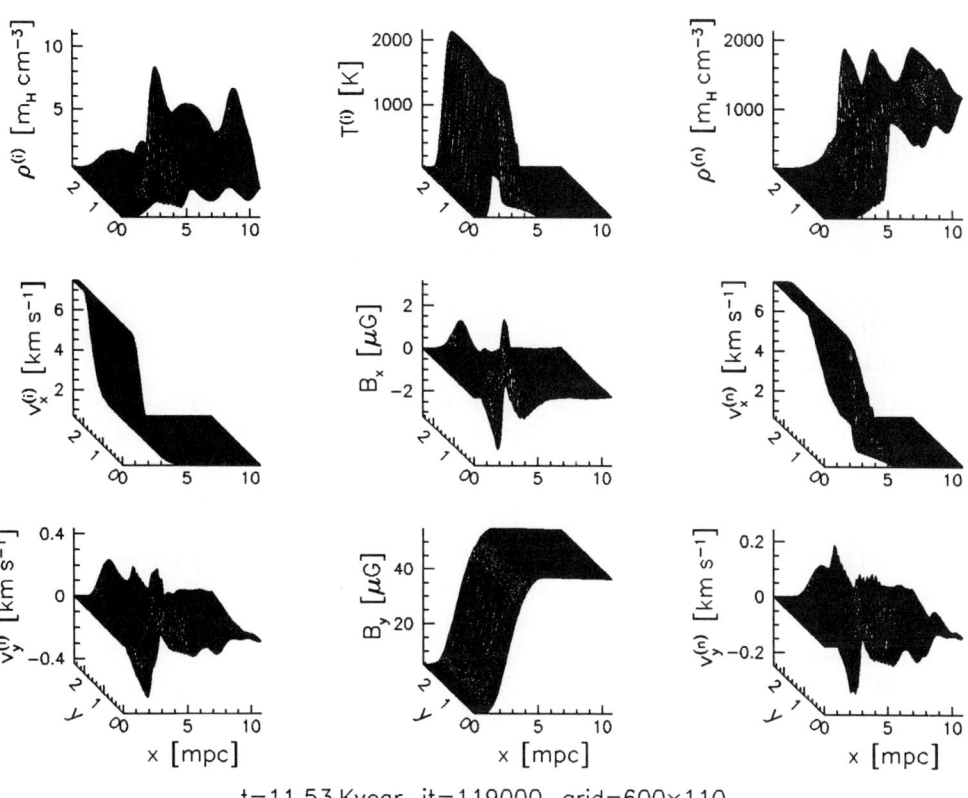

t=11.53 Kyear, it=119000, grid=600×110

FIGURE 3. The non-linear stage of the Wardle instability

A systematic search is done in the numerically accessible parameter space to determine the wave vector $\vec{k}_{max}$ and the growth rate s_{max} of the fastest growing modes. The neutral Alfvén number $A^{(n)} \equiv v_s/(B/\sqrt{4\pi\rho^{(n)}})$, and the angle θ_s between the shock normal and the upstream magnetic field determine the strength and obliqueness of the shock, as well as the the dimensionless parameters of the fastest growing mode. The other physical quantities, like the ambient neutral density, the fractional ionization, and the strength of the magnetic field set the global length and time scales for the shock. The simulations show a strong transient amplification of the initial perturbations, thus to some extent the amplitude and spectrum of the seed perturbations in the ambient ISM will contribute to the selection of the dominant growing mode, but the measured s_{max} and $\vec{k}_{max}$ values should give a reasonable estimate of the characteristic time and length scales of the Wardle instability.

For perpendicular shocks ($\theta_s = 90°$) both s_{max} and k_{max} are found to agree well with the predictions of the approximate analytic calculations. For oblique shocks $k_\parallel \equiv k \cos \phi_k$, the component of the wave vector lying in the plane of the magnetic field and the shock normal, is found to roughly determine the growth rate of the instability for a wide range of $\phi_k \equiv \tan^{-1}(k_\perp/k_\parallel)$. The results confirm and extend Wardle's linear analysis, which is restricted to $\phi_k = 0$. In strongly oblique shocks with $\theta_s \lesssim 45°$, there is some preference for $\vec{k}_{max}$ to lie outside the plane, i.e. $\phi_{max} > 0$, while "quasi-perpendicular" shocks with $\theta_s \gtrsim 45°$ have their fastest growing modes in the symmetry plane; $\phi_{max} \approx 0$. The results suggest that high density clumps formed by the Wardle instability will have relatively well defined diameters in the symmetry plane but the size distribution in the orthogonal direction may vary appreciably.

While relatively small grids (180×30) were sufficient to study the linear stage, the simulation of the non-linear evolution needed a much higher resolution (600×110), and consequently more CPU time, which was generously provided by the CRAY Research Inc on a CRAY Y-MP prior to this workshop. The preliminary results show saturation when the amplitude of the density perturbation is the same order as the density jump accross the shock front. Next the high density clumps become more elongated in the direction of the flow, finally the clumps detach from the shock front and slowly drift downstream (see Figure 3).

3. Conclusions

I have implemented a divergence free version of the FCT finite difference scheme to solve the coupled differential equations describing the dynamics of the two-fluid system, which contains the neutral gas, the ions and the magnetic field [6]. The code succesfully models C-type shock waves. The Wardle instability is studied in detail. The fastest growing modes are found for both perpendicular and oblique shocks in good agreement with the linear analysis. The saturation does not occur until the density contrast of the perturbations becomes comparable to the density jump over the shock front. In the non-linear stage, high density clumps form behind the shock, then detach from it and drift downstream.

This work was supported by NSF, the Hungarian Science Fundation, Cray Research de México, S. A. de C. V., and Universidad Nacional Autónoma de México. The simulations were performed with the CRAY-YMP of the Supercomputing Center at UNAM.

REFERENCES

[1] Book, D. L., Boris, J. P. & Zalesak, S. T. 1981, *Finite-Difference Techniques for Vectorized Fluid Dynamics Calculations*, ed. D.L. Book, Springer-Verlag, New York, 29.
[2] Chernoff, D.F. 1987, ApJ, 312, 143.
[3] DeVore, R.C. 1991, JCompPhys, 92, 142.
[4] Draine, B.T., Roberge, W.G., & Dalgarno, A., 1983, ApJ 264, 485.
[5] Smith, M.D., & Brand, P.W.J.L., 1990, MNRAS, 242, 495 (erratum: 244, 384).
[6] Tóth, G., 1993, ApJ, accepted for publication.
[7] Wardle, M. 1990, MNRAS, 246, 98.
[8] Wardle, M. 1991a, MNRAS, 250, 523.
[9] Wardle, M. 1991b, MNRAS, 251, 119.

MHD Experiments on the Thick Galactic Disk of Gas

By M. A. Martos[1], D. P. Cox[2]

[1] Department of Physics and Steward Observatory, University of Arizona, Tucson, AZ 85721, USA.

[2] Department of Physics, University of Wisconsin-Madison, 1150 University Avenue, Madison, WI 53706, USA.

We present simple models for the Galactic disk of gas, cosmic rays and magnetic field in hydrostatic equilibrium with gravity and parameters adequate to the solar neighborhood. The disk is thick, with a scaleheight of the order of 1 kpc. The stability of these structures, and their response to perturbations of a spiral density wave are studied in two and three dimensions for an isothermal gas. In the stability study, we examine the nonlinear evolution of undular and interchange modes. A model in which the temperature distribution is stratified was found stable. In a model with constant temperature throughout the thick layer, a direct consequence of the assumed thickness is the much increased timescale for the Parker instability in comparison with previous studies. The instability forms a periodic array of dense, thin sheets of gas extending perpendicular to the midplane. We stress, however, the mainly random topology of the magnetic field in the real Galaxy, which should restrict the Parker mechanism's range of operation to the local scales only. Our experiments on the response of the thick layer of gas to the spiral density wave perturbation show a vigorous and complex kinematics above the Galactic midplane. In 2-D, we explored two geometries for the magnetic field: with the field lines along the streamline, the Parker instability is excited but the dense sheets that are formed do not stand perpendicular to the midplane, but making an angle with it, and the shocks are strong; with the field lines perpendicular to the plane of motion the shocks are weak and, for a given amplitude of the spiral perturbation and signal speed, the flow is more disturbed at low relative speeds with respect to the spiral pattern, resulting in eddies and wave motion of large scale.

1. Introduction

The warm interstellar medium is like a galactic atmosphere, a thick disk of cosmic rays, magnetic field and gas extending to a height z above the Milky Way's midplane of $|z| > 1$ kpc. It has been rather slowly that the existence of this thick layer gained recognition over the past few years. Material at high z has been detected in neutral and ionized forms. The neutral component, with a scaleheight of roughly 500 pc, was revealed by 21 cm line studies [13] and observations of trace ions such as Ti II [5]. The thermal electron distribution, with a scaleheight of about 1500 pc, (assumed exponential) is implied by the analysis of pulsar dispersion measures [18]. The nonthermal counterpart is the thick disk of cosmic rays and magnetic field, revealed by the galactic synchrotron emission and Faraday rotation studies [3]. The system of gas, cosmic rays and magnetic field is strongly coupled, with an approximate equipartition between the turbulent, magnetic, and cosmic ray contributions to the total midplane pressure; the thermal pressure is small compared to each of those terms. The quantities displaying the nonthermal distribution drop slowly from $[z] = 0$ to

$[z] = 1.5$ kpc. Thus, throughout the layer, the magnetic pressure is a dominant form of pressure that confers rigidity and elasticity to the medium. As a consequence, thermal pressure balance may not have to be enforced between components of the interstellar medium (ISM), after all; and the inclusion of the thick layer of large nonthermal pressure could significantly alter previous analyses of various processes proposed to ocurr in the ISM, such as galactic fountains or superbubble blowout. All this has been reviewed by Cox [4]. The new picture described above was our motivation to undertake a study on how the effects of disturbances created – say – near the midplane, are propagated in the disk to influence the observed gas kinematics above the midplane. We start by considering relatively small amplitude waves, which is the subject of this paper. In section 2 we present simple models for the disk based on the observed local vertical distributions of gas and gravity; the hydrostatic equilibrium assumption forces the non-thermal distribution. The effect of perturbations on these hydrostatic structures was followed by performing simulations with the 3-D MHD code Zeus. The numerical procedures, and the preliminar results of these numerical explorations on the stability of possible structures and their response to a local perturbation simulating the spiral density wave in the stellar population are given in sections 3 and 4, and summarized in section 5.

2. Models

The condition for hydrostatic equilibrium in the disk, if all quantities depend on the height z only, can be written as an integral for the total pressure: $P(z) = -\int_z^b \rho K_z \, dz$, in which the outer boundary condition $P(b) = 0$ is adopted at $b = 5$ kpc. $\rho(z) = 1.27 m_H n(z)$ and $K_z(z)$ represent the density of interstellar matter and gravity from stars in the z direction, respectively, at the solar neighborhood, where the equation is assumed to be valid. To calculate the pressure integral, we start from estimates of the observed $\rho(z)$ and $K_z(z)$. Our expression for the number density is

$$n(z) = 0.6 e^{-\frac{z^2}{2(70pc)^2}} + 0.3 e^{-\frac{z^2}{2(135pc)^2}} + 0.07 e^{-\frac{z^2}{2(135pc)^2}} + 0.1 e^{-\frac{|z|}{400pc}} + 0.3 e^{-\frac{|z|}{900pc}} \text{cm}^{-3}. \quad (2.1)$$

The different terms correspond to the contributions of H_2, cold HI, warm HI in clouds, warm intercloud HI, and warm diffuse H II, respectively. Our expression for $K_z(z)$ provides a good fit to the gravity given by Bienaymé, Robin & Crézé [2] in the interval $0 < |z| < 2$ kpc

$$K_z = 8 \times 10^{-9} \left(1 - .52 e^{-\frac{|z|}{325pc}} - .48 e^{-\frac{|z|}{900pc}}\right) \text{cm s}^{-2}, \quad (2.2)$$

which is lower in magnitude and assumes less local dark matter than previous determinations, in agreement with more recent data [2,6,10]. We model the disk with two forms of pressure only: a thermal term, representing both the gas kinetic and cosmic ray contributions $P_t = n(z) K T_{eff}$, and a magnetic term $P_b = B(z)^2/8\pi$, allowing the effective temperature T_{eff} to be a function of z in general. Perturbations are imposed assuming quasi-isothermality ($\gamma = 1.01$) and flux freezing. The magnitude of the magnetic field $B(z = 0)$ is taken as $5\mu G$, which gives the correct P_b in the midplane [3]. The problem is initiated with the usual geometry of magnetic field lines parallel to the Galactic midplane. Self-gravity is neglected.

The numerical simulations were performed with the code Zeus, developed by M. Norman, D. Clarke (author of the 3.2 version we used), J. Stone and associated group at NCSA–Illinois. Zeus solves the set of nonlinear, coupled partial differential equations for a compressible, magnetized, perfectly conducting fluid in the usual ideal MHD approximation by finite differencing on an Eulerian mesh, fully explicit in time, with von-Neumann Richtmyer artificial viscosity to smear shocks. A detailed description of the algorithms in the code has been given elsewhere [23,24,25]. Using the adopted $K_z(z)$ and $n(z)$, the pressure integral is numerically solved. The extra condition $P(z) = P_b(z) + n(z)KT_{eff}$ applied at the midplane gives $T_{eff} = 10900K$. One of our models (model A), is defined by T_{eff} constant and independent of z. This condition determines $B(z)$. In a second model (model B), the magnetic pressure $P_b(z)$ is prescribed by

$$P_b(z) = 10^{-12} sech^2 \left(\frac{z}{800pc}\right), \qquad (2.3)$$

which results in a stratified effective temperature $T_{eff} = T_{eff}(z)$ that increases with height.

3. Stability

An exploration of the stability of the models naturally leads to a study of the Parker instability: the equilibrium of horizontal magnetic field lines in a stratified layer of gas subject to external gravity tends to be unstable against two modes of perturbation. One is the interchange mode, an overturning mode that exchanges flux tubes without distorting the straight field lines. The other is the undular mode, which waves the field lines inducing the gas to slide down the lines and collect in their troughs, while the crests of the lines become lighter and rise further. The Parker instability has been proposed to be an effective mechanism to form interstellar cloud complexes [16] with a spacing of a kpc, which in their analysis is consistent with the wavelength of the fastest growing modes of the instability; the spurs extending perpendicular to the Galactic midplane observed in radio continuum [20] and in HI 21 cm line [8], and other structures. Since the seminal work by Parker [17], the theory of the subject was established from linear analyses until recently, when progress in supercomputing allowed the first nonlinear, 3-D simulations of the instability by Matsumoto and Shibata [14]. In contrast to the latter study, our calculations account for the possibility of the sound speed and the Alfven speed being functions of z, and assume a much thicker disk and functional forms for $\rho(z)$ and $K_z(z)$ that are directly motivated by observations. In Cartesian coordinates (x, y, z), we take x in the direction of the initial field lines – the local azimuthal direction in the Galaxy –, y in the radial direction and z in the direction perpendicular to the Galactic midplane. The range of variation in the coordinates we spanned was up to 25 kpc in x, 2 kpc in y, with periodic boundary conditions; z was held in the interval $-2 < z < 2$ kpc with a zero gradient, free outflow boundary condition. Small amplitude (= 1% of the signal speed at most) velocity perturbations were imposed to the static equilibrium state of the form

$$V_z(x\ rmor\ y, z) \sim sin\left(\frac{2\pi x}{\lambda_x}\right)\ orsin\left(\frac{2\pi y}{\lambda_y}\right) \times cos\left(\frac{\pi z}{\lambda_z}\right) \qquad (3.4)$$

depending on whether the undular $V_z(x, z)$ or interchange $V_z(y, z)$ mode was under scrutiny. We had zones of 15 to 100 pc size across and Zeus was integrating about 70000 zones-cycles per second in a Cray Y-MP supercomputer. The grid was fixed and equally spaced except possibly in the z direction.

3.1. Results

Model A, a constant temperature model, was found readily unstable to undular modes. In 2-D, the growth rate increases with wavelength and is large for $\lambda_x \geq 3$ kpc. The timescale is a few times 10^8 yr, an order of magnitude larger than previous estimations that assume a much thinner disk. The instability forms a periodic array of dense sheets of gas perpendicular to the disk plane. The sheets are thin and contain the troughs of the curved magnetic field lines, extending from $z = 0$ to the maximum allowed z. The

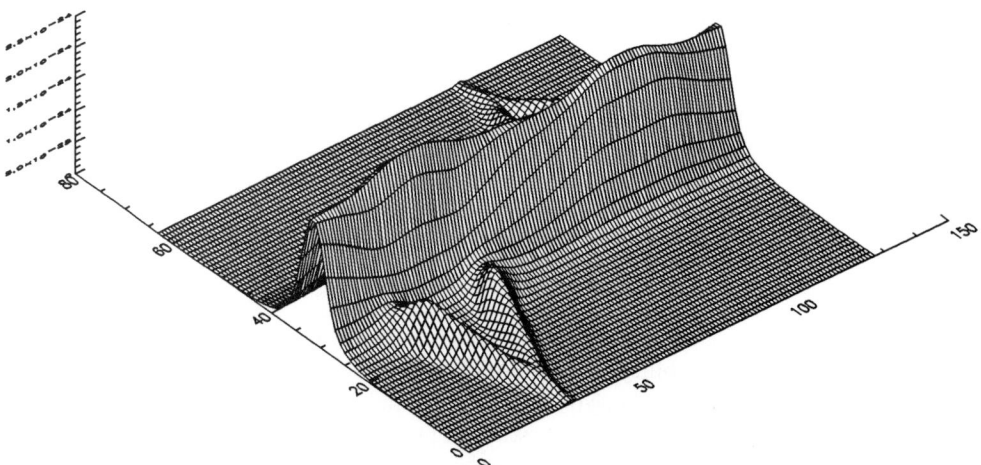

FIGURE 1. Density in the plane xz in g cm^{-3}. x (128 zones) varies from 0 to 6 kpc; z (64 zones) from -1.5 to 1.5 kpc.

array has an odd inversion symmetry with respect to the midplane which is characteristic of the most unstable mode of the Parker instability. When the flow converges onto each sheet supersonically from the sides as the gas slides down the field lines, it conferes a v-shaped double shock structure to the sheets, which is illustrated in Figure 1. In this regime we found that the sheets collapse to form compact, high concentrations of density, or "clouds", but their occurrence was rather exceptional in our experiments. In general the sheets stand perpendicular to the midplane and do not collapse over timescales of up to 10^9 yr, our limit because of numerical difficulties due to the large rarefactions between sheets. Although linear theory predicts that the array is indeed an equilibrium state of the system [11], the equilibrium is expected to be unstable [12]. What we found is that the sheets are a long lived phase. The perturbations wave the densest central disk in modes that are reminiscent of bending waves in the stellar population. The instability proceeds faster for 3-D interchange modes, but the growth rate is not much different that in 2-D. Short waves have the largest rates in these 3-D modes; they also result in the formation of dense sheets. Model B, the stratified temperature model, remained stable. This is a consequence of allowing the kinetic pressure to increase with z in the model, which makes the sheets wider and wider at higher z and no energy reduction is possible by clumping the gas into sheets.

4. Response to a Spiral Density Wave-like Perturbation

Early studies of this problem found that periodic solutions for the gas flow in Galactic differential rotation perturbated by spiral arms contain shocks [7,19]. The compression of gas by the shocks, with important consequences for star formation and observable signatures has motivated extensive theoretical work. A possibility is that the shocks are the triggering mechanism for the Parker instability [16]. The vertical structure of these large scale shocks has received less attention, despite HI 21 cm observations of vertical features and motions of rising and falling gas associated with the spiral arms [9,21,28]. Early nonlinear simulations of the gas flow allowing the extra degree of freedom in z did not confirm the vertical motions, but the shocks stood vertically the entire allowed z range and their extension was reduced if the effective temperature could increase with z [22,27]. Large scale vertical motions are to be expected in a thick Galactic disk of gas, in which gravity from midplane stars can reach large values. The bulk of restoring forces for gravity waves will come from high z. These motions are supressed in previous models, which assumed a thin layer of gas of 3 to 4 hundred parsecs, although the importance of gravity waves had been already anticipated [26]. Also, the breakdown of hydrostatic equilibrium at the shock will drive gas away from the Galactic plane if the dimensionality allows the gas to do so, expanding the scaleheight of the gas at the front. The transition will resemble the hydraulic jumps or bores occuring in incompressible fluids [15]. Here we extend previous analyses by allowing z motions and including the magnetic field. Using Zeus, we calculated the gas flow through the potential well of the spiral arm. The interaction is modeled as a local one, removing the usual assumption of periodicity. In 2-D, our Cartesian computing grid is placed at the solar location in the Galaxy and the WKB approximation is adopted, so that the relevant coordinate in the disk plane is x, measured perpendicular to the spiral; z is the vertical coordinate as above. Physical effects based on the radial dependences of K_z and differential rotation are deemed small over the grid and ignored. The gas enters the grid from the $x = 0$ edge, injected steadily with a speed V_e independent of z, and leaves it at the $x = 4$ kpc edge with a free outflow boundary condition which holds also for the $z = \pm 2$ kpc boundaries. We initialized grids of 128×64 to 256×256 zones with the hydrostatic distribution of model A and carried out two different sets of calculations, each corresponding to a given orientation of the field lines: i) along the initial gas flow, $\vec{B} = B(z)\hat{x}$; or ii) kept perpendicular to the plane of motion, $\vec{B} = B(z)\hat{y}$. The spiral perturbation is modeled fixed in the coordinate system (the pattern frame) as a potential well $U(x,z)$ which consists of one cycle of a cosine funcion in the variable r defined as the distance to a fixed point on the x axis representing the potential minimum within the influence zone $r \leq 1$ kpc. The amplitude of this perturbation introduces a new free parameter, V_p, which is the speed that a test particle would acquire if it fell to the deepest point in the well from a rest position outside the well. The speed V_e is the component transverse to the arm of the relative speed with which the gas encounters the spiral pattern. To minimize transients, the amplitude was varied linearly with time from 0 to its final value after a switch-one time of between 100 to 500 megayears. We studied the response for a range of parameters; for V_e, 5 to 30 km s^{-1} (which contains the probable local value inferred from diverse data and assumptions on the angular pattern speed, the galactocentric distance and at least a local value of the rotation curve); for V_p, we explored the values $V_p=15,20$ km s^{-1}, which correspond to a ratio of spiral to axisymmetric acceleration at the local standard of rest of 0.03 to 0.05. The signal speed c for magnetosonic waves is 12.5 km s^{-1} and 9.2 km s^{-1} for Alfven waves at the midplane in our model.

4.1. Results

Three parameters control the problem in 2-D: V_e, V_p and the signal speed c. Just like in 1-D [1], the shocks move in general with respect to the spiral perturbation. They are stationary only for a certain combination of the parameters: given V_p and c, there are values of V_e for which the shock remains stationary. Thus the shock front is locally determined; its strength is proportional to V_p. The results were not sensitive to the details of the perturbating potential. In all cases, the gas layer effectively thickens at the potential well, rising the density contours to higher z.

4.2. Case $\vec{B} = B(z)\hat{x}$

The spiral perturbation excites the Parker instability; the dense sheets do not extend perpendicular to the midplane in general, but in oblique angles. Because there is not a

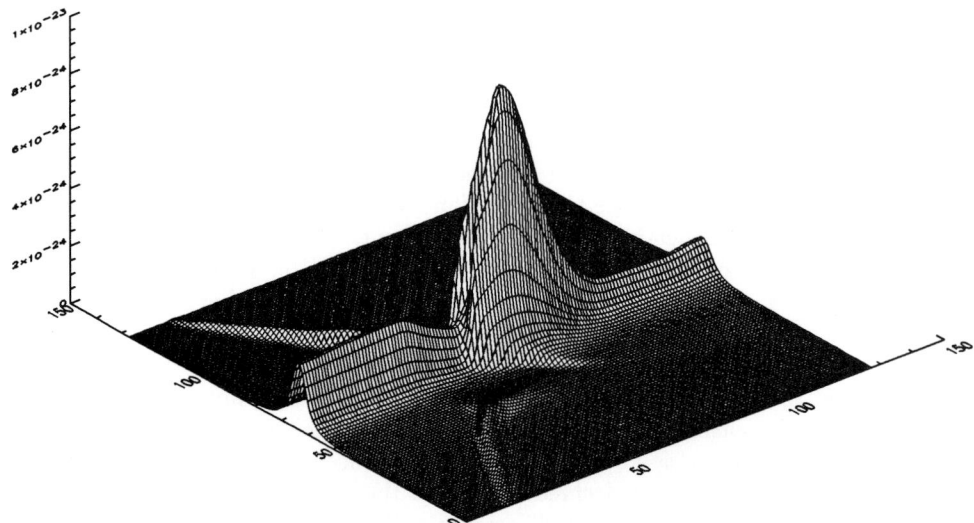

FIGURE 2. Density in the plane xz in g cm^{-3}. x (128 zones) varies from 0 to 4 kpc; z (128 zones) from -2 to 2 kpc. The line of density maxima (zone 64) corresponds to the Galactic midplane $z=0$.

well defined perturbating wavelength, the array is not perfectly periodic and the sheets show irregularities illustrated in Figure 2. The inversion symmetry with respect to the midplane is preserved, however. The shocks are strong, with compression factors of about 4 ($V_p=15$ km s^{-1}) to 7 ($V_p=20$ km s^{-1}) typically. The gas reaches velocities close to 40 km s^{-1} in the z direction and 70 km s^{-1} in the x direction as it slides down wide arches of magnetic lines.

4.3. Case $\vec{B} = B(z)\hat{y}$

Given c and V_p, for a high entry speed V_e (compared with c) the flow interacts weakly with the well and only accelerates through it, consequently decreasing the density in the well. For low V_e the gas slows down as it enters the well and tends to accumulate there. Although the perturbation assures supersonic flow, the shocks are transients that evolve into broad peaks which move downstream (high V_e) or upstream (low V_e) from the well, except for a certain interval of V_e which depends on the other parameters. The associated compressions never exceeded a factor of 2. The two regimes have distinct associated vertical motions. For fixed V_p and c, the low V_e regime ($V_e \leq 15$ km s^{-1}) is the one of important z motions, with frequent large scale eddies and reversed flow, (Figure 3) reaching 20 km s^{-1}, and up to 30 km s^{-1} in x at high z; the high V_e regime is one of negligible vertical motion: the gas simply accelerates past the well reaching horizontal speeds exceeding 60 km s^{-1} above the midplane for V_e between 20 to 30 km s^{-1}, for the 2 values of V_p we tried.

5. Summary and Discussion

Perhaps the most important caveat that should be borne in mind in drawing conclusions about the real Galaxy from all the Parker instability studies known to us, including the present one, has to do with the irrealistic assumption of straight magnetic field lines. The field is too disorderly to organize itself in a coherent large scale mechanism to form structure, but the instability should be effective locally, at some small but finite scale determined by other effects that supress the 3-D shortest waves of the dominant interchange modes, such as shear and self-gravity. The scale for relevant undular modes is too large,

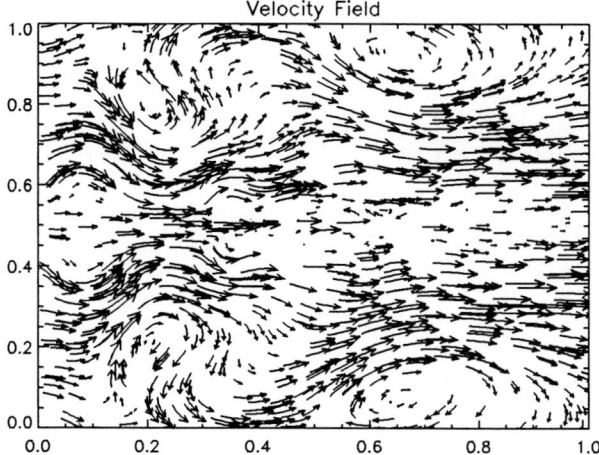

FIGURE 3. Velocity field in plane xz. The horizontal axis is the x axis, from 0 to 4 kpc. z varies from -2 to 2 kpc. The position of the potential well is the center of the mesh, which has $128 \times 128 zones$. The gas enters from the left (x=0) with V_e=5 km s^{-1}; V_p=20 km s^{-1}. Elapsed time is 400 megayears. The maximum vector length corresponds to about 15 km s^{-1}.

of at least 3 kpc for the thick disk of gas we find it represents better the observations; the timescale for the instability is $\sim 10^8$ yr, an order of magnitude larger than previously estimated and than the cosmic ray residence time, making a causal association between CR escape and the instability unlikely.

Interesting vertical motions of gas associated with midplane perturbations are to be expected at high z even in a regime of rather modest energy, as these experiments have shown. The spiral perturbation can accelerate gas radially to speeds of up to 65 km s^{-1} (at high z) without having to invoke the Parker mechanism, and create a complex kinematics of rising and falling motions extending to vast regions of the gaseous disk. The dense, thin extensions of material from the central plane into the galactic "atmosphere" could indeed have relevance to observed structures.

A more detailed analysis of the results of this work, and from additional 3-D experiments currently under way will be reported elsewhere. They include effects from thermal processes, differential rotation, and arbitrary angles between the magnetic field and the circumferential direction in different models for the system of gas, magnetic field and cosmic rays.

The numerical computations were performed in the Cray 2 and Y-MP supercomputers at NCSA-Illinois, and continue in the Cray Y-MP at UNAM-México. The authors are grateful to M. Norman, D. Clarke, R. Smith, L. Sparke, C. Goebel and B. Savage for fruitful conversations.

REFERENCES

[1] Baker, P. L. & Barker, P. K. 1974, AA, 36, 179.
[2] Bienaymé, O., Robin, A. C. & Crézé, M. 1987, AA, 180, 94.
[3] Boulares, A. & Cox, D. P., 1990, ApJ, 365, 544.
[4] Cox, D. P. 1989, *Structure and Dynamics of the Interstellar Medium*, ed. G. Tenorio-Tagle, M. Moles, and J. Melnick. (Berlin: Springer-Verleg), 500.
[5] Edgar, R. & Savage, B. 1989, ApJ, 340, 762.
[6] Einasto, J., Joeveer, M., and Saar, E. 1985, *IAU Symposium 117, Dark Matter in the Universe*, ed. G. R. Knapp and J. F. Kormendy (Dordrecht:Reidel), 243.
[7] Fujimoto, M. 1966, *IAU Symposium 29, Non Stable Phenomena in Galaxies*, ed. M. Arakeljan

(Academy of Sciences of Armenia SSR), 453.
[8] Heiles, C., 1984, ApJS, 55, 585.
[9] Kepner, M., 1970, AA, 5, 444.
[10] Kuijken, K. & Gilmore, G. 1989, MNRAS, 239, 605.
[11] Lerche, I., 1967a, ApJ, 149, 395.
[12] Lerche, I., 1967b, ApJ, 149, 553.
[13] Lockman, F., Hobbs, L., & Shull, J. 1986, ApJ, 301, 380.
[14] Matsumoto, R. & Shibata, K. 1991, *Numerical Astrophysics in Japan 2*, ed. S.M. Miyama and M. Nagasawa (NAO, Mitaka), 177.
[15] Moore, D. W. & Spiegel, E. A. ApJ, 154, 863.
[16] Mouschovias, T. Ch., Shu, F. H. & Woodward, P. R. 1974, AA, 33, 73.
[17] Parker, E. N. 1966, ApJ, 145, 811.
[18] Reynolds, R. 1989, ApJ, 339, L29.
[19] Roberts, W. W. 1969, ApJ, 158, 123.
[20] Sofue, Y. 1976, PASJ, 25, 207.
[21] Sofue, Y. & Tosa, M. 1974, AA, 36, 237.
[22] Soukup, J. E. $ Yuan, C. 1981, ApJ, 256, 376.
[23] Stone, J. M. & Norman, M. L. 1992a, ApJS, 80, 753.
[24] Stone, J. M. & Norman, M. L. 1992b, ApJS, 80, 791.
[25] Stone, J. M., Mihalas, D. & Norman, M. L. 1992, ApJS, 80, 819.
[26] Tosa, M. 1973, PASJ, 25, 191.
[27] Tubbs, A. D. 1980, ApJ, 239, 892.
[28] Weaver, H. & Williams, D.R.W., AAS, 17, 1.

Turbulence and the Interstellar Medium

By A. Pouquet

Observatoire de Nice, BP 229, 06304 Nice Cedex 4, France.
(pouquet@tailleferre.obs-nice.fr)

The plausibility of an analysis of existing data on the interstellar medium in terms of turbulence has been pointed out by several authors. However, many open questions remain, some of them concerning the structure (large scale versus small scale) of the magnetic field, as well as its origin, and its role in preventing gravitational collapse. A succint description of our present understanding of the main characteristics of turbulent flows, with either compressibility and/or magnetic fields included, is reviewed on the basis of recent numerical simulations of the basic equations. Some numerical models of small–scale flows are also described. Statistical quantities, as well as characteristic structures are briefly discussed. Also quoted, with strong bias, are some of the evidence that indicates that molecular clouds embedded in the interstellar medium are in a state of fully developed complex turbulence. Within that framework, the influence of several agents on the dynamics of the interstellar medium is discussed, in particular that of the slowing down of gravitational collapse by turbulence through either pressure effects, or through a coherent mechanism of energy transfer to small scales due to the formation of shocks, and leading to a marginal equilibrium at all scales, and the open questions when this phenomenology is extended to the MHD case.

---

1. Introduction

Molecular clouds observed in the disk of our Galaxy have a complex structure. The process of star formation within them is inefficient (as low as a few percent of the gas is transformed into stars), and the life time of these clouds is longer than the Jeans' analysis predicts, probably by a factor ten. Why? Turbulence is often invoked, but observations tell us that it is supersonic so that shocks, as well as vortices, are bound to play a role. Magnetic fields, observed on large scales and dynamically important, can also slow down gravitational collapse, but they are turbulent so that their underlying structure may also be an agent in the evolution of the cloud. Heating, in particular through stellar winds of previously formed stars and supernovæ explosions, and cooling must also be taken into account; they take place on time–scales significantly shorter than the hydrodynamical times, and yet turbulence is essential in the energetic balance of clouds, at least at the 1 kpc scale [45], and presumably at smaller scales as well.

This paper attempts a brief review of the features of what is presently known of complex turbulent flows that may be relevant in the context of the interstellar medium, this knowledge stemming mostly from a series of three–dimensional numerical simulations over the last decade; the supersonic case is summarily discussed in Section 2, together with the modelization of turbulent flows in the context of their numerization; a plausible phenomenology of the interaction of self–gravity and turbulence supported by two–dimensional numerical simulations for the neutral fluid and for the MHD case is described in Section 3; conclusions are given in Section 4.

There are numerous indications that the interstellar medium is turbulent, and extensive reviews have already appeared [42,11]. This paper centers on the molecular clouds of intermediate size (typically a few parsec across, containing of the order of 200 solar masses, and at a temperature of 20 K) embedded in a giant molecular clouds of the galactic disk. Within such clouds, supersonic velocities are common, with Mach numbers of the order of four or more. At the microscale, supernovæ explosions of previously formed stars have a disruptive effect on the cloud and inject back energy into the medium. Moreover, both through Zeeman broadening and polarization measures, a dynamically important large-scale magnetic field has been observed with both a uniform and a random component; the flow is sub-alfvenic ($v_0/v_A < 1$ with $v_A \sim B_0/\sqrt{\rho_0}$ the Alfvén velocity constructed on the mean magnetic field B_0 and mean density ρ_0 of the medium, and v_0 the *rms* velocity).

As pointed out for example in Scalo's review [42] as well as by Elmegreen [9] but in a different context, interstellar turbulence differs in many ways from classical incompressible turbulence, because of the presence of many complex phenomena, such as gravitational and thermal instabilities, galactic rotation, density fluctuations possibly fragmenting or coalescing, tidal effects, supersonic motions, energy input at various scales including small scales with SN explosions and stellar winds, magnetic fields, radiative transfer, interactions with interstellar dust grains , chemistry within shocks, ... If turbulence concepts were restricted to the classical Kolmogorov case (K41) – with a large–scale energy containing range, an intermediate inertial range through which kinetic energy flows to small scales at a constant rate $\varepsilon = dE^V/dt$ leading to a modal distribution $E^V(k) \sim \varepsilon^{2/3} k^{-5/3}$ where E(k) is the Fourier kinetic energy spectrum, and a dissipation range at small scales with a rapidly falling–off spectrum – they might not apply to the ISM.

However, there are examples both in meteorology and oceanography where injecting energy into the physical system at various scales may nevertheless lead to a K41 spectrum, or its variant for two–dimensional turbulence. In fact, it was shown that, given some restrictions on the rate of energy injection at the different scales, a Kolmogorov spectrum may still occur globally at all inertial scales [39]. Indeed, through interstellar scintillation, Armstrong *et al.* [1] have demonstrated the existence of a density spectrum $E^\rho(k) \sim k^{-5/3}$ on eighteen orders of magnitude of scales, a result that is being confirmed by more recent observations, closing in particular some of the gaps in the previous data (Spangler, private communication).

Furthermore, numerous studies of flows more complex than an incompressible neutral fluid have allowed a better understanding of turbulence, in particular in the context of weak turbulence interacting with waves, in the acoustic case [9] as well as for various plasmas conditions [50]. So that even though the *relevance* of the concepts of turbulence in the ISM is unquestionable, the precise way in which these concepts can be applied is quite open since, in the theoretical framework, we can only deal with more or less sophisticated variants of a phenomenological approach. Hence the important rôle that numerical simulations (and modelization) of astrophysical flows is playing.

In the absence of a theory of turbulence, one cannot define it precisely but one can attempt to describe it. *Non–linearity* is the key. By opposition to temporal chaos – which only involves a few spatial modes (at least three, such as in the Lorenz model) and which leads to complex temporal behavior – spatio–temporal turbulence involves a large number of interacting spatial scales as well. The dimension of the underlying attractor is large and may not saturate as the Reynolds number increases. The non–linear interactions among the modes lead to steep gradients, by opposition to solitons – which need not be considered turbulent and for which an exact balance between steepening and dispersive effects is reached. With this formation of small scales, enhanced dissipation occurs – modeled by eddy-viscosities – in either a finite time or a Reynolds–dependent time. This point is relevant for example in the context of the heating of solar and stellar corona by magnetic currents and is still quite open even in the 2D case: one of the difficulty involves the various relevant time scales due to the superposition of ideal and resistive instabilities (see [33], Figure 1). The power law $E(k) \sim k^{-m}$ is not necessarily a K41 law: m may depend on the physics. For example in incompressible MHD [20,23] $m = 3/2$, unless the correlations between the velocity and the magnetic field are strong.

Another striking feature of turbulent flows is the intermittency (small filling factor) of small–scale features, leading to non–gaussian probability distributions of velocity derivatives, see for example [48]. Such exponential wings have also been observed in MHD, including when the small–scale flow is not treated fully explicitly [26], and for supersonic flows [37].

There is numerous evidence for turbulence in the interstellar medium. The best known is Larson's law [24] that displays scaling laws between the velocity dispersion Δv, the scale of a cloud R and the density ρ, namely:

$$\Delta v \sim R^{1/2} \quad , \quad \rho \sim R^{-1} \ .$$

For Fourier spectra, the equivalent relations are $E^V(k) \sim k^{-1}$ and $E^\rho(k) \sim k^0$. Note that this implies that the kinetic energy is independent of scale. These laws have been confirmed by many observations since. Another evidence of turbulence is the fact that velocity histograms for several molecular clouds display exponential wings [11]: high ± values of the velocity are more probable than if the distribution were gaussian; this also occurs for natural flows and in the laboratory. Other evidence yet comes from the computation of the fractal dimension associated with molecular clouds through an evaluation of the scaling between perimeter and area [43] and through a wavelet analysis [18].

A characteristic feature of molecular clouds is the multiplicity of structures, and in particular their high degree of filamentation on several length scales. Filaments are also observed in 3D numerical simulations of turbulence. In the incompressible case, vorticity sheets form in the inviscid phase which then roll–up either through a Kelvin–Helmholtz instability or by a self–focusing mechanism, as first described by Neu [29], see also [34]. The flow is composed of a multitude of randomly oriented filaments which are long–lived structures [5,48]. What happens in the MHD case is not clear, since the coupling of current and vorticity may lead to competing phenomena [36]. In the case of an initially weak magnetic field, the structures that develop are intermittent [27] and filamentary [15] including in the slightly compressible convective case [30].

2. Numerical simulations of supersonic flows

2.1. *Three–dimensional computations*

Numerical simulations of decaying three–dimensional compressible flows have been performed using spectral methods on grids ranging from 64^3 points [13,21] to 128^3 [3], and more recently, using the PPM algorithm for supersonic flows, on a uniform grid of 512^3 points [37]. What those simulations show clearly is that (i) the shocks develop first, the vortices later; (ii) at early times, the production of vorticity sheets occur mostly in the vicinity of either curved or colliding shocks; (iii) at late times, a plethora of vortex filaments, interacting and curving around each other, is found; (iv) in the late subsonic phase, the energy spectrum follows approximately a $k^{-5/3}$ law, with a shallower buffer range at small scales, which may follow a k^{-1} law. Forced computations in [21] also find a $-5/3$ law at lower resolution by time–averaging the data. However it should be pointed out that, by comparing the energy spectra at different resolutions ranging from 64^3 to 512^3 grid points, it is clear that 512 grid points in each direction are insufficient to correctly resolve the inertial range: out of the 256 : 1 ratio in wavenumbers, less than a factor 10 is available for the range in which non–linear turbulent interactions take place un–impedded by dissipation processes or un–contaminated by large–scale eddies. Moreover, with the PPM code, the spectra have not yet converged at those resolutions.

The other prominent structures that develop, besides vorticity filaments, are density filaments, both at high and at low densities; low entropy patches stretch into filaments which are regions of high density (and vice–versa) because of small pressure fluctuations (since the original pressure fluctuations are dissipated by shocks).

The numerically–computed velocity, density and entropy fields stemming from the 512^3 computation are being used presently as a data–base to empirically model the ISM. The same analysis – in terms of column density and velocity–intensity spectra as well as histograms of moments of the velocity distribution – are compared for the computation and for recent observations in the ^{12}CO $J2:1$ transition in a cloud [12]. Broad features of observed spectra are recovered in the simulated ones, such as double peaks, asymmetries and non–gaussian profiles. Moreover it can be shown on specific examples that the high–velocity structures are associated with a few identifiable events, involving either shocks or strong vortices.

Of the few simulations that have been done in compressible MHD turbulence in two space dimensions, two simple facts emerge [44,35,40]: **(i)** There is, like in the non–MHD case, a critical Mach number above which the small–scale flow in the early–time regime is dominated by shocks and below which it is dominated by vortices as shown for example in the Table 1 below. The data stems from numerical simulations of the Orszag–Tang vortex [31] for various rms Mach numbers; the initial conditions consist of a large velocity eddy, at the center of which is located a magnetic X–point. Table 1 gives the ratio of the maximum current density to the maximum vorticity, and of the maximum to minimum density in the flow at a fixed time, of the order of a few eddy turn–over times, and for computations at a kinetic and magnetic Reynolds numbers of ~ 500; two clearly distinct regimes appear, with a transitional Mach number around 0.3, corresponding presumably to the occurence of shocks locally in space. **(ii)** The ratio of the longitudinal to solenoidal component of the velocity is higher in the MHD case than in the non–MHD case, because of magnetic pressure; **(iii)** expanding bubbles as well as filaments appear in the density field because of the local Joule heating at magnetic neutral X–points.

TABLE 1

Mach number	j_{max}/ω_{max}	ρ_{max}/ρ_{min}
0	2.83	1
0.2	2.78	2
0.5	1.75	14
1.0	1.40	60

Looking at scaling laws, it appears that there are several intermediate regimes that can be identified temporally, although it is not clear whether higher–resolution 2D runs would still behave similarly; such runs should certainly be done. If one looks for example at the scaling with Reynolds numbers R (assuming unit magnetic Prandtl number) of the total dissipation $2R^{-1}(<\omega^2> + <j^2>)$ – where $\omega = \nabla \times \mathbf{u}$ and $\mathbf{j} = \nabla \times \mathbf{b}$ are the vorticity and current density – the first temporal peak scales logarithmically with R whereas at late times after reconnection has taken place, it appears independent of R [33].

In three dimensions in the incompressible case, an inverse cascade of magnetic helicity $H^M =< \mathbf{a} \cdot \mathbf{b} >$ where $\mathbf{b} = \nabla \times \mathbf{a}$, takes place leading to large–scale force–free ($\mathbf{j}//\mathbf{b}$) fields, see [16] for a review; also [38]. In the subsonic case, Horiuchi & Sato [19] have shown that inverse transfer of H^M also occurs; however, the fully supersonic case has not been computed, and yet is highly relevant to the ISM. Does one obtain large–scale helical filaments that are magnetized? Are such filaments force–free? This latter question is of importance since in the case of quasi–force–free fields, the drift term of ambipolar diffusion becomes negligible (although in fact a self–consistent formulation of the complete equations including the ambipolar term should be used to perform the analysis of the behavior of magnetic helicity). Finally, it should be noted that three–dimensional twisted filaments have been observed for example in L204 [17], in L1641 in Orion [14] and in HI in high latitude high velocity clouds [47]; in the latter case, they seem to be controlled by the magnetic field, and with fine structures on several length scales.

2.2. Models of turbulence in the numerical framework

In view of the large Reynolds numbers of astrophysical flows, a direct numerical simulation including all dissipative terms with their proper (small) magnitude is not realistic. There are several ways around that problem. One can for example introduce explicitly in the equations the transport coefficients computed analytically, such as eddy–viscosities. However, such coefficients are numerous: besides eddy–viscosity, eddy–noise stemming from two small–scale eddies feeding through a beating mechanism the large scales, should also be considered; and for more complex flows involving compressibility, rotation, magnetic fields, anisotropies [49], ... the intricacy of such models grows fast. Another possibility is to resort to the $k - \epsilon$ modelization often used in aerodynamics [10]. Common to both is an attempt to keep a handle on the underlying physics.

Of a different nature is the numerical approach, which considers that the small scales do not have to be treated in detail, granted that properties of the flow such as steepness and monotonicity of profiles are kept. In this case, the Euler equations are used with proper numerical treatment of steep gradients; there are several examples of this approach at this conference.

Yet another method is to include dissipative terms explicitly, but in a modified way that leads to substantial savings in both CPU and memory. In that category, one finds the often–used hyperviscosity methods, linear or non–linear, where the Laplacian term in the Navier–Stokes equations is replaced by $\Delta^{2\alpha}$ with $\alpha \sim 8$ in two dimensions and $\alpha \sim 2$ in three dimensions (because of the lower resolution of 3D runs, α must be smaller to ensure that decay in the vicinity of the smallest wavenumber does not arise too abruptly). Such operators do not ensure positivity of the dissipation everywhere in space, but are only globally dissipative. This leads to non–monotonicity of profiles near shocks [32,33], which has to be removed through appropriate filtering. With $\alpha = 4$, they have been recently used for self–gravitating fluids [6,46].

A method introduced recently deals directly with wavenumbers: the simple remark is that for a computation on a grid of, say, 1024^3 points, roughly only a few millions are in the inertial range; the idea is then to eliminate modes in a systematic fashion by allowing non–linear interactions only on a restricted set of wavenumbers [45,26]; whereas in the latter paper modes are only eliminated in the dissipative range, in the former paper they are eliminated throughout. These methods have been tested in a variety of cases in 2D, for example for supersonic flows and MHD. and give satisfactory results at large scales, although much remain to be done to further test them.

3. The interaction of gravity and turbulence

3.1. Turbulence, pressure and shocks

Consider a cloudlet of size l ; it receives energy through gravitational contraction at a rate $\tau_{ff} \sim \sqrt{\rho_0}$ independent of l ; and it looses energy at a rate τ_{tr} by transferring it presumably to smaller scales. The ratio of those two times

$$\tilde{r} = \tau_{tr}/\tau_{ff}$$

will govern the behavior of the cloud: for small enough $\tilde{r}$, collapse is prevented, the critical value of $\tilde{r}$ being ~ 1. Once τ_{tr} is evaluated, the scale–dependence of $\tilde{r}$, as in a dispersion relation, gives the critical scale of onset of collapse. In the case of Jeans' analysis, $\tau_{tr} = \tau_{ac} = l/c_s$ where c_s is the speed of sound in the quiescent medium. From the dispersion relation that results, one concludes that scales larger than the Jeans' length L_J are unstable. Chandrasekhar [7] took into account the effect of turbulent pressure as well to oppose gravity; this can be reformulated as taking $\tau_{tr} = \tau_{NL} = 1/\sqrt{k^3 E^V(k)}$ where τ_{NL} is the eddy turn–over time of the turbulent flow evaluated in a Kolmogorov–like fashion and $E^V(k)$ the energy spectrum. Bonazzola et al. [4] noted that with the above expression of τ_{NL}, there will be an *inversion* of Jeans' criterium – i.e. now *small scales* would be unstable – when the energy spectrum is *steeper* than k^{-3}.

However we know from observations that $E(k) \sim k^{-2}$ for a wide range of scales. But in fact, in evaluating in a phenomenological way the time τ_{tr}, we should ideally take into account the fact that the flow is both supersonic (with eddies, acoustic waves and shocks interacting) and sub–alfvenic. Léorat et al. [25] showed that for a non–magnetized supersonic flow the transfer time can be simply evaluated as the characteristic time of shock formation: $\tau_{tr} = \tau_{shock} = L_0/\Delta_s u$ where L_0 is the large scale of the flow and $\Delta_s u$ the velocity jump in a shock. This time, like τ_{ff}, is independent of the size of the cloud: for τ_{shock} sufficiently small (*i.e.* for supersonic flows), all scales are in *marginal gravitational equilibrium* without collapsing. This was corroborated by a series of numerical simulations in 2D [25]. Moreover, along filamentary shocks, the density concentration, in particular for the radiative case, may be sufficient to render collapse possible again because locally the free–fall time becomes short, and clumps along filaments will emerge. Intermediate cases, when the cloud as a whole slowly collapses and there are numerous local regions with local clumps with a flat density spectrum may be the case of most interest for the ISM; indeed, Larson's relations do give a flat density spectrum.

3.2. The rôle of the magnetic field

What happens in the magnetic case? We can expect an effect of magnetic pressure in further stabilizing the cloud [41], as shown for example in several 2D computations [40]. On the other hand, when shocks form, the mechanism of marginal equilibrium quoted above may also come into play. In the compressible case, the evaluation of the transfer time in the collapse parameter $\tilde{r}$ should be based on the fastest mode, *ie* the fast magnetosonic wave with its associated Mach number $M_{sr} = v_0/\sqrt{c_s^2 + v_A^2}$ [40] which involves the combined effect of the compressibility of the flow and of the magnetic field. For a supersonic sub–alfvenic turbulence such as in the ISM, the two effects may in fact partially cancel each other, and the slowing–down of collapse may be reduced to simply a pressure effect, less efficient than the supersonic mechanism described above. However, when magnetic fields are not able to suppress or sufficiently prevent shock formation, then an extension to MHD of the phenomenology discussed in the previous section leads to a criterium for collapse that reads

$$\tilde{r} = \tau_{tr}/\tau_{ff} \sim (l/L_J)\, M_{sr}^{-1} \ .$$

Further studies are needed in this area, in particular in considering whether a strong magnetic field can suppress shock formation. Several observational features constrain the possible range of parameters, such as the occurence of a turbulence at the sub–alfvenic level; another constraint on the dynamics is the observed correlation between the magnetic field and the density namely $B \sim \rho^{1/2}$.

Finally, it should be noted that the correlations between velocity and magnetic field may play a role in the rate at which the collapse takes place. Let us define the correlation coefficient as $\rho_{vb} =< \mathbf{v} \cdot \mathbf{b} > / < \sqrt{v^2 + b^2} >$; when the dynamical interactions produce a strong v − b correlation, the non–linear terms in the MHD equations are weakened substantially ($\mathbf{v} \times \omega$ and $j \times \mathbf{b}$ balancing each other, and Ohm's law being negligible since $\mathbf{v}$ and $\mathbf{b}$ are either parallel or anti–parallel for maximal correlations). In the absence of gravity, two–dimensional computations on the Orszag–Tang vortex showed that the level reached by the correlation coefficient at a fixed time decreases logarithmically with Reynolds number, as shown in Table 2 below for $t = 8$ in units of the eddy turn–over time; the correlation coefficient ρ_{vb} is given normalized by its initial value of 0.5, and the highest Reynolds number is 2,500.

			TABLE 2			
ρ_{vb}	1.56	1.62	1.70	1.72	1.78	1.82
$-\ln \nu$	5.3	4.9	4.4	4.2	4.0	3.7

Table 2 gives the scaling with viscosity ν of the normalised correlation coefficient in a series of 2DMHD computations on the Orszag–Tang vortex. The growth of $|\rho_{vb}|$ also occurs in the presence of self-gravity. In one such case, the correlation coefficient decays from a normalised value close to zero to −70%, and the magnetic field is thus unable to stop the collapse, as exemplified for example on the density contrast ρ_{max}/ρ_{min} which reaches values 140 times larger than its initial value.

4. Conclusions

Turbulence can be looked at from the statistical view–point, when one uses as diagnostics power–law spectra, structure functions or histograms. Another striking feature of turbulent flows is the coherent long–lived structures that develop within them, such as localized vortices, shocks or density filaments, together with clumps. Involving forcing of the flow at various scales, to mimic large–scale shear and small–scale supernovae explosions, both for a neutral and a magnetic fluid, and the resulting emergence in some cases of a correlation in the pre–collapse stage between magnetic energy and density, as may be observed, are but a few examples of the many open problems to be addressed. Also, in MHD, the weakening of nonlinearities through the emergence of structures such as Alfvén waves or force–free fields may greatly affect transport properties of such flows: is there an inverse cascade of magnetic helicity in 3D–MHD in the supersonic case? And when ambipolar diffusion [28] is taken into account? How wide is the turbulent spectrum if waves predominate? Are they observable structural differences between gravitational collapse in the presence of a random magnetic field as opposed to a wave–dominated case?

Much remain to be done in order to understand the complexity of the interstellar medium. In particular, several parameter regimes occur, from the kiloparsec galactic disk to the dark cold dense cores at the 0.01 pc scale, from the coronal and hot diffuse phases, to the cold molecular clouds. The acknowledgement or denial of the rôle of turbulence in the ISM may be more of a semantic matter than anything else; if one understands "turbulence" à la Kolmogorov but *stricto sensu*, then indeed such concepts could not strictly apply. On the other hand we can also define an "interstellar turbulence" in much the same way as there is "plasma turbulence". One way to proceed is to dichotomize the problem: cut it into little pieces and study each piece of the puzzle. That there are filaments in the ISM is clear. That there are vortices may be more difficult to visualize but will appear progressively as unavoidable. Recent numerical studies [22] also point out to the important role played by vorticity. One challenge is to sort out, from all these physical mechanisms and their interplay (cooling and heating, transfer, thermal instabilities, turbulence, magnetic fields, rotation, ...), which one can be neglected and which one are prevalent. It is the ensemble of parametric studies that will lead us, by comparison with careful and detailed observations, to the right model–equation(s) for the ISM.

I am thankful to Edith Falgarone for a useful and friendly discussion. Partial support for this work comes from DRET contract 92–1202.

REFERENCES

[1] Armstrong J., Cordes J. & Rickett B. 1981, Nature **291** 561.
[2] Biskamp D., Welter H. & Walter M. 1990, PhysFluidsB, 2, 3024.
[3] Blaisdell G., Mansour N. & Reynolds W. 1991, Dep. Mech. Eng. Rep. **TF–50**, Stanford Univ.
[4] Bonazzola S., Falgarone E., Heyvaerts J., Perrault M. & Puget J.L. 1987, AA, 172, 293.
[5] Brachet M., Meneguzzi M., Vincent A., Politano H. & Sulem P. 1992, PhysFluidsA, 6, 2845..
[6] Broc A. 1993, Rapport de DEA, Univ. de Nice, Obs. de la Côte d'Azur.
[7] Chandrasekhar S. 1951, ProcRoySocA, 210, 26.
[8] Elmegreen B. 1993, in *Protostars and Planets* **III** 97, University of Arizona Press, Tucson.
[9] Elsässer K. & Schamel H. 1976, *Z. Physik* 23, 89.
[10] Erlebacher G., Hussaini M., Speziale C. & Zang T. 1992, JFluidMech, 238, 155.
[11] Falgarone E. & Philipps T. 1990, ApJ, 359, 344; and Falgarone E. 1991, in *Structure and Dynamics of the Interstellar Medium*, IAU Symposium, 120, G. Tenorio–Tagle, M. Moles & S. Melnick Eds, Springer; and Falgarone E., Puget, J. & Pérault M. 1992, AA, 257, 715.
[12] Falgarone E., Lis D., Philipps T., Porter D., Pouquet A. & Woodward, P. 1994, ApJ, submitted.
[13] W.J. Feireisen, W.C. Reynolds & J.B. Ferziger; 1981, Report TF-13, Stanford Univ.
[14] Fukui Y & Mizuno A. 1991, IAU Symposium 147, *Fragmentation in Molecular Clouds*, Grenoble, June 1990, F. Boulanger, G. Duvert & E. Falgarone Eds., Kluwer.
[15] Galloway D. & Frisch U. 1986, GeophysAstrophysFluidDyn, 36, 53.
[16] Hasegawa A. 1985, Advances in Phys, 34, 1.
[17] Heiles C. 1988, ApJ, 324, 321.
[18] Henriksen R. 1991, IAU Symposium 147, *Fragmentation in Molecular Clouds*, Grenoble, June 1990, F. Boulanger, G. Duvert & E. Falgarone Eds., Kluwer.
[19] Horiuchi R. & Sato T. 1988, PhysFluids, 31, 1142.
[20] Iroshnikov P. 1963, SovAstron, 7, 56.
[21] Kida Y. & Orszag S. 1990, JSciComput, 5, p. 1.; and p. 85.
[22] Klein R. 1994, This conference.
[23] Kraichnan R.H. 1965, PhysFluids, 8, 995.
[24] Larson R. 1981, MNRAS, 194, 809.
[25] Léorat J., Passot T., Pouquet A. 1990, MNRAS, 243, 293.
[26] Meneguzzi M., Politano H., Pouquet A. & Zolver M. 1994, JCompPhys, submitted.
[27] Meneguzzi M., Frisch U. & Pouquet A. 1981, PhysRevLett, 47, 1060.
[28] Mouschovias T. 1991, ApJ, 373, 169.
[29] Neu J. 1984, JFluidMech, 143, 253.
[30] Nordlund A., Brandenburg A., Jennings R., Rieutord M., Ruokolainen J., Stein R. & Tuominen I. 1992, ApJ, 392, 647.
[31] Orszag, S. & Tang C. 1979, JFluidMech, 90, 129.

[32] Passot T. & Pouquet A. 1988, JCompPhys, 75, 301.
[33] Passot T., Politano H., Pouquet A. & Sulem P.L. 1990, TheoCompFluidDyn, 1, 47.
[34] Passot T., Politano H., Sulem P.L., Angilella M. & Meneguzzi M. 1994, JFluidMech, submitted.
[35] Picone J. & Dahlburg R. 1991, PhysFluidsB, 3, 29.
[36] Politano H., Pouquet A. & Sulem P.L. 1994, in preparation.
[37] Porter D., Pouquet A. & Woodward P. 1994, PhysFluidsA, in press.
[38] Pouquet A., Frisch U. & Léorat J. 1976, JFluidMech, 77, 321.
[39] Pouquet A., Frisch U. & Chollet J.P. 1983, PhysFluidLett, 26, 877.
[40] Pouquet A., Passot T. & Léorat J. 1991, Advances in Turbulence, 3, 343; Eds. A.V. Johansson & P.H. Alfredsson, Springer-Verlag; and 1991, Symposium IAU 147, p. 101; F. Boulanger, G. Duvert & E. Falgarone Eds., Kluwer Dordretch.
[41] Pudritz R. 1990, ApJ, 350, 195.
[42] Scalo J. 1987, in *Interstellar Processes*, p. 349–394, D.J. Hollenbach & H.A. Thronson Jr Eds, Reidel.
[43] Scalo J. 1990, in *Physical Processes in Fragmentation and Star Formation*, R. Capuzzo-Dolcetta, C. Chiosi & A. Di Fazio eds., Kluwer Dordrecht.
[44] Shebalin M. 1988, Princeton Meeting on Compressible Flows, T. Birmingham Ed.
[45] Vazquez-Semadeni E. & Scalo J. 1992, PhysRevLett, 68, 2921.
[46] Vazquez-Semadeni E., Passot T. & Pouquet A. 1994, in preparation.
[47] Verschuur G. 1991, IAU Symposium 144, 93, H. Bleomer Ed.
[48] Vincent A. & Meneguzzi M. 1991, JFluidMech, 225, 1.
[49] Yoshizawa A. 1985, PhysFluid, 28, 3313.
[50] Zakharov V. & Sagdeev R. 1970, SovPhysDokl, 15, 439.

A Turbulent Model for the Interstellar Medium

By T. Passot[1,2], E. C. Vázquez-Semadeni[3], A. Pouquet[1]

[1] Observatoire de Nice, BP 229, 06304 Nice Cedex 4, France

[2] Mathematics Dept., University of Arizona, Tucson, AZ, 85721, USA

[3] Instituto de Astronomía, UNAM, Apdo. Postal 70-264, México, D. F. 04510, Mexico

We present results from two-dimensional numerical simulations of a supersonic turbulent flow with parameters characteristic of the interstellar medium at the 1 kpc scale in the plane of the galactic disk. The simulations include heating above a critical density, in order to model heating from stars at unresolved scales, as well as cooling and diffuse heating terms. The calculations show segregation of the ISM into a warm, diffuse phase at $\sim 10^4$ K and a cold, denser phase at $\sim 1.5 \times 10^3$ K, coexisting in nearly (but not exactly) pressure equilibrium. The model suggests a life cycle for the ISM and cloud formation in which turbulence generates high-density complexes (cloud complexes) by mass advection (ram pressure), while self-gravity and cooling allow the complexes to reach densities high enough to produce star formation, which in turn generates more turbulence. The cloud energetics and the equivalent equation of state of the flow are briefly discussed. The system also exhibits large-scale gravito-acoustic waves, equivalent to density waves in galaxies, which appear to be triggered by the initial transients.

1. Introduction

The interstellar medium (ISM) is a highly compressible turbulent flow (e.g. [4, 6, 15]) characterized by the availability of strong kinetic and thermal energy sources, efficient cooling mechanisms, and the ability to nonlinearly transfer energy among a wide range of scales. However, a comprehensive theoretical account incorporating both the relevant thermal processes and the turbulent dynamics has not been given. In particular, the problems of phase segregation of the ISM, that of maintenance of the turbulence, the mechanisms of cloud formation, and the origin of cloud virialization remain unanswered. The pioneering attempt of Chiang & Prendergast [2] to numerically model the structural properties of the ISM with a bi-fluid (gas + stars) model suggested that a "star formation instability" is responsible for the phase segregation of ISM. Their calculations, however, assumed a star formation rate (SFR) directly proportional to the gas density, while observational evidence suggests that a better approximation is that star formation (SF) turns on above a certain critical (column) density, e.g. [7, 16]; see also [17] and references therein. In this case, the equilibrium state for the SF instability does not exist. Thus, the origin of phase segregation in a medium with discontinuous SF is another important problem. Furthermore, neither Chiang & Prendergast nor subsequent papers [1, 14] analyzed the role of turbulence in the formation of structure in the ISM. On the other hand, work that has explicitly analyzed turbulence in the ISM [10, 12] has so far neglected thermal effects and considered forcing regimes not representative of the ISM.

In this paper we investigate some of the problems mentioned above by means of two-

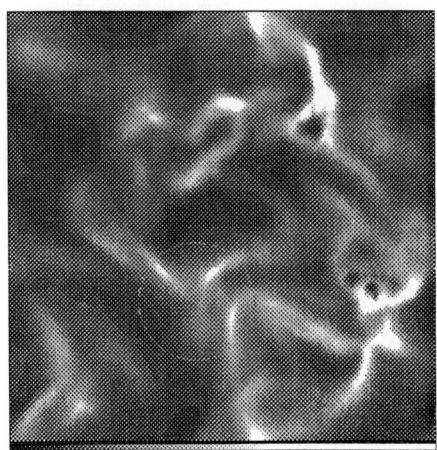 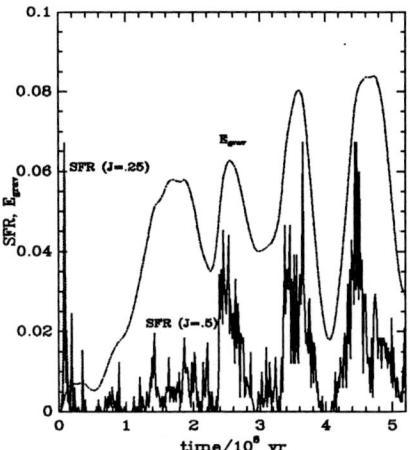

FIGURE 1. a) Density gray-scale image of the 512^2 run at $t = 1.9 \times 10^8$ yr. The highest density regions are produced by the turbulence generated by star formation, and have the strongest vorticity production. b) Evolution of the star formation rate for two 128^2 models, one with normal gravity $J = 0.5$ (solid line) and one with one-quarter the gravitational strength ($J = 0.25$) (dotted line, multiplied by 5). Gravitational energy for the run with $J = 0.5$ (dash-dotted line).

dimensional (2D) numerical calculations of supersonic turbulent flows at the kpc scale in the galactic plane. In an attempt to make the simulations more representative of the actual conditions in the ISM, the equations include model terms for heating effects from stellar sources, radiative cooling (assuming an optically thin medium), and large-scale shear mimicking galactic differential rotation shear.

However, as a first step we neglect magnetic fields, supernovae, and rotation effects (Coriolis and centrifugal forces), which will be considered elsewhere.

2. Numerical Model

The computations solve the following nondimensionalized equations:

$$\frac{\partial \rho}{\partial t} + \nabla \cdot (\rho \mathbf{u}) = \mu \nabla^2 \rho$$

$$\frac{\partial \mathbf{u}}{\partial t} + \mathbf{u} \cdot \nabla \mathbf{u} = -\frac{1}{M^2} \frac{\nabla P}{\rho} + \frac{1}{R_e} \nabla^8 \mathbf{u} + \mathbf{f}_s \quad (1)$$

$$\frac{\partial e}{\partial t} + \mathbf{u} \cdot \nabla e = -(\gamma - 1) e \nabla \cdot \mathbf{u} + \frac{\gamma}{R_e P_r} \frac{\nabla^2 e}{\rho} + \Gamma_d + \Gamma_s + \rho \Lambda$$

$$\nabla^2 \phi = \rho - 1$$

for the two-dimensional case with periodic boundary conditions. As usual, ρ is the density, $\mathbf{u}$ is the fluid velocity, e is the specific internal energy, P the pressure, and ϕ the gravitational potential. An ideal-gas equation of state $P = (\gamma - 1)\rho e$ is used. The diffuse heating Γ_d is taken as a constant in space and time. The stellar heating Γ_s is taken as a constant at each grid point $\mathbf{x}$ if locally $\rho > \rho_c$, $\nabla \cdot \mathbf{u} < 0$, and $t < t_{onset} + \Delta t_s$, and zero otherwise. Here, ρ_c is the critical density for star formation, t_{onset} is the time at which star formation

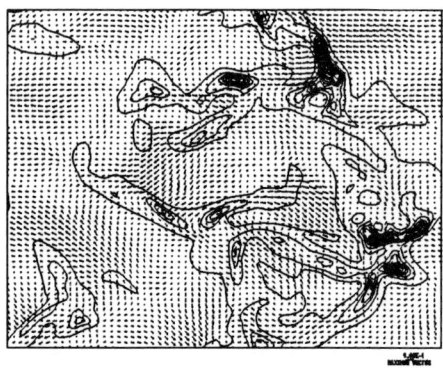 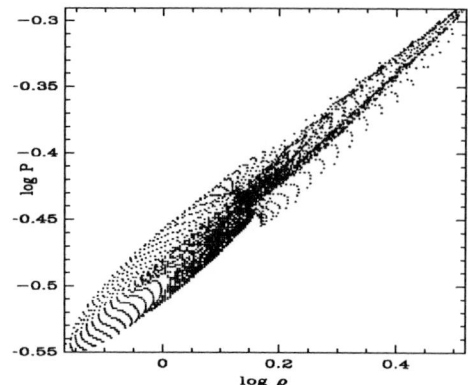

FIGURE 2. a) Density contours (subsampled at 128^2 resolution) and velocity vectors (at 64^2 resolution) for the 512^2 run at $t = 8 \times 10^7$ yr showing that clouds form by ram pressure. b) $\log P$ vs. $\log \rho$ for a cloudy but non-star-forming region in the 512^2 run at the same t as in (a). The slope of the region covered by the dots is approximately 0.38

turns on at position **x**, and Δt_s is the stellar lifetime, taken as a typical lifetime of OB stars (6×10^6 yr). This scheme allows us to monitor the instantaneous SFR as the fraction of points in the grid in which SF is being turned on at each timestep.

The temperature is related to the internal energy by $e = c_v T$, with c_v the specific heat at constant volume, and Λ is the cooling function, which we take as a piecewise power law in the temperature giving *no thermal instability* below 10^5 K between cooling and the diffuse heating. The cooling function is an approximation to that given by Dalgarno & McCray [3] and Raymond et al. [13], as used by Rosen et al. [14]. Finally, $\mathbf{f}_s$ is a large-scale sinusoidal shearing term, mimicking the effect of galactic differential rotation shear. Other parameters are the Mach number $M = u_o/c_s$, the Jeans number $J = L_o/L_J$, giving the number of Jeans lengths in the simulation, the mass-diffusion coefficient μ, and the Reynolds number R_e. A hyperviscosity scheme is used which employs a ∇^8 viscosity operator, confining viscous effects to the smallest resolved scales. The integration technique is pseudo-spectral, and the time stepping uses a third-order Runge-Kutta scheme. More details on the computations will be given elsewhere (Vázquez-Semadeni et al. [18]).

Several simulations have been performed. Low resolution runs (128^2) were used to test the effects of variations in the parameters, while the dynamics and structures in the flow are discussed on a 512^2 run. All runs were normalized to values of the variables typical of the warm phase of the ISM, and evolved for a few times 10^8 yr.

3. Results and Discussion

The calculations show segregation of the flow into a warm, diffuse phase at $\sim 10^4$ K and $\rho \sim 0.2$ cm^{-3}, and a cold, denser phase at $\sim 1.5 \times 10^3$ K and $\rho \sim 5$ cm^{-3}, coexisting in nearly (but not exactly) thermal pressure equilibrium, in agreement with results from previous workers [1, 2, 14]. A third component consists of the regions with active stellar heating ("HII regions"), which are formed in the densest parts of the cold phase, and expand rapidly until they reach the conditions of the warm phase (fig. 1.a). New SF is generally induced in the dense shells surrounding the expanding regions. SF thus self-propagates in a manner similar to the schemes of Gerola & Seiden [9] and Franco and collaborators (Franco [8], see Tenorio-Tagle, this volume).

The warm phase is characterized by a transonic behavior, with small density contrasts, low-level vorticity, and slow timescales, while the cold phase is characterized by a highly supersonic behavior, with high density contrasts, fast timescales, and strong production of vorticity. It should be noted that the parameters used are such that the thermal (heating *and* cooling) timescales are shorter than the dynamical timescales by factors ~ 0.01 to 0.1. Moreover, the equilibrium temperature T_{eq} (recall there is no thermal instability) between diffuse heating and cooling is a decreasing function of the density, due to the different functional forms of Γ_d (a constant) and Λ (a piecewise power law). Thus, the phase segregation of the flow can simply be understood as a consequence of the lower equilibrium temperature of the dense regions ("clouds") generated by the turbulence. Note that the temperature in the "clouds" in the simulations is too high compared to the actual temperatures of diffuse and giant molecular clouds in the ISM. This is due to a combination of two effects. First, the density contrast achieved by the simulations ($\rho_{max}/\rho_{min} \sim 50$) is rather low due to the limited resolution, causing the cooling to be comparatively inefficient. Secondly, molecular clouds are known to have strong self-shielding against the diffuse radiation field, which, however, has not been taken into account in the present calculations. (Preliminary calculations with a prescription for mimicking self-shielding have cloud temperatures as low as 30 K, although the prescription introduces other spurious effects which still need to be eliminated.) Because of these two problems, no self-gravitating clouds are formed in the simulations. It is found, however, that models without gravity do not sustain SF indefinitely. Instead, the SF activity and the turbulence decay rapidly (fig. 1.b), the system tending to a final quiescent state of thermal equilibrium between cooling and the diffuse heating. The model thus suggests a life cycle for the ISM and cloud formation in which turbulence generates high-density complexes by mass advection (ram pressure), while self-gravity and cooling allow the complexes to reach densities high enough to produce star formation, which in turn generates more turbulence. Furthermore, all processes appear to be essential in the maintenance of the cycle.

As a result of the fast thermal timescales and the monotonic decrease of T_{eq} with ρ, the thermal pressure in the dense phase is typically only 2-3 times that of the diffuse phase, and so the system is in near thermal pressure balance everywhere (except for expanding HII regions). Clouds thus appear to be formed and maintained by ram pressure from the surrounding medium (fig. 2.a). For quiescent regions devoid of SF, the flow behaves nearly as a polytropic gas, with a polytropic index $\gamma \sim 0.38$ (fig. 2.b) reminiscent of a value predicted by Elmegreen [5] in the framework of a model of cloud formation including several instabilities. A compilation of cloud properties by Myers [11] gives an effective value of $\gamma \sim 0.25$. However, for SF regions, no unique trend of P with ρ exists, due to the pressure variations caused by the stellar heating.

Although no self-gravitating clouds were formed in the simulations, gravitational to (turbulent + thermal) energy ratios ranging from 0.02 to 0.1 are found for a sample of five clouds in the simulation. The longest-lived clouds in the simulations have lifetimes $\sim 10^8$ yr, and are the closest to be in virial equilibrium. An important suggestion is then that individual clouds do not necessarily have a tendency to become virialized; it is only because nearly-virial clouds have longer lifetimes that they are more frequently observed.

Finally, a clear periodicity is observed in the star formation rate and in the global gravitational energy of the system (fig. 1.b). These have periods consistent with those expected for gravito-acoustic density waves in the system, which are the equivalent of spiral density waves in galaxies. These waves appear in all simulations, even if the initial conditions have uniform density and temperature, as long as all processes are incorporated. Thus, they appear to be triggered by any non-equilibrium condition initially present in the flow.

This work was partiallly supported by grants UNAM-DGAPA IN104092 and UNAM-CRAY SC000393.

REFERENCES

[1] Chiang, W.-H. & Bregman, J. N. 1988, ApJ 328, 427
[2] Chiang, W.-H. & Prendergast, K. H. 1985, ApJ 297, 507
[3] Dalgarno, A. & McCray, R. A. 1972, ARAA 10, 375
[4] Dickman, R. L., in Protostars and Planets II, ed. D. C. Black & M. S. Matthews (Tucson: Univ. of Arizona Press)
[5] Elmegreen, B. G. 1993, in Protostars and Planets III, ed. E. H. Levy & J. I. Lunine (Tucson: Univ. of Arizona Press)
[6] Falgarone, E. 1989, in Structure and Dynamics of the Interstellar Medium, ed. G. Tenorio-Tagle et al. (Berlin:Springer-Verlag)
[7] Franco, J. & Cox 1986, PASP 98, 1076
[8] Franco, J. 1992, in Star Formation in Stellar Systems, ed. G. Tenorio-Tagle et al. (Cambridge: Cambridge U.Press), 515
[9] Gerola, H. & Seiden, P. 1978, ApJ 223, 129
[10] Léorat, J., Passot, T., & Pouquet, A. 1990, MNRAS 243, 293
[11] Myers, P. C. 1978, ApJ 225, 380
[12] Passot, T., Pouquet, A., & Woodward, P. R. 1988, A&A 197, 228
[13] Raymond, J. C., Cox, D. P. & Smith, B. W. 1976, ApJ 204, 290
[14] Rosen, A., Bregman, J. N. & Norman, M. 1993, preprint
[15] Scalo, J. M. 1987, in Interstellar Processes, ed. D. J. Hollenbach & H. A. Thronson (Dordrecht: Reidel), 349
[16] Skillman, E. D. & Bothun, G. D. 1986, A&A 165, 45
[17] Vázquez-Semadeni, E. C. & Scalo, J. 1989, ApJ 343, 644
[18] Vázquez-Semadeni, E. C., Passot, T. & Pouquet, A. 1993, in preparation

The Hydrodynamics of Cloud Interactions

By Richard I. Klein[1], Christopher F. McKee[2]

[1] Department of Astronomy, University of California, Berkeley CA 94720, and Lawrence Livermore National Laboratory, Livermore CA 94550, USA.

[2] Departments of Physics and of Astronomy, University of California, Berkeley CA 94720, USA.

1. Introduction

The interaction of shock waves with interstellar clouds is a fundamental problem in interstellar gas dynamics. Shock waves are common in the interstellar medium (ISM) because radiative cooling is able to maintain the temperature of most of the gas in the ISM well below the temperatures characteristic of energetic events in the ISM, such as supernovae, stellar winds, bipolar flows, the creation of HII regions, or shocks associated with spiral density waves. An understanding of the physics of the interaction of shock waves with interstellar clouds is essential to understanding the evolution of the ISM as it is rent by shock waves from supernovae, stellar winds, cloud-cloud collisions and spiral density waves [6,19]. The interaction between shocks and interstellar clouds is central to a number of problems in interstellar gas dynamics. More generally, this interaction is a particular example of the interaction between a cloud and a surrounding medium in relative motion. The shock determines the manner in which the cloud is injected into the flow, the accompanying increase in pressure, and the Mach number of the flow past the cloud. Similarly, the physics of the cloud shock interaction is embodied in the essential physics of cloud-cloud interactions. As we shall see, much of the destructive effect of the cloud-shock interaction is associated with the postshock flow of the intercloud gas past the cloud. This is substantially true as well for cloud-wind and cloud-cloud interactions. Given the importance of the interaction of interstellar shocks with clouds for understanding the structure and the dynamics of the ISM, as well as the possible importance of the interaction as a means of triggering new star formation, the problem has been studied extensively. Despite much work on this important problem, the key questions have remained unresolved: 1) What is the rate and total amount of gas stripped from the cloud? What mechanisms are responsible? 2) What is the rate of momentum transfer to the cloud? That is, how long does it take for the cloud to become comoving with the shocked intercloud medium? 3) What is the appearance of the shocked cloud -its morphology, velocity dispersion, luminosity? 4) How is the interaction between the shock and the cloud affected if the shock in the intercloud medium is itself radiative? 5) Under what conditions will the shocked cloud become gravitationally unstable? 6) How does a magnetic field affect the evolution of the shocked cloud?

In order to address these questions, we have undertaken a comprehensive numerical study of the cloud shock problem [12-14] using for the first time high resolution adaptive mesh refinement hydrodynamics (AMR). In this paper we focus on the simplest case: a planar shock impacting an isolated, spherical cloud under the assumptions that radiation, magnetic fields, gravity and thermal conduction are negligible. We show that the problem is completely determined by two dimensionless parameters: the Mach number of the shock,

M and the density ratio between the cloud and the intercloud medium, χ. We shall demonstrate that the dependence on the Mach number can be scaled out for strong shocks, so the results depend primarily on the density ratio χ.

Our approach to the cloud-shock problem is formulated and the relevant timescales developed in §2. In §3 we briefly describe the AMR hydrodynamics and our approach to the analysis of the results. We describe the morphological evolution of shocked clouds and scaling with the Mach number M and the density ratio χ as well as study the effects of a radiative cloud by using a soft equation of state in §4. We discuss the cloud drag in §5 and the dynamics of vorticity in §6. In §7 we discuss a simple theory for cloud fragmentation that is borne out by our detailed calculations and in §8 we present results of 3D calculations [15]. In §9 we briefly discuss the application of our results to a shocked cloud in the Cygnus Loop and in §10 we discuss the importance of convergence studies for a variety of hydrodynamic cloud interactions and present a summary of our convergence analysis. This is discussed in detail in [16]. We also present a test challenge problem for multi-dimensional hydrodynamics in which shock interactions and shear flows are important. Preliminary results of cloud-cloud collisions using AMR are presented in §11 and previous studies of this problem are reviewed.

2. Formulation of Cloud-Shock Interaction

Consider a cloud in pressure equilibrium with an ambient medium density ρ_{io}. We shall focus on the case of a spherical cloud with radius ao. We assume that the cloud is approximately isothermal; since gravity and magnetic fields have been assumed to be negligible, the cloud will have a nearly uniform density ρ_{io}. The density contrast between the cloud and the intercloud medium is $\chi = \rho_{co}/\rho_{io}$.

The time for the shock in the intercloud medium to sweep across the cloud is $t_{ic} = 2a_o/V_b$, where V_b is the velocity of the blast wave in the intercloud medium. The characteristic time for the cloud to be crushed by the shocks moving into the cloud is a_o/V_s, where V_s is the shock velocity in the cloud. Since $V_s = V_b/\chi^{1/2}$, we define the cloud crushing time to be $t_{cc} = \chi^{1/2}a_o/V_b$. If the cross sectional area of the cloud A, were to remain constant as the blast wave accelerates the cloud then the characteristic drag time $t_{drag,o}$ for a strong shock is $t_{drag,o} = \chi^{1/2}t_{cc}/C_D$, where C_D is the drag coefficient. The cloud under goes a lateral expansion after being shocked [21] and the actual drag time $t_{drag,o}$ is considerably smaller, of the order a few times the cloud crushing time t_{cc}. After the blast wave has swept over the cloud, the shocked cloud is subject to both the Kelvin-Helmholtz (KH) and Rayleigh-Taylor (RT) instabilities. For $\chi \gg 1$, the time scale t_{KH} for the growth of the KH instability for perturbations of wavenumber k parallel to the relative velocity V_{rel} between the cloud and intercloud media is $t_{KH}^{-1} = kV_{rel}/\chi^{1/2}$ [3]. Thus, the KH growth time is comparable to the cloud crushing time, $t_{RT}/t_{cc} = (V_b/V_{rel})/ka_o$. The shortest wavelengths have the fastest growth, but longer wavelengths ($ka_o \sim 1$) are more disruptive. The RT growth time is also of the order of the cloud crushing time, $t_{RT}/t_{cc} = 1/ka_o$. The results suggest that the cloud will be destroyed in a time related to the cloud crushing time.

The final time scale of interest is the pressure variation time scale t_p. For a dense cloud in a Sedov-Taylor blast wave, McKee [20] found that $t_p = |\partial lnP/\partial t|^{-1} = 01.R_c/V_b$ where R_c is the distance of the cloud from the site of the explosion. We define small clouds to have $t_{cc} \ll t_p$ so that $a_o \ll 0.1R_c/\chi^{1/2}$. Hence the cloud is sufficiently small that the blast wave does not change significantly as the cloud is crushed and destroyed. In this paper we will study the interaction of shock waves with small clouds. We assume radiative cooling to be negligible, both in the intercloud medium and in the shocked cloud.

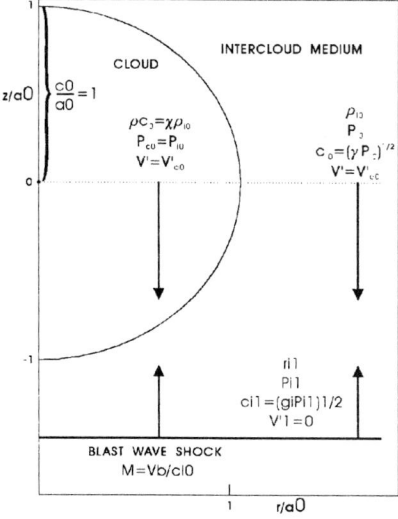

FIGURE 1.

3. Numerical Simulations

The numerical calculations have been described in [12,13] and will be described in some detail in [16]. We have used both 2D and 3D codes based on local adaptive mesh refinement with an underlying Godunov hydrodynamic scheme [2]. A rectangular grid is constructed that covers the computational domain (the level 1 grid). The AMR then uses a nested sequence of rectangular meshes to solve the Eulerian partial differential equations. In practice we have used three levels of grids with a factor of 4 refinement in each direction: thus a level 1 grid cell encompasses 256 level 3 cells in 2D. The decision to move to a higher level of refinement at any spatial location of the computational domain is based on using Richardson extrapolation to estimate the local truncation error. In addition to setting the refinement level by spatial error estimation, we can directly control the maximum refinement to level 3 to be in the cloud whereas intercloud material not near the cloud is restricted to lower levels of grid refinement. The code is two-fluid throughout the calculation. This greatly minimizes the numerical diffusion at the interface boundary of cloud and intercloud and permits us to accurately calculate the effects of shear instabilities as well as the extent of cloud-intercloud mixing. A no-slip boundary condition is applied at the interface. The AMR is applied to both fluids.

We consider a plane shock propagating along the z-axis at velocity V_b (Fig. 1). The cloud is centered on the z-axis. The calculation is carried out in the frame of the shocked intercloud gas, in which the cloud will eventually come to rest. We denote velocities in this frame by a prime; velocities measured in the frame of the unshocked gas, which is often the observer's frame, are unprimed. In the frame of the unshocked gas, the shock propagates in a positive z-direction. Initially, the cloud moves downward in the negative z-direction. In this paper we will consider the case for the which the ratio of the specific heats $\gamma = 5/3$ for both the cloud and intercloud medium. Other cases are considered in [14].

4. Evolution of the Shocked Cloud

Four stages can be identified in the interaction of a shock with a cloud [21]. There is (1) an *initial transient* when the blast wave first strikes the cloud, sending a shock into the cloud and a reflecting shock back into the intercloud medium. The reflecting shock settles into a standing bow shock in a time of order $a_o/V_b = t_{ic}/2$. The next stage is (2) *shock compression*: After a time of order t_{ic}, the flow around the cloud converges on the axis behind the cloud, producing a high-pressure reflected shock in the intercloud medium and driving a shock into the rear of the cloud [25]. The shocks compressing the cloud from the sides are weaker than those at the front and back of the cloud because the pressure is a minimum at the sides. The result is that the cloud is compressed into a thin pancake, with its axial dimension reduced about a factor of 2. The collision of the main shock propagating in from the front of the cloud with the shock coming from the rear produces yet greater compression. The (3) $re - expansion$ stage is initiated when the main cloud shock reaches the rear of the cloud, causing a strong rarefaction to be reflected back into the cloud and leading to an expansion of the shocked cloud downstream [25]. At the same time, the low pressure at the sides of the cloud compared to that on the axis causes the cloud to expand laterally. The lateral expansion continues a few cloud crushing times. The final stage is (4) *cloud destruction*, as instabilities and differential forces due to the flow of the intercloud gas past the cloud cause it to fragment.

4.1. *Evolution of the Cloud Density*

We will follow the evolution of the shocked cloud with a series of isodensity contours of cloud and intercloud matter at different times. In Figs. 2a-2h we note that only one half the cloud is calculated due to symmetry along the z coordinate. As the blast wave engulfs the cloud, it drives a weaker shock into the sides of the cloud. When this shock interacts with the main cloud shock, a third shock is created which deflects the gas that has passed through the side shock. These three shocks intersect at a triple point in the flow. At $t = 0.84t_{cc}$ in Fig. 2a we see evidence of the triple point as the confluence of three shocks, as seen in the sharp gradients within the cloud. The initial transient in the intercloud medium begins with the reflection of the blast wave from the cloud. After crossing the cloud, the intercloud shock converges behind the cloud at time $\simeq 0.94t_{cc}$, marking the end of the initial transient.

When the intercloud shock first converges on the axis, it does so at normal incidence, and a strong reflected shock is formed. Eventually, the reflected shock interacts with the incident shock to produce a Mach reflected shock that propagates along the axis. The Mach reflected shock extends from the cloud to the blast wave (Fig. 2b). At somewhat later time $t = 1.26t_{cc}$ a double Mach reflected shock forms with two triple points [11]. A powerful supersonic vortex ring forms just behind the double Mach reflection and is carried away from the cloud (fig. 2c). The high pressure which can be as large as $1.65\rho_{io}V_b^2$ drives a strong shock into the rear of the cloud, and this rear shock collides with the main cloud shock at a time $\sim 1.26t_{cc}$ (Fig. 2c). At $t = 1.68t_{cc}$, the back side of the cloud is RT unstable as low density material from the intercloud medium is accelerated into the cloud, and the cloud becomes axially flattened. The sides of the cloud are now KH unstable from the shear flow around the cloud boundary. When the shocks reach the cloud surface, rarefaction waves propagate back into the cloud and the re-expansion phase commences. At the point of maximum compression $t = 1.26t_{cc}$, the shocks have flattened the cloud to the point that the diameter along the axis is about half that in the transverse direction.

Cloud destruction proceeds with the re-expansion. The Richtmyer-Meshkov instability [22] contributes to this destruction. This instability, which is due to the impulsive acceleration of the cloud-intercloud boundary, grows linearly with time, rather than exponentially as does the RT instability. For this instability to be important, the cloud must have large surface perturbations. The perturbations on the surface of the cloud are provided by the mesh. We note at $t = 2.1t_{cc}$, (Fig. 2e) the RT and Richtmyer Meshkov instabilities

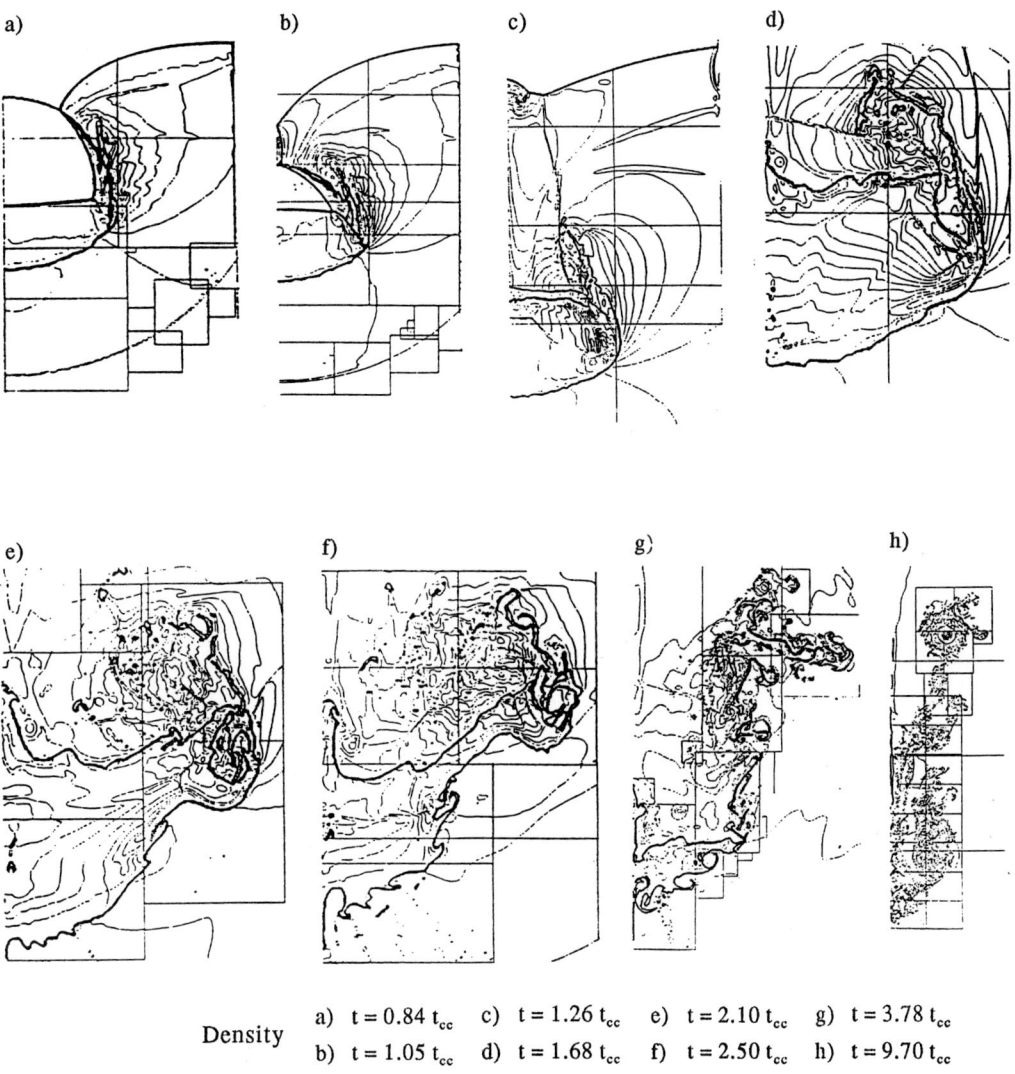

FIGURE 2.

Density
a) $t = 0.84\, t_{cc}$ c) $t = 1.26\, t_{cc}$ e) $t = 2.10\, t_{cc}$ g) $t = 3.78\, t_{cc}$
b) $t = 1.05\, t_{cc}$ d) $t = 1.68\, t_{cc}$ f) $t = 2.50\, t_{cc}$ h) $t = 9.70\, t_{cc}$

are starting to become noticeable on the front of the cloud. The dominant destruction mechanism of the cloud appears to be the KH and RT instabilities. These instabilities have growth time of order t_{cc} and the effects of the instabilities are evident by $2.5 t_{cc}$ (Fig. 2f). By $3.8 t_{cc}$, the cloud consists of a distorted, axially flattened core with a plume of fragments that contains over 70% of the mass of the cloud extending behind and to the side of the cloud (Fig. 2g). A prominent shear layer exists along this layer. Vortex rings appear along this layer and coincide with regions of severe cloud fragmentation. The vortex rings act as dynamic mixmasters further aiding in the fragmentation and mixing of the cloud. By $t = 9.7 t_{cc}$ (Fig. 2h), the cloud is completely fragmented. It occupies a volume with about twice the transverse dimension and five times the axial dimension of the initial cloud, with no single fragment having more then 2% of the initial cloud mass.

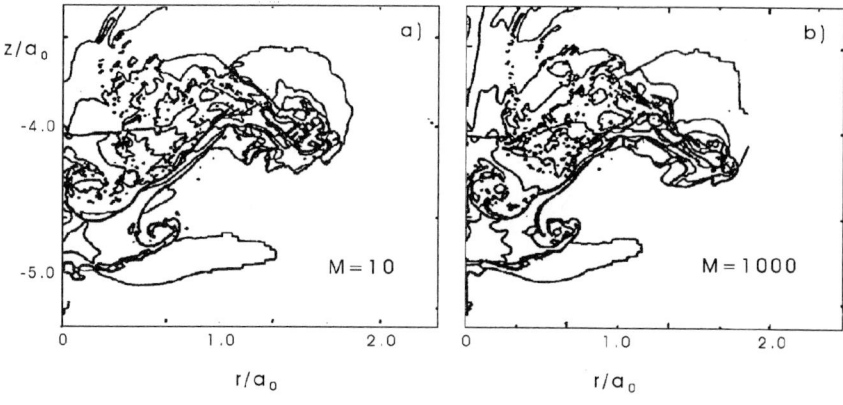

FIGURE 3.

4.2. Mach Scaling

The conditions behind a strong shock are virtually independent of the sound speed ahead of the shock, and this suggests that there should be a simple scaling among the cases run at different values of the Mach number $M = V_b/C_{io}$. We have shown that the hydrodynamic equations are invariant under the transformation $t \to tM$, $v \to v/M$, $P \to P/M^2$ with the position and density left unchanged [14]. For shocks for which $M \geq 10$, the postshock density and pressure are within a few percent to a fraction of a percent of their values at $M \to \infty$. If we can neglect terms of order M^{-2} in the jump conditions, the transformed state will be the same for all Mach numbers. Figure 3 shows the isodensity contours for the $M = 10$ and $M = 1000$ cases at time $t = 2.95 t_{cc}$. Although the cloud has undergone substantial distortion due to RT and KH instabilties, the agreement between the two cases is remarkable. We conclude that Mach scaling is valid [14].

4.3. Scaling with Density Ratio

To lowest order we expect the time evolution to scale with the cloud crushing time t_{cc} since the dynamical time scale of the shocked cloud and the instability growth times are all of order t_{cc}. This has been demonstrated [14] but calculations have revealed that there are weak, but significant, deviations from this scaling. These differences are attributed to the fact that the initial time scale for cloud drag does not scale with t_{cc} but rather with $\chi^{1/2} t_{cc}$; correspondingly, the velocity imparted to the cloud by the initial shock scales as $\chi^{1/2}$. As a result, the normalized drag time t_{drag}/t_{cc} increases weakly with χ. Because the dominant instabilities both depend on the relative velocity of the cloud and intercloud medium, the higher relative velocities at high χ lead to somewhat faster growth rates, thereby causing the normalized destruction time t_{dest}/t_{cc} to decrease weakly with χ.

4.4. Dependence on γ_c

We have considered the effect of varying the equation of state of the cloud. By setting $\gamma_c = 1.1$, we can see some of the qualitative effects of radiative losses in the shocked cloud. Softening the equation of state reduces the velocity of the cloud slightly and also leads to substantially greater compression. We have found that the transverse dimension is reduced by as much as a factor 2.0 and the axial dimension by as much as a factor 4.4, compared to factors of 1.2 and 2.3, respectively, for $\gamma_c = 5/3$. The re-expansion of the shocked cloud occurs with the characteristic sound speed of the shocked cloud, $0.22 V_b/\chi^{1/2}$, substantially slower than for $\gamma_c = 5/3$. As a result, the cross sectional area is significantly smaller than for $\gamma_c = 5/3$, which implies that the drag is less, the relative velocity remains high, and the instabilities are more violent.

5. Cloud Drag

If the cross section of the shocked cloud remained constant, then the cloud would become comoving with the postshock flow in a time $t_{drag,o} \simeq \chi^{1/2} t_{cc}$. In fact, the re-expansion of the shocked cloud substantially increases the drag and makes the drag time significantly less than this. By taking into account the variation of the cloud radius with time, [14] have shown that the cloud velocity obeys

$$\frac{V'_c}{V'_{co}} = \frac{1}{1 + \frac{9 C_D t}{8 \chi^{1/2} t_{cc}} \left[1 + \frac{1}{3} \left(\frac{C_c t}{V_s t_{cc}} \right)^2 \right]}, \qquad (t \leq t_m)$$

where t_m is the time at which the cylindrical radius of the cloud first reaches a value within 10% of its maximum value. After the cloud ceases its expansion, the cloud velocity obeys

$$\frac{V'_c}{V'_{co}} = \frac{1}{1 + \frac{9 C_D}{8 \chi^{1/3}} \left[\frac{t}{t_{cc}} + \left(\frac{C_c t_m}{V_s t_{cc}} \right)^2 \left(\frac{t}{t_{cc}} - \frac{2 t_m}{3 t_{cc}} \right) \right]}. \qquad (t > t_m)$$

This formulation of the cloud drag agrees reasonably well with the numerical results if the drag coefficient $C_D = 1$.

6. Vortex Dynamics

A striking aspect of the interaction of a shock wave with a cloud is the development of powerful vortex rings which play an important role in the destruction of the cloud. It is well known that in an ideal fluid, vorticity can not be produced so long as the fluid is isentropic (or, more generally, so long as there is a one to one correspondence between P and ρ [17]. Our problem is different in two respects: the flow is not isentropic, both because of the initial conditions and because of the presence of shocks, and the inevitable presence of some numerical viscosity due to the discretization of the partial differential equations means that the fluid is not ideal. There are two effects that can produce or destroy vorticity, a pressure gradient that is not aligned with the density gradient, and viscosity. It can be shown that the baroclinic term in the vorticity production equation is indeed responsible for the production of vorticity in the cloud-shock interaction [14]. In §10 we will discuss the effect of viscosity on the vorticity by examining the effects of different grid resolutions on the total circulation generated by the shocked cloud.

We have developed an analytic model for estimating the amount of vorticity produced by the shock-cloud interaction [14]. Here we will state the results they find for the dominant mechanisms. During the initial passage of a shock over a cloud they find the circulation generated is

$$\Gamma_{shock} \simeq -2.25 V_b a_o (1 - \chi^{-1/2}).$$

After the shock has swept over the cloud, vorticity continues to be generated by the baroclinic term and they find that in the subsequent post shock flow at the cloud-intercloud boundary, the circulation generated is

$$\Gamma_{post} = \frac{-9}{64} \left(\frac{\chi^{1/2} t_{drag}}{t_{cc}} \right) V_b a_o.$$

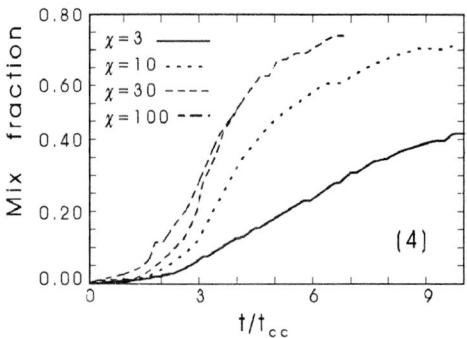

FIGURE 4.

The drag time t_{drag} is typically several times t_{cc}, and is considerably smaller than the initial drag time $t_{drag} \simeq \chi^{1/2} t_{cc}$. The circulation associated with the supersonic vortex ring in the intercloud medium that is produced behind the cloud is found to be

$$\Gamma_{ring} \simeq \frac{3}{4} V_b a_o.$$

Vorticity is produced in the cloud as well, but since the velocities in the cloud are smaller than those in the intercloud medium by a factor of order $\chi^{-1/2}$, the vorticity in the cloud is smaller by a similar factor. Excellent agreement has been found between the theoretical model of vorticity production and the numerical calculations indicating that we have indeed identifies the major sources of vorticity. We therefore conclude that the numerical viscosity does not have a significant effect on the calculation of the vorticity. We will show this to be consistent with the conclusion reached in §10 on the basis of convergence studies.

7. Cloud Fragmentation

The destruction of the shocked cloud is due to its fragmentation on both large and small scales. This fragmentation is driven by KH and RT instabilities, each of which has a characteristic growth time of order the cloud crushing time. The Richtmyer-Meshkov instability also contributes to fragmentation, but it is less important because it grows linearly rather than exponentially. We focus on two measures of the time it takes to destroy the cloud: the destruction time, t_{dest}, which focuses on large scale fragmentation, and the mixing time, t_{mix}, which focuses on small scale fragmentation.

A shocked cloud develops a core-plume structure, in which some of the cloud mass is concentrated near the axis while the remainder is stretched out behind the cloud in a plume (e.g., Fig. 2g). The destruction time is defined as the time at which the mass of the core has been reduced to a fraction $1/e$ of the initial cloud mass. Our results show that for density ratios in the range $10 \leq \chi \leq 100$, the destruction time is $t_{dest} \simeq 3.5 t_{cc}$. It is a feature of the KH and RT instabilities that, in the absence of viscosity, surface tension, and magnetic fields there is no minimum length scale. As a result we expect the fragmentation to occur down to the resolution of the grid. At this point, individual zones will contain both cloud and intercloud material. The fraction of the cloud mass in such zones is the "mix fraction", and the time at which the mix fraction reaches $1/2$ is the mixing time, t_{mix}. In Fig. 4 we show the mixing time as a function of time for different values of the cloud-intercloud density ratio. The mixing occurs more rapidly for larger χ. We attribute this to the higher relative velocity at higher χ which decreases the growth time of both the KH and RT instabilities, t_{KH} and t_{RT}. The velocity dependence of the KH and the

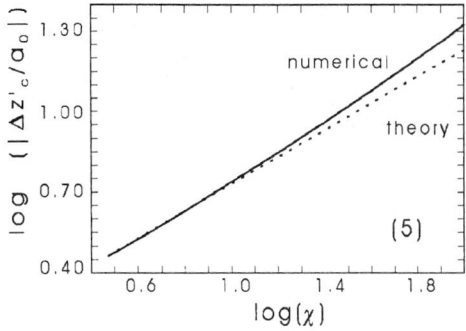

FIGURE 5.

RT growth times can be shown to scale as $1/v_{rel}$, so the growth time is reduced for higher velocities.

A simple hierarchical model for fragmentation by the KH instability can be developed by assuming that the cloud first fragments into a few large fragments of comparable mass, and that these in turn fragment into several fragments of comparable mass, and so on. Such a process leads to the most rapid possible mixing. Using our simple model for fragmentation, it is possible to calculate the distance the cloud travels during the time it takes the cloud to fragment to very small scales [14]; $\Delta z'_c \propto -\chi^{1/2} a_o$. As shown in Fig. 5, this $\chi^{1/2}$ scaling is well satisfied for $3 \leq \chi \leq 100$. At the largest χ the calculated value of $\Delta z'_c$ is about 25% larger than this. Rather than initially breaking up into several large fragments, at large χ the core appears to undergo stripping from its surface, which is a less effective mixing mechanism. This represents a change in the nature of fragmentation at large χ. We are currently studying the nature of this transition in fragmentation in 3D.

8. 3-D Calculations

We have performed detailed 3D AMR calculations of the cloud-shock interaction [15]. We will summarize here some of the key results in this paper. Our calculations were performed in a 3D quadrant of the full spherical domain using four-fold symmetry to permit increased resolution. Current calculations we are doing on the full sphere will investigate the consequences of assuming this symmetry. Our calculations were done with varying resolutions of 48, 72 and 96 zones in the initial cloud radius. The initial size of the 3D volume was chosen to contain the full evolution of the cloud out to $t = 5t_{cc}$. A fixed grid code would require in excess of 50 million zones to match this resolution; AMR required less than 6 million zones at late times in the calculation, but considerably fewer zones at earlier times. In Figs. 6a and 6b we display volume rendered images of the cloud density at times $3t_{cc}$ and $4.2t_{cc}$. These images were constructed from the full 3D data set by using a ray trace through the data with an appropriate opacity transfer function to bring out the details of the 3D internal structure of the cloud.

We have compared these calculations in 3D with axisymmetric 2D calculations using the identical resolution; 96 zones in an initial cloud radius at the highest grid level resolution. We calculate integral properties of the cloud by taking appropriate moments of the density, velocity, velocity dispersion, pressure and luminosity, cf. [14]. We find that in 3D, the mean compression of the cloud is about 30% higher than the mean compression found in 2D because non-axisymmetric interaction of cloud fragments can occur which results

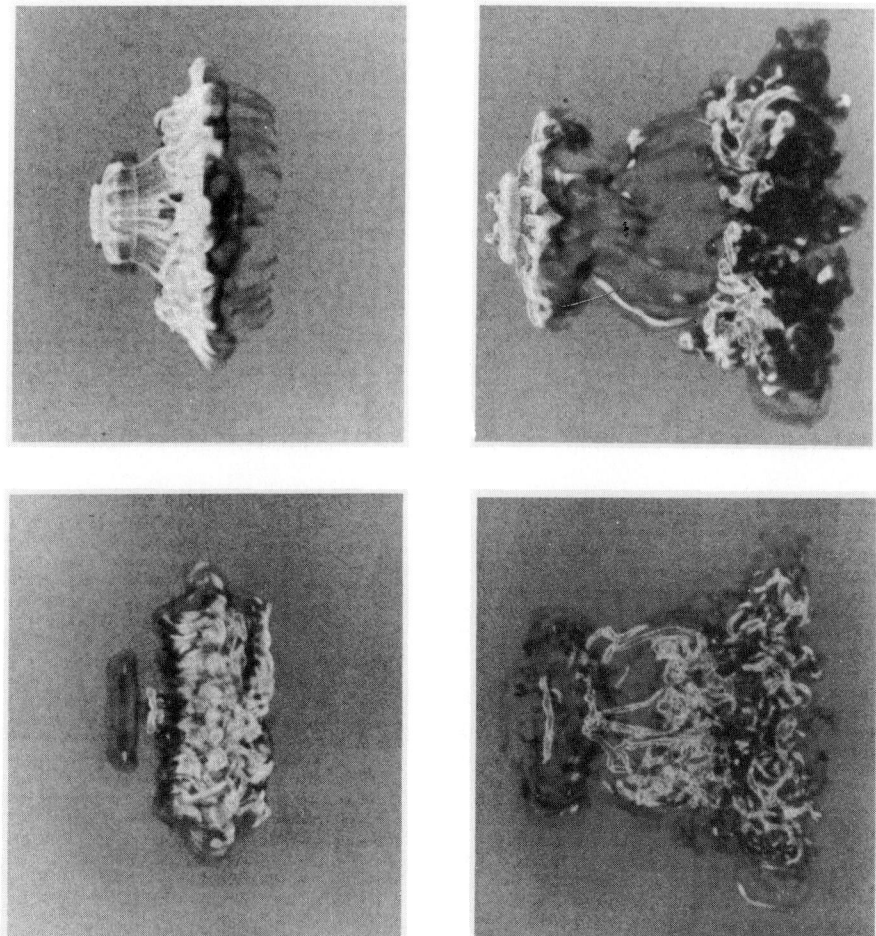

FIGURE 6.

in the increase in cloud density. In 3D, the mean axial velocity of the cloud is ~ 40 % larger than 2D. The higher velocity is most likely due to the higher density, corresponding to a smaller cross section. In Fig. 6a, the cloud undergoes severe azimuthal instabilities resulting in a fluted cloud structure. This structure is more highly fragmented and less coherent than in 2D at a comparable time. The axial and radial components of the dispersion velocity is essentially the same in 2D and 3D as is the mean cloud pressure. The luminosity of the cloud in 3D is however about twice what it is in 2D due to the increase in cloud density. The fragmentation of the cloud is significantly more severe in 3D than in 2D due to azimuthal bending mode instabilities in vortex rings [24]. We find in 2D that at $t = 4.2 t_{cc}$, the mass remaining in the core of the cloud is ~ 32 % of the initial cloud mass while the mass remaining in the back side of the cloud is ~ 37 % of the initial cloud mass. In 3D the mass remaining in the core of the cloud is comparable (29 % of the initial cloud mass), but in the back side of the cloud (Fig. 6b) significant fragmentation results in many smaller fragments, none more massive than 5 % of the initial cloud mass.

Figures 6c and 6d show the volume rendered magnitude of the vorticity at $t = 3 t_{cc}$ and

$4.2t_{cc}$. We note the appearance of hairpin vortices and twisted vortex tubes characteristic of turbulent flow. At $t = 3t_{cc}$ the coherent vortex rings apparent in our 2D calculations have undergone severe non-axisymmetric instability and begin to break into smaller less coherent vortex structures with evidence of a coherent vortex ring remaining near the front of the cloud. The flow appears to be fully turbulent at $t = 4.2t_{cc}$. The coherent vortex rings found in 2D are clearly unstable in 3D and this results in more effective mixing in 3D. We have calculated the energy spectrum of the turbulent flow by taking fast Fourier transforms of the flow field. We find that $E(k) \propto k^{-2.0 \pm 0.1}$. The spectrum may indeed be dominated by the presence of shocks which would yield a k^{-2} spectrum.

9. Application to a Shocked Cloud in the Cygnus Loop

Recently, new H_α images of the Cygnus Loop in the eastern region obtained by [7] have provided direct evidence for the interaction of a blast wave from a supernova remnant with an isolated interstellar cloud. They suggested that their data could be understood in terms of a shock-cloud interaction based on our earlier predictions [12]. Here we make a very brief comparison of their observations with our calculation of a RradiativeS cloud shock, where we have modeled the radiative losses in the cloud by using $\gamma_c = 1.1$ in the cloud. For a more detailed comparison see [14]. Using estimates for the blast wave velocity and the initial cloud radius Fesen estimates the elapsed time that the interaction began, which when corrected for the motion of the cloud gives an observed age of about 4700 yr, or $1.9t_{ic}$(obs) where t_{ic}(obs) is the intercloud crossing time inferred from an adopted initial cloud radius and a blast wave velocity. We compare the observed cloud to our calculation at a similar time, $t/t_{ic} = 2$. Our calculations show that the intercloud shock interacts at a kinked angle behind the cloud and reattaches through a Mach disk. The observations show that the shock front also shows evidence of kinking and reattachment at the same relative locations as in the calculations, with some evidence of a Mach disk. Our calculations predict that after the shock has swept over the cloud, it produces a strong shear surface along the cloud boundary. The resulting KH instabilities and vortex rings stretch, distort and fragment the cloud, producing arm-like features that are swept back downstream. Interacting shocks within the cloud result in a highly flattened cloud core attached to the distorted, back swept arms. This morphology is clearly seen as well in the H_α observations.

Although our predictions for the intercloud shock structure and its separation from the cloud and our overall predictions of the cloud morphology are in good agreement with the observations for a cloud with density ratio equal to 10, recent X-ray observations by Graham (1993, private communication) are not consistent with a spherical cloud. We suggest that the cloud is elongated along the line of sight [14], but the overall dynamics are expected to be similar to that for a spherical cloud.

10. Convergence Studies

The accuracy of a finite difference solution to the partial differential equations governing the physics of fluid dynamics can be assessed by an proper study of the convergence properties of the solution. For problems involving complex, highly non-linear flows, such as those described in this paper as well as many of the talks delivered at this conference, the notion of convergence itself becomes a complex issue. For finite difference schemes in particular, appropriate convergence analysis gives a measure of the truncation error remaining in the solution. For smooth particle hydrodynamics (SPH) the relationship between convergence and truncation error in the equations is more complicated. Nevertheless, *it is crucial for* proper convergence tests to be performed of all relevant quantities that are important to the physics of a given problem. These tests must be performed to look at convergence properties as a function of time for different grid, or in the case of SPH, particle resolutions. Only with establishment of convergence properties, can one be certain that both the method and the resolution used are appropriate to resolve the essential physics and take seriously the quantitative outcome of the results.

In this section, we briefly describe a few of the results of careful convergence studies we have made for the physics of cloud-shock interactions. For the details of our broader analysis the reader is referred to Klein, Colella & McKee (1994, in preparation). We introduce the notion that convergence must be established for relevant quantities as a function of time as well as resolution. This becomes particularly important because the sense of convergence at a particular resolution over some temporal range may change significantly at a later time. We shall also offer a test challenge problem for the entire community to assess how well their particular codes will do with a reasonably difficult problem involving shock interactions and shear layers originating from slip lines in the flow, phenomena that have appeared repeatedly in many of the talks in this conference. If a specific technique such as SPH cannot obtain reasonably accurate results for this problem, then using this technique on astrophysical gas dynamic problems for flows for which shock interactions and hydrodynamic instabilities (such as KH and RT) are important to the outcome is inappropriate. Either the method is not appropriate for the problem, or the resolution used is not adequate to capture the essential physics. We briefly describe the results from three detailed investigations of convergence. We first examine the convergence of global quantities, cf. [14], which are integrated moments over the entire volume of cloud material. In Figs. 7a, b and c we plot the circulation, a direct measure of the vorticity, the mean density in the cloud and the radial dispersion velocity in the cloud as a function of time in units of the cloud crushing time, for grid resolutions ranging from 30 to 240 grid zones in an intitial cloud radius. We find in general that 120 grid zones (R_{120}) is necessary for ~ 10 % convergence, but 60 grid zones (R_{60}) gives errors in some cloud global properties as well as morphological differences. Although R_{60} gives a reasonable convergence for early time (before t_{cc}) this resolution is quite inadequate at later times in the evolution of the cloud. Finally, in 7d we examine the convergence properties of mixed mass; that fraction of the cloud mass that is mixed in with intercloud matter. This gives us important information on the rate mixing for processing cloud material back into the intercloud medium. Here we see that calculations made with R_{60} give reasonably converged results over the entire time for cloud destruction. The results of these calculations and more detailed investigations lead us to conclude that at least 120 zones per cloud radius is necessary to capture the essential physics of cloud-shock interactions and obtain high accuracy in all the global properties of the cloud. For the more difficult case of radiative clouds where compressions can be significantly larger, higher resolution may be necessary.

10.1. *2-D Test Challenge*

Many problems of importance in astrophysical gas dynamics include as fundamental phenomena the appearance of interacting shock waves, slip sheets, leading to vorticity production and shear flows resulting in KH instabilities. Such phenomena are of particular importance in the cloud-shock problem we have discussed as well as cloud-cloud interactions where these phenomena play a role in the evolution of the cloud as well as the advent of eventual star formation. Modeling these phenomena accurately is essential to any hydrodynamical numerical scheme that has the objective of capturing the physics of such flows. We stress the importance of these phenomena by offering a test challenge problem to the full community of fluid dynamic simulators to test their individual numerical meth-

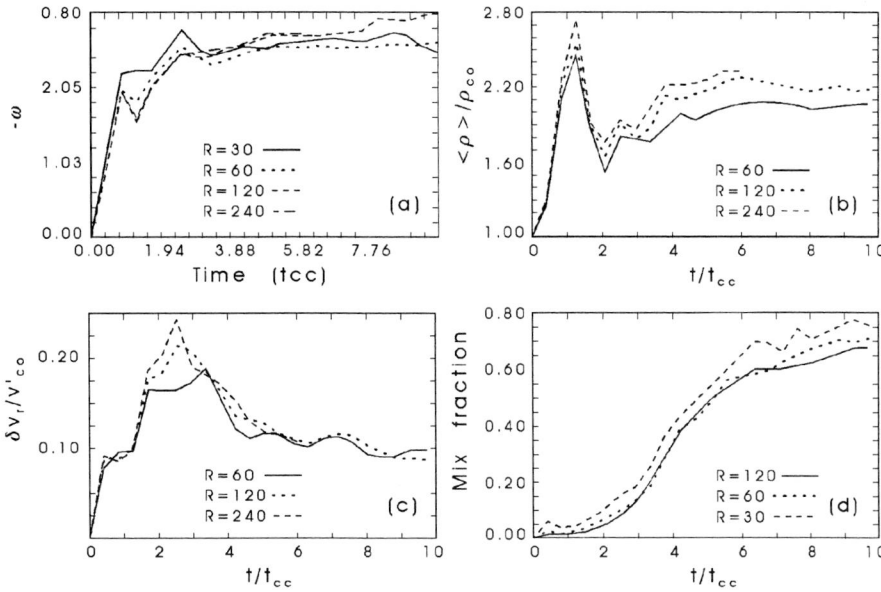

FIGURE 7.

ods against. Although some of the more sophisticated hydrodynamic methods (i.e. piece wise parabolic method, Godunov, AMR) have "cut their teeth" on the challenge problem, and published results, there has been no evidence to date that more recent entries such as SPH have attempted to validate results to multi-dimensional calculations where complex shocked heated flows and shear flow instabilities might be important.

In the interests of establishing credibility for multi-dimensional hydrodynamic calculations of complex nature, we consider the problem of double Mach reflection of oblique, strong shock waves. This problem has a long experimental history, cf. [1,8,23], in addition to a well documented numerical study, [5,9]. The flow is set up experimentally by driving a strong shock down a tube which contains a wedge. When the shock encounters the wedge the flow becomes a comlex shock reflection which is self-similar and depends upon the wedge angle, the Mach number of the shock and adiabatic index of the gas. The problem is most easily set up in a 2D cartesian geometry by initializing a Mach 10 shock in air ($\gamma = 1.4$) which initially makes a $60°$ angle with a reflecting wall. The preshock fluid has a density of 1.4 and pressure of 1. The postshock values follow from the R-H shock conditions. Since the planar shock initially makes a $60°$ angle with the x-axis, this is the equivalent of a shock moving down a wedge that is inclined $30°$ with respect to the horizontal. By initially inclining the shock with respect to the x-axis, we avoid the difficulty of putting in inclined boundaries and explicitly including the wedge. As a result of the inclination, and the high Mach number, the shock reflects as a double Mach reflection forming two Mach stems with two contact discontinuities. The reflected shocks adjoin to the Mach stems with the formation of two triple points. A well pronounced slip line forms and results in a jet of dense material along the reflection wall which culminates in a vortex sheet roll-up in the flow. The problem and it's set-up are well described in [26]. The solutions are given for a variety of numerical schemes. In simulating this important problem, it is crucial to correctly get the strength and positions of the Mach stems, the angle subtended by the two triple points from the point where the double Mach reflected shock rejoins the reflecting x-axis, the location of the slip line, the formation of the jet on the reflection axis and the vortex roll-up in addition to the correct morphology of the double reflected shock. If the essence of this problem cannot be captured by the numerical method chosen, or if the resolution necessary to obtain an accurate solution is prohibitive, then results obtained with the method for astrophysical problems for which the phenomena are present, must be looked upon with great suspicion. In this case it may be advisable to employ a different numerical approach.

11. Cloud-Cloud Collisions

Recent attempts to study gravitational instability in collisions of interstellar clouds leading to possible triggering of binary and multiple star formation have used smooth particle hydrodynamics, e.g. [4,10,18]. As a rule, these calculations have used very few particles to model an otherwise complex multi-dimensional problem. In particular, [18] use 4000 particles in a 3D simulation of the interaction of equal and unequal size cloud collisions. In a similar 2D simulation [10] use 4300 particles and [4] use 4000 particles in 3D to attempt to follow the fragmentation stages of an ensemble of clouds dissipating turbulent energy by shocks through many orders of magnitude in density. In all cases it is clear that the calculations are substantially under resolved. None of these studies have made any attempt to do a proper convergence analysis to study the sensitivity of the results to particle resolution. As a result, it is impossible to assign any degree of accuracy to the final outcome of the simulated collisions. Given our experience with detailed resolution studies of the cloud shock problem, it is clear that many of the same processes are involved in cloud-cloud collisions; i.e. complex shock interactions and shear flow instabilities leading to ablation and fragmentation of clouds. Without the appropriate resolution in particles, it becomes impossible to resolve the ablative instabilities resulting from the cloud interactions, and their effect on cloud destruction. We have begun to study the problem of cloud-cloud collisions with high resolution AMR in both 2D and 3D. Here we present preliminary results without gravity but including detailed atomic cooling to demonstrate the complexity of the interaction and the sensitivity to grid resolution. We have sufficient resolution to capture the ablative instabilities, but we point out the calculations are still not converged using 80 grid cells per cloud radius. We first consider the collision of two clouds each with 33 $M_\odot$; cloud temperature $T = 55$ K and relative velocity 10 km s^{-1} corresponding to a relative Mach number of 13. These preliminary AMR calculations have 80 grid cells per initial cloud radius, although we have made calculations with up to 320 cells per radius to establish convergence properties. The AMR permits us to refine cells only in the region of the clouds for the entire interaction, a huge economy in overall computation time over fixed grid codes. After 3.27×10^{13} s the clouds have collided on the mid-plane perpendicular to the collision axis producing a substantial compression. At 1.72×10^{14} s the coalesced clouds undergo collapse back unto the collision axis. This occurs because the high compression during initial collision results in substantial cloud cooling near the axis. The cooling causes a pressure vacuum resulting in the re-expansion of the cloud back onto the axis. At this time (Fig. 8a) we see the development of the RT instabilities on the cloud surface. At late times, 4.95×10^{14} s (Fig. 8b) the clouds have reached a near equilibrium with substantial compression on the collision axis and significant fragmentation due to the KH instability. Most of the mass is concentrated near the mid plane. To demonstrate the effects of grid resolution, we show the simulation at a comparable time (Fig. 8c), but with 40 grid cells per initial cloud radius. The final distribution of mass is dramatically different. The effect of the KH instability is not nearly as significant, but large amounts of the mass are now distributed much farther away from the collision plane. Thus a modest factor of 2 in resolution can result in a significant difference. It is only with appropriate

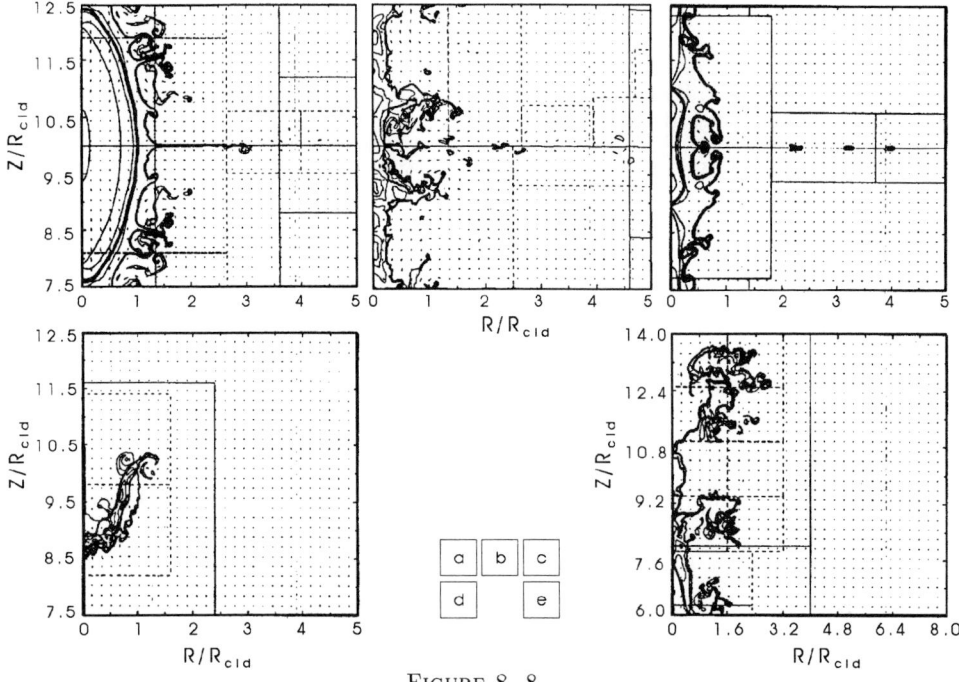

FIGURE 8. 8

resolution study that we have been able to demonstrate that several hundred grid cells per cloud radius were necessary model the collision due to the large compressions obtained in these models.

As another illustration we show the results of cloud-cloud collisions with two unequal mass clouds. The clouds initially have 33 $M_\odot$ and 9.77 $M_\odot$ respectively. At $t = 3.85 \times 10^{13}$ s (Fig. 8d) the clouds have collided and the smaller mass has been overtaken and "absorbed" by the larger cloud. We see that as the larger cloud overtakes the smaller mass there is significant shearing on the cloud-intercloud interface resulting in a KH instability on the outside surface of the larger cloud. The interaction is similar to the cloud shock interaction and the interaction of a cloud with a wind. Evidence for this instability is completely lacking in the SPH calculations that have far too few particles to represent the instability and accurately account for the cloud ablation. At late time, the instability is fully developed (Fig. 8e). Cloud mass is highly ablated and distributed far from the plane of collision. Here we show a portion of the total mass distribution.

To summarize, we find that only with appropriate resolution can we hope to capture the essential physics of instability and ablation in the cloud-cloud collision problem. The fragmented, unstable morphology we find is missing from the SPH calculations. Any conclusions about final mass distribution or fragmentation based upon such calculations must be held as very tentative until such convergence studies are performed.

We thank T. Woods for useful discussions. The work of RIK was performed under the auspices of the U.S. Department of Energy by Lawrence Livermore National Laboratory under Contract W-7405-Eng-48. CFM was supported by a grant from the NSF, AST92-21289. Both RIK and CFM have benefited from the support of a NASA grant to the Center for Star Formation Studies.

REFERENCES

[1] Ben-Dor, G. & Glass, I.I., 1979, JFluidMech, 92, 459.
[2] Berger, M. J. & Oliger, J. 1984, JCompPhys, 53, 484.
[3] Chandrasekhar, S. 1961, Hydrodynamic and Hydromagnetic Stability (New York: Dover).
[4] Chapman, S. Pongracic, H., Disney, M., Nelson, A., Turner, J. & Whitworth, A. 1992, Nature, 359, 207.
[5] Colella, P. & Glaz, H.M., 1984, 9th Intl. Conf. on Numerical Methods in Fluid Dynamics, (Berlin: Springer-Verlag).
[6] Cox, D. P. & Smith, B. W. 1974, ApJ, 189, L105.
[7] Fesen, R. A., Kwitter, K. B. & Downes, R. A. 1992, AJ, 104, 719.
[8] Fursenko, A. A., Golovizin, V. P., Komissaruk, V. A., Mende, N. P. & Zhmakiu, A. I. 1981, Leningrad Phys. Tech. Inst. No. 709.
[9] Glaz, H. M., Colella, P. Glass, I. I. & Deschambault, R. L., 1985, ProcRSocLondon, 13.
[10] Habe, A. & Ohta, K. 1992, PASJ, 44, 203.
[11] Hornung, H. 1986, AnnRevFluidMech, 18, 33.
[12] Klein, R. I. McKee, C. F. & Colella, P. 1990, *The Evolution of the Interstellar Medium*, ed. L. Blitz (San Francisco: Astron. Soc. of the Pacific), 117.
[13] Klein, R. I., Colella, P. & McKee, C. F. 1992, *Advances in Compressible Turbulent Mixing*, ed. W. Dannevik, A. Buckingham & C. E. Leith (US Government Printing Office), 452.
[14] Klein, R. I., McKee, C. F. & Colella, P. 1994, ApJ, Jan. 1.
[15] Klein, R. I., McKee, C. F. & Bell, J. 1994, in preparation.
[16] Klein, R. I., Colella, P. & McKee, C. F. 1994, in preparation.
[17] Landau, L. D. & Lifschitz, E. M. 1959, *Fluid Mechanics*, (Reading: Addison Wesley).
[18] Lattanzio, J. C., Monaghan, J. J. Pongracic, H. & Schwarz, M. P. 1985, AA, 215, 125.
[19] McKee, C. F. & Ostriker, J. P. 1977, ApJ, 218, 148.
[20] McKee, C. F. 1988, *Supernova Remnants and the Interstellar Medium*, ed. R. S. Roger & T. L. Landecker(Cambridge: Cambridge University Press), 205.
[21] Nittman, J., Falle, S. & Gaskell, P. 1982, MNRAS, 201, 833.
[22] Richtmyer, R. D. 1960, ComPureApplMath, 13, 297.
[23] Shirouzu, M. & Glass, I.I., 1982, Univ. of Toronto, UTIAS Report No. 264.
[24] Widnall, S. E., Bliss, D. B. & Tsai, C. Y. 1974, JFluidMech, 66, Part 1, 35.
[25] Woodward, P. R. 1976, ApJ, 207, 484.
[26] Woodward, P. R. & Colella, P. 1984, JCompPhys, 54, 115.

Dynamical evolution of HII regions powered by strong stellar winds

By F. Comerón

Departament d'Astronomia i Meteorologia, Universitat de Barcelona, Av. Diagonal, 647, 08028 Barcelona, Spain

An approach to the champagne phase of HII regions is presented based on the thin shell approximation and considering the shocked stellar wind from the central star, rather than photoionization, as the dominant effect determining the disruption of the parental molecular cloud.

1. Introduction

The classical theory of HII regions describes the evolution of a photoionized sphere of gas in a uniform surrounding interstellar medium. More realistic modelling of the initial scenario of massive star formation takes into account the fact that stars are born in dense molecular clouds which eventually get disrupted by the mechanical energy input resulting from stellar activity [1,7,8]. Modelling of the disruption of the cloud-intercloud interface has shown in some detail the gas dynamics associated with the so-called *champagne phase*, in which high velocities can be reached when the high pressure gas of the HII region expands into the tenuous intercloud medium. In this way, a much better agreement between theory and observations is found [14].

The ultraviolet flux is not the only energy output of an early type star that dramatically alters the physical conditions of the interstellar medium around it. Strong stellar winds efficiently inject momentum and energy, giving rise to expanding shells of gas [12], and substantially modify the classical evolutionary scenario for HII regions [2]. The morphology of compact cometary HII regions, completely embedded in dense molecular material, has been successfully modelled on the basis of an O-type star supersonically moving through the molecular gas and generating a bow shock ahead of it [10,11]. Including the effects of the stellar winds, in addition to the photoionization, is a key point in understanding the diversity of morphologies of HII regions [4,13].

We have carried out several simplified simulations of the evolution of the HII region as it reaches the cloud-intercloud boundary, as a first approach to a modified view of the champagne phase, in which it is the high pressure of the hot coronal gas produced by the shocked stellar wind, rather than that of the warm, dense photoionized gas, that is the driving agent of the cloud disruption.

2. Outline of the dynamical evolution

We will assume that the exciting star is embedded in its parental molecular cloud at a depth greater than the Strömgren radius of its HII region. Five major stages can then be distinguished in the evolution:

1 - The ionizing photons fill the Strömgren sphere in a time of order of the recombination time ($\sim 10^3$ yr for an ambient density $n_H = 100$ cm^{-3}). The wind from the star drives a shock into the photoionized material and soon it forms a dense expanding shell separating the hot bubble of shocked stellar wind gas from the ionized, but mechanically nearly undisturbed, outer medium.

2 - The dense shell reaches a thickness large enough so as to completely trap the ionizing radiation of the central star. This happens before the shell has reached the initial Strömgren radius, in a time of several 10^4 yr for a cloud density of 100 cm^{-3}, that is, less than the dynamical timescale of the Strömgren sphere. Given the high density of the ionized gas outside the shell, recombination and cooling of the gas take place very quickly. The expansion of the shell now proceeds within a cool neutral medium.

3 - The expanding shell reaches the cloud-intercloud boundary and the hot shocked gas inside the bubble blows out to the intercloud medium (champagne phase). Upon reaching the strong density gradient, the dense shell is accelerated outwards under the push of the low density, high pressure gas inside the cavity, and becomes Rayleigh-Taylor unstable [3,9]. The destruction of the shell releases the ionization front which progresses into the intercloud gas.

4 - A new dense, quickly expanding shell forms between the hot gas of the bubble and the intercloud medium. Eventually, the volume density and the thickness of this new shell can grow high enough so as to trap again the ionization front. Nevertheless, the recombination rate in the intercloud medium is much lower than in the cloud, and it remains ionized for a time of order 10^6 yr or more. This only happens for the earliest type stars, whose mechanical output- to- ultraviolet luminosity ratio is higher than for later type stars [5]; shells blown by stars of later spectral types deccelerate to subsonic expansion velocities before their density and thickness can trap all the ionizing photons.

5 - Finally, the shell expanding into the cloud medium becomes subsonic. The age of the bubble when this happens is greater the earlier the spectral type of the star is, and for the earliest stars this last stage takes place once the star has exploded as supernova, thus modifying the entire dynamics of the region.

The relative importance of stellar winds and ionizing radiation is best described by the parameter $\xi = S_*^2/(L_*^3 n_0)$, where S_* is the stellar luminosity shortwards of the Lyman limit, L_* is the mechanical power of the stellar wind and n_0 is the ambient number density. In terms of ξ, several relevant ratios at the turnoff of the Strömgren sphere outer to the shell have simple expressions:

1 - Expansion velocity to sound speed at turnoff:

$$\frac{v_b}{c_s} = 0.31 \left[\frac{\beta_2^2}{(\mu m_h)^3 c_s^9}\right]^{1/5} \left(\frac{1}{\xi}\right)^{1/5}$$

2 - Bubble radius to Strömgren radius at turnoff:

$$\frac{r_b}{r_s} = 2.17 \left[\frac{\mu m_h c_s^3}{\beta_2^{2/3}}\right]^{2/5} \xi^{2/15}$$

3 - Pressure inside the bubble to ambient pressure at turnoff:

$$\frac{p_b}{p_{ext}} = 0.07 \left[\frac{\beta_2^2}{\mu^3 m_h^3 c_s^9}\right]^{2/5} \frac{1}{\xi^{2/5}}$$

4 - Age of the bubble to the expansion timescale of the photoionized region at turnoff:

$$\frac{t_b}{t_s} = 4.13 \mu m_h \frac{c_s^3}{\beta_2^{2/3}} \xi^{1/3}$$

(In these formulae, β_2 is the recombination coefficient to levels other than the fundamental, and μm_H is the mean mass per particle).

Values of these quantities for different spectral types in a medium of $n_0 = 100$ cm^{-3} are given in Table 1:

Table 1

Sp. type	ξ (cm^3 s erg^{-3})	v_b/c_s	r_b/r_s	p_b/p_{ext}	t_b/t_s
O8	$4.1 \cdot 10^{-14}$	7.1	0.27	35.5	0.02
B0	$4.5 \cdot 10^{-13}$	4.4	0.37	13.6	0.05
B2	$1.6 \cdot 10^{-11}$	2.1	0.60	3.3	0.17

3. Results

The simulations we have carried out follow the evolution of the dense shell and the ionization front through the five phases indicated above. The dynamical evolution of the shell is based on the Kompaneets (thin shell) approximation (see e.g. ref. [6], and other references therein). A cloud with density $n_H = 100$ cm^{-3} and temperature 20 K, in pressure equilibrium with an intercloud medium at 8500 K, has been assumed in all cases. We have simulated the evolution for different sets of parameters L_*, S_*, corresponding to stars with spectral types between O8 and B2, located at distances between 1.5 and 4 Strömgren radii from the edge of the cloud.

The champagne phase which initially releases the ionization front into the intercloud medium starts at a time of a few times 10^5 years after turn on of the exciting star. For stars of spectral type B0 or earlier, the fast shell which expands into the intercloud medium has grown to a surface density and thickness large enough to trap the ionization front inside it again at an age of 10^6 years or less after turn on. For these stars, the ionization front runs along with the expanding shell until the latter becomes subsonic. Our simulations are stopped when the expansion velocity drops to twice the sound speed in the intercloud medium, which happens after some $5 \cdot 10^6$ years for a O8 star, $2 \cdot 10^6$ years for a B0 star and $8 \cdot 10^5$ years for a B2 star, the precise values depending on the initial depth of the star inside the cloud.

The main difference introduced by the stellar wind relative to the photoionization-driven champagne phase has to do with the early evolution after the ionized shell reaches the bubble. At these times, the pressure inside the bubble is still high enough to keep the dense, ionized shell from expanding, filling up the cavity surrounding the star and expanding into the intercloud medium. Disruption of the cloud-intercloud interface is thus followed by an outflow of coronal gas, rather than of dense HII. High velocity, dense HII also appears in the stellar wind model; a part of it is associated to the remnants of the Rayleigh-Taylor unstable shell, while the other is the intercloud medium collected on the new shell expanding into it. Both the velocity and density structure of this material should be markedly different from the results obtained for the photoionization-driven champagne phase.

Position-velocity maps can be readily made from our results for various viewing geometries, which should allow comparison to actual spectral line observations.

Acknowledgements: The author wishes to thank Dr. J. Torra for valuable discussions. This work has been supported by the DGICYT under contract PB91-0857 and by the CICYT under contract ESP93-1020-E.

REFERENCES

[1] Bodenheimer, P., Tenorio- Tagle, G., Yorke, H.W. 1979, ApJ, 233, 85.
[2] Breitschwerdt, D. 1987, in *Circumstellar Matter*, IAU Symp. 122, I. Appenzeller and C. Jordan (eds.), D. Reidel Publ. Co.
[3] Chevalier, R.A. 1976, ApJ, 207, 872.
[4] Churchwell, E. 1990, Astr. Ap. Rev., 2, 79.
[5] Leitherer, C., Robert, C., Drissen, L. 1992, ApJ, 401, 596.
[6] Mac Low, M.-M., Mc Cray, R. 1988, ApJ, 324, 776.
[7] Tenorio-Tagle, G. 1979, AA, 71, 59.
[8] Tenorio-Tagle, G., Yorke, H.W., Bodenheimer, P. 1979, AA, 80, 110.
[9] Tenorio-Tagle, G., Bodenheimer, P., Różyczka, M. 1987, AA, 182, 120.
[10] Van Buren, D., Mac Low, M.-M. 1992, ApJ, 394, 534.
[11] Van Buren, D., Mac Low, M.-M., Wood, D.O.S., Churchwell, E. 1990, ApJ, 353, 570.
[12] Weaver, R., McCray, R., Castor, J., Shapiro, P., Moore, R. 1977, ApJ, 218, 377.
[13] Yorke, H.W. 1986, ARAA, 24, 49.
[14] Yorke, H.W., Tenorio-Tagle, G., Bodenheimer, P. 1983, AA, 127, 313.

Line formation during the disruption of molecular cloud cores by photoionization

By J. A. Rodriguez-Gaspar[1], G. Tenorio-Tagle[1], J. Franco[2]

[1] Instituto de Astrofísica de Canarias, 38200-La Laguna, Tenerife, Spain.

[2] Instituto de Astronomía–UNAM, Apartado Postal 70-264, 04510 México D.F., México.

We have solved the line transfer problem for hydrogen recombination lines and forbidden lines of oxygen in expanding H II regions. We calculate their profiles along several lines of sight during the disruption of molecular cloud cores by photoionization. The density, velocity, and ionization structure of spherically symmetric models with an initial power law density distribution, $\rho \propto r^{-w}$, were used to calculate the source function. For steep density gradients ($w \geq 1.5$), the gas moves at supersonic speeds and generates shocks caused by the rapid expansion of the ionized cores. Collisionally ionized oxygen is found to be a good tracer of the hot ($T > 3 \times 10^4$ K) and fast moving shocked gas. Hydrogen emission, on the other hand, footprints mainly the slower photoionized gas with $T \sim 10^4$ K. The observational signature of the cloud core disruption is apparent in the calculated total line profiles integrated through the spherical nebula and in the radial surface brightness distribution.

1. Introduction

Here we continue with the general study of the evolution of spherical H II regions started by Franco et al. [1,2], where clouds with an initial power-law density distribution, $\rho \propto r^{-w}$, are photoionized by a luminous O star. They derived (analytically and numerically) the time evolution of the expanding gas in different geometries and derived the critical exponent, w_{crit}, above which the ionization front (IF) cannot be slowed down by geometrical dilution and recombinations. For $w > w_{crit}$, the IF continues as a supersonic R-type front and is able to ionize the whole cloud. In the formation phase, the value of the critical exponent depends on the assumed initial conditions (i.e., on the ionizing flux of the star, and on the peak density and size of the cloud core) but it has a fixed value, $w = 1.5$, during the expansion phase. For $w > 1.5$ the whole cloud becomes ionized by a supersonic weak R-type IF and the gas acquires the same temperature everywhere ($T \sim 10^4$ K). The pressure gradient follows the density gradient and all the cloud is set into motion at the same time ("the Champagne flow"). The acceleration of the ionized matter decreases with the radius and the inner core expands faster than the outer parts. Thus, the expansion of the core generates a strong shock. Depending on the shock strength we have two differents regimes: for $w \leq 3$ ("slow regime") the shock speed is about constant, and increases rapidly with time for $w > 5$ ("fast regime"). The evolution then depends on the value of the exponent w: the core disruption time and the maximum speed reached by the gas are functions of the density gradient. Here we present the hydrodynamics and line transfer calculations for hydrogen and oxygen lines from spherical H II regions during the disruption of molecular cloud cores.

2. Numerical Results

We have performed simulations for two extreme cases of the exponent, $w = 1.5$ and 5.0. In order to avoid extremely low densities, a constant low-density medium (LDM) was added as background gas with a value of $n_H = 10^{-3}$ cm^{-3}. In each of the calculated

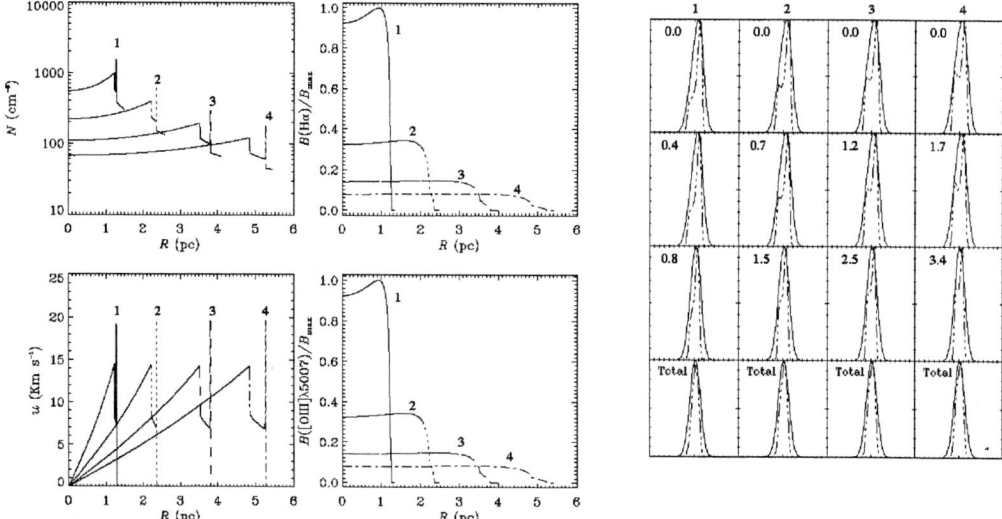

FIGURE 1. $w = 1.5$. On the left the calculated hydrodynamical models and the surface brightness distributions in erg cm^{-2} sec^{-1} strad^{-1} are shown for four different evolutionary times: $t_1 = 4.9 \times 10^4$, $t_2 = 9.1 \times 10^4$, $t_3 = 1.5 \times 10^5$ and $t_4 = 2.0 \times 10^5$ yrs, and on the right the normalized line profiles of Hα (solid line) and $\lambda 5007$ (dash line) as functions of the velocity along the LS (the LS impact parameters in parsecs are on the upper left corner and each tick mark on the x axe represents 10 km s^{-1} matching the range from -90 to 90 km s^{-1}).

cases, the line profiles were integrated through two different *lines of sight* (LSs) using a *pencil beam* with unity diameter. The results are shown at four different evolutionary times. In addition, the surface brightness and the total line profiles (integrated through the spherical nebula) are also displayed.

• $w = 1.5$ (Figure 1). This is the critical exponent above which the IF cannot be trapped inside the cloud. It corresponds to the evolution of a weak D-type IF with a leading shock preceding it which moves with a velocity $\sim 2c_i$ (c_i is the sound speed in the ionized gas) into the neutral gas. For $w < 1.5$, a neutral interphase grows between the two fronts due to the neutral gas accumulated behind the shock, but for $w = 1.5$ the IF is able to ionize the entering into the shock and the neutral shell cannot be formed (the IF is located just behind the shock in the four models). The surface brightness plots in $\lambda 5007$ and Hα show an expanded core which resembles a "bright globule" with similar fluxes in both lines.

The $\lambda 5007$ line profiles are strongly affected by dust absorption. As time proceeds, the average density in the expanded core decreases and so does the absorption caused by dust. Nonetheless, the $\lambda 5007$ shows an increasingly wider hump or a nearly splitted line profile. On the other hand, Hα does not show blends or splittings at all. It is almost constant over all LSs (the beam impact parameter, b, is indicated in parsecs at the upper left corner) with a slight broadening and a positive displacements only along the innermost LSs. The displacements coincide with those of $\lambda 5007$ and have a value of less than 10 km s^{-1}. It is also possible to distinguish slight asymmetric profiles due to dust absorption on gas receding from us (negative speeds). The total line profiles, integrated through all the spherical cloud, do not show displacements nor asymmetries. The $\lambda 5007$ line is

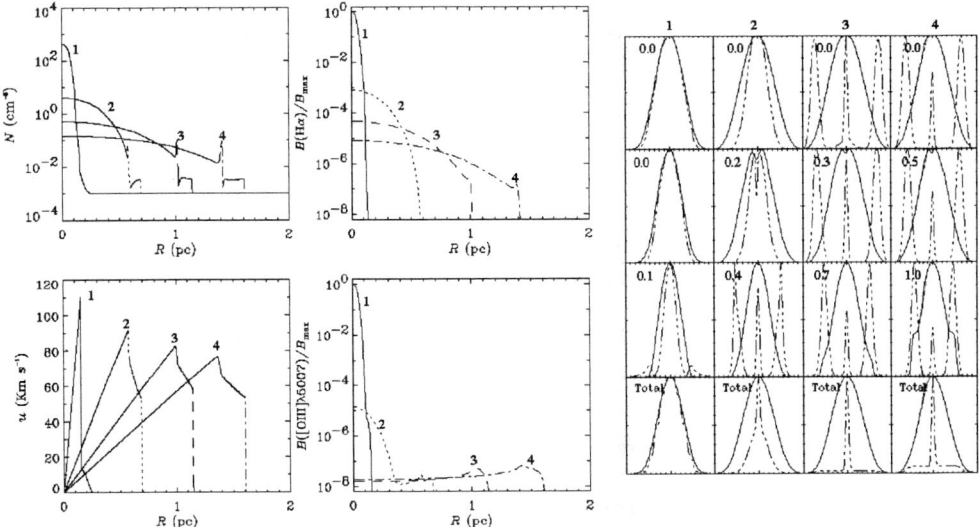

FIGURE 2. $w = 5$. The evolutionary times are: $t_1 = 1.7 \times 10^3$, $t_2 = 6.7 \times 10^3$, $t_3 = 1.2 \times 10^4$ and $t_4 = 1.8 \times 10^4$ yrs.

broadened up to 15 km s^{-1} (this range matches the bulk mass velocity) while its thermal FWHM is only 7 km s^{-1}. Hα presents its thermal broadening (FHWM$\sim$ 20 km s^{-1}).

- $w = 5.0$ (Figure 2). This case is well within the "fast" regime, where the shock acceleration is very strong and has important effects on the structure of the expanded core. In particular, the core density cannot relax to uniform values and presents a centrally condensed profile. The temperature is roughly constant up to the shock position, where it rises above 10^5 K. The gas flow can reach supersonic speeds ($\sim$ 110 km s^{-1}) in less than 10^3 yr as a consequence of the steep pressure gradient. The shock soon moves into the LDM sweeping it into a growing shell, while the core cools due to its quasiadiabatic expansion.

The Hα brightness distributions have a central maximum with a smooth fall and with a negligible halo contribution while the λ5007 lines evolve into a rim pattern with a significant halo contribution ($\sim$ 10% of the maximum). The oxygen lines are about three orders of magnitude weaker than Hα due to the low density hot rim that produces this emission. The line emission has differents sources depending on the gas temperature. The λ5007 is mainly generated in the shocked and hot high speed gas surrounding the core while HII emission comes from the expanding cooler gas of the core. The Hα profiles are broadened and the width increases slightly with time due to the background swept up matter which produces an important contribution at high speeds. The λ5007 profiles initially single broadened profiles, evolve into three components. The central component is produced by the high density and cool core with the corresponding FWHM very small ($\sim$ 5 km s^{-1}). The other components are the approaching and receding features of the shocked hot gas; they appear displaced about 70 km s^{-1} with a FWHM $\sim$ 15 km s^{-1}. The Hα line profiles from unresolved sources are also very broad. For λ5007 they show the same evolution into a three components structure.

3. Conclusions

The hydrodynamical calculations described here were performed with the code described in [3], and served as input data for the line transfer calculations. The case with $w = 5$ is interesting because it shows how a cloud core can be disrupted in a short interval of time ($\Delta t < 10^4$ years) and the flow can reach supersonic speeds under the action of the pressure gradient alone. The same hydrodynamical solution applies to a fully ionized condensation regardless of the location of the photon source.

The disruption via photoionization of high density condensations within molecular clouds leads to a number of distinct signatures depending on the original density fall off of the condensations. Some of these can be summarized as follows: i) The surface brightness for the "champagne" cases ($w > 1.5$) decreases abruptly when the cloud core is rapidly expanding but the $\lambda 5007$ emission remains always lower than the Hα counterpart. For larger values of w, the brightness distribution of $\lambda 5007$ changes into a rim profile with the maximum matching the shock location. While the brightness in Hα remains centrally condensed. ii) The $\lambda 5007$ line profiles show a large splitting, as large as the averaged speed of the shocked gas, with an additional minor contribution from the relaxing core at lower speeds (e.g. the central components in Figure 2).

This work was partially supported by DGAPA-UNAM through the grant IN103991, a CRAY R&D grant, and the EEC grant CI1*-CT91-0935. Some of the simulations were performed with CRAY/YMP of the Supercomputing Center at UNAM.

REFERENCES

[1] Franco J., Tenorio-Tagle, G. & Bodenheimer P. 1989, RevMexAstAstrof, 18, 65.
[2] Franco J., Tenorio-Tagle, G. & Bodenheimer P. 1990, ApJ, 349, 126.
[3] Tenorio-Tagle G., Bodenheimer P., Lin D. N. C. & Noriega-Crespo A. 1986, MNRAS, 221, 635.

Highly Collimated Outflows in Two and Three Dimensions

By P. E. Hardee[1], D. A. Clarke[2]

[1] Department of Physics & Astronomy, University of Alabama, Tuscaloosa, AL, USA

[2] Harvard-Smithsonian Center for Astrophysics, Cambridge, MA, USA

A three dimensional numerical simulation of a highly collimated outflow is compared to a two dimensional simulation with identical parameters. Both simulations were performed on a Cartesian grid using the ZEUS astrophysics MHD code. In two dimensions an equilibrium slab jet and in three dimensions an equilibrium cylindrical jet were established completely across the computational grid in pressure balance with the surrounding medium. The additional degree of freedom in three dimensions leads to spatially rapid filamentation of the flow. The resulting disruption of highly collimated flow in three dimensions is considerably more rapid than in two dimensions.

1. Introduction

A large number of effectively two dimensional numerical simulations of highly collimated outflows have been performed in the past few years [6]. The two dimensional numerical simulations have proven very useful in establishing the qualitative behavior of highly collimated outflows in situations of astrophysical interest, e.g., the jets associated with extragalactic radio sources. The numerical simulations have shown that the Kelvin-Helmholtz stability properties of the flow are crucial to understanding the resultant morphology. In three dimensions a linearized analysis of the time dependent fluid equations suggests that the additional degree of freedom in three dimensions results in a much more unstable system as a consequence of additional unstable modes that do not exist in two dimensions.

2. Two Dimensional Numerical Simulation

Figure 1 is a gray scale density image from a simulation [5] of a two dimensional slab jet of half-width R, at dynamical time $\tau = t(a_x/R) = 18$, which was initially established across the computational grid in equilibrium with an external medium ten times more dense than the jet and with sound speed a_x. A slab jet is spatially resolved along two Cartesian axes and is infinite in extent in the third dimension. The jet flow speed is 3 times the internal sound speed and 9.5 times the external sound speed. The simulation was performed on a 320 x 600 zone computational grid and forty zones spanned the jet width. A small amplitude sinusoidal oscillation of frequency $\omega R/u = 0.1$ was imposed at the origin to break the symmetry (u is the flow speed). In the simulation a sinusoidal oscillation of wavelength 14.5R grows in amplitude disrupting continuous flow about 28R from the origin, and a lobe of jet material forms.

3. Three Dimensional Numerical Simulation

3.1. Two Dimensional Slices

Figure 2 contains a gray scale axial density slice and transverse density slices from a simulation [4] of a three dimensional cylindrical jet at the same dynamical time as the slab jet shown in Figure 1. The axial slice through the three dimensional data cube provides the most direct comparison with the two dimensional simulation. The cylindrical jet was

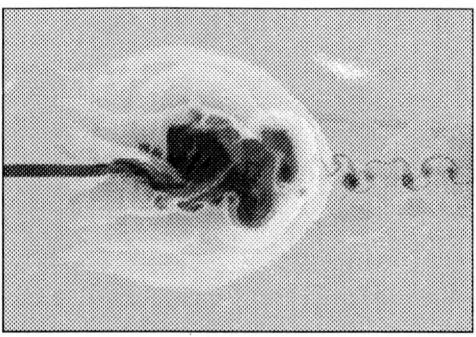

FIGURE 1. A sinusoidal oscillation excited by the sinusoidal perturbation at the origin disrupts continuous flow in a slab jet simulation. Patterns to the right of the growing lobe are remnants of the original flow which was established across the grid. Image is 90 x 60 jet radii.

initially established across the computational grid with exactly the same conditions as the slab jet shown in Figure 1. The three dimensional simulation was performed on a 100 x 100 x 300 zone computational grid and twenty zones spanned the jet diameter. A periodic precession with frequency identical to that of the sinusoidal oscillation used in the slab jet simulation has been imposed at the origin to break the symmetry. Continuous flow is disrupted about 20 jet radii from the origin and a lobe of jet and entrained material has formed. The transverse slice at axial distance of 20R shows the flow splitting into twin filaments. Filamentation is largely responsible for the rapidity of disruption when compared to the slab jet. The filamentation leads to entrainment of external material and considerable shock heating. The difference in transverse expansion between slab and cylindrical jets is largely the result of the different volume dependence on radius.

3.2. Three Dimensional Ray Integrations

A line of sight integration through the three dimensional data cube provides the closest approximation to what might be observed on the plane of the sky. Figure 3 is a gray scale rendering of line of sight integrations of the temperature (top) and velocity divergence (bottom) that shows the three dimensional structures at dynamical time $\tau = 14$. The temperature image shows both a large scale helical twist of wavelength 24R induced by precession at the origin and cross sectional elliptical flattening at an axial distance of about 12R. Bright strands in the velocity divergence show regions of compression, indicative of shocks. Elliptical flattening in the temperature image is associated with the widely spaced twin filaments in the velocity divergence image which mark the outer extent of the long axis of the elliptical distortion. Accompanying the large amplitude elliptical distortion is a weaker third filament indicative of a weaker triangular distortion of the jet cross section. At larger distance this triple surface filamentation leads to jet trifurcation. At this distance the jet loses collimation, and bright filaments wrap tightly around a growing lobe. Fine scale surface filamentation exists close to the origin but does not reach the large amplitude of the double and triple surface filamentation associated with elliptical and triangular jet distortion, so does not show up in the velocity divergence or temperature image.

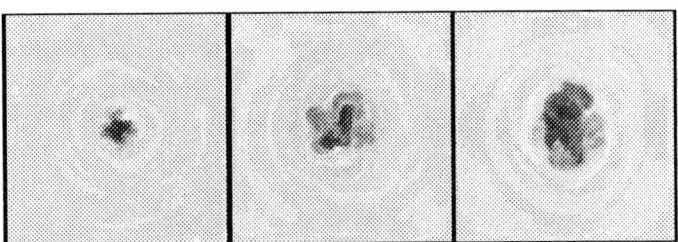

FIGURE 2. a) Axial density slice with dimension 60 x 26 jet radii. Patterns to the right of the growing lobe are remnants of the original flow which was established across the grid. b) Transverse density slices at axial distance 10R, 20R and 30R with dimension 26 x 26 jet radii.

4. Kelvin-Helmholtz Instability

Considerable theoretical work on the stability of two and three dimensional flows has been performed in the past few years [1]. Solutions to the dispersion relations obtained by linearizing the time dependent fluid equations for slab and cylindrical geometries are shown in Figure 4. The solutions describe the propagation and growth of the fundamental (f) sinusoidal wave perturbation and accompanying internal sinusoidal perturbations (r1,r2,r3,...) to the slab jet, and of helical (f1), elliptical (f2), triangular (f3) and higher order fundamental wave perturbations to the cylindrical jet for parameters appropriate to the numerical simulations. On the cylindrical jet the fundamental wave modes correspond to helical twisting of the jet, elliptical distortion and bifurcation of the jet, triangular distortion and trifurcation of the jet, etc.. All distortions twist around the jet circumference. Jets become more stable as the Mach number increases, and spatial growth lengths scale as the Mach number times the jet radius. The maximum spatial growth rates are greater on the cylindrical jet than on the slab jet, and it appears that the elliptical (f2) and triangular (f3) modes and the accompanying double and triple filamentation are largely responsible for the rapid disruption of the cylindrical jet. Higher order modes (fine scale surface filamentation seen in the simulation closer to the origin) do not disrupt the jet in spite of their faster maximum growth rate because they are confined to the jet surface. The linear analysis reveals that the effect of perturbations falls off as distance away from the jet surface like $(r/R)^{n-1}$ where n is the mode number [3].

5. Conclusion

It is clear that the additional degree of freedom and additional unstable wave modes in three dimensions affect the dynamics of highly collimated flows significantly and lead to decollimation somewhat more rapidly than might have been anticipated. On a positive note the structures seen in the simulation correspond to structures predicted by the linear theory

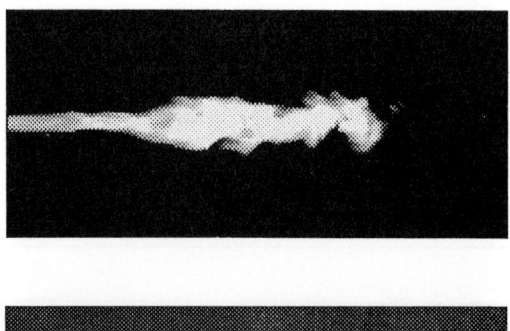

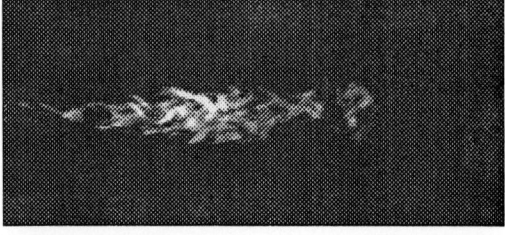

FIGURE 3. a) Temperature image at top shows large scale jet distortions. b) Velocity divergence image at bottom shows filamentary compressions associated with jet distortion and with the boundary layer around the growing lobe.

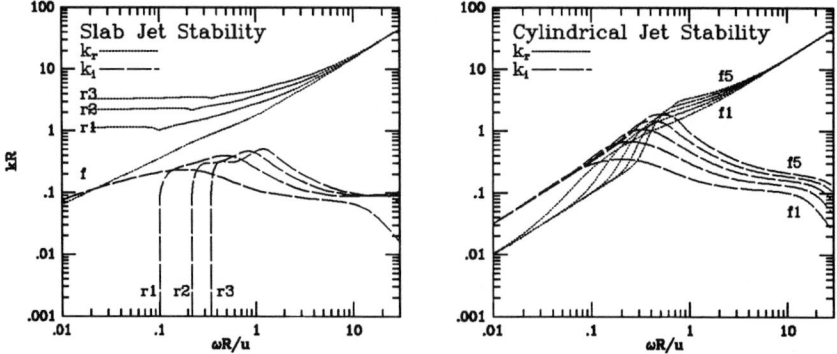

FIGURE 4. Dispersion relations describing spatial growth of non-axisymmetric Kelvin-Helmholtz unstable modes on slab and cylindrical jets.

and resemble structures observed in extragalactic radio jets. Helically twisted filaments seen on the surface of the jet in the numerical simulation, e.g., the velocity divergence image in Figure 3, appear similar to twisted filamentary structures observed in a radio image of the jet in M87 between the nucleus and knot A [7]. Beyond knot A the radio image suggests a sinusoidal oscillation (projected helical twist) in the plane of the sky and also shows a narrowing and subsequent broadening of the jet (projected elliptical distortion). The filaments wrap around the jet as in the simulation and are relatively tightly wrapped beyond knot A. The tighter wrapping of filaments beyond knot A as in the simulation may occur in a shear layer around a more rapidly moving jet core. However, the spatial scale of powerful extragalactic jets (FR II radio sources) requires a much more stable outflow than that of our present simulation. For example, the jet in Cygnus A[2] shows an inferred helical twist and apparent bifurcation over almost half its considerable length. Thus, our present simulation might more appropriately represent a low power outflow such as the jet in M87 (FR I radio sources). Mechanisms providing a stabilizing influence include much higher initial Mach numbers, or jet expansion which enhances stability through adiabatic cooling (increased Mach number) and through an increase in the jet radius (increased scale length), and magnetic fields (magnetic tension). Additionally, high power extragalactic jets propagate through a lobe that is of lesser density than the jet itself and this may provide a stabilizing influence.

This work was supported by NSF grant AST-8919180 and EHR-9108761 to the University of Alabama, and by NSF grant AST-9148279 to Harvard University.

REFERENCES

[1] Birkinshaw, M. 1991, in *Beams and Jets in Astrophysics*, ed. P. A. Hughes, (Cambridge:Cambridge), p.278.
[2] Dreher, J. W., Carilli, C. L. & Perley, R. A. 1987, ApJ, 316, 611.
[3] Hardee, P. E. 1983, ApJ, 269, 94.
[4] Hardee, P. E. & Clarke, D. A. 1992, ApJ Let, 400, L9.
[5] Hardee, P. E., Cooper, M. A., Norman, M. L. & Stone, J. M. 1992, ApJ, 399, 478.
[6] Norman, M. L. 1993, in *Astrophysical Jets*, eds. D. Vurgarella, M. Livio & C. P. O'Dea, (Cambridge:Cambridge), in press.
[7] Owen, F. N., Hardee, P. E. & Cornwell, T. J. 1989, ApJ, 340, 698.

Numerical Simulations of Stellar Outflows

By F. Rubini[1], G. Manzini[2], S. Lizano[3], C. Giovanardi[4]

[1] Dipartimento di Astronomia e Scienza dello Spazio, Università di Firenze, 50125-I, Firenze

[2] CERFACS, 47, Avenue G. Coriolis, 31057 Toulouse Cedex

[3] Instituto de Astronomia, UNAM, Apdo. Postal 70-264, México, D. F. 04510, Mexico

[4] Osservatorio di Arcetri, Largo E. Fermi 5, 50125 Firenze, Italy

We present numerical simulations of a collimated highly supersonic jet injected in a molecular cloud at rest. We show that, for nonuniform clouds with a density structure $\rho \propto r^{-2}$, the passage of the bow shock establishes a velocity field similar to one observed in molecular outflows.

1. Introduction

Recent developments in the observation and interpretation of stellar outflows reveal a close link between stellar jets and molecular outflows. Essentially, they would not be separate phenomena, but a transfer of energy would take place from the stellar jet to the external matter, producing fast molecular winds, with the observed characteristic lobe-structure. In this picture a key role is played by the bow shock generated as the jet enters the cloud with highly supersonic Mach numbers. The aim of our simulations is to show that the velocity pattern of the molecular cloud outflow is driven by the bow shock, in contrast with the predictions based on a mixing-layer mechanism (see e.g. [1]). Along these lines, Raga and Cabrit [5] have recently studied a simple analytic model in which the molecular outflow corresponds to the gas entrained in the wake of a bowshock created by an internal jet working surface. Numerical simulations, in fact, show that the mixing layer influences just a small region close to the jet (e.g. [4]) whereas the velocity field created in the matter at rest, when traversed by the bow shock, fills a region comparable with the observed lobes. This effect, due to non-linear interactions, can be studied by solving the complete set of Euler equations. To this purpose we used a numerical code which implements modern high-resolution shock-capturing techniques and prevents the formation of spurious numerical oscillations even if the flow is characterized by high gradients and discontinuities.

The simulations are characterized by pressure and temperature ratios between the jet and the external cloud and the Mach number. We also explore the effect a jet being injected into a non uniform cloud.

2. Computational Results

We have investigated the mechanism of interaction between an axial supersonic jet and the surrounding molecular outflow, regardless of their formation mechanism (e.g. direct ejection from the star or from an accretion disk) and of their collimation properties (e.g. effects of shocks or jet instabilities). In particular, we address the following question: can the bow-shock be the main responsible of the energy transfer from the jet to the

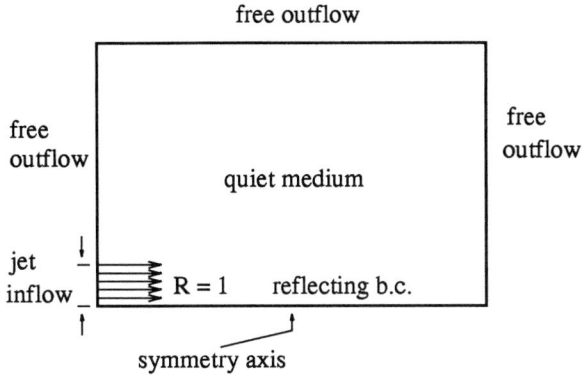

FIGURE 1. Sketch of the integration domain. In the following simulations the length of the box is 1 pc and the width is 0.5 pc.

outflow? In order to give an answer we have solved the compressible Euler equations by means of numerical shock capturing techniques, able to take into account the shock-like discontinuities in the solution. For a description of the code see [2].

Fig. 1 schematically shows the integration of domain, with the corresponding boundary conditions. A collimated highly supersonic jet is continously admitted through an orifice of radius R into the quiet gas inside the rectangular box. The width of the inflow section is assumed to be the lenght scale in our calculations. The southern boundary conditions corresponds to the axis of simmetry, hence, only the upper half of the jet is calculated. In the other boundaries a free flow boundary condition is imposed. In the simulations presented below, the length of the box is 1 pc and the width is 0.5 pc.

In our simulations we fix the ratio of the jet and cloud temperature, $T_j/T_c = 16$, and the Mach number, $Mach = 35$, and changed the ratio of the jet and the cloud pressure $K = P_j/P_c$. The small scale structures cannot be resolved in this simulations but a global and straightforward view of the external cloud is possible. We considered jets moving into uniform and nonuniform clouds. For a more extense discussion of the effects of varying the physical parameters, see [3].

Fig. 2 shows a $Mach = 35$ overexpanded jet, with $K = 0.5$, moving into a uniform cloud. The time of the simulation and the velocity of the lowest contour are shown in the left plot. The pictures shows the jet propagating from the left to the right. The right picture puts in evidence the bow-shock propagating inside the molecular cloud, the left picture shows the magnitude of the velocity field that is established immediately after its passage. The absolute values of this velocity field are still too small with respect to the observed ones. In fact, velocity field quickly decreases from 200 Km/sec, in the inner beam, to a few Km/sec quite close to the axis. The external matter stays essentially at rest, and no real molecular lobe is present.

Things work a little better in Fig. 3, representing an underexpanded jet with $K = 10$ and the same Mach number. The time of the simulation and the velocity of the lowest contour are shown in the left plot. When we compare this case with the previous one, we see that this jet is much faster. This is due to the higher inner density. In both the simulations, in fact, we assume an inner jet temperature 16 times the external one, so that the inner density increases with the pressure ($n_j = \frac{K}{16} n_c$). This means that the outer molecular cloud is a weaker obstacle, and the head of the jet can travel faster.

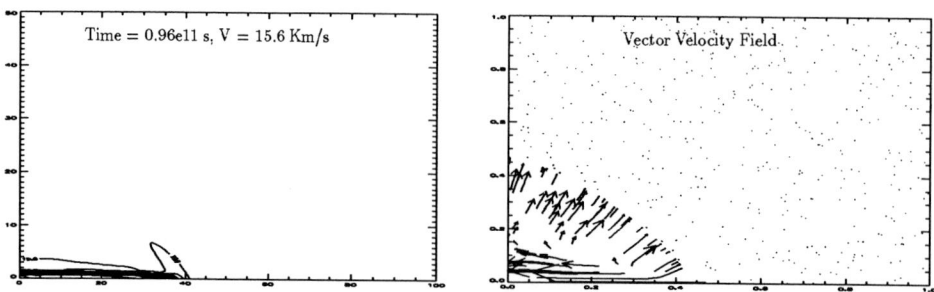

FIGURE 2. The velocity contours on the left and the velocity vectors on the right are shown for the case Mach=35 and $K = 0.5$ for a jet entering a uniform cloud.

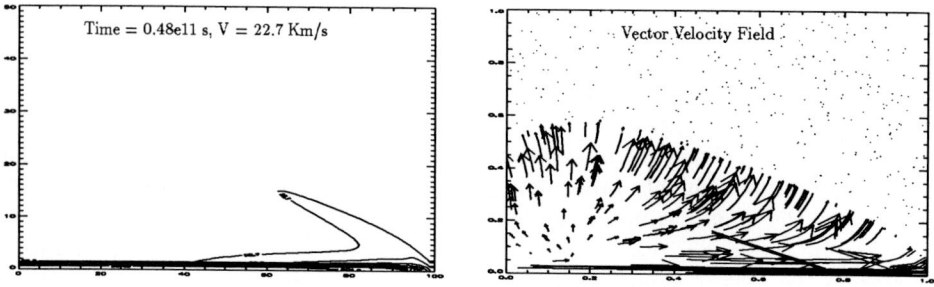

FIGURE 3. The velocity contours on the left and the velocity vectors on the right are shown for the case Mach=35 and $K = 10$ for a jet entering a uniform cloud.

A dramatically new scenario emerges when a non uniform cloud is considered. We assumed an external density field that decreases with power of the longitudinal distance, $\rho \propto r^{-2}$, due to the self gravity of the cloud. In this density field less and less the external matter works against the bow-shock and the working surface of the jet. As a consequence, the velocity of the head of the jet, that is the velocity of the bow-shock, increases and for the same reason, the momentum transfer to the particles at rest increases. In Fig. 4 and 5 both the underexpanded ($K = 0.5$) and the overexpanded ($K = 10$) simulations show a post-shock velocity field in which over a wide, lobe-like region the computed absolute values are in a very good agreement with the observed ones. The time of the simulation and the velocity of the lowest contour are shown in the left plot.

3. Conclusions

Our numerical simulations show that as soon as the jet impacts the external medium, the bow-shock begins to propagate into the cloud transferring energy and momentum to the shocked particles of the rest matter. This mechanism produces long range effects and the velocity field of the shocked particles fills a lobe-like region whose size is of order 10^{18} cm, after a time comparable to the time-life of the jet, $\sim 10^3$ yrs. The shape of such a region and the mean value of the velocity field, of the order 50 Km/sec, agree with the observed data.

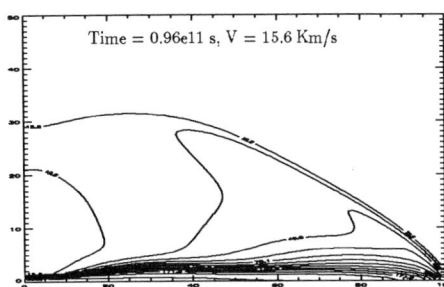 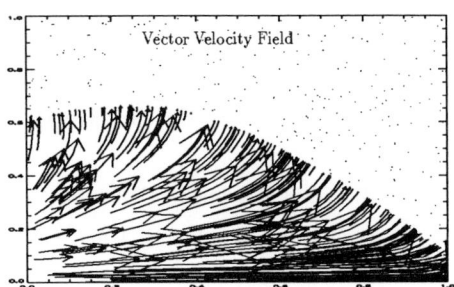

FIGURE 4. The velocity contours on the left and the velocity vectors on the right are shown for the case Mach=35 and $K = 0.5$ for a jet entering a cloud with density $\rho \propto r^{-2}$.

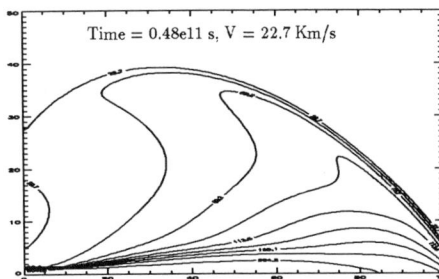 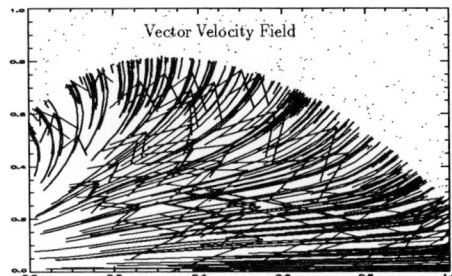

FIGURE 5. The velocity contours on the left and the velocity vectors on the right are shown for the case Mach=35 and $K = 10$ for a jet entering a cloud with density $\rho \propto r^{-2}$.

It seems therefore, that the bow-shock alone could be responsible of the momentum transfer from the jet to the molecular cloud that produces the observed CO outflows.

REFERENCES

[1] Cantó, J. & Raga, A. C. 1991, ApJ, **372**, 646.
[2] Manzini, G. 1994, Ph. D. Thesis, Université Paul Sabatier, Toulouse, France.
[3] Manzini, G. & Rubini, F. 1994, AA, submitted.
[4] Norman, M. L., Smarr, L., Winkler, K.-H. A. & Smith, M. D. 1982, AA, **113**, 285.
[5] Raga, A. & Cabrit, S. 1993, AA, **278**, 267.

Variable Velocity Jets: Internal Working Surfaces

By S. Biro[1], A. C. Raga[2]

[1] Department of Astronomy, University of Manchester, Manchester, M13 9PL, UK.
[2] Mathematics Department, UMIST, P.O. Box 88, Manchester M60 1QD, UK.

Work in progress on the detailed study of "Internal Working Surfaces" is presented. Prompted by our simulations of stellar jets with a periodic variation in the magnitude of the velocity, we have begun the study of the evolution and properties of a single knot or internal working surface. The previous results are reviewed and the present status of our work is described.

1. Introduction

Stellar Jets from young stars are strings of Herbig-Haro (HH) objects with varying degrees of alignment. These HH objects are small knots bright in emission lines characteristic of the cooling regions behind shocks. Several of these HH jets have interesting kinematic properties. The knots in the jet HH34 have highly supersonic velocities [5]. Position-velocity diagrams of HH46/47 [9] present velocity changes (which are a large fraction of the jet velocity) along the body of this jet, as well as sharp velocity drops at the brighter knots. Finally, high resolution mapping of jets such as HH111 [10] reveal several of the knots to be bow-shaped, similar to the "head" of the jet. These observational results can be interpreted in terms of a time-dependence in the ejection of the jet material. Specifically it would seem to indicate multiple events of increased ejection velocity.

Wilson [12] presented numerical simulations of jets (in the extragalactic context) with a periodic, supersonic, variation in the magnitude of the velocity. A double shock structure forms inside the jet beam (one to brake the faster gas upstream and one to accelerate the slower gas downstream). Because they are similar to the working surfaces at the heads of jets, these shock structures have been called "internal working surfaces" (IWS). Different aspects of this type of flow have been modeled analytically and numerically. Analytic models making predictions about both the dynamical and emission properties of the flow have been calculated [6-8]. One-dimensional simulations including detailed atomic processes have been performed [4], while in 2D a steady state calculation with radiative cooling of a single IWS has been made [3], and in 3D a time-dependent but quasi-adiabatic calculation has been presented [11].

We have done both a simulation of a complete jet with supersonic, periodic variations in the magnitude of the velocity, and of a single IWS. In §2 we briefly describe the results of the full jet simulation, which is discussed in more detail in [1]. In §3 we show a progress report of the zoom-in on a single IWS.

2. The complete jet simulation

For all the results that will be shown here we have assumed a compressible, inviscid, non-adiabatic gas of cosmic abundance and have used our 2D axysymmetric Eulerian Flux-Vector-Splitting [2] code, which includes a non-equilibrium cooling function.

The initial conditions for the full jet simulation are as follows. The simulation was run on a 600 x 200 point grid. The surrounding medium is initially uniform and stationary. The jet gas has a particle density of $n_j = 7$ cm^{-3}, is overpressured ($P_j/P_e = 10$) and supersonic ($V_j=100$ km s^{-1} at $T_j = 10^4$ K). At the point of injection, the velocity is given a periodic (sinusoidal) variation with a supersonic amplitude: $v(t)=100 + 50 \sin(2\pi t/\tau)$ km s^{-1}. Where $\tau = 380$ yrs. is the period of the variation.

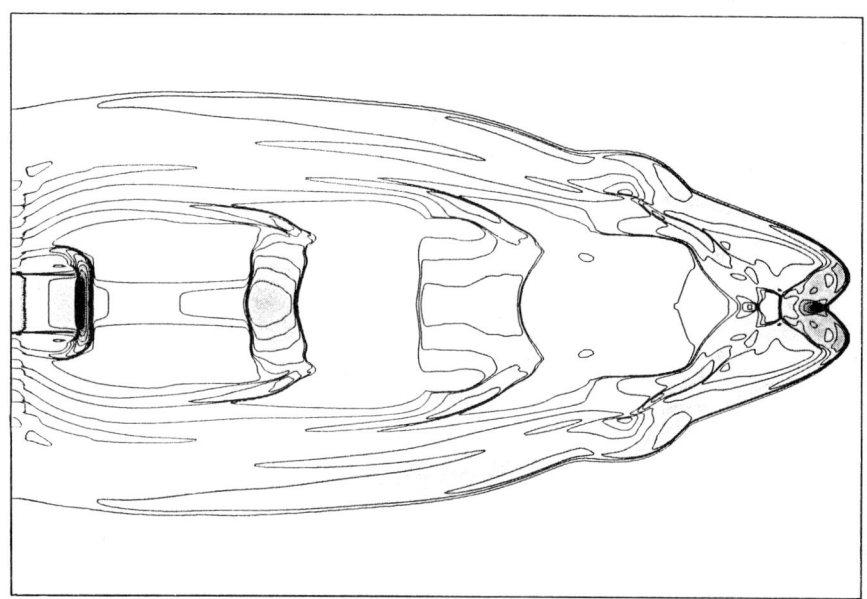

FIGURE 1. The pressure structure of the complete jet at the final time (1900 yrs.) of the simulation. Isobaric contours are logarithmically spaced. Greyscales are added to aid the eye. The places where the contours are "piled-up" indicate the loci of shocks. The head of the jet and three complete IWS (at different phases of their evolution) can be seen. A fourth one is running into the head. Taken from [1].

Figure 1 shows a plot of the resulting pressure structure. The properties of one IWS as it moves away from the source were measured. The size of the IWS (defined as the distance between the two shocks along the axis of the jet, see Figure 1) is proportional to the distance from the source, in agreement with the properties predicted analytically for an isothermal IWS [1]. The total emission of the knot in the [SII]6717+31 and Hα lines peaks very near the source (with Hα peaking first) and then decreases rapidly. At large distances from the source Hα is twice [SII]. Maps of emission (both in Hα and in [SII]) were constructed. At several angles of projection the knots have bow-shapes similar to the observations. Profiles of intensity as a function of position in both lines were made and were found to be spatially offset indicating that they come from different regions of the IWS.

These results prompted us to zoom in on the behavior of one single knot or IWS. Simulations of a steady-state, high resolution single knot have been done [3]. Our time-dependent calculation further expands on this approach.

3. A single Internal Working Surface

In order to "zoom-in" on a single IWS, we have assumed that at large enough distances from both the source and the head the spatial structure of the jet is periodic. In other words, we assume that two consecutive knots will be at very similar stages of their evolution. If this is the case, then the simulation of a single IWS with periodic boundary conditions in the up- and down-wind directions is equivalent to the whole jet, and permits greater resolution, both temporal and spatial.

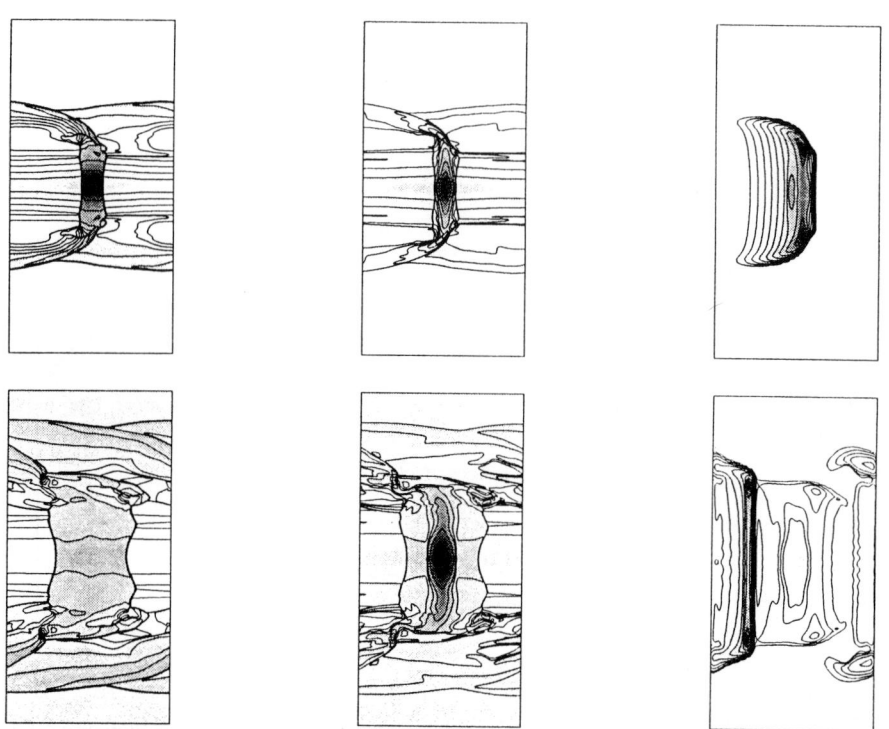

FIGURE 2. Temporal evolution of the IWS. Pressure (first column), density (second) and Hα emission (third) at two times: 150 (first row), 300 (second) yrs. Contours and greyscales are combined to aid the eye.

The setup for this simulation consists of a square grid of 450 × 450 grid points with a reflection condition on the top and bottom walls, and periodic on the left and right. At the initial time, a jet is defined in the lower 56 grid points (equivalent to a radius of 5 ×10^{15} cm). The jet has an initial particle density of $n_j = 700$ cm^{-3}, is highly overpressured ($P_j/P_e = 50$) and is given a sinusoidal velocity profile along the x (horizontal) direction of the grid such that material is coming in from left and right and meeting at the centre of the grid: $V(x) = 50 \cos(2\pi x/x_{max} - \pi/2)$ km s^{-1}. In order to simulate the relative motion of the knot in the ambient medium, we have given the ambient medium a negative velocity of $V_e = -100$ km s^{-1}.

Figure 2 shows the time evolution of the knot in pressure, density and Hα emission. Very soon the two-shock structure forms. Material flows out the side at a large rate initially (pushing a small bowshock into the surrounding environment) which at later times slowly decreases resulting in the collapse of the bow. Here we can see how well the cooling structure is resolved. Also we see how the jet material ends up almost completely (and quite quickly) in a cocoon around the jet beam. It can be seen that most of the emission initially comes from the knot itself, but at later stages of the evolution of the knot it originates in the wings of the bow shock.

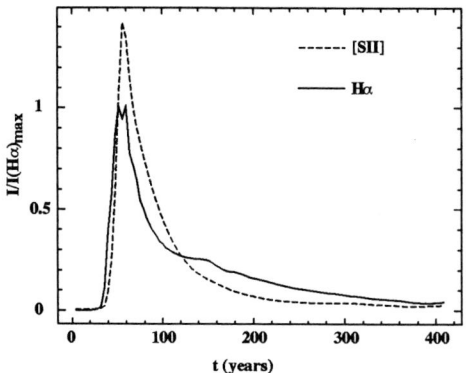

FIGURE 3. Total Hα and [SII] emission of the IWS as a function of time.

A very detailed history of the emission from the IWS is obtained from this simulation. At every time step, the emission integrated over the whole knot in the lines Hα and [SII] was calculated. Figure 3 shows the time-evolution of these lines. In both cases, the emission peaks very soon (or equivalently very near the source) and then drops off rapidly. Hα is stronger than [S II] at later times.

4. Summary

We have reviewed our previous results of a simulation of a stellar jet with a periodic, supersonic variation in the magnitude of the ejection velocity. In this work, information was obtained about the time-evolution of the size, shape and emission properties of an IWS. There is qualitative agreement between these results and both observations and the predictions of models. We argue that greater resolution of single IWS is necessary in order to obtain reliable predictions of observable quantities, and then show preliminary results of such a simulation. From the temporal evolution we see two regimes of lateral ejection of material. The Hα emission maps show how emission comes mostly from the knot at early times and from the bow later. A detailed time-evolution of the emission in Hα and [SII] is in progress. Further work will enable us to understand better the formation of such IWS.

This work was partially supported by UNAM, Manchester University, and ORS.

REFERENCES

[1] Biro, S. & Raga, A. C. 1993, ApJ, submitted.
[2] van Leer, B. 1982, Lect. Notes in Phys., 170, 507.
[3] Falle, S. A. E. G. & Raga, A. C. 1993, MNRAS, 261, 573.
[4] Hartigan, P. & Raymond, J. 1993, ApJ, submitted.
[5] Heathcote, S., & Reipurth, B. 1992, AJ, 104, 2193.
[6] Kofman, L. & Raga, A. C. 1992, ApJ, 390, 359.
[7] Raga, A. C., Cantó, J., Binette, L. & Calvet, N. 1990, ApJ, 364, 601.
[8] Raga, A. C. & Kofman, L. 1992, ApJ, 386, 222.
[9] Reipurth, B. 1989, *Low Mass Star Formation and Pre-Main Sequence Objects*, ed B. Reipurth, ESO, 247.
[10] Reipurth, B., Raga, A. C., & Heathcote, S. 1992, ApJ, 392, 145.
[11] Stone, J. M. & Norman, M. L. 1993, ApJS, 80, 753.
[12] Wilson, M. J. 1984, MNRAS, 209, 923.

Accretion Disks in Astrophysics

By S. K. Chakrabarti

Department of Astronomy and Astrophysics, The University of Chicago, and International Center for Theoretical Physics, Trieste.

Accretion disks are temporary storehouses of matter around compact objects, such as white dwarfs, neutron stars and black holes. A steady disk allows a controlled release of the gravitational energy of the infalling matter in the form of radio waves to high energy gamma rays. In this review, we briefly present a unifying view of various accretion disk models existing in the literature, and also discuss the current status of the numerical simulation works.

1. Introduction

Accretion disks are important ingradients in many astrophysical systems which involve mass transfer from one object or set of objects to another. There are ample evidences in the literature that accretion disks exist in systems ranging from CVs, LMXBs in small scale, to AGNs and Quasars in large scale. In this paper, I will present a unifying view of the theoretical models and numerical simulation works in accretion disks. The subject is extensively studied and cannot possibly be covered quite squarely in a few short pages. Keeping in mind that the present conference is on *Numerical Astrophysics*, I shall try to emphasize on the role that is played by the theoretical models in guiding numerical works in this field. Towards the end, I will lay down my personal view on some suggestive directions in order that the numerical work can contribute more fruitfully to the understanding of the physical processes around compact objects.

2. Developments in Analytical and Numerical Works

Though analytical models are always welcome, they tend to describe stationary properties of an accretion disk. There are increasing evidence that most of the astrophysical systems which are believed to harbor accretion disks are also variable. Occasionally the same system shows variabilities in completely different timescales (an example being the X-Ray flickering and optical micro-variabilities in Blazers). Sometimes, stability analysis helps to sort out which of the models might be applicable, but such a procedure is often very difficult and scopes are limited. Theoretical models investigate very different *physical* processes in detail (e.g., radiative transfer) in an accretion disk (which is why there are more than two dozen models in the literature), but fail to study detailed dynamical behavior. The numerical simulations, on the other hand, have been very successful in obtaining the *dynamical* behavior (e.g., presence of various instabilities, shock waves, etc.), though many of the important physical ingradients are missing (which is why there are only less than half a dozen disk simulation results available). In fact, the role of these two directions have been complementary and it is possible that either of them can never be totally independent. Though it might seem obvious that the results from numerical models should correspond more closely to the observations and therefore they must play a major role, so far, the aim and scope of this approach has been found to be limited as well. Only hope is that in future, one can include more and more physical processes in the code so that both the physics and the dynamics emerge from the same simulation. In certain situation, the exact nature of a physical process (such as, viscous process which is essential for transporting angular momentum outwards to enable matter to fall in) may be illusive and the theoretical input to numerical works has been meagre. This has resulted in the fact that most of the disk simulations have no 'physical' dissipation. They are energy and angular momentum conserving, and show violent behavior such as strong shock waves, contrary to some of the more popular theoretical disk models which do not have these discontinuities. Below, we shall follow the developments in these two directions.

2.1. Analytical Models

Various theoretical models of accretion disks are present in the literature for more than two decades. Most of these models, if not all, are distinguished by the equation of state (EOS) of matter that is used in the disk and whether or not the infall velocity is assumed significant compared to the rotational velocity. For example, consider the radial momentum equation of matter in a non-magnetized disk falling on a Newtonian star:

$$\frac{1}{\rho}\frac{dP}{dr} + v\frac{dv}{dr} - \frac{l^2(r)}{r^3} - F(r) = 0 \qquad (1)$$

Here, $F(r)$ is the gravitational acceleration $(-GM/r^2)$ and other symbols are usual. In the popular model of thin accretion disks [31, 39], pressure P and advection velocity v are considered to be negligible. The remaining two terms show that the angular momentum distribution is Keplerian. In other words, for accretion to take place, matter must lose significant angular momentum to be able to be Keplerian as it falls toward the star. This requirement fixes the accretion rate $\dot{M}$ of the flow in this model,

$$\dot{M}[l(r) - l_{in}] = 4\pi z r^2 f_\phi = -4\pi z r^3 \eta \frac{d\Omega}{dr}. \qquad (2)$$

When the pressure is non-negligible, it changes the angular momentum distribution as first shown by Maraschi, Raina & Treves [22]. If the advection velocity can still be ignored, the distribution of angular momentum becomes,

$$l(r) \propto [r^3 F(r) + \frac{r^3}{\rho}\frac{dP}{dr}]^{1/2} \qquad (3)$$

Clearly, therefore, where ever the pressure gradient term is positive(negative), the angular momentum is higher(lower) than Keplerian. This is the essence of thick accretion disk models. If the disk is radiation pressure supported, one has radiation torus [21, 33] and if the disk is ion pressure supported, one has ion torus [36]. In these models viscosity is negligible and accretion is pressure driven rather that viscous driven. Disk assumes the shape of a doughnut having all the properties of a torioidal star with maximum pressure and temperature in its center. It is obvious that more complex models of disks can be built by choosing more sofisticated EOS of the disk matter, which includes physical processes ranging from partially ionized matter to pair production, and often different EOS at different parts of the disk.

Since a black hole accretion is necessarily transonic, the flow velocity is expected to be very high, particularly close to the black hole and one can no longer ignore the advection term in Equation (1). This consideration gave rise to another class of disk, called a transonic disk [6, 20, 23, 27, 32]. Here, the flow could be pressure and/or viscous driven, though the topology of the solutions changes dramatically as viscosity is added [6, 23]. A natural extension of this model includes shock waves in disks which exploits the fact that there are multiple 'saddle'-type critical points in the flow. Theories of shock waves in the context of accretion on stars [18, 24, 40] are long present in the literature; presently they are extended to include accretion onto black holes (see, Chakrabarti [6] and references therein).

In the solutions which include shock waves, the flow first passes through the outer sonic point and after a shock transition, it passes through the inner sonic point before disappearing into the black hole. At the shock location, appropriate Rankine-Hugoniot conditions must be satisfied if the flow is non-dissipative, otherwise dissipative conditions must be used [6]. Among the important results which emerges out of these studies, one finds that (a) For a given set of initial parameters, the shock location is not unique [6, 15], but only at one location the shock is stable [13, 28]; (b) Properties of the entire flow, including that of the shock is predictable from the boundary conditions alone; (c) The stable shock becomes weaker and disappears altogether as the viscosity is increased [8]; (d) There is a range of parameters (such as energy and angular momentum) for which shocks form in the disk; (e) For a given set of initial parameters, there is a solution which does not include a shock; (f) In the case of an adiabatic flow, the shock-free solution has lesser entropy than the final solution which includes a shock transition; (g) In the case of an isothermal flow, the energy of the shock free solution is more than that in the final solution which includes a shock. These findings are very significant as they propose a unifying view of the accretion disks. This incorporates two extreme disk models into a single framework: for inviscid disks, strong shocks are produced, and for disks with high enough viscosity, the stable shock disappears altogether and angular momentum distribution can become Keplerian. Details of these solutions are discussed in [6,13].

In most of the analytical studies of the transonic disks and the disks which include shock waves, it is assumed that the flow motion is governed by the Newtonian equations, but the potential due to the central object is not $\propto 1/r$ but $\propto 1/(r - r_g)$, where r_g is the Schwarzschild radius of the central black hole. This so-called Paczyński-Wiita [33] pseudo-Newtonian potential preserves all the salient features of a Schwarzschild black hole and yet simplifies analytical work dramatically. Recently, this potential is successfully generalized to mimic geometry around a Kerr black hole [10].

So far, only axisymmetric disk models are considered. In the presence of non-axisymmetric perturbations, such as a binary companion or a massive object that is passing by, non-axisymmetric spiral waves are launched into the disk which steepens to produce spiral shock waves. In the literature, only self-similar analytical models are present [7, 42] and the results generally agree with more realistic numerical simulations.

2.2. Numerical Results

Because of the reasons I elucidated in the beginning of this Section, almost all the numerical simulations tend to have no explicit viscosity present [11, 16, 17, 25, 37, 38, 45] except for a case [14] where the so called α-viscosity [39] and radiative transport are also added. That is why, ironically, there is no simulation which reproduces the popular model of a thin, Keplerian disk. Rather, the simulations show stationary/non-stationary shock waves as important ingradients. It is possible that as the accretion rate changes at the outer boundary, various viscous processes could be turned on and off, and disks may change from Keplerian to shocked ones.

The first numerical attempt to study the behavior of matter around black holes was made fifteen years ago [45]. An Eulerian, fully general relativistic, first-order backward space difference technique was used. The spatial resolution was low and the solution was evolved only till $\sim 100 GM/c^3$. Considering the limited computational resource available during that period, it was certainly a bold step. It was shown that the large angular

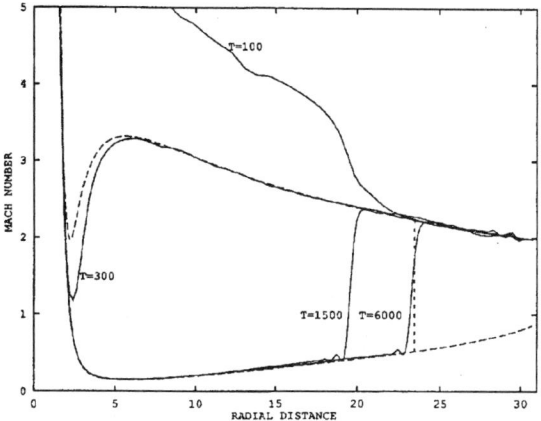

FIGURE 1. An example of a thin, accretion disk solution which includes a standing shock wave. Solid curves are simulation results at various times (as indicated), whereas the dashed curves and lines are those predicted by the analytical work.

momentum accretion is accompanied by shock waves which expand outwards. This code was later improved upon, with number of grid points as well as the evolution time orders of magnitude higher. A series of very important simulations were made with this code to show that thick accretion disks can indeed form in inviscid flows [16-17]. These simulations also confirm the results of Wilson [45] that non-steady shock waves are formed which travel outward. From the post-shock flow, a very strong wind is generated which is hollow in nature which 'hugs' the funnel wall. Due to inviscid nature of the simulation, centrifugal force kept the flow away from the axis of symmetry.

Most of the understanding of the shock formation in accretion disks around black holes has emerged [6] *after* these simulations were made. This had the effect that these simulation results were not interpreted adequately. For the same reason, in these simulations no example of any standing shock waves was provided, probably because the parameter space for which standing shocks might form was not known. In a series of simulations [13, 25], using Smoothed Particle Hydrodynamics (SPH) code written in axisymmetric co-ordinate system (and using Paczyński-Wiita [33] potential), the effects of preceeding theoretical background could be clearly seen. Several examples were presented which include *standing shock waves* in accretion disks and winds. The shocks are found to form *exactly* where they are predicted, particularly in an axisymmetric 1-D thin flow (for which accurate theoretical models could be constructed). In thick 2-D inflow simulations, shocks are found to form slightly outside the locations predicted by a 1.5-Dimensional model. This is probably due to turbulent pressure in the flow just behind the shock and/or inaccuracy of the 1.5D model which ignores the vertical component of velocity altogether. These simulations also find strong wind formation (as in the finite element methods discussed above), and the wind is found to become supersonic with finite distance. Several simulations are carried out using parameters which do not predict standing waves, and in these cases, no standing shocks are found either; after a transient phase, flow remains on the supersonic branch. These simulations verify all the assertions made in the theoretical work, such as the stability of the shock waves. They disprove assertions made earlier [1] that multiplicity of sonic points implies bi-periodicity. Impossibility of bi-periodicity could be understood by the fact that entropy of the flow passing through two sonic points are completely different

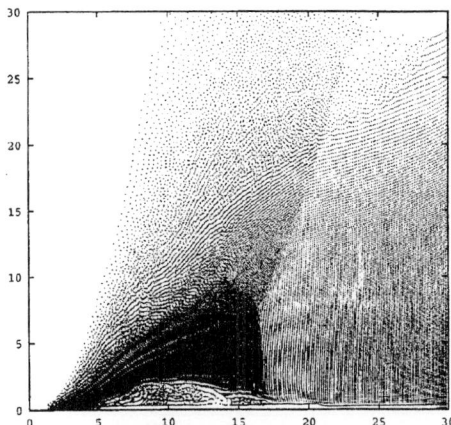

FIGURE 2. X-Z plot of the particle locations in the SPH simulation of the two dimensional accretion disk which includes a standing shock wave.

and a flow with a steady outer boundary condition cannot allow such instability. These simulations also disprove that any arbitrary shock transition (see, Fig. 9b of Wallinder, Kato & Abramowicz [44]) is also not allowed, no matter how intuitively obvious it may look.

Fig. 1 shows an example [13] of the simulation of a thin accretion disk which includes standing shock waves. Mach number of the flow is plotted against the radial distance (in units of the Schwarzschild radius of the central black hole). Solid curves are simulation results and the dashed curves are the supersonic and subsonic branches respectively. Two vertical dashed lines indicate locations of the analytically predicted shock transitions, the outer one being stable. After a transient phase, a shock forms close to the inner edge which then travels outward till it reaches the outer stable shock location predicted by analytical work. The specific energy and the specific angular momentum were chosen to be $\mathcal{E} = 0.011c^2$ and $\lambda = 3.8GM/c$ respectively.

Figs. 2-4 show results of a 2-Dimensional simulation [25] at $T = 700GM/c^3$. Here, 60,000 particles are used (in each quadrant). Fig. 2 shows the particles and the standing shock at $X \sim 16$. Note the presence of the oblique shocks also. The specific energy and angular momentum used were $0.006c^2$ and $3.3GM/c$ respectively.

Fig. 3 shows the contours of constant Mach number. Note that the flow, subsonic after the shock, becomes supersonic very close to the hole. Also, the wind which is originated subsonically on the disk surface, becomes supersonic very soon. Fig. 4 shows the contours of constant temperature (labelled in geometric units). At the shock location, the temperature of the flow becomes high and the velocity very low thus satisfying all the conditions of a thick accretion disk. The contours in the immediate vicinity of the post-shock region resemble that of a thick accretion disk. The shock heating causes a strong pressure gradient term which pushes matter out of the disk to form a strong wind.

As a passing remark, I like to mention here that the 1-Dimensional transonic disk solution which also includes a standing shock [13] is an well understood theoretical problem. It has more physical input than a simple Bondi flow. The agreement using 1-D SPH code has been so excellent (Fig. 1) that one feels confident in the 2-D SPH simulation results (Fig. 2-4), where the theoretical results are not available. Since it is always advisable to test a code with a problem which is as close to the original problem as possible, I personally believe that this shock problem in cylindrical geometry can be used as a *test problem for*

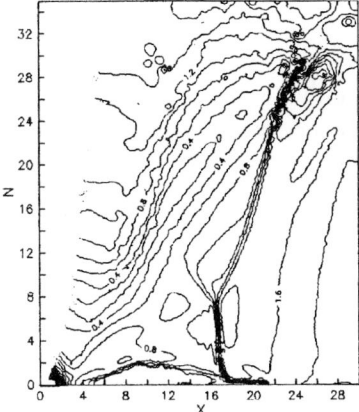

FIGURE 3. Equal spaced contours of constant Mach number in the X-Z plane. Flow becomes subsonic at the shock and subsequently becomes supersonic before disappearing into the hole. Note that close to the funnel, Mach no. of wind rises with Z.

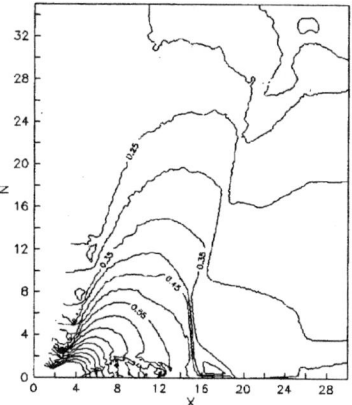

FIGURE 4. Equal spaced contours of the temperature of the flow. Note that the contours in the immediate vicinity of the post-shock region resemble that of a thick accretion disk since the temperature becomes very high and the infall velocity low.

any disk simulation around a black hole. It should not be enough for a code to just pass some test shock-tube problems in Cartesian grid alone.

Disk simulations which include α-viscosity and radiative transport, show completely different behavior [14] than what is shown above. Evolution of a Keplerian disk shows that it collapses into a thin sheet on the equatorial plane, as is expected since radiation pressure supported α disks are known to be unstable in viscous and thermal timescales.

Some very exciting time dependent solutions have been presented in this conference which are related to the formation of the star and the bipolar outflows from massive protostellar disks [43]. In these systems, the disks could be self-gravitating and the effects of self-gravity on the structure and stability of accretion disks need to be included also. Time *independent* work in this regard is present in the literature [5, 29]. The later solutions show interesting deviations of the Keplerian orbits from those in non-gravitating case. These results might affect star formation rates and the nature of the bipolar outflows, since the self-gravity of the disk can result in more 'saddle'-type sonic points and multiple shock waves [6].

An important set of simulations, which pertains to the formation of the non-axisymmetric shock waves, are carried out mainly by Kyoto group [11, 38] and others [37]. Here, in addition to the central object, a binary companion is also added. Non-axisymmetric modes amplify and steepen to produce spiral shocks inside the disk. It was shown that spiral shocks may remove angular momentum rather efficiently and it is possible to achieve an effective $\alpha \sim 10^{-2}$. Non-stationary shock waves, can also produce variabilities at different timescales [11-12]. Most of the simulations are carried out for thin accretion disks. Shocks can become weaker as the disk thickens (see Figs. 2 - 4 for axisymmetric case) and one needs to extend these simulation to a full 3-Dimensional accretion problem.

3. Magnetized Disk Models

It is usually believed that the origin of bipolar outflows and jets are related to the properties of accretion disks. Outflows and jets can extract angular momentum from disks very efficiently (especially when there is no binary companion), and thereby aid accretion process. Since various properties (such as velocity, kinematic luminosity, etc.) of an outflow must be directly related to the properties of the underlying disk, theoretical disk models as well as numerical simulations must reproduce both the disk and the jet *simultaneously*. There are a few theoretical models which achieve this in the context of bi-polar outflows [34] and in the context of radio jets [2, 4, 9, 19]. In the latter case, magnetohydrodynamic equations are solved inside as well as outside of the disk and the solutions are matched on the disk boundary. So far, no satisfactory numerical work is present which successfully produce collimated and accelerated outflows/jets *which are originated* from magnetized disks, even though separate simulations of magnetized disks and jets are present in the literature. Some of the simulations have discovered new insights into old problems – examples being the Balbus & Hawley instability [3] and the detailed instabilities at the working surface of a radio jet [30].

4. Concluding Remarks

There are increasing observational evidences that accretion disks in nature are not just thin or thick, slim or fat, smooth or shocked, twisted or centrifugally exhausted over their entire lifetime. For example, almost simultaneous correlated variabilities at multiple wavelengths, or the observed soft X-rays in AGN spectra, or the time variation of asymmetric splitting of line emissions in some AGNs are indicative of the fact that the underlying disk cannot be a simple Shakura-Sunyaev type thin disk. Similarly, ideal thick disks cannot be the only solution either, as high energy emissions come only from the region deep inside the funnel which are completely shadowed at any moderate inclination. Thus, a single model cannot work also. What shape and form a disk might assume depends on the energy, angular momentum and the quantity matter supplied, as well as the dominant viscous mechanism that may be prevalent. We have noted earlier that if the viscosity is low, the disk could have strong shock waves, whereas for high viscosity, shocks may disappear and disks become smooth. This view has a unifying character, but details are still to be worked out using both the approaches we discussed in this review.

So far, in my opinion, results of the numerical simulation of accretion disks have not made what one might call definite breakthroughs. However, it has potential as well as avenues to do so. In principle, it can answer the most difficult problem of all: who feeds the monstar? How does matter is supplied to the central black hole in AGNs and

Quasars? Is it due to the possible bar-inside-bar-inside-bar... instability [41] or striping out of closely passing stars [35] or accretion of winds from nearby star claustars as probably is the case in our own galactic center? The accretion from the large scale region ($\sim$ few Kpc) to the small scale region ($\sim 10^{-4}$pc) is to be understood, if possible, through a *single* simulation. In the *same* simulation, the emergence and propagation of jets/outflows from the small distances to the large distances is to be understood as well! A crucial input from the theorists should be to formulate the problem in a manner so that mathematical complexity is reduced without sacrificing essential physics. In the past years, one such important input which has saved a large amount of computing time has been to replace the full general relativistic equations by equations in flat geometry with pseudo-Newtonian potential [10, 33]. This potential naturally goes over to Newtonian form at a large distance, and therefore a single code could be used both at large and small scales. With the advent of various highly efficient methods (such as adaptive mesh technique) as well as faster super-computers, it may be possible to realize this *super-project* in future. For the time being, however, various component problems (such as star-disk interaction [26], effects of viscous and radiative transport, nucleosynthesis close to a black hole, etc.) could be tried out.

This work was carried out while the author was on leave from Tata Institute of Fundamental Research (Permanent Address). The work was supported by a grant NAG 5-1485 with the University of Chicago.

REFERENCES

[1] Abramowicz, M. A. & Zurek, W. 1981, ApJ, 246, 314.
[2] Blandford, R. D. & Payne, D. G., 1982, MNRAS, 199, 883.
[3] Balbus, S. A. & Hawley, J. F. 1991, ApJ, 376, 214.
[4] Camenzind, M. 1989, *Accretion Disks and Magnetic Fields in Astrophysics*, G. Belvedere (ed.), Kluwer, Dordrecht, 129.
[5] Chakrabarti, S. K. 1988, J. Astrophys. & Astron., 9, 49.
[6] Chakrabarti, S. K. 1990, *Theory of Transonic Astrophysical Flows*, World Scientific, Singapore.
[7] Chakrabarti, S. K., 1990, ApJ, 362, 406.
[8] Chakrabarti, S. K. 1990, MNRAS, 243, 610.
[9] Chakrabarti, S. K. & Bhaskaran, P. 1992, MNRAS, 255, 255.
[10] Chakrabarti, S. K. & Khanna, R. 1992, MNRAS, 256, 300.
[11] Chakrabarti, S. K. & Matsuda, T. 1992, 390, 639.
[12] Chakrabarti, S. K. & Wiita, P. J. 1993, ApJ, 411, 602.
[13] Chakrabarti, S. K. & Molteni, D. 1993, ApJ, (Nov. 10th).
[14] Eggum, G. E., Coroniti, F. V. & Katz, J. I. 1987, ApJ, 323, 634.
[15] Fukue, J. 1987, PASJ, 39, 309.
[16] Hawley, J. F., Smarr, L. L. & Wilson, J. R., 1984, ApJ, 277, 296.
[17] Hawley, J. F., Smarr, L. L. & Wilson, J. R., 1985, ApJ Suppl., 55, 211.
[18] Holzer, T. E. & Axford, W. I. 1970, ARAA, 8, 31.
[19] Königl, A. 1989, ApJ, 342, 208.
[20] Liang, E. P. T & Thomson, K. A. 1980, ApJ, 240, 271.
[21] Lynden-Bell, D. 1978, Phys. Scripta, 17, 185.
[22] Maraschi, L., Reina, C. & Treves, A. 1976, ApJ, 206, 295.
[23] Matsumoto, R., Kato, S., Fukue, J. & Okazaki, A. T. 1984, PASJ, 36, 71.
[24] McCrea, W. H. 1956, ApJ, 124, 461.
[25] Molteni, D., Lanzafame, G. & Chakrabarti, S. K., 1994, ApJ, (to appear).
[26] Molteni, D., Gerardi, G. & Chakrabarti, S. K., this volume.
[27] Muchotrzeb, B. 1983, Acta Astron., 33, 79.
[28] Nakayama, K. 1992, MNRAS, 259, 259.
[29] Nishida, S., Eriguchi, Y. & Lanza, A. 1992, ApJ, 401, 618.
[30] Norman, M., this volume.

[31] Novikov, I. D. & Thorne, K. S. 1973, *Black Holes*, C. DeWitt and B. DeWitt (eds.), Gordon and Breach, New York.
[32] Paczyński, B. & Bisnovatyi-Kogan, G. 1981, Acta Astron. 31, 1.
[33] Paczyński, B. & Wiita, P. J. 1980, AA, 88, 23.
[34] Pudritz, R. E. & Norman, C. A. 1986, ApJ, 301, 571.
[35] Rees, M. J., 1989, *Big Bang, Active Galactic Nuclei and Supernovae*, S. Hayakawa & K. Sato (Eds.), Universal Academy Press, Inc. Tokyo, Japan.
[36] Rees, M. J., Begelman, M. C., Blandford, R. D. & Phinney, E. S. 1982, Nature, 295, 17.
[37] Różyczka, M. L. & Spruit, H. C. 1989, *Theory of Accretion Disks*, NATO, ASI, F. Mayer et al. (Eds.), Kluwer, Dordrecht, 341.
[38] Sawada, K., Matsuda, T. & Hachisu, I. 1987, MNRAS, 224, 307.
[39] Shakura, N. I. & Sunyaev, R. A. 1973, AA, 29, 179.
[40] Shapiro, S. L. & Salpeter, E. E. 1975, ApJ, 198, 671.
[41] Shlosman, I., Begelman, M. C. & Frank, J. 1990, Nature, 345, 679.
[42] Spruit, H. C. 1987, AA, 184, 173.
[43] Yorke, H., this volume.
[44] Wallinder, F., Kato., S. & Abramowicz, M. A. 1992, AA Rev., 4, 79.
[45] Wilson, J. R. 1978, ApJ, 173, 431

Orbitally-Modulated Emission Line Profiles from Non-Keplerian Accretion Disks

By I. G. Martínez-Pais

Instituto de Astrofisica de Canarias, E 38200 La Laguna, Spain

Under the assumption of a vanishingly small pressure and a small viscosity accretion disk in a close binary system, gas streamlines are calculated via an approximate analytical method in the framework of the restricted three body problem. An outer limit for the disk is found, defined by the last non-intersecting orbit, in agreement with previous works. Once the velocity field is determined, a series of line profiles are numerically computed, under the assumption of a $r^{-\alpha}$ emissivity law, for different orbital phases and for a range of mass ratios.

1. Introduction

In his early work, Paczynski [3] constructed a simple dynamical model for accretion disks in close binary systems. His main assumption consisted in considering that pressure and viscosity forces are negligible against the gravitational forces of both the primary and secondary stars. Under this assumption, and also rejecting the effect of the interaction between the gas coming from the secondary and the disk itself, the gas streamlines are identical to simple periodic orbits in the restricted three body problem. The author searched numerically for simple non-intersecting periodic orbits confined to the orbital plane, finding just one family in which all the orbits were symmetrical with respect to the line joining the two stars. The most external orbits depart significantly from circular, a critical orbit always existing being the largest non-intersecting one. This critical orbit defines the outer limit of the pressure-free accretion disk. As far as I know, even rough line profile calculations have never been carried out in the framework of this simple model.

2. Calculation of the orbits

A polar coordinated system (r, θ) was adopted, centred at the primary component, and with the polar axis rotating with the two stars and passing through the center of the secondary. As usual, the values of the sum of the masses of both components, the orbital separation and the orbital angular velocity were taken as unity. To calculate the orbits, an analytical approximate method was used, first outlined by Huang [1]. Following the results of Paczynski [3] the orbits were assumed to be simple, periodic and symmetrical with respect to the line joining the two components. If, in addition, one demands that $\theta(t=0) = 0$ and $\lim_{\mu \to 0} r = r_0$; $\lim_{\mu \to 0} \theta = \omega t$; i.e. in the limit of low secondary mass (μ) the orbits are circular, it is easy to show that the equation of the orbits can be expressed as a power series of μ which explicitly depends on t:

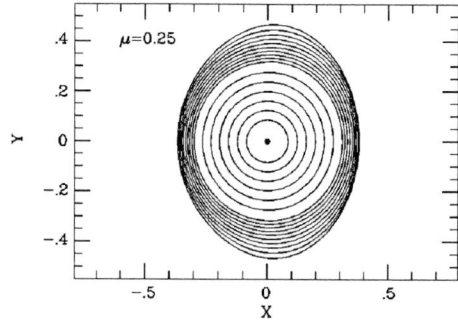

FIGURE 1. Some of the simple periodic orbits around the primary component of a binary system having $\mu = 0.25$ calculated in the restricted three body problem.

$$r = r_0 + \mu \sum_{k=0}^{\infty} a_k \cos k\omega t + \mu^2 \sum_{k=0}^{\infty} c_k \cos k\omega t + ... \quad (1.a)$$

$$\theta = \omega t + \mu \sum_{k=1}^{\infty} b_k \sin k\omega t + \mu^2 \sum_{k=1}^{\infty} d_k \sin k\omega t + ... \quad (1.b)$$

If, following a perturbative method, one substitutes these expressions into the infinitesimal body's equation of motion and retains only zero order terms in μ, one finds:

$$(\omega + 1)^2 = \frac{1-\mu}{r_0^3} \quad (2)$$

which is the third Kepler's law. Substituting Equation (1) in the equation of motion and retaining first order terms in μ it is possible to calculate both a_k and b_k (they are all zero except $k \leq 3$). Once the first order coefficients have been determined it is possible to calculate, following the same procedure, the second order ones, c_k and d_k (which are non-zero only for $k \leq 6$) and so on. All the coefficients depend only on r_0 and ω so, bearing in mind Equation (2), each orbit is determined using the value of r_0 only (or alternatively ω) [1]. Coefficients were calculated up to the second order in μ.

The characteristics of the calculated orbits are qualitatively compatible with those of Paczynski. In particular, they are non-intersecting up to a maximum value r_{0_M}, which defines the outer limit of the disk. Figure 1 shows some of the calculated orbits for $\mu = 0.25$. In order to compare the results obtained in this work with those obtained by Paczynski [3], the relative differences ($\Delta r/r$) for some parameters defining the shapes and sizes of the last non-intersecting orbits (where the differences should be greater) have been calculated. The parameters for comparison are: $R_1 = r(\theta = 0)$; $R_2 = r(\theta = \pi)$ and the maximum distance inside the orbit to the primary, $R_{\max}$. In Figure 2 the relative differences are represented versus μ, showing that for $\mu \leq 0.4$, which is a good range for Cataclysmic Variables, relative errors in R_1 and R_2 are ≤ 5 % and ≤ 2 % respectively, and ≤ 7 % for the maximum radius $R_{\max}$ (these relative errors show the effect of truncation of the series in μ).

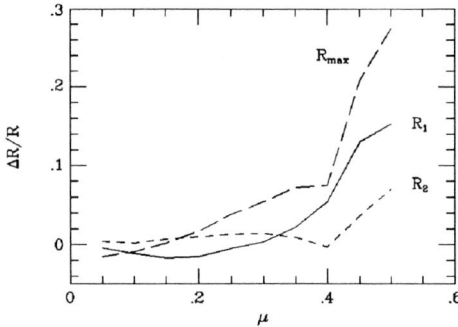

FIGURE 2. Relative differences between the results of this work and those of Paczynski [3] for the parameters $R_1 = r(0)$, $R_2 = r(\pi)$ and $R_{\max}$ of the last non-intersecting orbit.

3. Profile computations

In order to perform the profile calculations, two orbits were adopted defining the inner and outer limits of the disk, the corresponding values of r_0 being 0.01 and r_{0_M}, respectively. A velocity step for the line sampling was also defined in each case to be $\Delta v = 0.1 \times (1-\mu)^{1/2}$ i.e. about 10^{-2} times the maximum velocity in the line. A number of orbits were then calculated with values of r_0 between the limits just defined and having a step Δr_0 whose corresponding velocity step (for $\theta = 0$) was $0.2 \times \Delta v$. Once the orbits were calculated, the velocity field was determined by dividing each orbit into azimuthal sections, the difference in the projected velocity over the $\theta = 0$ direction, between two consecutive sections, being again $0.2 \times \Delta v$. The velocity of each section is then calculated as $u(r,\theta) = \sqrt{\dot{r}^2 + (r\dot{\theta})^2}$. The direction of the velocity was defined by the angle, ψ, between the velocity vector and the line joining the pole with the center of the section, namely, $\tan\psi = (r\dot{\theta})/\dot{r}$.

To calculate the profile line corresponding to a particular orbital phase ϕ, a projected velocity $v(r,\theta,\phi)$ was assigned to each azimuthal section given by $v(r,\theta,\phi) = -u(r,\theta) \cos(\phi + \theta + \psi)$. Then a partial profile $P_p(v,\phi)$ was calculated for each orbit assuming an $r^{-\alpha}$ emissivity law:

$$P_p(v,\phi) = U \Delta R \frac{r \Delta \theta}{u \sin \psi} R^{-\alpha} \qquad (3)$$

with $R = r(\theta = 0)$ and $U = u(r,0)$. Here ΔR is half of the radial distance between the two orbits adjacent to that considered in the azimuthal section corresponding to $\theta = 0$, and $\Delta \theta$ is the azimuthal extension of the section under consideration. A value of 1.5 was adopted for α, which seems to be appropriate for Cataclysmic Variables [2,4]. In order to take into account the local broadening due to turbulence, this partial profile was cross-correlated with a Gaussian [4] having $\sigma = K_t \bar{u}$, $\bar{u}$, being the calculated mean velocity along the whole orbit. The value of K_t was taken to be 0.2 [4]. Once all the partial profiles were calculated they were added, yielding the total line profile $P(v,\phi) = \sum_p P_p(v,\phi)$. The inclination angle of the system was not taken into account in the calculations because, in the framework of this model, its effect would simply be that the velocity scale would be multiplied by a factor $\sin i$ and the intensity scale by a factor $\cos i$.

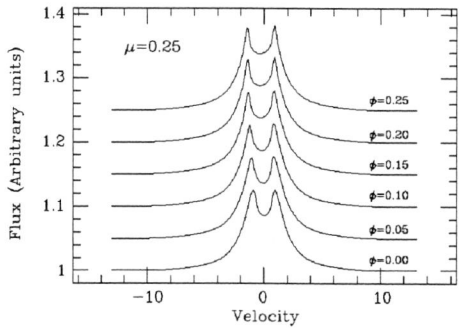

FIGURE 3. Line profiles corresponding to orbital phases $\phi = 0.0, 0.05, 0.10, 0.15, 0.20$ an 0.25 for the case $\mu = 0.25$.

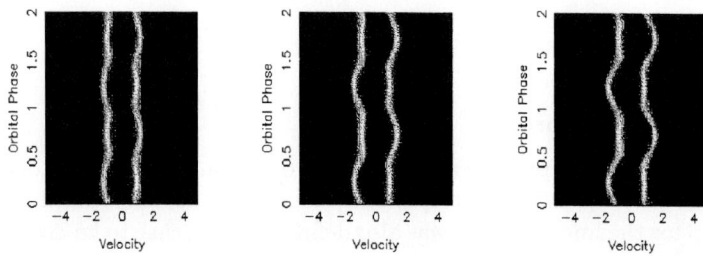

FIGURE 4. Greyscale images of the orbital evolution of the lines corresponding to three different values of μ: 0.1, 0.2 and 0.3.

4. Results

The profiles calculated exhibit qualitatively similar orbital behaviour for all the values of μ. As an example, in Figure 3 the profiles obtained with $\mu = 0.25$ are presented. Only profiles corresponding to orbital phases from $\phi = 0.0$ up to $\phi = 0.25$ are shown since for a phase $0.25 + \phi$ the profile is identical to that having phase $0.25 - \phi$, and for phase $0.5 + \phi$ it is symmetrical to that having phase $0.5 - \phi$. The most significant variations are those of the relative peak intensity and the widths of the peaks. No significant variation with the orbital phase of either the equivalent width or the width of the full line is present. Figure 4 shows three greyscale images of the evolution of the line for different values of μ. A distortion in the radial velocity curve of both line components is visible, which is due to the variation with the orbital phase of the separation of the doubling.

REFERENCES

[1] Huang, S.-S. 1967, ApJ, 148, 793.
[2] Marsh, T.R, Horne, K., Schlegel, E.M., Honeycutt, R.K. & Kaitchuck, R.H. 1990, ApJ, 364, 637.
[3] Paczynski, B. 1977, ApJ, 216, 822.
[4] Stover, R.J. 1981, ApJ, 248, 684.

Numerical Simulation of Co-planar Star-Disk Interaction

By D. Molteni[1], G. Gerardi[1] S. K. Chakrabarti[2]

[1] Istituto di Fisica, Via Archirafi 36, 90123 Palermo, Italy.

[2] Tata Institute of Fundamental Research, Bombay 400005, India.

We simulate interactions between a thin accretion disk around a massive black hole and an orbiting compact star or a black hole. We use the Smoothed Particle Hydrodynamics technique for this purpose. We take into account the following main processes: 1) loss of energy and angular momentum of the companion due to gravitational wave emission, 2) gain of angular momentum of the companion due to its accretion of the disk gas in the super-Keplerian region. We show that the orbit is stabilized against the loss of gravitational waves and the accretion rate onto the primary black hole is modulated on the time scale of the orbital period of the companion. Our simulation confirms the suggestion of Chakrabarti [1] that binary black holes can be kept in stationary orbits. We discuss possible observational signatures of such a system.

1. Introduction

There is increasing evidence from the spectra of AGNs and Quasars that accretion disks around the central engine may not be of the simple, thin, Keplerian type. In these systems, the angular momentum of the accretion disk close to the hole could be almost constant ($l \sim 4GM_1/c$), particularly when the viscosity of the disk is low but the pressure is sufficiently high so that the infall time is much shorter compared with the viscous timescale (in which the angular momentum is transported). In these 'pressure driven' disks, there is a region where the flow is super-Keplerian [1]. A black hole or a neutron star (which is on an instantaneous Keplerian orbit) orbiting inside this region, will accrete mass *as well as* angular momentum from the disk matter. Recently, Chakrabarti [2] suggested a theoretical possibility that a compact companion inside the super-Keplerian region of an accretion disk can have a stationary orbit as a result of the balance between angular momentum loss due to the gravitational wave emission and the corresponding gain from the disk matter accreted onto the companion. Assuming the companion to be on a circular orbit and assuming the Newtonian expression for the loss of energy and angular momentum, Chakrabarti [2] finds the expression for the radius of the equilibrium orbit to be given by,

$$r_{eq} = 15.63 \frac{M_1^{1/21}}{[\alpha \dot{M}_2]^{2/7}} (\frac{\mu}{M_\odot})^{4/7} (\frac{M}{M_\odot})^{8/21} (\frac{10^8 M_\odot}{M_1}) \qquad (1)$$

in units of the Schwarzschild radius of the primary ($r_g = 2GM_1/c^2$). Here, $4\alpha GM_1\dot{M}_2/c$ is the specific angular momentum transport rate (α being the efficiency factor), M_1 and M_2 are the masses of the components, μ is the reduced mass, $M = M_1 + M_2$ and $\dot{M}_2$ is the accretion rate onto the secondary. As an illustration, we choose $M_1 = 10^8 M_\odot, \alpha = 0.05$ and $\dot{M}_2$ as the critical rate onto the secondary. If we choose, $M_2 = 1 M_\odot$, we obtain,

$r_{eq} = 23 r_g$. Chakrabarti [2] also provides examples of realistic disk models where stable solutions are obtained.

Though the physical process just described is correct, the actual system could be more complex, since many important aspects of a star-disk interaction are ignored. In the present paper, we report the preliminary results of a numerical simulation of this system of two orbiting black holes with an accretion disk around the more massive one. Our results confirm the suggestion by Chakrabarti [2]. We find that the accreted disk material arrests indefinite shrinking of the companion orbit against the loss due to the gravitational wave emission. Beside this effect, we also find the spiral shock formation in the disk (first reported here using SPH method) and modulation of accretion rate onto the primary.

2. Model Assumptions

We assume that the central gravitating object as well as the companion are Schwarzschild black holes. Since the evolution due to the loss of gravitational waves is very slow, to save computing time, we start the simulation by launching the companion *inside the disk* in a circular orbit with Keplerian angular momentum. The goal would be to see whether the companion falls onto the central black hole, or gradually reaches its equilibrium orbit. The angular momentum loss of the companion is calculated by changing the velocity components (v_x and v_y) by the amount as appropriate from GW emission. Matter entering the companion is absorbed and its x and y momenta components are transferred to the companion. In this way, the companion receives energy and angular momentum.

To simulate the gaseous disk flow we use the SPH (Smoothed Particles Hydrodynamics) method. We have tested this method to simulate accretion disks and obtained excellent confirmation with theoretical predictions [1, 3]. The SPH uses an interpolating method, i.e. the fluid properties are locally approximated by the sum of interpolating functions, "pseudoparticles". The scheme is Lagrangian since the particles move according to the equations of motion derived with simple criteria by the continuous fluid dynamic equations. It has many attractive features that makes its use particularly appealing: (a) Its Lagrangian character permits an easy description of different fluid motion; (b) it is essentially grid free which allows computer memory saving, specially for 3-D problems with large voids in the integration domain; (c) it conserves energy, linear and angular momentum quite accurately; (d) it is easy to implement, the only time consuming part is the search of interacting neighbours. For a detailed description of the method see Monaghan [5]. We used Paczyński-Wiita potential [6] to simulate the black holes.

3. Numerical Results

We consider a thin disk confined only on the equatorial plane. We inject perfect gas with polytropic index $\gamma = 4/3$ at the outer edge of the disk with specific angular momentum $3.8 G M_1/c$ and specific energy $0.0075 c^2$. With this angular momentum, the stationary solution gives a super-Keplerian zone between $r = 2.3 - 4.1 r_g$; we launch the companion of mass $M_2 = 0.001 M_1$. We used a higher mass companion; if the stabilizing effect is found to be present for such a an object it will be also present for low mass ones. When a pseudo-particle reaches $r = 1.5 r_g$, it is assumed to be absorbed by the hole and is removed. Similarly, a particle is assumed to be accreted by the companion when its distance from it is less than $0.05 r_g$ and it is gravitationally bound to the companion.

Fig. 1 shows the radial distance of the companion as a function of time in units of GM_1/c^3. Initially, the distance increases very rapidly but eventually it reaches at a stable orbit at $r \sim 3.83 r_g$. The small oscillations are possibly due to the fact that the orbit

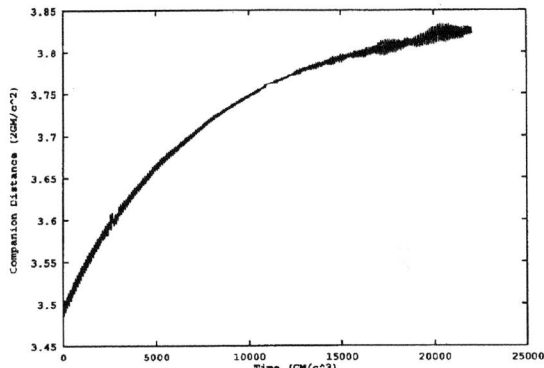

FIGURE 1. Variation of the radial distance of the companion as a function of time. The companion settles down at a stable orbit after gaining more angular momentum from the disk, than what is lost due to gravitational waves.

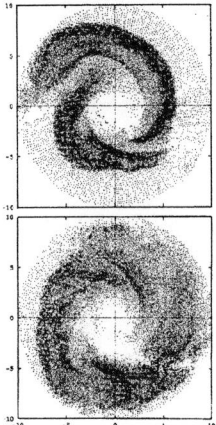

FIGURE 2. X-Y plots of the particle location at $T = 2750 GM_1/c^3$ (upper box) and $2800 GM_1/c^3$ (lower box) indicating substantial variation in timescales on the order of the orbital period.

becomes slightly elliptic. The presence of the non-axisymmetric perturbation due to the companion inside the disk causes non-axisymmetric density waves to develop which further steepens close to the central hole and produce non-axisymmetric shock waves. A non-axisymmetric shock wave dissipates angular momentum from the disk. This could be the reason why the companion orbit oscillates, and why the accretion rate onto the central hole is strongly modulated. Fig. 2 shows the particles in the simulation (on an average $40,000$) at two different times. Fig. 3 shows the variation of the accretion rate (in arbitrary units) with time. In these simulations, the injection radius was $r = 10 r_g$. The mass of the companion was $M_2 = 0.01 M_1$, and the specific angular momentum was $3.98 GM_1/c^2$.

The study of star-disk interaction is important for several reasons. A binary compact system is a steady emitter of gravitational waves which could be detected on earth. Secondly, the X-ray flickering and optical micro-variabilities observed in AGNs and Quasars

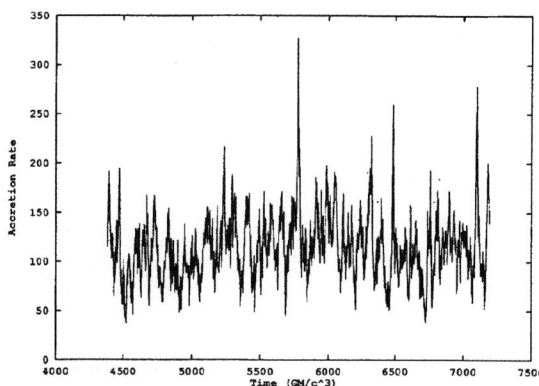

FIGURE 3. Vatiation of the accretion rate with time

could be partly due to the star-disk interactions. The X-rays/γ-rays emitted by the companion could be gravitationally lensed by the central hole, causing periodic variability. The details of our results will be published elsewhere [4].

This work was supported by the MPI.

REFERENCES

[1] Chakrabarti, S. K., (this volume).
[2] Chakrabarti, S. K., 1993, ApJ, 411, 610.
[3] Chakrabarti, S. K. and Molteni, D., 1993, ApJ, (Nov. 10th).
[4] Molteni, D., Gerardi, G. & Chakrabarti, S. K., 1993, (in preparation).
[5] Monaghan J. J., 1985, Comp. Phys. Repts., 3, 71.
[6] Paczyński, B. & Wiita, P. J., 1980, AA, 88, 23.

The Stability of Circumstellar Disks

By Shoken M. Miyama[1], Taishi Nakamoto[1], Nobuhiro Kikuchi[2]
Shu-ichiro Inutsuka[2], Ken'ichi Kobayashi[3], Taku Takeuchi[3]

[1] Division of Theoretical Astrophysics, National Astronomical Observatory, Mitaka Tokyo 181, Japan

[2] Department of Astronomy, University of Tokyo, Bunkyo-ku, Tokyo 113, Japan

[3] Department of Astronomical Science, the Graduate University for Advanced Studies, Mitaka Tokyo 181, Japan

The evolution of an infinitesimally thin, self-gravitating circumstellar disk is calculated numerically using SPH. Linear analyses of the stability of the circumstellar disk are made and the growth rate of the unstable mode is given. To test the reliability of our SPH code, numerical computations of the growth of small amplitude perturbations are compared with the results of linear theory and good agreement is obtained. Several models are computed and the nonlinear growth of an unstable, non-axisymmetric mode is studied. The rate of angular momentum transfer by these unstable modes is also discussed.

1. Introduction

Recent observational and theoretical studies suggest that solar-like pre-main-sequence stars pass through both protostellar and T Tauri stages. In both stages, a rotating gaseous disk is thought to exist around the star. In this paper we investigate the gravitational stability of such disks. The protostar, which preceeds the T Tauri stage, has a gaseous envelope. Because the envelope contains dust, completely absorbing the visual light from the center, the protostar is observed in the infrared (IR) continuum. While the central region cannot be observed directly, hot circumstellar disks have been proposed to explain the observed near-IR spectrum [1,3]. Computational studies of star formation also suggest that disks might form around stars [7]. Since gas from the envelope accretes onto the disk, the disk might become unstable in the course of its evolution.

In the case of a T Tauri star, the gaseous envelope which had encircled the protostar disappears because the gas has accreted onto the central region. The star can be observed directly in the visual regime. T Tauri stars have been studied extensively in recent years [4,6,8,22]. Spectra have been obtained ranging from ultraviolet to radio wavelengths. Almost all T Tauri stars have an IR excess and blue-shifted oxygen lines. In order to explain these phenomena, optically thick disks have been suggested, where the dust component radiates in the IR and also obscures the red-shifted oxygen component in the stellar wind. In order to explain the IR and radio continuum, the radius and the mass of the disk are estimated to be about 100AU and $0.01 - 1.0 M_\odot$ [4,6], respectively. The uncertainty in the disk mass comes primarily from uncertainties in the frequency dependence of the dust opacity.

Gaseous disks around two T Tauri stars have recently been observed. The first was seen in CO toward GG Tau using the Nobeyama radio telescope [23]. A double-peaked feature in the CO spectrum indicates the presence of a rotating gaseous disk. Similar observations were made of DM Tau [10]. The structure of such rotating disks was also observed by the Nobeyama Millimeter Array [11,20] and the size found to be about 500 AU.

The disk around a young star (T Tauri star or protostar) is considered to be a protoplanetary disk and hence its investigation is important for the study of planetary formation as well as star formation. From both observational and theoretical studies it is probable that a massive disk (*i.e.*, a gravitationally unstable disk) would be made in the course of star formation [17]. The stability of such disks, however, has not been throughly investigated. It is expected that if non-axisymmetric (spiral) modes develop, they would play an important role in angular momentum transfer in the disk [2]. In this paper we concentrate on such a gravitational instability in the circumstellar or protoplanetary disk.

Our main goals in this paper are to address the following questions:

(*a*) What is the critical mass of a stable disk against gravitational instability? What is the time scale of perturbation growth?

(*b*) What is the fate of an unstable disk? Is it possible that a companion star or a brown dwarf might be born from the disk?

(*c*) How effectively is angular momentum transferred by the unstable mode? How large is the amplitude of the nonlinear mode? [The amplitude $\delta\sigma$ is important to find because the gravitational torque follows at once from $T_{torque} \propto Gr^3(\delta\sigma)^2$].

In this paper, using two-dimensional SPH (smoothed particle hydrodynamics), we investigate the evolution of these circumstellar disks.

2. Strategy for Investigating the Evolution of Circumstellar Disks

Our purpose is to study nonlinear processes in the evolution of circumstellar disks. Hence we need numerical simulations in multiple dimensions. Using SPH, we can do three-dimensional simulations. Before calculating models, however, we must test the reliability and resolution of such simulations in the case of differentially rotating disks. One of the best ways to perform these tests is to compare numerical computations of the small amplitude perturbation growth with the results of linear perturbation analyses. It is difficult, however, to solve the eigenvalue equation of linear perturbation theory for a three-dimensional disk with self-gravity. Therefore we examine two-dimensional problems first. Below, we list the sequence of steps in out study of disk stability.

1) We use linear perturbation theory to find the gravitational stability of an infinitesimally thin disk [12]. Such solutions have been given by several authors [5,18,21], for the $m = 1$ case. We note, however, that these authors used unperturbed models in which the surface density has a power-law dependence on the distance from the central star and there are two boundaries with finite densities at finite radii. In these models the gravitational force must be infinite at the boundary because of the infinitesimally thin disk assumption. We cannot use models with such singularities for our numerical simulations. As shown below, we choose an appropriate unperturbed model as the initial condition for our numerical simulation.

2) We simulate the evolution of the thin disk [24] by two-dimensional SPH.

3) We compare the growth of the small amplitude perturbations predicted by numerical simulation with the growth predicted by linear perturbation theory. After we confirm the reliability of our code, we calculate many models with various surface densities, ratios of disk mass to central star mass, and various temperature structures.

4) Next we will make three-dimensional simulations for geometrically thin disks. A comparison between the results of the two and three-dimensional computations will identify the features arising from the three-dimensional geometry [13].

5) Finally, we plan to make simulations for the disk evolution that include viscosity, radiation loss, and other physical processes.

3. Results of Linear Perturbation Theory

In this section we give the results of a linear perturbation theory analysis for an infinitesimally thin disk [12]. The basic equations are written in cylindrical coordinates. The continuity equation and equations of motion for the fluid in the disk are

$$\frac{\partial \sigma}{\partial t} + \frac{1}{r}\frac{\partial r\sigma v_r}{\partial r} + \frac{1}{r}\frac{\partial \sigma v_\phi}{\partial \phi} = 0, \tag{1}$$

$$\frac{\partial v_r}{\partial t} + v_r\frac{\partial v_r}{\partial r} + \frac{v_\phi}{r}\frac{\partial v_r}{\partial \phi} - \frac{v_\phi^2}{r} = -\frac{1}{\sigma}\frac{\partial P}{\partial r} - \frac{\partial \psi}{\partial r} - \frac{\partial \psi_c}{\partial r}, \tag{2}$$

$$\frac{\partial v_\phi}{\partial t} + v_r\frac{\partial v_\phi}{\partial r} + \frac{v_\phi}{r}\frac{\partial v_\phi}{\partial \phi} + \frac{v_\phi v_r}{r} = -\frac{1}{r\sigma}\frac{\partial P}{\partial \phi} - \frac{1}{r}\frac{\partial \psi}{\partial \phi} - \frac{1}{r}\frac{\partial \psi_c}{\partial \phi}, \tag{3}$$

where P and σ are integrated pressure and surface density, respectively. The Poisson equation for the infinitesimally thin disk is

$$\Delta\psi = 4\pi G\sigma\delta(z). \tag{4}$$

The potential of the central star, ψ_c, with mass M_c, is given by

$$\psi_c = -\frac{GM_c}{|\vec{r} - \vec{r}_c|}. \tag{5}$$

The equation for the position of the central star, $\vec{r}_c$, is

$$\frac{d^2}{dt^2}\vec{r}_c = -\frac{\partial \psi}{\partial \vec{r}_c}. \tag{6}$$

In this paper, for the sake of simplicity, we assume the polytropic relation between the integrated pressure and the surface density, i.e., $P = K\sigma^{1+1/N}$, where K and N are constants. In this paper we assume $N = 1.5$. The physical variables are expanded around the unperturbed state as

$$\vec{v} = \vec{v}_0 + \vec{v}_1 = (0, r\Omega_0(r)) + \vec{v}_1, \quad P = P_0 + P_1,$$

where quantities with subscript 0 represent unperturbed values. The linear quantities which have subscript 1 are assumed to have the form $exp(-i\omega t + im\phi)$, where m is the azimuthal wave number. At the boundary where the unperturbed surface density vanishes, we take the free boundary condition. That is, the Lagrange change of pressure ΔP, should vanish at the boundary, i.e., $\Delta P = P_1 + \xi_r dP_0/dr = 0$, where ξ_r is the Lagrange displacement in the r direction. Finally we obtain integral-differential eigenvalue equations for ω.

3.1. Unperturbed State

The initial, unperturbed configuration for the nonlinear simulation is a rotating disk with equilibrium between the centrifugal force, pressure gradient, self-gravity of the disk and gravity of the central star,

$$r\Omega^2 = \frac{1}{\sigma_0}\frac{dP_0}{dr} + \frac{d\psi_0}{dr} + \frac{GM_c}{r^2}. \tag{8}$$

We specify the unperturbed state in the following way:
(a) The surface density distribution is given by $\sigma_0(r)$.

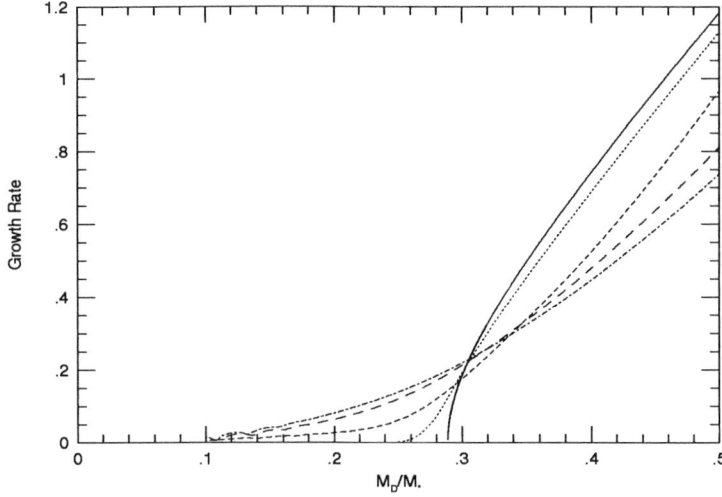

FIGURE 1. Growth rate given by the linear perturbation theory as a function of the mass ratio of the disk to the central star. In this case, $R_{out}/R_{in} = 5$ and $c_s/r\Omega(r_0) = 0.1$. The curves represent growth rates with m=0(solid), m=1(dotted), m=2(dashed), m=3(long dashed) and m=4(long & short dashed).

(b) Using $\sigma_0(r)$, we calculate the pressure gradient $(1/\sigma_0)dP_0/dr$, using the polytropic relation and the self-gravity potential ψ_0.

(c) The angular velocity is obtained from equation (8).

In this paper we use the following normalizations: $r_0 = 1$, $\sigma_0(r_0) = 1$ and $\Omega_0(r_0) \equiv [GM_c/r_0^3]^{1/2} = 1$, where r_0 is the radius where maximum pressure occurs. Our model for the circumstellar disk is very simple, but it includes the essential features of a circumstellar or protoplanetary disk. In this model the surface density is given in the range $[R_{in}, R_{out}]$ by

$$\sigma_0(r) = \left[\frac{2r - 1 - Cr^2}{r^2(1-C)}\right]^{3/2} \tag{9}$$

where

$$C = 4R/(R+1)^2 \quad \text{and} \quad R = R_{out}/R_{in}. \tag{10}$$

For a given value of R, the inner (R_{in}) and outer (R_{out}) radii are obtained by solving $\sigma_0(r) = 0$. Our models are characterized by the ratio of the disk mass to the central mass, $q = M_d/M_c$, and the ratio of the sound speed at r_0 to the corresponding rotation speed, $\hat{c}_s = c_s(r_0)/r\Omega(r_0)$.

In Figure 1 we show the results of linear perturbation theory, i.e., the growth rate of unstable modes as a function of mass ratio q. According to the *local* stability analysis by Toomre, the stability criterion is obtained from Toomre's Q value by

$$Q = \frac{c_s \kappa}{\pi G \sigma_0} = 1, \tag{11}$$

where κ is the epicyclic frequency. For the unperturbed state used here, the value of Q changes in the disk. If we take the criterion as $Q_{min} = 1$, where Q_{min} is the minimum value of Q in the disk, the criterion for gravitational instability is $q = 0.27$ for the model in Figure 1. Therefore, in the region of Fig. 1 where $q > 0.27$, the unstable mode is considered to be a gravitational mode. In Figure 2, the perturbed surface density σ_1 is shown for the case of $q = 0.5$ and $m = 2$ for the model shown in Figure 1. Here, in

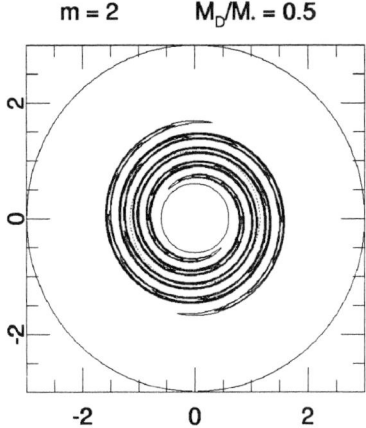

FIGURE 2. Eigenfunction of an unstable mode. The perturbed surface density is given for m=2, $M_d/M_c = 0.5$, $R_{out}/R_{in} = 5$ and $c_s/r\Omega_0(r_0) = 0.1$.

contrast, the region of $q < 0.27$, has unstable modes with $m > 1$. We consider these modes to be hydrodynamically unstable modes strengthened by self-gravity in the differentially rotating disk [9,19].

4. Results of Two Dimensional Nonlinear Simulation

4.1. *Two-Dimensional SPH*

Before showing results of the nonlinear simulation, we would like to explain briefly our two-dimensional SPH, which is almost the same as that given in previous papers [14-16]. In this case, however, the surface density is given by

$$\sigma(\vec{x}) = \sum_j W(\vec{x} - \vec{x}_j; h_j) \quad (12)$$

where W and h are the kernel function and the smoothing length. In our model we use a Gaussian kernel, $W = m/(\pi h^2)exp(-x^2/h^2)$, where m is the mass of one particle. In the case of an infinitesimally thin disk, the source term in the Poisson equation has a delta function as in equation (4). Hence, the gravitational potential at $\vec{x}$ from j-th particle is slightly different from that of the three-dimensional problem, in particular,

$$\psi_j(\vec{x}) = -2\pi G \int_0^\infty \vec{x}' W(\vec{x}' - \vec{x}_j; h_j) d\vec{x}' \int_0^\infty J_0(k|\vec{x}'|) J_0(k|\vec{x}|) dk = -Gm\frac{\sqrt{\pi}}{h} I_0(s) e^{-s}$$

where

$$s = \frac{(\vec{x} - \vec{x}_j)^2}{2h_j^2},$$

and J_0 and I_0 are the Bessel function and the modified Bessel function, respectively.

4.2. *Test Computations*

In this subsection we show results of the test problem, where the perturbation growth is computed by our SPH code. For this purpose, the initial condition consists of the equilibrium configuration along with the perturbation found from the linear theory given in section 2. Using our code, we calculate the growth of such perturbations and compare them

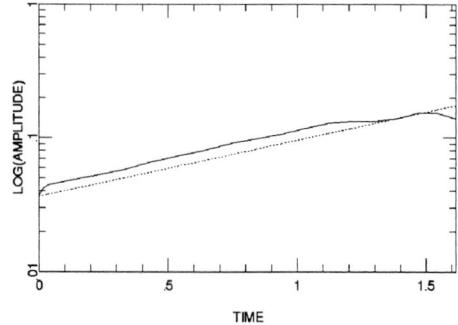

FIGURE 3. The numerical computation using 10000 SPH particles for the growth of perturbations in a self-gravitating disk. The solid curve represents the amplitude of the $m = 2$ mode, C_2, as a function of time. This is the case of $R = 5$, $C_s/r_0\Omega(r_0) = 0.1$ and $q = 0.5$. For comparison, the results using linear perturbation theory are shown by the dashed line.

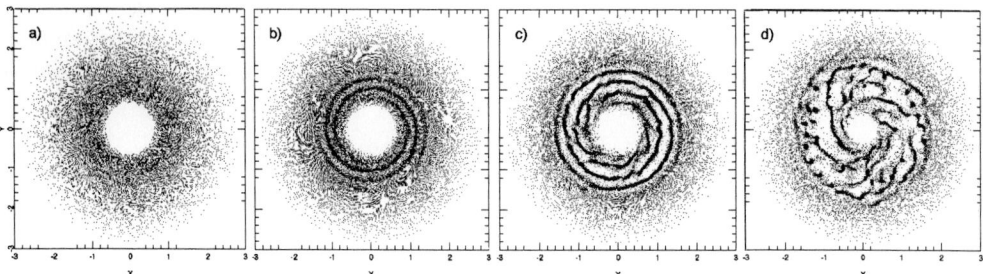

FIGURE 4. The numerical simulation of the evolution of a self-gravitating disk in differential rotation with 40000 particles. The disk model is $R = 5$, $C_s/r_0\Omega(r_0) = 0.1$ and $q = 0.5$. (a) $t = 0$, i.e., the initial state, (b) $t = 1.4$, (c) $t = 3.8$ and (d) $t = 5.3$.

with the linear result. In order to construct the equilibrium configuration, particle positions are adjusted by a relaxation method until the difference between the density calculated by SPH and the analytic density of equation (9) is less than 1%. The perturbation is added by changing the equilibrium position ($\vec{x}_0$) and velocity ($\vec{v}_0$) of each particle by $\vec{x} = \vec{x}_0 + \epsilon\vec{\xi}(\vec{x}_0)$ and $\vec{v} = \vec{v}_0 + \epsilon\vec{v}_1(\vec{x}_0)$, where $\vec{\xi}$ and $\vec{v}_1$ are the Lagrange displacement and velocity, respectively, which are eigenfunctions obtained by solving the perturbation equation as shown in section 2. In our computation we fix the amplitude of the perturbation at $\epsilon = 0.001$ which corresponds to a 4% perturbation in the surface density.

To observe the growth of the perturbation we calculate the Fourier amplitudes,

$$C_m(r,t) = \sqrt{A_m^2 + B_m^2}, \quad \text{where} \quad \sigma(r,\phi,t) = \sum_{m=0}(A_m \cos m\phi + B_m \sin m\phi). \quad (13)$$

In Figure 3, the time variation of the amplitude C_2 is given along with the result of the linear perturbation analysis. We find good agreement. The difference between the growth rate ω obtained by solving the linear theory and $(1/C_m)dC_m/dt$ from our SPH computations is less than 5% for $q = 0.5$, nearly 5% for $q = 0.4$ and about 10% for $q \leq 0.2$ as long as $C_m < 0.1$. Therefore we conclude that our code can reliably represent the linear growth of perturbations in a differentially rotating system.

4.3. Nonlinear Growth

In this subsection we present results of the nonlinear simulation. We would like to show only the dependence of the evolution on the mass ratio q, fixing the other disk parameters at $R = R_{out}/R_{in} = 5$ and $c_s/r_0\Omega(r_0) = 0.1$. We use 40000 particles and calculate models for q =0.1, 0.2, 0.3, 0.4 and 0.5 with values of Q = 2.24, 1.23, 0.92, 0.74 and 0.63, respectively. We use two different initial conditions for the perturbation. The first is the same condition as given in the previous subsection, i.e., the equilibrium state perturbed by the eigenfunction of the linear theory. The second initial condition is the equilibrium state with a *random* perturbation. In Fig. 4 the evolution of the unstable disk is shown for the same model as in Fig. 3. Figure 4a represents the initial data with the perturbation of the linear theory. At the stage of Fig. 4b, the amplitude C_2 is 0.08 at the maximum. It is similar to that of Fig. 2, i.e., it is still in the linear growth regime. After $t = 1.4$, the arm of the $m = 2$ mode begins to fragment and at the final stage of computation the disk fragments into many pieces. If the mass of the central star is assumed to be $1M_\odot$, the mass of each fragment is on the order of a brown dwarf mass (i.e., $m < 0.08M_\odot$).

We have done simulations for many models; the results are mainly determined by the minimum value of Toomre's Q-value. If $Q_{min} < 1$ for the initial model, the disk fragments into many pieces in a few rotation periods. If the number of fragments is large, then the mass of each fragment is much smaller than the disk mass. But if $Q_{min} > 1$, the amplitude of the initial perturbation grows and saturates to a finite value. Then the disk does not fragment and there are several non-axisymmetric modes.

4.4. Angular Momentum Transfer

We would like to show how effectively the specific angular momentum is redistributed in the disk by the unstable, non-axisymmetric mode. We define the mass distribution function in terms of specific angular momentum. That is, $m(j)dj$ is a mass whose specific angular momentum is in the range $[j, j+dj]$. Then we define the change of angular momentum in the disk by

$$\eta(t) = \frac{1}{M_d} \int |m(j) - m_0(j)| dj$$

where m_0 is the initial mass distribution as a function of the specific angular momentum. If the evolution of the disk is completely axisymmetric then $\eta(t)$ is always zero.

In the case of $q = 0.4$, i.e., $Q_{min} = 0.74$, as the amplitude of the perturbation grows the value of $\eta(t)$ increases exponentially with time. In the case of $q = 0.2$, i.e., $Q_{min} = 1.23$, the unstable mode grows but eventually saturates. Hence the value of η changes linearly as a function of time. In this case, the time scale of the angular momentum transfer is longer that 10 rotation periods at the maximum pressure point.

5. Summary

(a) We investigate the linear stability of differentially rotating disks with self-gravity. Our models show that the disks are always unstable even for disks with small ratios of disk mass to central star mass. The unstable mode is characterized by the minimum value of the Toomre's Q-value ($Q = c_s\kappa/\pi G\sigma_0$) in the disk. For the case of disks with $Q_{min} < 1$, the growth rate of the unstable mode is very high and on the order of the angular velocity of the disk. We consider this unstable mode to be gravitational. For disks with $Q_{min} > 1$, the growth rate is lower and we think the instability comes from a hydrodynamical instability in the differentially rotating system. We do not find the SLING mode which is reported by Adams et al. [5].

(b) We test the reliability of our numerical code by computing the evolution of the disk. We find good agreement between the results of the linear stability analysis and those of the numerical simulation.

(c) We compute nonlinear evolutionary processes of self-gravitating disks. When $Q_{min} < 1$, the disks fragment into many pieces. The mass of a fragment is on the order of a brown dwarf. When $Q_{min} > 1$, a non-axisymmetric mode appears but the disk does not fragment. In this case we find that the $m = 1$ mode is not dominant. We conclude that the value of Q_{min} is a good parameter to predict the evolution of the disk.

(d) The unstable mode can transfer angular momentum in the disk even when $Q_{min} > 1$. In that case the transfer rate is almost constant. Nevertheless, we must compute more models with various Q-values to determine the general theory of angular momentum transport by such non-axisymmetric waves.

REFERENCES

[1] Adams, F. C. & Shu, F.H. 1986, ApJ,, 308, 836.
[2] Adams, F. C., Emerson, J. P. & Fuller, G. A. 1990, ApJ, 357, 606.
[3] Adams, F. C., Lada, C. J. & Shu, F.H. 1987, ApJ, 312, 788.
[4] Adams, F. C., Lada, C. J. & Shu, F.H. 1988, ApJ, 326, 865.
[5] Adams, F. C., Ruden, & Shu, F. H. 1989, ApJ, 347, 959
[6] Beckwith, S. V. W., Sargent, A. I., Emerson, J. P., Harris, S., Mathieu, R., Benson, P. J. & Jennings, R. E. 1986, ApJ, 307, 337.
[7] Bodenheimer, P., Yorke, H. W., Rozyczka, M. & Tohline, J. E. 1990, ApJ, 355, 651
[8] Cohen, M., Emerson, J. P. & Beichman, C. A. 1989, ApJ, 339, 455.
[9] Goodman, J. & Narayan, R. 1988, MNRAS, 231, 97,
[10] Handa, T., et al.., 1993, ApJ, submitted.
[11] Kawabe, R., Ishiguro, M., Omodaka, T., Kitamura, Y. & Miyama, S. M., 1993, ApJ, 404, L63.
[12] Kikuchi N. & Miyama, S. M. 1993, ApJ, to be submitted.
[13] Laughlin et al. 1993, this volume.
[14] Miyama, S. M., Hayashi, C. & Narita, S. 1984, ApJ, 279, 621
[15] Miyama, S. M., 1992, PASJ, 44, 193
[16] Nagasawa, M. Nakamura, T. & Miyama, S. M. 1988, PASJ, 40, 691
[17] Nakamoto, T. & Nakagawa, Y. 1994, ApJ, in press
[18] Noh, H., Vishniac, E. T. & Cochran, W. D. 1991, ApJ, 383, 372
[19] Papaloizou, J. C. B. & Pringle, J. E. 1984, MNRAS, 208, 721
[20] Saitoh, M., et al.., 1993, ApJ, submitted.
[21] Shu, F. H., Tremaine, S., Adams, F. C., & Ruden, S. P. 1990, ApJ, 358, 495
[22] Strom, K. M., Strom, S.E., Edwards, S., Cabrit, S. & Skrutskie, M.F. 1989, AJ, 97, 1451.
[23] Strutskie, M. F. et al., 1993, ApJ, 409, 422.
[24] Tomley, L., Cassen, P., & Sleiman-Cameron, T., 1991, ApJ, 382, 530

Instabilities in Protostellar Disks

By Gregory Laughlin, Peter Bodenheimer

UCO/Lick Observatory, University of California at Santa Cruz, Santa Cruz CA, 95064

A hydrodynamic calculation which models the collapse of the pre-solar nebula from a dense 1 $M_\odot$ molecular cloud clump is presented. The collapse yields a massive equilibrium disk surrounding an equally massive protostellar core. The disk density and temperature profiles indicate that it is predisposed towards non-axisymmetric instabilities. We follow the development of these instabilities by evolving the system in 3–D using SPH. We find that the disk is prone to a series of spiral instabilities with primary azimuthal mode numbers $m=1$ and $m=2$. The torques induced by these spiral structures elicit material transport of angular momentum and mass through the disk, re-adjusting the surface density profile toward more stable configurations. The net effect of the instabilities on the particular disk examined here can be characterized by an effective viscosity $\nu = \frac{2}{3}\alpha c_s^2/\Omega$ with $\alpha \approx .03$. The effective viscosity in a disk lying closer to stability would most likely be somewhat lower than this.

1. Introduction

A remarkable feature of the present-day solar system is that the sun contains the vast majority of the mass, whereas the planets (primarily Jupiter) contain nearly all of the angular momentum. Rotation data from young T-Tauri stars indicate that the bulk of this separation must have been accomplished during the first 10^6 years after the solar nebula collapsed from its parental molecular cloud core [2].

It is important to examine how the inward transport of mass and the outward transport of angular momentum occurred in the early solar nebula. A full understanding of this process will be an integral part of the broader picture of star and planetary formation.

Several mechanisms for eliciting the transfer of mass and angular momentum within the requisite million year timescale have been suggested. Although magnetic effects are important in the hottest, innermost regions of the core-disk structure [3], it is likely that the two most important agents of transport through the bulk of the disk are turbulent viscosity and gravitational torques [6, 7]. Here, we summarize some of our recent efforts to simulate the effects of these two transport processes in the pre-solar nebula. We start by describing a 2–D axisymmetric hydrodynamic collapse calculation which begins from a initial configuration similar to that of a cold dense molecular cloud core, and ends up as an evolving, self-gravitating viscous disk orbiting a nascent protostar. We then outline how non-axisymmetric instabilities within the disk can be examined by tracing the 3–D evolution of the configuration with SPH, and then conclude by discussing how the net effect of this non-axisymmetric evolution can be modeled as the result of viscous diffusion acting on the disk's surface density profile. A detailed description of all these points is forthcoming [5].

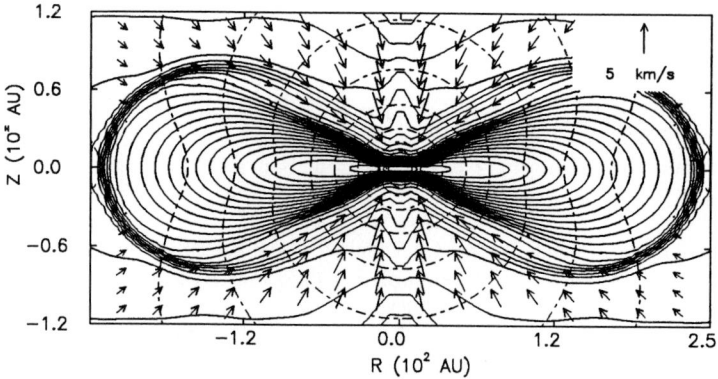

FIGURE 1. Structure of the 2-D model 40,000 years after the onset of collapse. Density contours (*solid lines*) are separated by $\Delta \log \rho = .2$, with $\log \rho_{min} = -15.22$. Temperature contours (*dashed-dotted lines*) are separated by $\Delta \log T = .1$, with $\log T_{min} = 1.506$. Arrows depict the gas velocity at the positions of the tips; the velocity scale is shown in the upper right of each frame.

2. Axisymmetric Collapse

Our axisymmetric finite difference calculations begin with an initially spherical, centrally condensed clump of molecular hydrogen gas containing 1 $M_\odot$ distributed so that $\rho \propto r^{-2}$. The cloud radius is 4×10^{16} cm, it is rotating with $\Omega = 4.4 \times 10^{-13}$ s^{-1}, and the temperature is an initially uniform 20K. These conditions are similar to those which are thought to exist in the central regions of magnetically supported molecular cloud cores shortly before they decouple from the magnetic field and begin to collapse [9]. A 2nd order eulerian numerical scheme [10] is used to evolve the system. We employ four explicit nested grids which resolve length scales spanning nearly three orders of magnitude [14]. Gray radiative transfer is included in the calculation as an implicit substep to the hydrodynamics. We use an α type viscosity law with $\alpha = .01$ [11]. We approximate the accretion luminosity from material falling onto the unresolved protostellar core by injecting a constant flux of 25 $L_\odot$ into the innermost grid zone. Further details of the code and justifications for the approximations which were used can be found in Yorke et al. [13].

In Fig. 1, the resultant core-disk structure is shown 40,000 years after the onset of collapse. At this time, slightly more than 0.5 $M_\odot$ has collected in the core, and a little less than 0.5 $M_\odot$ has been incorporated into a thick equilbrium disk. The remaining few percent of the orignal cloud mass is still collapsing in the outer envelope. The surface density profile of the disk is considerably smoother than the profile obtained with an inviscid calculation started from the same initial conditions [13]. The minimum Toomre Q value ($Q = c_s \kappa / \pi G \sigma$) in the disk is 1.3, indicating a likely susceptibility to non-axisymmetric instabilities [1]. The axisymmetric simulation was pursued through 70,000 years, at which time the core mass had increased to .6 $M_\odot$. The mass flux into the core at the end of the simulation was 1.87×10^{-6} $M_\odot$ yr^{-1}, typical of the rate required to clear out the disk within a million years.

3. 3-D Non-axisymmetric Calculations

To study the growth of spiral instabilities in our version of the pre-solar nebula, we constructed a 3-D disk containing 25,000 equal mass particles which were distributed to follow the density and temperature contours of the model shown in Fig. 1. This system was evolved for 5000 years (10 dynamical timescales) using a descendant of the TREESPH

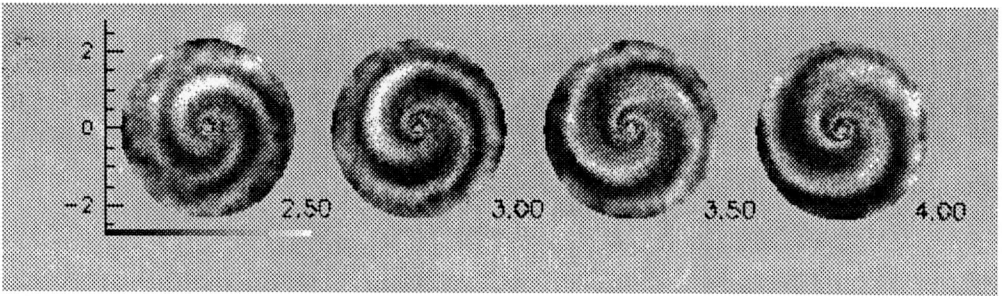

FIGURE 2. Spiral instabilities in the surface overdensity distribution of the 3–D SPH simulation described in the text. The number given at the lower right hand corner of each snapshot is the elapsed time in units of $T_{dyn}=477$ yrs. The axis at the left hand side is scaled in units of 100 AU. The grayscale bar runs from 50 % underdensity *black* to 50 % overdensity *white*.

code [4]. The central protostellar core was represented by a single massive particle, which was allowed to accrete disk particles straying too close to the center. The momentum and energy of the accreted particles are transfered to the core particle. In order to approximate the ram pressure of the remnant infalling envelope, particles straying beyond the intitial disk radius of 230 AU had their outward velocities set to zero, yielding an absorbing outer boundary condition. A locally isothermal equation of state was employed, in which the radial temperature distribution was held fixed throughout the run.

The result of this, and other related simulations, is a vigorous cycle of spiral instabilities with primary azimuthal mode numbers m=1 and m=2. The spirals, which often contain local overdensities in excess of 100%, develop from a random spectrum of initial perturbations averaging 2–3% in each global fourier component. In our upcoming paper, we show that it is very unlikely that the bulk of their growth in this gaseous disk is simply due to SWING amplification of the initial noise [12]. Figure 2 depicts four successive snapshots of the disk at different phases of the instability cycle.

After 5000 years of evolution, the strengths of the modes associated with the spiral instabilities have decreased considerably. Examination of the disk shows that a significant amount of angular momentum and mass transport has occurred, and that the minimum Q in the disk has increased to $Q \approx 1.45$ from its original value of 1.3. Additional simulations with configurations having the same initial density profile, but with adjusted radial temperature distributions indicate that the minimum Q value in these disks must lie below $Q \approx 1.5$ in order for the perturbations to grow into the non-linear regime. Uniformly lowering the disk temperature so that the minumum Q value drops from 1.3 to 1.15 unleashes a very active succession of spiral waves which rapidly rearrange the disk surface density distribution in the direction of stability. A further decrease of the minumum Q value to 1.0 causes the disk to fragment within a single dynamical timescale.

Certainly, our 3–D evolutionary calculations indicate that the detailed non-linear behavior of spiral waves within a protostellar disk is a complex phenomenon. We would like to have a simple theory which describes the overall effect that the kaleidoscope of instabilities has on the structure of the disk itself. It has been suggested that the net *long term* effect arising from the development of non-axisymmetric structures may amount to a diffusive process acting on the disk surface density profile [8]. If this is indeed the case, then the

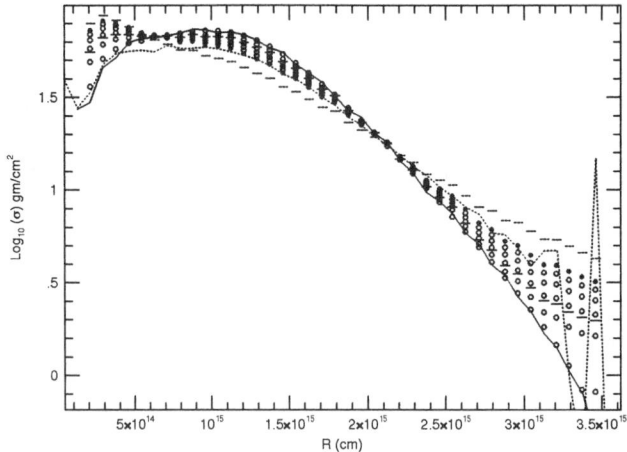

FIGURE 3. Comparison of the surface density redistribution in the 25,000 particle standard–case simulation with solutions of the viscous diffusion equation. *Solid line*: initial surface density distribution for SPH model. *Dotted line*: surface density distribution for SPH model at $T = 10T_{dyn}$. *Small open circles*: solution to the viscous diffusion equation at $T = 0,2,4,6,$ and $8T_{dyn}$ for $\alpha=.03$. *Small filled circles*: solution to viscous diffusion equation at T=10 T_{dyn} for $\alpha = .03$. *Dotted bars*: Solution to viscous diffusion equation at T=10T_{dyn} for $\alpha=.06$. *Solid bars*: Solution to the viscous diffusion equation at $T = 10T_{dyn}$ for $\alpha=.01$.

governing diffusion equation for the surface density distribution, $\sigma(r)$, will be

$$\frac{\partial \sigma}{\partial t} = -\frac{1}{r}\frac{\partial}{\partial r}((\frac{\partial \Omega r^2}{\partial r})^{-1}\frac{\partial}{\partial r}(\sigma \nu_g r^3 \frac{\partial \Omega}{\partial r}))$$

Figure 3 shows the result of evolving our initial surface density profile with this equation for the same 5000 year interval covered by the 3–D SPH simulation. We have assumed that the effective viscosity ν_g resulting from the gravitational torques generated by the instabilities can be parameterized by the standard α prescription ($\nu_g = \frac{2}{3}\alpha c_s^2/\Omega$) with a value of $\alpha=.03$, and as evidenced by the figure, this viscosity provides a surprisingly good fit over most of the disk. (The density spike beyond 200 AU results from mass piling up against the outer absorbing boundary and falling victim to the Toomre axisymmetric ring instability.)

The possibility that *intrinsic* viscosity in the SPH method itself might be making a significant contribution to the surface density evolution shown in Figure 3 has not escaped us. In the forthcoming paper we describe several tests which were done with the aim of isolating the degree to which numerical effects contribute to the derived α value. Briefly, an SPH simulation modeling the same configuration, but using 12,000 particles, showed only slightly more viscous evolution than the 25,000 particle case described here, whereas the same simulation with 6500 particles displayed an effective viscosity characterized by an α value $\approx .05$, and a considerably more ragged density profile. These and other tests appear to indicate that a major fraction of the disk evolution in the 25,000 particle simulation resulted from the presence of non-axisymmetric structures rather than numerical artifacts; the effective viscosity due to gravitational torques in our model protostellar disk provides significant mass and angular momentum transport on a timescale of approximately 100,000 years.

In all probability, however, the effective value of α in a real protostellar disk will be lower than the value of .03 derived here, and the transport times will be longer. Genuine gaseous disks are of course immune to numerical viscosity, but more importantly, our initial disk profile was probably more unstable to non-axisymmetric instabilities than a true-to-life disk would be. Our simulations have shown that the disk evolves rather quickly away from unstable configurations towards ones in which the Q value is increased and the effective viscosity is decreased. It is easy to imagine a process in which a much more subtle succession of spiral waves with a smaller effective viscosity transports angular momentum and mass through a disk that always lies very close to stability.

This work was supported in part by a travel and computer time grant to GL from Cray Research de México, S. A. de C. V. and Universidad Nacional Autónoma de México. It was also funded through a special NASA astrophysics theory program which supports a joint Center for Star Formation Studies at NASA/Ames Research Center, UC Berkeley and UC Santa Cruz. PB acknowledges a grant of computer time from the San Diego Supercomputer Center.

REFERENCES

[1] Adams, F.C., Ruden, S.P., & Shu, F.H. 1989, ApJ, 347, 959
[2] Bertout, C. 1989, ARAA, 27, 351
[3] Hayashi, C. 1981, ProgrTheoPhysSuppl, 70, 35
[4] Hernquist, L., & Katz, N. 1989, ApJS, 70, 419
[5] Laughlin, G. & Bodenheimer, P. 1993, ApJ, submitted
[6] Larson, R.B. 1984, MNRAS, 206, 197
[7] Lin, D.N.C. & Papaloizou, J. 1985, in *Protostars and Planets II*, ed. D.C. Black & M. S. Matthews (University of Arizona Press: Tucson), p. 981
[8] Lin, D.N.C. & Pringle, J. E. 1987, ApJ, 320, L87
[9] Lizano, S., & Shu, F. 1989, ApJ, 342, 834
[10] Różyczka, M. 1985, AA, 143, 59
[11] Shakura, N.I., & Sunyaev, R.A. 1973, AA, 24, 337
[12] Toomre, A. 1981, in *The Structure and Evolution of Normal Galaxies*, ed. S.M. Fall & D. Lynden-Bell (Cambridge: Cambridge University Press), p. 111
[13] Yorke, H.W., Bodenheimer, P., & Laughlin, G. 1993, ApJ, 411, 274
[14] Yorke, H.W., & Kaisig, M. 1993, ComputPhysComm, in press

The Dynamics of Massive Protostars and their Photoionized Disks

By H. W. Yorke, A. Welz

Institut für Astronomie und Astrophysik, Am Hubland, D-97074 Würzburg, Germany

A 2D radiation hydrodynamic code using multiply nested grids has been applied to the problem of the formation and early evolution of massive stars and photoionization of circumstellar disks. The basic hydrodynamic code is explicit (except for the equation of energy conservation). Gray radiation transfer is treated in the flux-limited diffusion approximation and solved implicitly in a separate substep. The gravitational collapse of a slowly rotating, density-peaked molecular fragment leads to the formation of a circumstellar disk surrounding a central source. During subsequent evolutionary phases, the interaction of the circumstellar disk with the hydrogen-ionizing flux of the hot central star and with its stellar wind are calculated using a modified version of the code. Both the direct stellar radiation and the transfer of diffuse hydrogen-ionizing photons produced by direct recombinations into the ground level are considered. It is shown that hydrodynamic processes alone can induce conical outflows ($v \sim 40$ km s^{-1}) without invoking magnetic fields and without postulating an intrinsic stellar wind from the central source. Numerical models including stellar winds of sufficiently low mechanical luminosity evolve in a similar manner, except for the addition of the high velocity outflowing wind confined and focussed along the pole.

1. Introduction

The formation of circumstellar disks appears to be a common side effect of the star formation process. Optically thick disks – diagnosed by excess emission at $\lambda \lesssim 10$ μm – are found in around 30% to 50% of young pre-main sequence low mass ($M \lesssim 3$ M$_\odot$) stars [3,16]. No main sequence stars appear to have such massive, optically thick disks. However, evidence of infrared excess emission consistent with the existence of optically thin disks – analogous to the structures surrounding Vega and β Pictoris – is found for young stars of all masses. Presumably such disks are the direct descendents of originally massive, optically thick disks. Non-detection of massive disks surrounding massive stars can be interpreted as a correspondingly short time scale for their disk evolution.

The detailed structure of these disks influences the early evolution of young stars. Massive disks provide a temporary reservoir of material with specific angular momentum too large to be directly accreted by a central object. When angular momentum is transported outwards, some of this material can be accreted. The relatively high densities in these disks provide the environment in which dust grains can further coagulate and evolve [12,21]. This affects not only the growth of planetisimals and the planet formation process directly but also the opacity of the disk material, which in turn affects the energetics in the disk and the disk's appearance. Finally, the disks are closely linked with the existence of outflows. In the case of high mass stars ($M \gtrsim 10$ M$_\odot$) the existence of a massive disk can be expected to strongly affect the formation and early evolution of ultra-compact and compact HII regions.

1.1. The computational problem

Previous numerical studies of the formation and early evolution of disks surrounding young (proto-)stars have concentrated on the low mass case; see *e.g.* the introductions in [28], hereafter YBL; [2,14]. On the one hand, this case pertains directly the early solar nebula and the formation of planets around our Sun. On the other hand there are several difficulties (principally numerical) associated with the formation and early evolution of intermediate and high mass stars. First of all radiation pressure and relative dust/gas drift can strongly affect the star formation process for cloud masses $M_{cl} \gtrsim 10$ $M_\odot$, *c.f.* [26,27], and may even be the primary cause of an absolute upper mass limit. Use of frequency-dependent opacities and self-consistent radiation transport in 1D together with a realistic model for grain evolution has resulted in a basic understanding of the relevant physics, but up to present only the spherically symmetric, non-rotating collapse has been treated in detail. The importance of a careful treatment of radiation transfer for $M \gtrsim 10$ $M_\odot$ has been substantiated in these calculations.

Self-consistent multidimensional radiation transfer calculations strain the capabilities and speed of present day computers and have only been attempted for special configurations und with poor spatial and/or angular resolution in 2D [5,6]. Radiation hydrodynamical calculations of collapsing rotating clouds which require solving the radiation transfer problem on the order of 10^3 to 10^4 times have therefore invoked approximations, such as the grey Eddington-approximation (*c.f.* [4,20]) or grey flux-limited diffusion (*c.f.* YBL; [3]). Fortunately, radiation acceleration is only able to reverse the inward flow in collapsing $M_{cl} = 10$ $M_\odot$ spherically symmetric clouds in rather late evolutionary phases [26], after most of the material has accreted onto the central object. Thus one can feel confident with the grey flux-limited diffusion (FLD) approximation at least during the initial collapse and disk formation phases of a 10 $M_\odot$ rotating clump, from which we expect a central core of several $M_\odot$ to form.

A second difficulty in calculating the evolution of collapsing clumps is the role played by magnetic fields. How do magnetic fields eventually decouple from the dusty molecular gas and allow a protostellar clump to collapse? To what extent are they initially dragged along with the collapsing gas and to what extent and at what time are they restrengthened by a dynamo effect in the disk and/or the central protostar? Are magnetic fields responsible for the winds and outflows observed in young objects? Are they responsible for the redistribution of angular momentum which occurs during these early phases?

Lizano & Shu [11] and Fiedler & Mouschovias [7] have calculated the quasi-magnetostatic contraction of non-rotating molecular clumps, but dynamical phases have not yet been considered. In a series of papers Stone & Norman [17,18,19] have presented an explicit version of a 2D magnetohydrodynamic code which after generalization to 3D MHD could be applied to such phases, provided sufficient computational power is available.

The problem is further complicated when the central source is luminous and hot enough to ionized its immediate surroundings. Presumably this phase is accompanied by a stellar wind. Thus, the computer code to be applied to the massive star formation problem will have to be able to deal with ionization/recombination and heating/cooling in regions of strong shocks.

Finally, there is a large range of length and time scales involved. The radius of the final star is $\sim 10^{11}$ cm, whereas the typical observed size of an ammonia core is $\sim 10^{17}$ cm. This resolution problem can be handled in one space dimension by adaptive grid techniques [24,25,23] or by joining together solutions applicable in particular regions of the protostar. In 2 or 3 dimensions the problem becomes more difficult. In an explicit code, the time step is restricted by the Courant-Friedrichs-Lewy (CFL) condition. Therefore, even

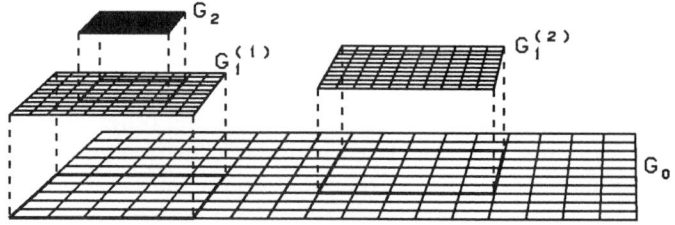

FIGURE 1. Grid structure for four nested 2D grids at three ($L = 2$) grid levels with a refinement ratio $r = 2$.

if the central star is not resolved and the smallest grid cell has size 0.1 AU, the time step becomes $\lesssim 5 \times 10^5$ s, four million times smaller than the free-fall time of a cloud starting at a density of 10^{-18} g cm^{-3}. Adams & Shu [1] used approximate analytical techniques in their models of protostars, rather than a full hydrodynamic treatment. Tscharnuter [20] used an implicit 2-D code with a radially moving grid, in order to resolve the inner regions of a protostar down to a scale of 10^{10} cm. However, the time step still became so short that he was unable to follow the collapse beyond the point where 0.07 M$_\odot$ had collected in the central core and the extent of the disk was only several AU.

1.2. Photoionized disks

In this communication we concentrate on the hydrodynamical evolution of a massive circumstellar disk under the influence of ionizing radiation and a stellar wind from the central source. The initial conditions (temperature, density, velocity) are taken from the results of 2D radiation hydrodynamical collapse calculations of an initially $\rho \propto r^{-2}$ density-peaked, uniformly rotating 10 M$_\odot$ protostellar cloud; see [29,30]. Preliminary results of hydrodynamic calculations of the photoionization of disks with the same initial conditions considered here are discussed in [32]. However, these early calculations did not take the diffuse field of recombination photons into account, which are able to ionize hydrogen.

The problem of photoionization of infinitely thin disks including the diffuse ionizing field has been addressed in [8], using a combination of analytical and semi-analytical methods. An important result of this investigation is a collection of simple formulae for calculating mass loss rates from the disks and disk destruction time scales over a wide range of disk and stellar parameters.

2. The computational code

The code used by YBL is based on the 2D axisymmetric code written by Różyczka [13] which was improved in the following manner: 1) rotation, 2) self-gravity, 3) FLD radiation transfer were included, and 4) multiply nested grids were implemented to deal with the spatial resolution problem. Many basic features of the code, including 1) and 2) are discussed in detail in [17,18]. The implementation of 3) and 4) into hydrodynamic codes is discussed in [31].

2.1. Multiply nested grids

In Fig. 1 we give an example from Yorke & Kaisig [31] of the 2D grid structure possible with multiply nested grids. The discretized finite difference equations for the hydrodynamics, radiation transfer, Poisson equation, etc. are to be solved on each of the four grids shown, using boundary conditions obtained from the next coarser grid. Boundary conditions for

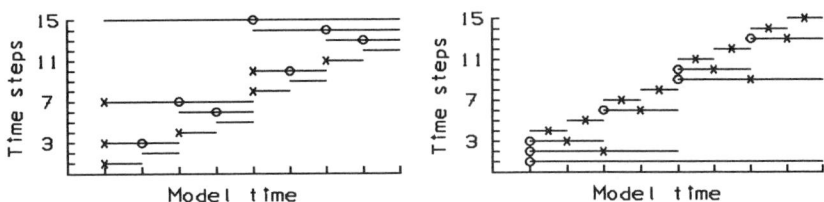

FIGURE 2. Two possible integration sequences for multiply nested grids with r=2 and L=4: starting with the finest grid (left) and starting with the coarsest grid (right). Symbols indicate the updating of interior values in a particular grid using data from its subgrid (circles) or updating of grid boundary values using data from its next coarser grid (crosses).

the coarsest grid are those imposed on the problem. Information is also transferred from the fine grids to the coarser grids at specified points in model time (see Fig. 2).

Different time step increments Δt_l are used at each grid level $l = 0, 1, 2$, which must satisfy the condition:

$$\Delta t_l \leq \min(\Delta t_l^{CFL}, \frac{\Delta t_{l-1}^{CFL}}{r}, \frac{\Delta t_{l-2}^{CFL}}{r^2}, ..., \frac{\Delta t_0^{CFL}}{r^l}) \quad , \tag{2.1}$$

where $l = 0$ corresponds to the coarsest grid level. Δt_l^{CFL} is the CFL restriction of the time step necessary for the grid level l and r is the refinement ratio between neighboring grid levels. Because r time steps are necessary on a particular grid for each coarser grid's time step, the total number of time steps necessary to advance the coarsest $l = 0$ grid by one and all finer grids up to the same time is:

$$N = 1 + r + r^2 + \cdots + r^L = r^{L+1} - 1 \quad , \tag{2.2}$$

where L is the finest grid level. For the examples shown in Fig. 1 ($L = 2$) and Fig. 2 ($L = 3$) the number of integration steps is $N = 7$ and $N = 15$, respectively.

Time-centered outer boundary conditions can be used in the scheme depicted at the right of Fig. 2, whereas the interior of a grid can be replaced by time-centered data from a subgrid when the fine grids are integrated first. A more detailed description of grid interaction including boundary flux corrections is given in [31]. The interested reader is also referred to [2].

2.2. Numerical treatment of photoionization

In conjunction with ionization/recombination we treat the gas component as hydrogenic. The fate of ionizing photons emitted directly by the central source is calculated along 241 radial directions and all relevant quantities are interpolated onto the 4*124×124 cylindrical (R, Z) grid (4 nested grids, each with 124 grid points in both the R and the Z directions; see Fig. 3). Along each line of sight direction we integrate the equation describing the transfer of radially moving photons:

$$\frac{dr^2 F_{\text{Lyc}}}{dr} = -(\sigma_{\text{Lyc}} + \sigma_{\text{dust}}) r^2 F_{\text{Lyc}} \quad , \tag{2.3}$$

where F_{Lyc} is the radial energy flux of hydrogen-ionizing Lyman continuum photons and σ_{Lyc}^{-1} ($\sigma_{\text{dust}}^{-1}$) is the mean free path for absorption of these photons by ionization of hydrogen (by dust). When the degree of ionization $x = n_e/n = n_p/n$ is known (n_e, n_p and n are the electron, proton and hydrogen particle densities, respectively), the absorption coefficient

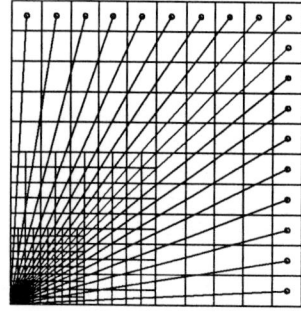

FIGURE 3. Grid structure for 3 nested 11 × 11 self similar grids showing the strategy of choosing 19 radial lines of sight to compute the transfer of stellar ionizing photons. Only one quadrant of the cylindrical (R, Z) grid need be considered because of the assumed axial symmetry and symmetry with respect to the equatorial plane. Note that at least one line of sight passes through each grid cell.

TABLE 1. Model parameters for calculations of photoionized disks. The last two columns are the latest evolutionary time calculated and the mean mass loss rate from the disk.

case	$\frac{\dot{M}_{\text{wind}}}{10^{-8}\,M_\odot\,\text{yr}^{-1}}$	$\frac{u_{\text{wind}}}{\text{km s}^{-1}}$	$\frac{L_{\text{wind}}}{L_\odot}$	$\frac{S}{10^{46}\,\text{s}^{-1}}$	$\frac{t}{\text{yr}}$	$\frac{\dot{M}_{\text{evap}}}{10^{-8}\,M_\odot\,\text{yr}^{-1}}$
A	2.0	60	0.006	0.35	240	11
B	2.0	60	0.006	1.74	250	24
C	2.0	60	0.006	7.08	575	40
D	0.5	600	0.15	0.35	25	27
E	0.5	600	0.15	7.08	49	50
F	10.0	600	2.97	0.35	41	> 7

σ_{Lyc} can be determined from the ionization cross section κ_{Lyc} of neutral hydrogen:

$$\sigma_{\text{Lyc}} = n(1-x)\kappa_{\text{Lyc}} \quad , \tag{2.4}$$

The subscript 'Lyc' is used to denote all quantities pertaining to Lyman continuum photons ($h\nu_{\text{Lyc}} \geq 13.6\,\text{eV}$) from the star. In general weighted mean values depending on the spectral distribution in this wavelength range are necessary.

Ionizing photons resulting from recombinations to the ground state are treated using the flux-limited diffusion approximation (see below). Time dependent ionization/recombination and heating/cooling is considered, whereby collisions of atoms and molecules in the neutral regions with dust grains determine the gas temperature. In the ionized regions the main source of heating is the excess energy of ionizing photons; the primary cooling mechanism is inelastic collisions of ions with electrons. We use a simplified form of the cooling function (fixed degrees of ionization for all elements except hydrogen in the ionized region).

We solve the following parabolic partial differential equation for FLD diffuse radiation transfer implicitly in a separate substep

$$\left(\frac{\partial u_1}{\partial t}\right) = -\nabla \cdot \mathbf{F}_1 + \epsilon_1 \,, \tag{2.5}$$

where u_1 is the energy density of diffuse hydrogen-ionizing photons, $\mathbf{F}_1$ is the corresponding radiative flux, and ϵ_1 is the energy production rate of these photons:

$$\epsilon_1 = h\nu_1 \alpha_1 n_e n_p = h\nu_1 \alpha_1 n^2 x^2 \tag{2.6}$$

The subscript '1' is used to denote all quantities pertaining to $h\nu_1 = 13.6$ eV photons produced by direct recombination of hydrogen into the $n = 1$ ground level; α_1 is the corresponding recombination coefficient. (*c.f.* [15] for a description of symbols).

In the framework of the FLD approximation (*e.g.* [10]) $\mathbf{F}_1$ is approximated by:

$$\mathbf{F}_1 = -\frac{c}{\sigma_1} \lambda \nabla u_1, \tag{2.7}$$

where the flux-limiter λ is given by

$$\lambda = \frac{1}{S}(\coth S - \frac{1}{S}) \quad \text{with} \quad S = \frac{|\nabla u_1|}{\sigma_1 u_1}. \tag{2.8}$$

σ_1^{-1} is the mean free path for absorption of these photons.

We note that for the two extremes, $S \to 0$ (optically thick) and $S \to \infty$ (optically thin), the correct limits for $\mathbf{F}_1$ are obtained:

$$\mathbf{F}_1 \to \begin{cases} -\frac{c}{3\sigma_1} \nabla u_1 & \text{for } S \to 0 \quad \text{(diffusion limit)} \\ c u_1 & \text{for } S \to \infty \quad \text{(streaming limit)} \end{cases} \tag{2.9}$$

A discussion of our method of numerical solution is given in [31].

To complete this description, we formulate the equations necessary to solve for the degree of ionization x:

$$\frac{dx}{dt} = (1-x)\left(\frac{cu_1\kappa_1}{h\nu_1} + \frac{F_{\text{Lyc}}\kappa_{\text{Lyc}}}{h\nu_{\text{Lyc}}}\right) - nx^2\alpha, \tag{2.10}$$

where d/dt is time derivative in a coordinate system moving with the gas. α is now the total recombination coefficient, including direct recombinations into the ground $n = 1$ level.

3. Numerical Results

3.1. *The evolution of a 10 $M_\odot$ collapsing cloud*

Yorke et al. [31] consider the fate of a 10 $M_\odot$ gravitationally unstable molecular cloud fragment of diameter ~ 2700 AU. The cloud is assumed to be originally density-peaked with $\rho \propto r^{-2}$ and to rotate as a solid body ($\Omega = 5 \times 10^{-12}$ s^{-1}). These calculations show how a warm, quasi-hydrostatic disk is formed. Whatever flows into the central cell is assumed to accrete onto a central hydrostatic core.

The growth of the central core virtually ceases at 2.68 $M_\odot$ (under the assumption of no angular momentum transport on such a short time scale). The evolution is quasi-stationary at this point — the equilibrium disk continues to grow in size and mass as material from the outer regions of the clump fall inward with too much angular momentum to be accreted onto the core. The disk is encased in an accretion shock front several scale heights from the symmetry plane which is similar in many respects to a bow shock. The temperature contours are still nearly spherical.

At the end of this computation the central unresolved region with 2.68 $M_\odot$ is found to be unstable to non-axisymmetric perturbations. In addition, the inner portions of the disk containing most of the mass are unstable according to the local Toomre criterion, implying that here also non-axisymmetric perturbations will lead to the partial breakup of the inner disk. Assuming that gravitational torques will subsequently transport angular momentum out of this region (*c.f.* [9]), a central core of 8.4 $M_\odot$ with a stable disk of

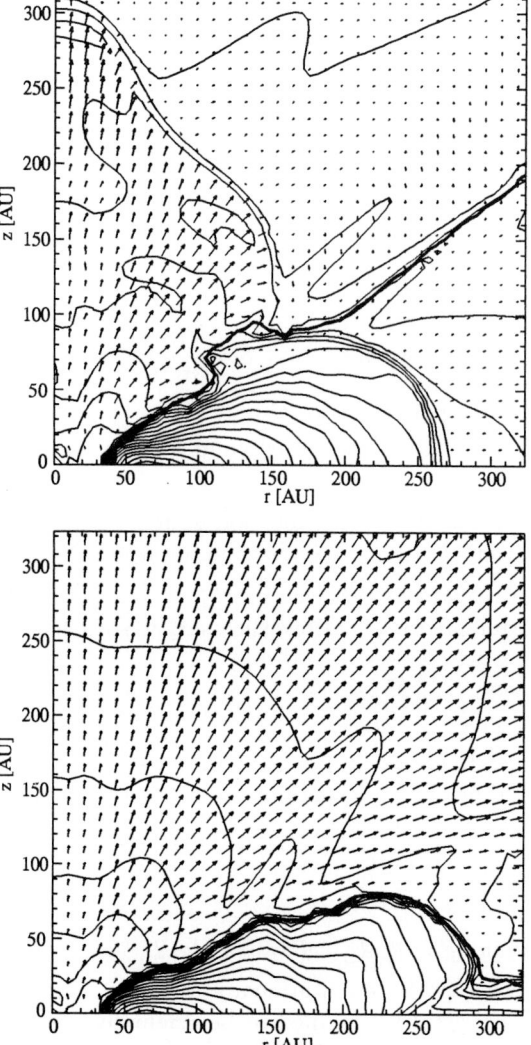

FIGURE 4. Structure of 10 $M_\odot$ case C (see Table 1) 33 yr (top) and 575 yr (bottom) after a central source of ionizing photons and a weak stellar wind has turned on. Density contour lines are separated by $\Delta \log \rho = 0.5$, velocity arrows are normalized by their maximum value 46 km s^{-1}. The thick solid line denotes the location of the ionization front which separates the neutral disk material from the warm ($T \approx 10^4$ K) outflowing ionized gas.

1.6 $M_\odot$ should result. This constitutes the initial conditions for subsequent calculations of the disk's interaction with the central source.

3.2. Photoionization of circumstellar disks

Once a massive protostar ($M \gtrsim 8$ $M_\odot$) reaches hydrogen burning temperatures and densities, it produces hydrogen-ionizing photons copiously. The star can begin to ionize and heat its immediate surroundings. In addition, mass outflow from the star can be expected. Such models are excellent candidates for explaining ultra-compact HII regions.

In Fig. 4 we show the density and velocity structure and the location of the ionization front for case C at two evolutionary times. A central ionizing flux 7×10^{46} s^{-1} and a rather weak stellar wind ($\dot{M}_W = 2 \times 10^{-8}$ $M_\odot$ yr^{-1}, $v_W = 60$ km s^{-1}) were assumed (see Table 1). We find that ionized material leaves the inner grid at a rate $\dot{M}_{HII} \approx 4 \times 10^{-7}$ $M_\odot$ yr^{-1} at a velocity of $v_{HII} \approx 40$ km s^{-1}. The disk becomes noticeably flattened because of the increased external pressure. There is no trace of the accretion shock front after ≈ 200 yr.

For cases with stronger stellar winds the mass loss rate from the disk appears to be higher (*c.f.* case E) with the exception of case F (see Table 1). For this case the powerful wind is able to significantly compress the disk material and it becomes more difficult to ionize. Also, there are fewer diffuse photons because the low density wind fills a larger percentage of the volume in the vicinity of the disk. However, in contrast to case C a steady state situation has not been reached. There is evidence in Fig. 5 that the strong wind is interacting directly with the disk via a Helmholtz ("wind-water wave") instability. There should be some mixing at the disk-wind boundary which could contribute to the dissipation of the disk. A detailed description of these results is forthcoming [22].

4. Conclusions

From the numerical calculations discussed here we envisage the following sequence of events for the formation of ultra-compact HII regions. Disk formation is rapid ($10^3 - 10^4$ yr), when a 10 $M_\odot$ density-peaked molecular clump becomes unstable. The initial disk has the structure of an "accreting accretion disk" as it continues to accrete material from the outer regions of the clump. Although it is technically a "thin disk" (the density scale height at radius $R \approx 0.1\,R$), the equilibrium region extends to about 10 scale heights above the equator, where the accretion shock front is located.

The hydrostatic core and inner parts of the disk are unstable with respect to non-axisymmetric perturbations; presumably the core mass grows within several 10^4 yr to a significant percentage of the original clump's mass. The pre-main sequence contraction time scale is also short (several 10^4 yr), so that main sequence activity – hydrogen-ionizing radiation together with a hot stellar wind – can begin even while the remaining disk continues to accrete material from the still collapsing clump. The result is a BN-type source with both inflow and outflow. The disk can be "photoevaporated" on a time scale of several 10^5 yr.

This work was supported by the DFG under grants Yo 5/6-1,2. We thank the Nuclear Research Facility in Jülich (HLRZ) for a grant of computer time. Computing facilities of the University of Würzburg (RZUW) and the Leibniz Computing Center (LRZ) in Munich were also used.

REFERENCES

[1] Adams, F. C., Shu, F. H. 1986, ApJ, 308, 836.
[2] Berger, M. J., Colella, P. 1989, JCompPhys, 82, 64.
[3] Beckwith, S., Sargent, A. 1993, *Protostars and Planets III*, ed. E. H. Levy, J. Lunine, (Tucson: Univ. of Arizona Press), 521.
[4] Bodenheimer, P., Yorke, H. W., Różyczka, M., Tohline, J. E. 1990, ApJ, 355, 651.
[5] Boss, A. P. 1988, ApJ, 331, 370.
[6] Dent, W. R. F. 1988, ApJ, 325, 252.
[7] Efstathiou, A., Rowan-Robinson, M. 1990, MNRAS, 245, 275.
[8] Fiedler, R. A., Mouschovias, T. Ch. 1992, ApJ, 391, 199.
[9] Hollenbach, D. J., Johnstone, D., Shu, F. H. 1993, *Massive Stars: Their Lives in the Interstellar Medium*, ed. J. P. Cassinelli, E. B. Churchwell, ASP Conf. Ser., Vol. 35, 26.
[10] Laughlin, G., Bodenheimer, P. 1993, this volume.
[11] Levermore, C. D., Pomraning, G. C. 1981, ApJ, 248, 321.
[12] Lizano, S., Shu, F. H. 1989, ApJ, 342, 834.
[13] Morfill, G. E., Tscharnuter, W., Völk, H. J. 1985, *Protostars and Planets II*, ed. D. C. Black, M. S. Matthews (Tucson: Univ. of Arizona Press), 493.
[14] Różyczka, M. 1985, A&A, 143, 59.
[15] Shu, F. H., Adams, F. C., Lizano, S. 1987, ARAA, 25, 23.
[16] Spitzer, L., Jr. 1978, *Physical Processes in the interstellar Medium*, (New York: Wiley).
[17] Strom, S. E., Edwards, S., Skrutskie, M. F. 1993, *Protostars and Planets III*, ed. E. H. Levy, J. Lunine, (Tucson: Univ. of Arizona Press), 837.

[18] Stone, J. M., Norman, M. L. 1992a, ApJS, 80, 753.
[19] Stone, J. M., Norman, M. L. 1992b, ApJS, 80, 791.
[20] Stone, J. M., Norman, M. L. 1992c, ApJS, 80, 819.
[21] Tscharnuter, W. 1987, A&A, 188, 55.
[22] Weidenschilling, S. J., Cuzzi, J. N. 1993, *Protostars and Planets III*, ed. E. H. Levy, J. Lunine, (Tucson: Univ. of Arizona Press), 1031.
[23] Welz, A., Yorke, H. W. 1993, in prep.
[24] Winkler, K.-H. A., Norman, M. L. 1986, *Astrophysical Radiation Hydrodynamics*, ed. K.-H. A. Winkler, M. L. Norman, (Dordrecht: Reidel), 71.
[25] Winkler, K.-H. A., Newman, M. J. 1980a, ApJ, 236, 201.
[26] Winkler, K.-H. A., Newman, M. J. 1980b, ApJ, 238, 311.
[27] Yorke, H. W. 1984, *Star Formation Workshop, Occ. Rep. Roy. Obs. Edinburgh No.*, 13, ed. R. Wolstencroft, (Edinburgh), 63.
[28] Yorke, H. W., Krügel, E. 1977, A&A, 54, 183.
[29] Yorke, H. W., Bodenheimer, P., Laughlin, G. 1993a (YBL), ApJ, 411, 274.
[30] Yorke, H. W., Bodenheimer, P., Laughlin, G. 1993b, *Star Formation, Galaxies and the Interstellar Medium*, ed. J. Franco, F. Ferrini, G. Tenorio-Tagle, (Cambridge: Cambridge Univ. Press), 127.
[31] Yorke, H. W., Bodenheimer, P., Laughlin, G. 1993c, ApJ, in prep.
[31] Yorke, H. W., Kaisig, M. 1993, CompPhysComm, in press.
[32] Yorke, H. W., Welz, A. 1993, *Star Formation, Galaxies and the Interstellar Medium*, ed. J. Franco, F. Ferrini, G. Tenorio-Tagle, (Cambridge: Cambridge Univ. Press), 239.

Three-Dimensional Fragmentation Calculations of Protostar Collapse

By P. Bodenheimer

UCO/Lick Observatory, Board of Studies in Astronomy and Astrophysics, University of California, Santa Cruz, CA 95064, USA.

Hydrodynamical three-dimensional numerical calculations of the collapse and fragmentation of rotating protostars are reviewed. Recent calculations have made some progress in explaining the general properties of observed binary and multiple systems. However, a large parameter space must be explored, the physics that determines the actual initial conditions must be clarified, and the numerical resolution and accuracy of existing numerical codes must be significantly improved before achievement of this goal can be realized.

1. Introduction

The protostellar collapse phase is a challenge to treat numerically. The initial state is a dense interstellar cloud with a density of about 10^{-19} g cm^{-3}, while the end product of collapse is a star with a mean density of 10^{-1} g cm^{-3}. Thus, calculation of the collapse involves a change of 18 orders of magnitude in density and a change of scale from 10^{17} cm to 10^{11} cm. Furthermore, a shock wave forms at the edge of the central stellar core or at the surface of a surrounding disk. In one space dimension, with assumed spherical symmetry, the numerical problem can be handled with an adaptive grid [46] or by reduction of the problem to ordinary differential equations under the assumption of steady-state accretion [40]. However, rotational effects become important during the later stages of the collapse, resulting in the formation of disks or binary systems, so the problem actually must be calculated in 2 or 3 space dimensions. Treatment of the full range of densities then becomes extremely difficult. Existing 3-D calculations resolve at best three orders of magnitude in length scale. In this talk I will concentrate on the earlier phases of protostar collapse, during which the isothermal assumption is generally valid, in order to describe calculations regarding fragmentation of the protostar into binary and multiple systems. The appropriate length scale is ≈ 0.05 pc. In a later talk Harold Yorke will discuss the collapse of individual fragments, in two space dimensions with radiation transfer, in which the calculations are resolved down to scales of a few AU.

The study of several important issues underlies the three-dimensional calculations:

1) How are the general observed properties of binary and multiple systems explained? How can we understand the distribution of binaries according to period, the distribution of secondary masses for a given primary mass, and the distribution of orbital eccentricities?

2) How is the frequency of binary systems explained? What are the initial conditions that lead to a binary or multiple system rather than a single star with a planetary system?

3) What is the formation mechanism? Is there more than one? Some of the possibilities include fragmentation during protostar collapse, fragmentation of a disk through gravita-

tional instability, cloud-cloud collisions in the interstellar medium, capture, or fission. Is hierarchical fragmentation important? Or are the observed hierarchical multiples formed by capture through interactions in N-body systems?

In this talk I will not consider all possible mechanisms for the origin of binaries but will concentrate on the fragmentation process during the collapse of a rotating protostar.

2. Observational Background

Several recent reviews of the general properties of pre-main-sequence and main-sequence binary systems have appeared [1, 4, 10, 25, 49]. To summarize: (1) about 70% of all main-sequence stars in the F to G range of spectral types have one or more companions; (2) the period range runs from $0 < \log P < 9$, where P is given in days and where the median period is about 180 yr; (3) for long-period systems the secondary masses for a given mass primary are distributed approximately as the normal initial mass function, but for short period systems the secondary mass distribution is somewhat flatter in the sense that there is a tendency toward more nearly equal-mass companions; (4) in general there is a wide range in eccentricities, except that in very close systems (P < 7 to 19 da, depending on age [25]) the orbits are circular; and (5) there is considerable evidence that binary systems form during the very earliest evolutionary phases of a star. Support for this last point comes from the following facts: first, the binary frequency among pre-main-sequence stars is now suspected to be even greater than that of main-sequence stars [16,35], second, some of the youngest stars known, those near the birth line in the HR diagram, are known to be binaries, and third, binary systems in the protostellar phase are now being discovered. A good example of an interesting protostellar binary is IRAS 16293 - 2422 [44,47], with a separation of 750 AU and evidence for circumbinary disk material. For these reasons, the protostar collapse phase should be investigated with regard to the origin of binaries.

3. Standard Test Case

The typical initial condition in the core of a molecular cloud that is thought to lead to star formation involves a particle density of H_2 molecules of 10^5 cm^{-3}, a temperature of 10 K, a radius of 0.05 pc, and a uniform angular velocity of 3×10^{-14} s^{-1}. If we define α_i and β_i as the initial ratios of thermal energy and rotational energy, respectively, to the absolute value of the gravitational potential energy, then these particular initial conditions correspond to $\alpha_i = 0.41$ and $\beta_i = 3.3 \times 10^{-3}$. If the cloud is to collapse, the product $\alpha_i \beta_i < 0.2$, and if the cloud is to fragment, $\alpha_i \beta_i < 0.12$ [27]. Both of these criteria apply to clouds of initially uniform density and temperature that remain isothermal during collapse.

The general picture of low-mass star formation [38] is that molecular cloud material in the density range $10^3 - 10^4$ cm^{-3} is supported against collapse by magnetic and turbulent effects. Some regions evolve to higher densities, losing angular momentum through magnetic braking [15], and gradually diffusing across magnetic field lines through plasma drift [39]. When cloud cores approach the above physical state, turbulent, rotational and magnetic effects have become relatively unimportant, and the line widths observed in the cores have become essentially thermal. At some point gravitational collapse begins, and the general problem is to follow the collapse from these or similar initial conditions.

A standard test case was set up a number of years ago for the purpose of comparing the results of different codes [12]. It involves a mass of 1.0 $M_\odot$, a radius of 3.2×10^{16} cm, an $m = 2$ azimuthal perturbation in density with an amplitude of 50% superimposed on a uniform background density of 1.44×10^{-17} g cm^{-3}, a uniform angular velocity of 1.6

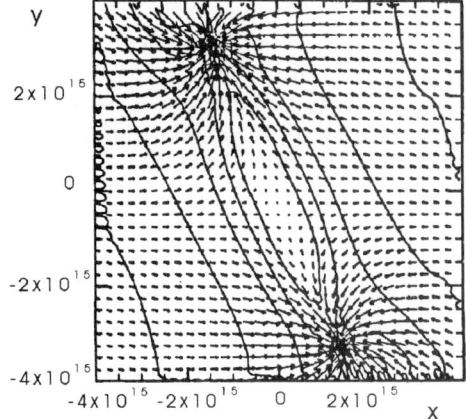

FIGURE 1. Standard test case after a collapse time of 8.02 ×10^{11} s. Solid curves: logarithmically spaced contours of equal density in the (x, y) plane; log ρ_{min} = −14.46 and log ρ_{max} = −11.23. Arrows give velocities with a maximum value of 1.2 ×10^6 cm s^{-1}. From ref. [13].

×10^{-12} s^{-1}, and a temperature of 10 K. The above-defined parameters turn out to be α_i = 0.25 and β_i = 0.20. Several different codes obtain the same result, namely that a binary forms on an eccentric orbit with components having about 0.1 of the mass of the original cloud. The highest-resolution calculation of this case [13] gives excellent agreement with earlier results, demonstrates that the outcome is independent of numerical resolution, and shows that the bar that forms between the binary pair is accreted onto the binary (Fig. 1).

4. What is the correct initial condition?

The standard test case is not a realistic initial condition because both theoretical and observational evidence indicates that molecular cloud cores are centrally condensed. Available observations [31,48] indicate that the cores have density distributions corresponding to power laws $\rho \propto r^{-p}$ where $1.5 < p < 2$. Theoretical calculations show that before collapse sets in molecular cloud material is magnetically dominated and that given regions contract in quasi-hydrostatic equilibrium with the time scale determined by the rate of plasma drift across magnetic field lines. Detailed two-dimensional models [24,42] show that at the time the central regions begin to collapse they have density distributions with $\rho \propto r^{-2}$ in the equatorial plane and along the polar axis. Observations also suggest [30] that the cores are prolate in shape rather than spherical, with typical axial ratios of 0.5 – 0.6. Accordingly, Boss [11] has carried out calculations for a number of values of α_i and β_i with an initial density distribution that is mildly centrally condensed, with a ratio of 20 between the central density and the density at the outer edge. The initial models have a bar-like structure, with unequal axes in the (x, y) plane (the z axis is the rotation axis), in accordance with observations. Although fragmentation still occurs, the region of the (α_i, β_i) plane is smaller than it would be for a cloud of uniform density. For example, if the ratio of axes in the (x, y) plane is 2:1, the region where fragmentation occurs is limited to $\alpha_i < 0.45 - 0.36\beta_i$. Clouds near the borderline of the region of instability to fragmentation tend to collapse to a bar-like configuration; then if α_i is reduced, a bar forms, then fragments into a binary.

However, the observations suggest a greater degree of central condensation than that assumed by Boss, although it is true, as he claims, that the central regions of the observed clouds are not resolved. Three-dimensional collapse calculations on a Cartesian grid with 64^3 zones have been performed by Myhill & Kaula [32] with initial density profiles $\rho \propto r^{-1}$ and $\rho \propto r^{-2}$ and with 10% $m = 2$ initial density perturbations. For initial values $\alpha_i = 0.16$ and $\beta_i = 0.17$ they find that if the initial rotation is uniform, there is no fragmentation, just a single central density maximum. However if differential rotation is assumed, with the angular momentum distribution of the uniformly rotating sphere of uniform density, fragmentation into a binary occurs for both initial density profiles. An independent analytic calculation [43] shows that the singular isothermal sphere ($\rho \propto r^{-2}$) is stable against fragmentation.

Thus it is clear that a centrally condensed density distribution suppresses fragmentation, although initial differential rotation may solve the problem. However, there is no observational evidence that differential rotation exists. The observations of Goodman et al. [18] are analyzed in terms of a fit to a uniformly rotating model; they generally are consistent with such a model, but the observations are not accurate enough to discern differential rotation even if it were present. Theoretical models of the pre-collapse contraction including magnetic braking and ambipolar diffusion [42] show that in fact the dense cores arrive at the point of collapse with uniform rotation in their central regions. We thus have the dilemma that, for example in the Taurus-Auriga star formation region, we have a high proportion of binary systems [17,23], yet the cores that form the stars have $\rho \propto r^{-2}$ and are probably uniformly rotating, so they should not have been unstable to fragmentation. Another possible solution to this problem is suggested in the next section.

5. Further Numerical Calculations

A wide variety of three-dimensional calculations, employing many different initial conditions, has been performed to investigate the question of fragmentation of interstellar clouds and protostars. The available parameter space is so large, however, that the set of existing calculations should be considered as a random sampling rather than a systematic search through this space. In this section we provide a few examples, but we do not attempt a complete review of all existing calculations.

The observed elongated structure of some interstellar clouds has prompted investigations of the collapse of isothermal finite cylinders [3,22,36]. Two fragments generally form on the axis of the cylinder, but near its ends, as long as $J_0 = 1/\alpha_i < 2$. More recent 3-D SPH calculations by Nelson & Papaloizou [33] have treated the related problem of a prolate spheroid, non-rotating and isothermal, with various axis ratios. The initial density was either homogeneous or perturbed, and the particle number was 2500 - 5000. Fragmentation is expected to occur when the mass per unit length along the major axis is greater than a critical value $m_{crit} = 2C_s^2/G$, where G is the gravitational constant and C is the sound speed. When the cloud collapses toward its major axis, fragmentation tends to occur near the center of the cloud where the mass per unit length is greatest.

Boss [9] has carried out a finite difference calculation of the collapse of a rotating cloud with moderate central density concentration with initial parameters $\alpha_i = 0.26$, $\beta_i = 0.16$. An $m = 2$ initial perturbation with 10% amplitude was applied. The numerical grid had 51 by 45 by 64 zones in spherical coordinates. The cloud first collapses to a ring-like structure in the equatorial plane, then two density peaks develop in the ring, each of which splits into two fragments. This result is probably the effect of a high-order mode, rather than hierarchical fragmentation, which is likely to occur only on a much smaller scale after the fragments have collapsed to a point where rotation again dominates. Klapp, Sigalotti, & de Felice [21] have done essentially the same calculation, with a similar resolution to that of Boss, except for the fact that a small initial velocity field is present. The results show about six fragments along a bar-like structure at the center; the details of the outcome are quite different from those of Boss, and they do depend on the numerical resolution.

An SPH calculation with 13,000 particles by Monaghan & Lattanzio [29] treats a cloud of mass 10^4 $M_\odot$ and a radius of 12.6 pc, $\alpha_i = 0.30$ and $\beta_i = 0.47$. The initial model contains random density fluctuations with amplitudes of 14%. Molecular cooling is applied as the cloud collapses, and the temperature decreases from 70 K to 5 K. The calculation employs a variable smoothing length that decreases with time but remains constant in space. About 6 gravitationally unstable knots form in a ring-like configuration; they are identified with the ring of HII regions in W49A [45]. Without cooling this configuration does not fragment because the initial $\alpha_i \beta_i = 0.14$, above the limit for isothermal collapse.

Monaghan [28] has performed more recent calculations starting from uniform-density ellipsoids which have a velocity field that is a linear function of the coordinates. The vorticity is assumed to be parallel to the total angular momentum. Even if the angular momentum is fixed, the outcome depends sensitively on the vorticity distribution, in particular, on the degree of shear. If initial shear is present, ellipsoids collapse to long filaments which are unstable to fragmentation, but if there is no initial shear (uniform rotation), the filament becomes wound up and has not fragmented after 1.5 free fall times. The results are consistent with the presence of filamentary structures in star-forming regions, and these calculations show how such filaments can be formed even without magnetic fields.

Miyama [26] has done a series of SPH collapses of rotating clouds with a polytropic equation of state $P = K\rho^\gamma$. Random initial perturbations are assumed on a cloud with uniform density and uniform rotation, and the criterion for fragmentation is examined for various values of γ as a function of α_i and β_i. An analytic stability analysis combined with the numerical results shows that the marginal condition for fragmentation is given by $\alpha_i \beta_i^{4-3\gamma} =$ constant. If applied to the adiabatic collapse of a gas of molecular hydrogen, which would represent the later, optically thick phase of protostar collapse, $\gamma \approx 7/5$ and the criterion beomes $\alpha_i = 0.09\beta_i^{0.2}$. If $\beta_i = 0.1$ then instability occurs only if $\alpha_i < 0.06$, indicating that much lower values of α_i are required in the adiabatic collapse than in the isothermal collapse for fragmentation to occur. The question that arises is whether such initial conditions are likely to be realized, because if they are not, it becomes very difficult to explain the origin of close binary systems by fragmentation during the collapse. Very similar fragmentation criteria had previously been obtained [19,20] based on the stability properties of very rapidly rotating, flattened disk-like equilibria.

An alternative point of view regarding fragmentation and the formation of multiple systems has been proposed by Pringle [34]. He states that formation of such systems by collapse of an initially static cloud with a density distribution $\rho \propto r^{-2}$ "is not a plausible option", since both fragmentation and fission are suppressed. The alternative is that collapse is induced by cloud-cloud collisions with a typical collision velocity equal to the sound speed. The effect of even a mild collision on a cloud can induce rapid reduction of the Jeans mass and collapse if the densities are in the range $\rho = 10^{-22} - 10^{-23}$ g cm^{-3}, where molecular and grain cooling occurs [41]. With regard to the magnetic field problem, Pringle suggests that the gas flows preferentially along field lines until a critical mass is attained. The impulse typically produces a binary system in a wide, highly elliptical orbit, but subsequent interactions between the components and their circumstellar disks could reduce the semimajor axis.

This suggestion has been subjected to numerical tests by Chapman and collaborators [14]. The hydrodynamics (including cooling) of the collisions are calculated with an SPH code and the gravity with a tree code. Early published calculations used 4000 particles, although this number has been substantially increased in more recent work. The results show, for example, that if clouds of 75 $M_\odot$ with radii of 2×10^5 AU collide at Mach 5, the gas is compressed to high density, and binary fragmentation occurs with the fragments accompanied by rotationally supported disks. Interactions between the disks and fragments leads to stable multiple systems. Another example treats the collision of an ensemble of 20 subclouds, each of 4 $M_\odot$ and a radius of 0.2 pc, with relative velocities of Mach 1.7. They merge to form a dense layer, and on a smaller scale the layer fragments into a quadruple, two of whose components later merge to form a triple with orbital separations 1200 and 7500 AU. More details of these calculations are presented elsewhere in this volume.

A whole series of calculations has been carried out with an SPH code for the fragmentation of initially elongated cylindrical clouds [5,6,7,8]. Zinnecker [49] first suggested that such prolate clouds, with rotation about an axis perpendicular to the axis of the cylinder, could fragment. The detailed calculations, which use 1000 to 8000 particles in an isothermal gas, show that in such a system two fragments generally form, which can later subfragment into a hierarchical multiple system. Disks were observed to form about the fragments, and in some cases disk fragmentation was also observed. Calculations were also performed in which there was assumed to be rotation about the axes both parallel to and perpendicular to the axis of the cylinder. As a result, hierarchical multiple systems were formed in which the long-period and short-period orbits were not co-planar, in general agreement with observations of such systems. Also, if slight density variations are introduced into the initial configuration along the axis of the cylinder, binaries can be formed with a range of mass ratios. For example, a density variation by a factor of 1.5 along the cylinder produces a binary with mass ratio 1:10.

Several interesting features emerged as a result of these calculations, which were carried out over a wide range of parameter space. (1) Orbits tend to be elliptical, but accretion of surrounding material onto the secondary as it forms has a tendency to equalize the masses and circularize the orbit. As the mass ratio decreases, the eccentricity tends to decrease. (2) Disk fragmentation, that is, formation of a binary companion out of a disk, can be induced by the tidal effect of a nearby companion. (3) The effect of a companion is also to induce mass and angular momentum transport through tidal effects on the disk of the primary. (4) The fragmentation process, along with angular momentum transport by gravitational torques, can reduce the angular momentum of fragments by up to two orders of magnitude as compared with that of the parent cloud. Thus these processes go in the direction of explaining the origin of the closer binary systems. However, because the initial scale is large (1 pc) and the numerical resolution is low, the multiple systems that are formed have relatively long periods and orbital separations greater than 50 AU. (5) If the numerical resolution is improved, the resulting mass ratios change.

These calculations further illustrate the fact that a wide variety of outcomes can result from fragmentation processes combined with disk formation. Examples include simple binary formation, destruction of one of the binary fragments by tidal dissipation, collision of a fragment with the disk of another fragment, leading to coalescence, fragmentation in the spindle that connects the primary fragments in the case where rotation is perpendicular to the axis of the cylinder, disk fragmentation into a hierarchical multiple, and fragmentation into a multiple from a bar-like configuration.

Burkert & Bodenheimer [13] have improved the spatial resolution considerably, treating the case of an isothermal collapse. They use a fixed Eulerian Cartesian grid which is not rezoned during the calculation. The main grid has 64^3 equally spaced zones. Up to 3 levels of nested subgrids can be installed, with a factor 4 improvement in linear resolution per subgrid. For example, the first subgrid generally has 64 by 64 by 32 zones in the x, y, and z directions, respectively, and is centered on the origin of the main grid. In general, the subgrids can be placed in arbitrary locations. The standard hydrodynamic equations are solved, with operator splitting and a second-order van Leer scheme for advection. A

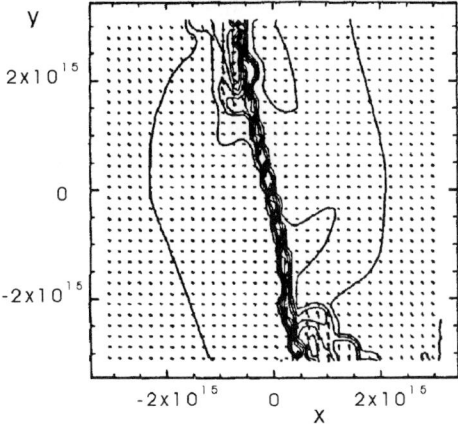

FIGURE 2. Density contours in the (x,y) plane after 1.36×10^{12} s for a highly resolved grid-based calculation ($\alpha_i = 0.26$, $\beta_i = 0.16$). The 15 contours are logarithmically spaced; log $\rho_{min} = -17.46$ and log $\rho_{max} = -11.06$. The maximum velocity is 1.16×10^6 cm s^{-1}. From ref. [13].

number of numerical tests have been performed, including one-dimensional shock tubes, spherical pressureless collapse, the explosion of a supernova in a uniform medium, collapse to a rotating equilibrium disk, and the standard fragmentation test case (§3).

A high-resolution calculation with 3 subgrids was performed for a cloud of mass 1 M$_\odot$, a radius of 5×10^{16} cm, a temperature of 10 K, a uniform angular velocity of 7.2×10^{-13} s^{-1}, and a uniform initial density with an imposed 10% $m = 2$ perturbation. The values of α_i and β_i were 0.26 and 0.16, respectively. The initial collapse results in a disk-like structure which proceeds to fragment into a binary in an elliptical orbit, with each component having about 5% of the mass of the initial cloud. There is evidence for disk-like structures around the components. A narrow bar forms between the main fragments, and after a short time it fragments also, into nine small pieces, each with about 0.3% of the initial cloud mass (Fig. 2). This result is found to be independent of the grid resolution. A calculation by Max Ruffert [37] from exactly the same initial conditions with a completely independent ppm-based code, also with a nested grid structure, gives a very similar result. Thus the fragmentation of the bar seems to be a robust result. The system of fragments that is produced is undoubtedly unstable, and there is a possibility that at least one low-mass fragment can be ejected and could evolve to be a free-floating brown dwarf. The calculations had to be stopped, however, when the fragments became unresolved even on the finest grid and when the isothermal approximation began to break down in the densest regions. Thus further calculations are required, in which the 11 fragments are treated as an N-body system in a dissipative medium, so that their further evolution can be determined. More realistic initial density stratifications are also required for the collapse calculations.

6. Conclusions

The main conclusion that can be reached from the set of numerical studies just discussed is that many problems remain, both physical and numerical. (1) Typical observed conditions in the cores of molecular clouds indicate central density concentration and uniform rotation. Under these conditions fragmentation is suppressed. Possible solutions to the problem of binary formation include disk fragmentation [2], differential rotation [32], or induced star formation [34]. (2) Many three-dimensional isothermal calculations produce fragmentation into binary or multiple systems, generally of rather long periods. But are the initial conditions realistic? A very large number of parameters is needed to describe the initial state, so how do we know which region in a large parameter space is the correct one? Further studies of how molecular clouds evolve to produce bound subcondensations are necessary to address this question. (3) The simulations do not produce close binaries, in particular, those with separations less than 1 AU. However, many possibilities remain to be explored, including generation of fragments with low α_i which later can subfragment, orbital decay of longer-period systems, interactions of fragments with disks, or N-body interactions and captures in tight systems of fragments. (4) The initial condition of an elongated cloud with arbitrary rotation axis shows some promise in explaining mass ratios, formation of hierarchical systems through disk fragmentation, and non-coplanarity in multiple systems. But the numerical resolution in the calculations published so far is not sufficient to justify all conclusions that have been made, and the initial densities in the clouds have been assumed to be uniform. (5) Existing calculations have shown that the effect of numerical resolution on the results is significant, and that small changes in initial conditions can also have significant effects on the results. Significant effort is needed to isolate real physical effects from numerical effects in such calculations. A more rigorous standard test case is needed which requires a high degree of numerical resolution. Finally, it is evident that the application of intense computer power is required before the aim of explaining the properties of existing binary systems can be attained.

This work was supported in part through a special NASA astrophysics theory program which supports a joint Center for Star Formation Studies at NASA-Ames Research Center, UC Berkeley, and UC Santa Cruz, and in part by a grant of computer time from San Diego Supercomputer Center.

REFERENCES

[1] Abt, H. 1983, ARA&A, 21, 343.
[2] Adams, F., Ruden, S., & Shu, F. 1989, ApJ, 347, 959.
[3] Bastien, P. 1983, A&A, 119, 109.
[4] Bodenheimer, P., Ruzmaikina, T., & Mathieu, R. 1993, in *Protostars and Planets III*, eds. E. Levy & J. Lunine (Tucson: University of Arizona Press), p. 367.
[5] Bonnell, I., Arcoragi, J.-P., Martel, H., & Bastien, P. 1992, ApJ, 400, 579.
[6] Bonnell, I., & Bastien, P. 1992, ApJ, 401, 654.
[7] Bonnell, I., & Bastien, P. 1993, ApJ, 406, 614.
[8] Bonnell, I., Martel, H., Bastien, P., Arcoragi, J.-P., & Benz, W. 1991, ApJ, 377, 553.
[9] Boss, A. P. 1991, Nature, 351, 298.
[10] Boss, A. P. 1993, in *Realm of Interacting Binary Stars*, eds. J. Sahade, G. McCluskey, & Y. Kondo (Dordrecht: Kluwer), p. 355.
[11] Boss, A. P. 1993, ApJ, 410, 157.
[12] Boss, A. P., & Bodenheimer, P. 1979, ApJ, 234, 289.
[13] Burkert, A., & Bodenheimer, P. 1993, MNRAS, 264, 798.
[14] Chapman, S., Pongracic, H., Disney, M., Nelson, A., Turner, J., & Whitworth, A. 1992, Nature, 359, 207.
[15] Ebert, R., von Hoerner, S., & Temesvary, S. 1960, *Die Entstehung von Sternen durch Kondensation diffuser Materie* (Berlin: Springer), p. 311.
[16] Ghez, A., Neugebauer, G., & Matthews, K. 1992, in *Complementary Approaches to Double and Multiple Star Research*, eds. H. A. McAlister & W. I. Hartkopf (San Francisco: Astronomical Society of the Pacific), p. 1.
[17] Ghez, A., Neugebauer, G., & Matthews, K. 1993, AJ, in press.
[18] Goodman, A. A., Benson, P. J., Fuller, G. A., & Myers, P. C. 1993, ApJ, 406, 528.
[19] Hachisu, I., & Eriguchi, Y. 1985, A&A, 143, 355.
[20] Hachisu, I., Tohline, J., & Eriguchi, Y. 1988, ApJ Suppl., 66, 315.
[21] Klapp, J., Sigalotti, L, & de Felice, F. 1993, A&A, 273, 175.
[22] Larson, R. B. 1972, MNRAS, 156, 437.

[23] Leinert, Ch., Weitzel, N., Zinnecker, H., Christou, J., Ridgeway, S., Jameson, R., Haas, M., & Lenzen, R. 1993, A&A, in press.
[24] Lizano, S., & Shu, F. 1989, ApJ, 342, 834.
[25] Mathieu, R. 1994, ARA&A, in press.
[26] Miyama, S. M. 1992, Pub. Astr. Soc. Japan, 44, 193.
[27] Miyama, S. M., Hayashi, C., & Narita, S. 1984, ApJ, 279, 621.
[28] Monaghan, J. 1993, preprint.
[29] Monaghan, J., & Lattanzio, J. 1991, ApJ, 375, 177.
[30] Myers, P. C., Fuller, G. A., Goodman, A. A., & Benson, P. J. 1991, ApJ, 376, 561.
[31] Myers, P. C., Fuller, G. A., Mathieu, R. D., Beichman, C. A., Benson, P. J., Schild, R. E., & Emerson, J. P. 1987, ApJ, 319, 340.
[32] Myhill, E., & Kaula, W. M. 1992, ApJ, 386, 578.
[33] Nelson, R., & Papaloizou, J. 1993, MNRAS, in press.
[34] Pringle, J. 1989, MNRAS, 239, 361.
[35] Reipurth, B., & Zinnecker, H. 1993, A&A, in press.
[36] Rouleau, F., & Bastien, P. 1990, ApJ, 355, 172.
[37] Ruffert, M. 1994, in preparation.
[38] Shu, F. H., Adams, F. C., & Lizano, S. 1987, ARA&A, 25, 23.
[39] pitzer, L. Jr. 1978, *Physical Processes in the Interstellar Medium* (New York: John Wiley & Sons), §13.3.
[40] Stahler, S., Shu, F. H., & Taam, R. E. 1980, ApJ, 241, 637.
[41] Tohline, J. E., Bodenheimer, P., & Christodoulou, D. M. 1987, ApJ, 322, 787.
[42] Tomisaka, K., Ikeuchi, S., & Nakamura, T. 1990, ApJ, 362, 202.
[43] Tsai, J., & Bertschinger, E. 1989, Bull. A. A. S., 21, 1089.
[44] Walker, C., Carlstrom, J., & Bieging, J. 1993, ApJ, 402, 655.
[45] Welch, W. J., Dreher, J. W., Jackson, J. M., Terebey, S., & Vogel, S. N. 1987, Science, 238, 1550.
[46] Winkler, K.-H, & Newman, M. J. 1980, ApJ, 236, 201.
[47] Wootten, A. 1989, ApJ, 337, 858.
[48] Zhou, S., Evans, N. J., Butner, H. M., Kutner, M. L., Leung, C. M., & Mundy, L. G. 1990, ApJ, 363, 168.
[49] Zinnecker, H. 1989, in *Low Mass Star Formation and Pre-Main-Sequence Objects*, ed. B. Reipurth (Garching: ESO), p. 447.

Collapse and Fragmentation of Magnetized Cylindrical Clouds

By Kohji Tomisaka

Faculty of Education, Niigata University, 8050 Ikarashi-2, Niigata 950-21, Japan.
(tomisaka@ed.niigata-u.ac.jp.)

Gravitational collapse of a cylindrical cloud is studied by MHD simulations. An infinitely long cloud with axial B-field, initially in magnetohydrostatic equilibrium, is considered. Perturbations with small amplitudes are magnified by the gravitational instability. The most unstable mode predicted by a linear perturbation analysis is found to grow preferentially, even from numerical noise, with a growth rate in agreement with the linear theory. At the start of the collapse, the density-enhanced region has an elongated, i.e. prolate spheroidal shape. As the collapse proceeds, the high density fragment begins to contract mainly in the direction parallel to the B-field. Finally, in the case of a non-magnetized cloud, a spherical core is formed, whereas for a magnetized cloud a dense oblate disk forms. The radial size of this disk is proportional to the initial characteristic density scale height in the r-direction. During the collapse, a slowly contracting inner dense fragment is formed ($\lesssim 10\%$ of the mass) inside of an outer fast contracting disk-envelope. From consideration of the Jeans mass and magnetic critical mass, it is concluded that such fragments formed in a long cylindrical cloud cannot be supported against self-gravity and will eventually collapse.

1. Introduction

There are several pieces of observational evidence which suggest that the processes of star formation for low mass and high mass stars are different. This seems to be related to the difference between the collapse of supercritical clouds and subcritical clouds [8]. Supercritical clouds have no stable magnetohydrostatic configuration and hence undergo dynamical collapse. On the other hand, subcritical (or magnetohydrostatic) clouds evolve slowly due only to plasma drift and/or magnetic braking (although, if the cloud becomes supercritical through decrease in the magnetic flux at the center of the cloud or loss of angular momentum, it too begins to collapse dynamically). However, *this is true only when the static equilibrium of the cloud is stable.* If the time scale of some dynamical instabilities is shorter than the evolutionary time scale, the cloud may begin a dynamical contraction or fragmentation before it reaches the condition for the supercritical cloud. We meet this situation in cylindrical clouds.

Linear analyses of the gravitational instability show that a cylindrical isothermal cloud is unstable to a perturbation whose wavelength is longer than a critical length. The most unstable perturbation has a wavelength of $\lambda_{\max} \simeq 20 c_s/(4\pi G \rho_c)^{1/2}$ and an e-folding time of $\tau_{\max} \sim 3(4\pi G \rho_c)^{-1/2}$, where ρ_c represents the density on the axis of the cylinder [3]. This growth timescale is comparable to that for ambipolar diffusion [5].

2. Model and Numerical Method

The initial configuration assumed is a magnetohydrostatic equilibrium of a cylindrical isothermal cloud, which is uniform in the z-direction. Assuming that the magnetic pressure is proportional to the thermal pressure, i.e., $B_z^2/8\pi = \alpha c_s^2 \rho$, the distribution of the density is specified by two parameters: the center-to-surface density ratio, $F \equiv \rho(0)/\rho_s =$

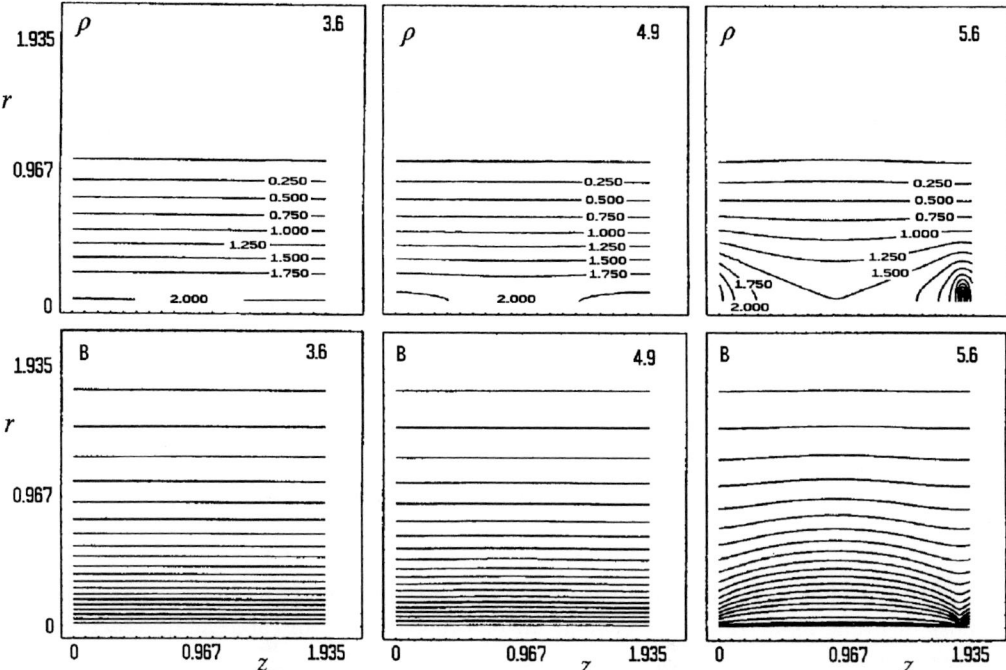

FIGURE 1. The time evolution of density (upper panel) and magnetic fields (lower panel) for model M. The z-axis and r-axis are oriented horizontally and vertically, respectively. The numbers attached to the density contour lines represent the logarithm of the density: $\log_{10}\rho$. The contour interval is taken to be constant $= 0.25$. Three snapshots at times (a) $t = 3.6$, (b) $t = 4.9$ and (c) $t = 5.6$ are shown.

$\rho(0)/(p_{\text{ext}}/c_s^2)$, and the above α. In this paper, results for models with $F = 100$ and $\alpha = 1$ (magnetic), $\alpha = 0$ (non-magnetic model) are shown.

Cylindrical symmetry is assumed and the basic equations are ideal MHD with an isothermal equation of state and the Poisson equation. The unsteady MHD equations are solved with the Monotonic Scheme and the Constrained Transport method, and the Poisson equation by the Conjugate Gradient Squared (CGS) method preconditioned by the modified incomplete LU decomposition (MILUCGS) (for original references, see [9]). On the upper ($z = l_z$) and lower ($z = 0$) boundary, a periodic boundary condition is applied, while on the outer r- boundary a fixed boundary condition is used.

3. Results

Model M: For a magnetized cloud, the parameters $\alpha = 1$ and $F = 100$ are chosen. The size of the calculated region is taken as $l_z = 1.93H$ (where $H = c_s/(4\pi G\rho_s)^{1/2}$), which coincides with the wavelength which has the maximum growth rate in the gravitational instability [4]. At $t = 0$, the density is perturbed as $\rho \longrightarrow \rho[1 - 10^{-2}\cos(2\pi z/l_z)]$.

At $t = 3.6$ (Figure 1a) there is no prominent fragmentation, while by $t = 4.9$ (Figure 1b), a high density region elongated in the z-direction appears. The shape of this high-density region is prolate, with major axis coinciding with the symmetry (z-) axis. This agrees with the result of linear analyses [3,4]. The wavelength of the most unstable perturbation as predicted from linear analyses is $\simeq 2H$. Also, the global shape of the high-density region in Figure 1b is the same as that expected from an eigenfunction of the linear analysis.

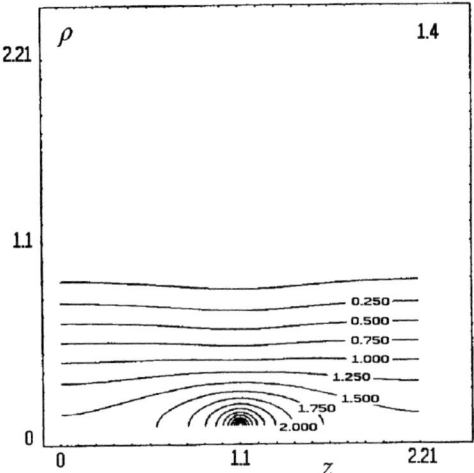

FIGURE 2. The time evolution of a non-magnetic cloud at $t = 1.4$.

From this figure it is apparent that perturbations with wavelengths shorter than that of the most unstable mode do not appear at all before the most unstable mode reaches the non-linear regime.

After the stage depicted in Figure 1b, collapse proceeds along the symmetry axis as well as in the radial direction. The final structure is a disk contracting towards the center, as seen in Figure 1c. Since gas falls most easily along the magnetic field lines, the plane of this disk is perpendicular to the symmetry axis of the cylinder. The magnetic field lines get squeezed by the effect of radial contraction near the disk (lower panel). Since this configuration is unstable against the Parker instability ([6] and references therein), the gas accumulates by flowing along the magnetic field lines.

The speed of contraction in the z-direction is about twice that in the radial direction. In the central region a high density core is formed onto which mass continues to accrete. Eventually, near the core, the infall speed exceeds the local isothermal sound speed. It is concluded that a typical product of self-gravitational instability in a long cylindrical cloud is a sequence of oblate spheroidal disks, separated by intervals of $\lambda_{\max}$.

Model N: This model corresponds to a non-magnetic isothermal cylinder with $F = 100$. The low density ($\simeq 10^{1.5}$) envelope has a prolate spheroidal shape (see Figure 2). In contrast, the high density central region appears to be almost spherical. This is due to the lack of the lateral magnetic restoring force that was present in Model M above.

Non-Magnetic Core: When the density ratio, F, becomes large, the hydrostatic configuration reaches Chandrasekhar's singular isothermal sphere solution [1] $\rho_{\mathrm{sing}} = c_s^2/2\pi G r^2$, or in a non-dimensional form: $\rho_{\mathrm{sing}}/\rho_s = 2/(r/H)^2$. The density in the core seen in Figure 2 is everywhere greater than ρ_{sing}, so the mass here is larger than the Jeans mass and the core should continue to contract.

Magnetic Core: The mass-to-flux ratio at the center of the cloud, $M/(\Phi/G^{1/2})$, plays a crucial role in dividing the clouds into super- and subcritical clouds. That is, considering a cloud whose mass is larger than the non-magnetic critical mass (i.e. $> M_{\mathrm{cr}} = 1.18 c_s^4 p_{\mathrm{ext}}^{-1/2} G^{-3/2}$), then the cloud has no equilibrium solutions when the mass-to-flux ratio is larger than $1/(2\pi)$, while for smaller values of this ratio the cloud has at least one hydrostatic solution.

Since the mass-to-flux ratio remains constant along the contraction unless the MHD

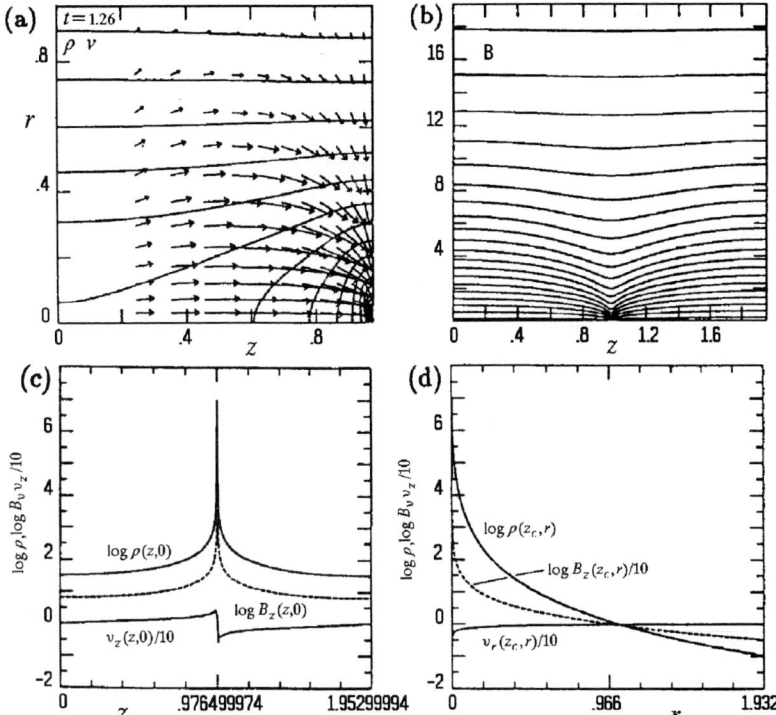

FIGURE 3. The structure of the fragment of model CM at time $t = 0.705$. Panel (a) shows the density and velocity fields and (b) shows the magnetic field lines. Panels (c) and (d) show cuts along the z-axis and r-axis respectively.

condition is broken, the mass-to-flux ratio *in the fragment* can be written using quantities from the initial stage as

$$\frac{M}{\Phi/G^{1/2}} \simeq \frac{\rho_c l_z G^{1/2}}{B_c} = \frac{\rho_c^{1/2} l_z G^{1/2}}{(4\pi\alpha)^{1/2} c_s}, \qquad (3.1)$$

where l_z represents the z-length of the numerical box and is taken to be equal to the wavelength of the most unstable perturbation, $\simeq 20 c_s/(4\pi G \rho_c)^{1/2}$. This ratio can be written approximately as $M/(\Phi/G^{1/2}) \simeq 1.59/\alpha^{1/2}$. Thus, *only when the magnetic field is extremely strong* i.e. $\alpha \gtrsim 100$, will the magnetized fragment be *subcritical*. The contraction of the core cannot be stopped by the magnetic field strength that is normally assumed, $\alpha \simeq 1$.

In conclusion, the cores formed in the fragments of the cylindrical cloud continue to contract until other effects not considered here work to support the fragments, for example, the equation of state of the gas changes from an isothermal to a harder one. It is shown that even if the initial cylindrical cloud has a magnetohydrostatic configuration, the contracting disk formed by the gravitational instability may be supercritical.

4. Further Evolution

To examine the structure of the contacting central core in more detail a larger calculation with much finer resolution has to be performed.

Model CM has the same physical parameters as model M but has about 4.28 times the spatial resolution. The evolution can be traced until the maximum density reaches about 10^6, that is, 10^4 times the initial central density. In Figure 3, the structure at time $t = 0.705$ for model CM is shown. As can be seen in Figures 3c and 3d, the speed of the infalling gas increases as it approaches the core and it is found that the maximum infall speed increases as the collapse proceeds. The maximum infall velocity in the z-direction at the time shown in Figure 3c agrees with that found from the asymptotic solution for spherical isothermal collapse by Larson [2] and Penston [7], that is, $v_z = 3.28 c_s$. A high density core with a thickness of about $10^{-2} H$ (i.e., $|z - z_c| \lesssim 10^{-2} H = 10^{-1} c_s/(4\pi G \rho_{c\ \text{init}})^{1/2}$) is formed, inside of which the gas velocity decreases as it approaches the center of the core, z_c. In contrast, in the r-direction, very smooth distributions of density, magnetic field strength and velocity are seen. This numerical result agrees qualitatively with a calculation by Hanawa, Nakamura & Nakano (private communication) using a different numerical scheme.

By the end of the calculations, the slowly contracting thin dense core seems to be detached from the outer inflow region (defined by $|z - z_c| \gtrsim 10^{-2}$). The fraction of the initial cloud mass which forms this static high density disk is approximately 6–10%, and if stars form from this material this would also be the ratio of mass in protostars to the mass of the parent cloud.

This work was supported in part by Grants-in-Aid from Ministry of Education, Science & Culture (04233211, 05217208).

REFERENCES

[1] Chandrasekhar, S. 1939, An Introduction to the Study of Stellar Structure (Chicago: University of Chicago Press) §22.
[2] Larson, R. B. 1969, MNRAS, 145, 271.
[3] Nagasawa, M. 1987, Prog. Theor. Phys. 77, 635.
[4] Nakamura, F., Hanawa, T, & Nakano, T. 1993, PASJ in press.
[5] Nakano, T. 1979, PASJ, 31, 697.
[6] Parker, E. N. 1979, Cosmical Magnetic Fields (Oxford: Oxford University Pr.), §13 & §22.
[7] Penston, M. V. 1969, MNRAS, 144, 457.
[8] Shu, F. H., Adams, F. C., & Lizano, S. 1987, ARAA, 25, 23.
[9] Tomisaka, K. 1993, submitted to ApJ.

Effect of Deceleration on the Gravitational Instability of Shocked Gas Layers

By T. Yoshida[1], A. Habe[2]

[1] Department of Physics, Ibaraki University, Mito 310, Japan.

[2] Department of Physics, Hokkaido University, Sapporo 060, Japan.

We have performed two-dimensional hydrodynamic calculations to investigate the effect of deceleration on the gravitational instability of isothermal shocked gas layers. We calculate the time evolution of the decelerating shocked layer with a corrugation-type perturbation. We find that there is the most unstable wave number. In this case, the density perturbation grows rapidly due to the decelerating shock instability t_{cr}, where this is the time when the self-gravity of the layer becomes equal to its deceleration. After t_{cr}, the density perturbation continues to grow rapidly due to the gravitational instability. In this way, the deceleration accelerates the gravitational instability. The gravitational collapse occurs in half of the free fall time t_{ff} of the preshock gas. On the other hand, if the density perturbation does not grow enough in t_{cr}, the gravitational collapse occurs in t_{ff} or does not occur. We discuss the conditions for fast gravitational collapse.

1. Introduction

The stability of shocked gas layers has been investigated with view to modeling sequential star formation in OB associations [2]. This problem is also important for the formation of high-redshift objects [3]. The shocked gas layer can be formed by comological H II regions in the intergalactic medium, whose radiation sources may be quasars, young galaxies, or Population III stars [5]. The shocked gas layer formed by the H II region decelerates as the shock propagates into the ambient medium, since the column density is increasing. Linear analyses of the self-gravitating shocked layer [1,4,6] show that the deceleration is important for the stability. If the perturbations grows to a nonlinear stage due to the deceleration shock instability, the layer may be distorted and the gravitational instability will be affected. We need the fully-nonlinear analysis to investigate the effect of the deceleration on the gravitational instability.

In this paper, we perform two-dimensional numerical hydrodynamic calculations of shocked layers to investigate the final fate of the self-gravitating and decelerating shocked layer and to explore the typical size and time scale of the instability.

2. Models and Methods

As the initial condition, we set the hydrostatic equilibrium solution of the plane-parallel isothermal gas layer in the decelerating frame, which is bounded on one side by the ram pressure P_s of an isothermal shock and on the other side by thermal pressure P_t[1]. We take the following parameters: preshock density $\rho_E = 4.8 \times 10^{-22}$ g cm^{-3}, sound velocity $c_s = 1\,\mathrm{km\,s^{-1}}$, and initial column density of the layer $\sigma(t_i) = 1.5 \times 10^{-2}$ g cm^{-2}. The initial position of the shock front in the rest frame is taken to be $Z_s(t_i) = \sigma(t_i)/\rho_E = 10$ pc. The initial ratio of the deceleration to the average self-gravity of the layer is 3.8. The thermal pressure at the trailing surface of the layer is assumed to vary with distance from UV sources as $P_t \propto Z_s^{-\epsilon}$. We assume that $\epsilon = 1.5$, which is appropriate for the expansion law of H II regons.

We simulate the evolution of an isothermal decelerating shocked layer with self-gravity by performing numerical hydrodynamic calculations in $x - z$ Cartesian coordinates. We use a frame comoving with the shock front. Instead of solving the overall gas flow, we solve only the gas flow in the shocked gas layer.

We initially displace the position of the shocked layer in the z-direction from the unperturbed position. We assume six different wavelengths for the initial perturbations. These dimensionless wave numbers κ are 0.3, 0.9, 1.5, 3.0, 6.0, and 12.0, where $\kappa = \lambda_{ff}/\lambda$, and $\lambda_{ff} = c_s/\sqrt{(G\rho_E)} = 5.7\,\mathrm{pc}(c_s/1\mathrm{km\ s}^{-1})(n_E/100\,\mathrm{cm}^{-3})^{-1/2}$. We assume that the amplitude a is 5.7×10^{-2} pc for all the initial perturbations. In all cases, we use 401 grid cells in z-direction and 65 grid cells in the x-direction. We use a flux-split Eulerian scheme with second-order accuracy in space and first-order in time. In the x-direction, we assume a periodic boundary condition. We solve the Poisson equation, using a FFT algorithm.

3. Numerical Results

We can separate the numerical results into two classes: gravitationally unstable ($\kappa = 0.3$, 0.9, and 1.5) cases and stable ones ($\kappa = 3.0$, 6.0, and 12.0).

3.1. Gravitationally unstable cases

This class is further separated into fast ($\kappa = 0.9$ and 1.5) and slow ($\kappa = 0.3$) collapse cases. We give the results of the fast collapse cases in Figure 1a, which shows the number density and the velocity in the layer for $\kappa = 1.5$ at $t = 0.6 t_{ff}$. In Figure 1, the shock front is at the upper side of the layer and the trailing surface is at the lower side of it. Figure 1b shows the results of the slow collapse case ($\kappa = 0.3$ at $t = 0.84 t_{ff}$).

These two cases differ in the location of the gravitationally collapsing part. Figure 1b shows that the gravitationally collapsing part is in the middle of the layer. On the other hand, for $\kappa = 1.5$ the gravitationally collapsing part goes ahead of the average position of the shock front, as shown in Figure 1a. This difference in the locations is because the deceleration of the layer for $\kappa = 1.5$ is larger than that for $\kappa = 0.3$ when the column density grows rapidly.

3.2. Gravitationally stable cases

In these cases, the tangential flow generates a density perturbation faster than in the gravitationally unstable cases. However, this density perturbation does not result in the gravitationally unstable and diminishes with time.

We clearly see these results in Figure 2. Figure 2a shows the time evolution of the maximum relative column density of models for $\kappa = 0.3$, 0.9, and 1.5. The time when the maximum relative column density reaches unity for $\kappa = 0.3$ is about twice that for $\kappa = 0.9$ and 1.5. Figure 2b shows the time evolution of the relative column density at the initial leading part for $\kappa = 3.0$, 6.0, and 12.0. Although the column density perturbation grows in the early stage, gravitational collapse does not occur.

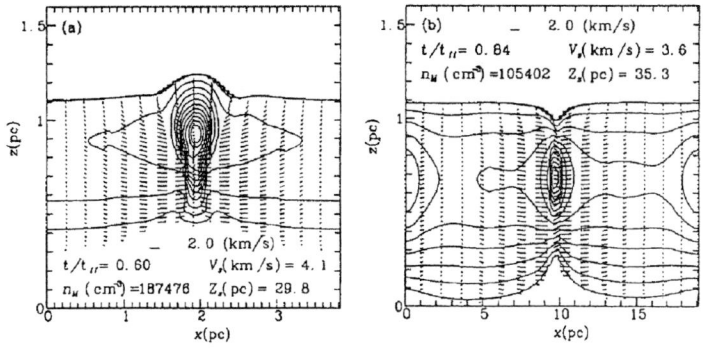

FIGURE 1. a) Number density contours and the velocity fields for $\kappa = 1.5$ at $t = 0.6t_{ff}$. b) $\kappa = 0.3$ at $t = 0.84t_{ff}$.

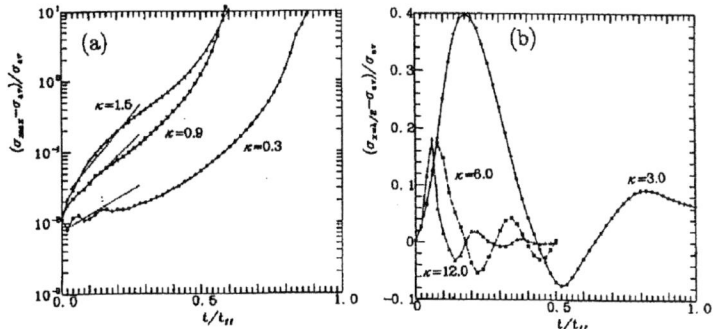

FIGURE 2. a) The time evolution of the maximum relative column density for $\kappa = 0.3, 0.9$ and 1.5. Each dashed line shows the growth rate, which is obtained from linear analysis, of the decelerating shock instability at the initial time t_i. b) The time evolution of the relative column density at the initial leading part $x = \lambda/2$ for $\kappa = 3.0, 6.0$ and 12.0.

4. Conditions for fast gravitational collapse and conclusion

Our numerical results show that there is a most unstable wave number and also that there is a maximum unstable wave number between $\kappa = 1.5$ and $\kappa = 3.0$. We discuss these wave numbers, comparing our numerical results with results of linear analyses [4].

In Figure 3a, the solid line shows the growth time τ_D of the decelerating shock instability at the initial time t_i with κ and the dashed line shows the growth time of the gravitational instability τ_G. Figure 3b shows the time scales τ_D and τ_G at $t = t_{cr}$, which is the time when the average self-gravity in the layer is equal to the deceleration. In our model, t_{cr} is $0.12t_{ff}$. Figure 3b shows that the maximum unstable wave number given by our numerical results is as large as κ of the layer at $t = t_{cr}$. We suggest that gravitational collapse occurs for $\kappa < \kappa_G(t = t_{cr})$.

Next we discuss the most unstable wave number and the condition of fast gravitational collapse. Our numerical results show that for $\kappa = 0.9$ and 1.5 the density perturbations grow well within $t = t_{cr}$ due to the deceleration of the layer and after t_{cr} the gravitational collapse occurs rapidly. On the other hand, the linear analyses show that for $\kappa = 0.9$ and 1.5 τ_D at t_i is shorter than $t_{cr} = 0.12t_{ff}$ (Figure 3a) and that τ_G at $t = t_{cr}$ is comparable to t_{cr} (Figure 3b). We conclude that when the wave number of the perturbation satisfies the

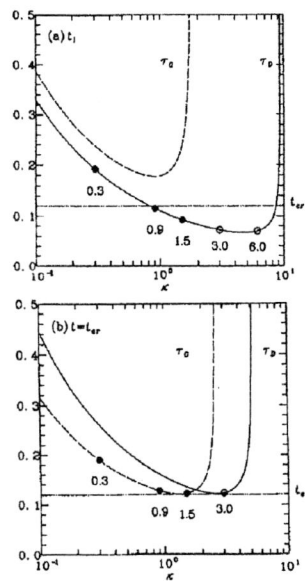

FIGURE 3. The growth time of two instabilities for the dimensionless wave number κ, which are obtained from linear analyses. The solid line shows the growth time of the decelerating shock instability τ_D. The dashed line shows the growth time of the gravitational instability τ_G. a) t_i, b) $t = t_{cr}$.

condition $\tau_D(t_i) \leq t_{cr}$ and $\tau_G(t = t_{cr}) \leq t_{cr}$, the decelerating shock instability accelerates the gravitational instability and the gravitational collapse occurs in half of t_{ff}.

REFERENCES

[1] Elmegreen, B. G. 1989, ApJ, 340, 786.
[2] Elmegreen, B. G. & Lada, C. 1977, ApJ, 214, 725.
[3] Madau P. & Meiksin, A. 1991, ApJ, 374, 6.
[4] Nishi, R. 1993, Prog.Theor.Phys., 87, 347.
[5] Shapiro, P. R. & Giroux, M.L. 1987, ApJ, 321, L107.
[6] Vishniac, E. T. 1983, ApJ, 274, 152.

The Formation of a "Protocluster"

By J. A. Turner[1], A. S. Bhattal[1], S. J. Chapman[1], M. J. Disney[1], H. Pongracic[1,2], A. P. Whitworth[1]

[1] Department of Physics and Astronomy, University of Wales, College of Cardiff, Cardiff, PO Box 913, CF2 3TH, UK

[2] Research Center for Theoretical Astrophysics, School of Physics, University of Sydney, N.S.W. 2006, Australia

Observations of the Orion B cloud by Lada et al. [4] in the near-infrared indicated that $\sim 96\%$ of the sources detected were clustered into 4 distinct regions. It appears, therefore, that star formation occurs predominately in a clustered mode. We present a simulation of the collision of two massive elements of a Giant Molecular Cloud (GMC) in which the outcome is a region of high stellar density containing ~ 20 protostellar discs. We speculate that this region has similar properties to one of the clusters of infrared sources observed by Lada et al. [4].

1. Introduction

Detailed observations of the Orion B cloud using CS reveals 42 cores within the GMC [5]. These cores are distributed in two regions to the north and south of the cloud. Most of the cores have masses of less than 100 $M_\odot$, with only 5 having a mass of 200 $M_\odot$ or greater. The same area was observed in the near-infrared at 2.2 μm [4]. As a result ~ 700 infrared sources were found; after correcting for background stars it was estimated that $\sim 96\%$ are clustered into 4 distinct regions. These clusters are associated with known star formation regions, NGC 2023, 2024, 2068 and 2071, and with 4 of the 5 densest cores found in the CS survey. The smallest cluster is associated with NGC 2023; it is 0.6 pc in size and contains 21 sources. The largest cluster is associated with NGC 2024; it is 2.0 pc in size, and contains 309 sources.

The large changes in density that occur when regions of interstellar gas ($n \sim 10^2$ cm^{-3}) are converted into stars ($n \sim 10^{24}$ cm^{-3}) requires a numerical code with a large dynamic range. We use a self-gravitating (tree-code [3]) hydrodynamic (SPH with variable resolution [6]) code.

We present here a numerical simulation of a cloud-cloud collision and the subsequent formation of a protocluster. The plots represent a grey-scale of the logarithm of the column density; black represents the highest column density, and white represents the lowest column density. Each figure is also over-layed with contours of the logarithm of the column density. Some figures show only part of the computational domain. The numbers in parenthesis after each reference to a figure indicate the zoom factor of the plot.

All our simulations were performed on the Cray Y-MP at the Rutherford Appleton Laboratory, and analysed using X-based packages developed at Cardiff.

2. Initial Conditions

The simulation starts with two identical clouds each having mass 750 $M_\odot$ and radius 6.5 pc. Each cloud represents a small region of a GMC. The clouds are in detailed hydrostatic balance, have sub-Jeans mass, and are surrounded by a hot rarefied intercloud medium. A Mach-5 collision occurs at an impact parameter of 5 pc — Figure 1a(1). When the gas is compressed, it cools radiatively. The gas has a barotropic equation of state which corresponds to its having isothermal sound speed $a_o \sim 0.6$ km s^{-1} for $n_o < 300$ cm^{-3} (i.e.

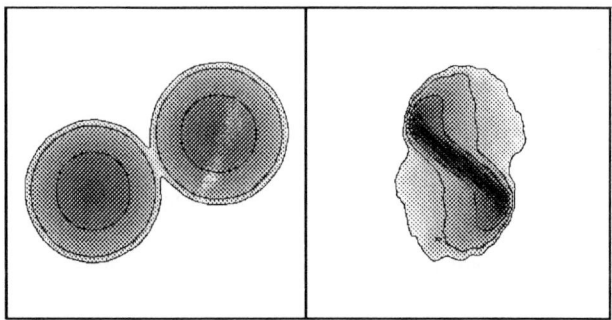

FIGURE 1. (a). The initial conditions ($Z_f = 1$, where Z_f is the zoom factor relative to the initial conditions): each cloud has mass 750 $M_\odot$ and radius 6.5 pc. They collide at Mach-5 and with a 5 pc offset. (b). The clouds during ($t = 6.38 \times 10^6$ years) the collision ($Z_f = 1$) — a compressed layer has formed.

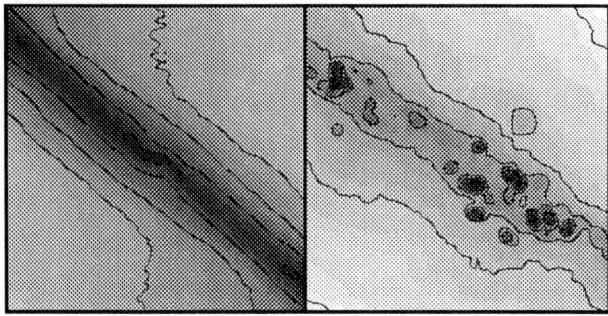

FIGURE 2. (a). Close-up of the compressed layer ($Z_f = 8$) at the same time as Figure 1b. (b). A further close-up of the central part of this compressed layer ($Z_f = 35$). The layer has fragmented into a number of protostellar discs. There is a high interaction rate.

in the unshocked clouds) and $a_s \sim 0.2$ km s^{-1} for $n_s > 10^4$ cm^{-3} (*i.e.* in the shocked layer); in between these densities it cools approximately according to $a \propto n^{-1/3}$. We use $\sim 10^5$ particles.

3. Results

Shortly after the start of the simulation a shock develops at the collision interface. This shock proceeds through the cloud at a steady rate. As a result a compressed layer, bounded by two shocks, is formed — Figures 1b(1) and 2a(8). Fragmentation occurs whilst confinement of the layer is dominated by ram pressure (*i.e.* well before confinement of the layer becomes dominated by self-gravity). This results in fragments whose separations are greater than the thickness of the layer and whose masses are large; typically ~ 10 $M_\odot$ —Figure 2b(35). See Chapman *et al.* "The formation of OB subgroups" (this volume) for a discussion of why the layer fragments while it is still confined by ram pressure, and why the fragment masses are large.

Subsequent fragmentation produces ~ 20 protostellar discs— Figure 3(35). The proto-

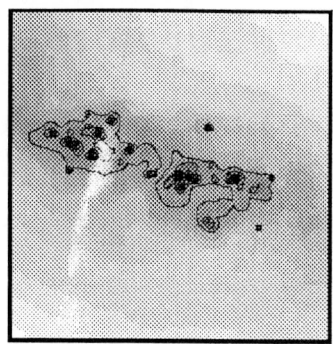

FIGURE 3. The same region as Figure 2b is displayed, but we are now at a later time ($t = 6.74 \times 10^6$ years; $Z_f = 35$). There are ~ 20 protostellar discs within this region. There is still a high interaction rate at this stage, which is as far as the simulation was taken due to its computational expense, so the final outcome is unknown.

stellar discs have masses in the range $4 - 15$ $M_\odot$, and radii of $0.001 - 0.010$ pc. The size of the region shown in Figure 3 is 0.8×0.8 pc^2.

4. Discussion

The collapse of each protostellar disc in the simulation is halted by rotation. The disc radii are comparable with the values inferred from observations of discs around young stellar objects in star formation regions such as Taurus-Auriga and Orion [1].

With its high protostellar density and similar size, the results of this simulation are similar to the cluster of infrared sources associated with NGC 2023 as detected by Lada et al. [4].

When the simulation was stopped the effect of the accretion of material onto the protostellar discs was only just becoming important. Accretion of material can induce rotational fragmentation and thereby produce more protostellar discs (see [2] and Chapman et al. "Two Formation Mechanisms for Binary (Multiple) Protostars in Shocked Interstellar Gas Layers" - this volume). However, there may also be a number of mergers or ejections which reduce the eventual numbers of stars that populate the cluster.

We are unable to follow the evolution of this "protocluster" any further due to limitations of computer time. We therefore cannot predict if it will survive in something like its present form, whether mergers and ejections will seriously deplete the eventual number of stars present, or whether the eventual number of stars within the cluster will grow via the rotational fragmentation of the protostellar discs shown in Figure 3.

REFERENCES

[1] Beckwith, S. V. W., Sargent, A. I., Chini, R. S. & Güsten, R., 1990, A J, 99, 924.
[2] Chapman, S. J., Pongracic, H., Disney, M. J., Nelson, A. H., Turner, J. A. & Whitworth, A. P., 1992, Nature, 359, 207.
[3] Hernquist, L., 1987, Ap J Suppl, 64, 715.
[4] Lada, E. A., DePoy, D. L., Evans, II, N. J. & Gatley, I, 1991, Ap J, 371, 171.
[5] Lada, E. A., 1990. PhD thesis, University of Texas at Austin.
[6] Monaghan, J. J., 1992, ARAA, 30, 543.

The Formation of Hierarchical Binary Systems in Turbulent GMCs

By S. J. Chapman[1], A. S. Bhattal[1], M. J. Disney[1], H. Pongracic[1,2], J. A. Turner[1], A. P. Whitworth[1]

[1] Department of Physics and Astronomy, University of Wales, College of Cardiff, Cardiff, PO Box 913, CF2 3TH, UK

[2] Research Center for Theoretical Astrophysics, School of Physics, University of Sydney, N.S.W. 2006, Australia

We investigate the effects of internal motions within a GMC by modelling a region of a GMC (henceforth a subcloud) as being made up of 20 randomly moving elements (henceforth sub-subclouds). Inelastic collisions between these sub-subclouds results in a high density core at the position where a few of the sub-subclouds collided. Within this core, at first a hierarchical quadruple system and then, due to a merger, a hierarchical triple system is formed.

1. Introduction

Observations of both main sequence and pre-main sequence stars suggest that most stars are binary or multiple, *e.g.* [1] and Bodenheimer, this volume. Therefore any star formation theory must account for the occurence of multiple systems. Massive stars form predominantly in Giant Molecular Clouds (GMCs) such as those seen in Orion. These GMCs have been resolved into a hierarchy of clouds within clouds on scales ranging down to ~ 0.01 pc. This clumpiness is thought to have been generated by turbulent motions within the GMC.

We present here a model of a turbulent GMC. We follow its subsequent evolution, leading to the formation of a stable triple system, *i.e.* three protostellar discs in orbit about each other (see [2]). The plots represent a grey-scale of the logarithm of the column density; black represents the highest column density, and white represents the lowest column density. Each figure is also over-layed with contours of the logarithm of the column density. Some figures show only part of the computational domain. The numbers in parenthesis after each reference to a figure indicate the zoom factor of the plot.

The large changes in density that occur when regions of interstellar gas ($n \sim 10^2$ cm^{-3}) are converted into stars ($n \sim 10^{24}$ cm^{-3}) requires a numerical code with a large dynamic range. We use a self gravitating code (tree code [3]) hydrodynamic (SPH with variable resolution [4]) code.

All our simulations were performed on the Cray Y-MP at the Rutherford Appleton Laboratory, and analysed using X-based packages developed at Cardiff.

2. Initial Conditions and Input Physics

The GMC is envisaged as an ensemble of 20 subclouds each having a radius of 1 pc and a mass of 100 M$_\odot$. We model just one such subcloud, by treating it in turn as an

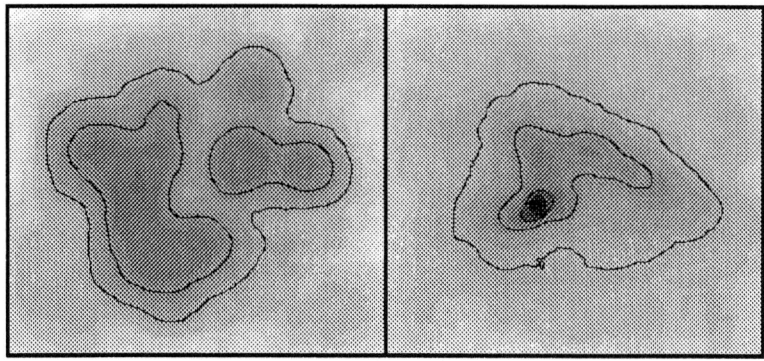

FIGURE 1. See text for details.

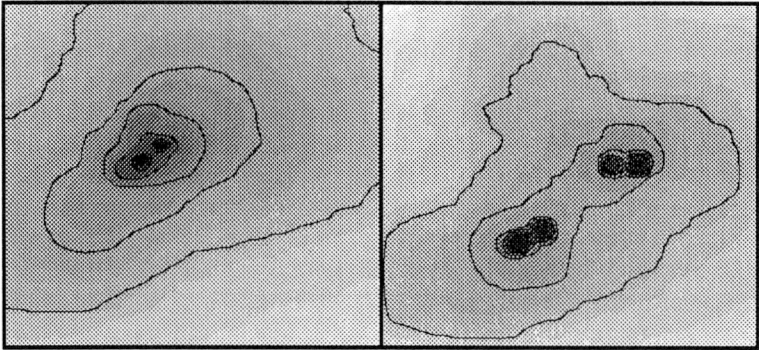

FIGURE 2. See text for details.

ensemble of sub-subclouds, each having a mass 4 $M_\odot$, a radius 0.2 pc, mean separation 0.6 pc, and moving randomly within the subcloud. The whole ensemble of sub-subclouds is approximately virialized. The regions between the sub-subclouds is occupied by 20 $M_\odot$ of a hot rarefied gas. The initial conditions are shown in Figure 1a(1). We use $\sim$ 4000 particles.

The gas has a barotropic equation of state which corresponds to its having isothermal sound speed $a_o \sim 0.6$ km s^{-1} for $n_o < 300$ cm^{-3} (*i.e.* in the unshocked clouds) and $a_s \sim 0.2$ km s^{-1} for $n_s > 10^4$ cm^{-3} (*i.e.* in the shocked layer); in between these densities it cools approximately according to $a \propto n^{-1/3}$.

3. Results

Collisions between sub-subclouds are highly inelastic, and their bulk kinetic energy is quickly dissipated — the dense core that results is shown in Figure 1b(1). In Figures 2a(4) and 2b(16) we zoom in at the point when the first condensations form. As can be seen, there are four in total. On closer inspection all are rotationally supported discs with their rotation axes orientated in different directions. They are orbiting one another in a hierarchical manner (*i.e.* two relatively close binary systems in orbit about each other).

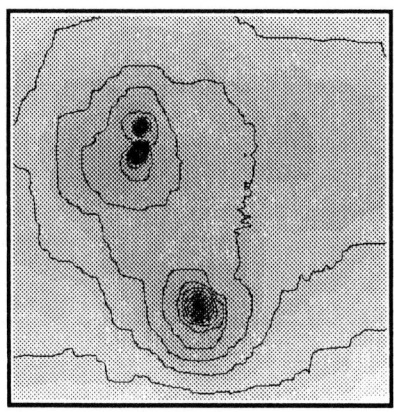

FIGURE 3. See text for details.

After a number of orbits one of the pairs merges to form a single disc surrounded by a spherical halo of accreting material. We display in Figure 3(16) the final proto-triple. The surviving close binary pair has completed six orbits, and appears to be stable. Over 80% of the mass in Figure 3 is in the three protostellar discs, and so the orbits are unlikely to decay. The third component has also completed an orbit which is roughly circular. The masses of the close pair are 11.2 and 4.9 $M_\odot$, and they have diameters 600 and 150 AU. The third protostellar disc has a mass 8.8 $M_\odot$, and diameter 400 AU with its spherical halo extending out to 800 AU. The close pair have a separation of 1200 AU, while the third protostellar disc has a separation of 7500 AU from the close pair.

REFERENCES

[1] Bodenheimer, P., 1992. In: Evolutionary processes in interacting binary stars, p. 9, eds Kondo, Y., Sistero, R.F. & Polidan, R.S., (Dordrecht: Kluwer Academic Publishers).
[2] Chapman, S.J., Pongracic, H., Disney, M.J., Nelson, A.H., Turner, J.A. & Whitworth, A.P., 1992, Nature, 359, 207.
[3] Hernquist, L., 1987, ApJS, 64, 715.
[4] Monaghan, J.J., 1992, ARAA, 30, 543.

Two Formation Mechanisms for Binary (and Multiple) Protostars in Shocked Interstellar Gas Layers

By S. J. Chapman[1], A. S. Bhattal[1], M. J. Disney[1], H. Pongracic[1,2], J. A. Turner[1], A. P. Whitworth[1]

[1] Department of Physics and Astronomy, University of Wales, College of Cardiff, Cardiff, PO Box 913, CF2 3TH, UK

[2] Research Center for Theoretical Astrophysics, School of Physics, University of Sydney, N.S.W. 2006, Australia

We present numerical simulations illustrating two mechanisms of binary formation. In the first mechanism, protostellar discs form independently and then capture one another to form binary and multiple systems. In the second mechanism, a single protostellar disc becomes rotationally unstable due to the accretion of matter with high specific angular momentum, and breaks up into two or more pieces. Both mechanisms occur for a wide range of initial conditions.

1. Introduction

Observations of both main sequence [1] and pre-main sequence [6,3] stars suggest that most are binary or multiple (see the article by Bodenheimer in this volume).

An attractive model of single low mass, star formation is that due to Shu, Adams & Lizano [7]. These authors consider the evolution of a magnetically supported cloud. Neutral particles within the cloud are not directly supported by the magnetic field and gradually percolate toward the center. Once enough mass has accumulated at the center of the cloud the magnetic support felt by the ions cannot prevent overall collapse. This theory accounts for many of the properties of single stars in TT associations. However, on its own it cannot explain the observed multiplicity of main sequence and pre-main sequence stars.

Pringle [5] suggests that the collapse of a molecular cloud to form protostellar discs may be caused by an external impulse. He argues that the ideal case is one in which the impulse is sufficient enough to provoke collapse without leading to the complete destruction of the cloud. A low-Mach collision of molecular clouds, along with efficient cooling decreases the local Jeans mass and length and thereby promotes fragmentation.

We use a self-gravitating (tree-code [2]) hydrodynamic (SPH with variable resolution [4]) code to test the Pringle's [5] hypothesis. We use clouds that are in detailed hydrostatic balance, have sub-Jeans mass, and are surrounded by a hot rarefied intercloud medium.

We present here some numerical simulations of two identical clouds colliding off-axis at Mach-5. Each cloud has mass 750 $M_\odot$ and diameter 13 pc. The gas has a barotropic equation of state which corresponds to its having isothermal sound speed $a_o \sim 0.6$ km s^{-1} for $n_o < 300$ cm^{-3} (i.e. in the unshocked clouds) and $a_s \sim 0.2$ km s^{-1} for $n_s > 10^4$ cm^{-3}

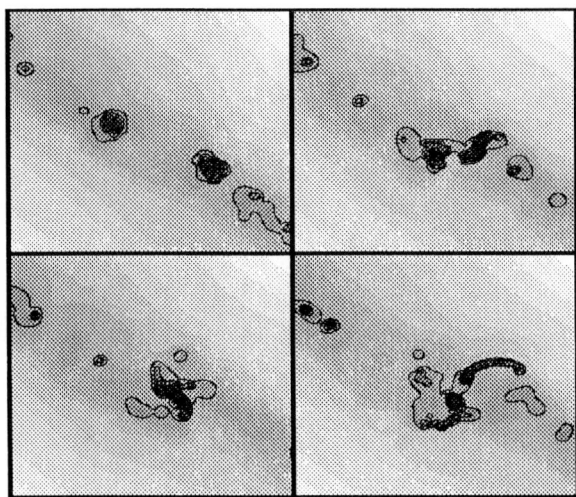

FIGURE 1. See text for details.

(*i.e.* in the shocked layer); in between these densities it cools approximately according to $a \propto n^{-1/3}$. We use $\sim 10^5$ particles.

The plots represent a grey-scale of the logarithm of the column density; black represents the highest column density, and white represents the lowest column density. Each figure is also over-layed with contours of the logarithm of the column density. All the figures are zoomed in by a factor of 35 with respect to the whole computational domain.

We have found that two instabilities play fundamental rôles in the formation of multiple protostellar discs.

2. Prompt Initial Fragmentation and Capture

The first instability is thermal, and causes the fragmentation of the dense layer formed during the collision. There is a tendency for the layer to break up into filaments (see the article by Chapman *et al.* "The Formation of OB Subgroups" in this volume) and then groups of protostellar discs condense out of these filaments. These discs have typical masses of ~ 10 M$_\odot$, and radii of ~ 0.001 pc. Because the protostellar discs fall toward one another along the filaments, and because the discs can be quite extended, there is a high capture rate.†

Figure 1a (upper left) shows the layer side-on. The layer has broken up into a number of fragments. The two largest are rotationally supported discs, and are travelling toward one another. Figure 1c (upper right) shows them just before periastron. Figure 1b (lower left) shows them at periastron. Most of their transverse kinetic energy is dissipated via a disc-disc interaction. Figure 1d (lower right) shows them in orbit around one another. Figure 2 shows the captured pair at a much later time (they are the two largest of discs shown). At this point they have completed two orbits. However, their subsequent evolution is about

---

† There is also a relatively high merger rate, but a single protostellar disc is not necessarily the outcome of such an event. Most merged protostellar discs break up via the second mechanism — see later.

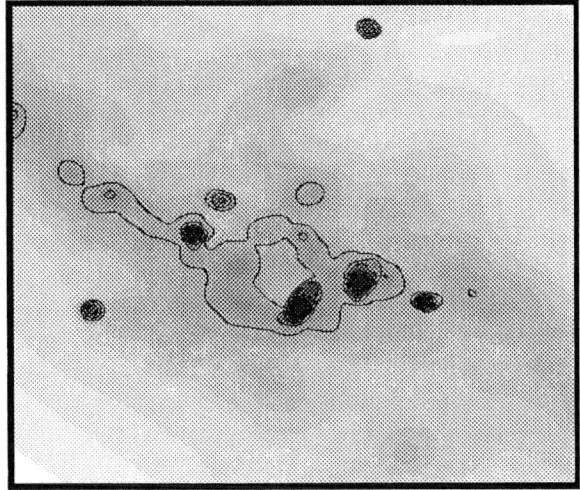

Figure 2. See text for details.

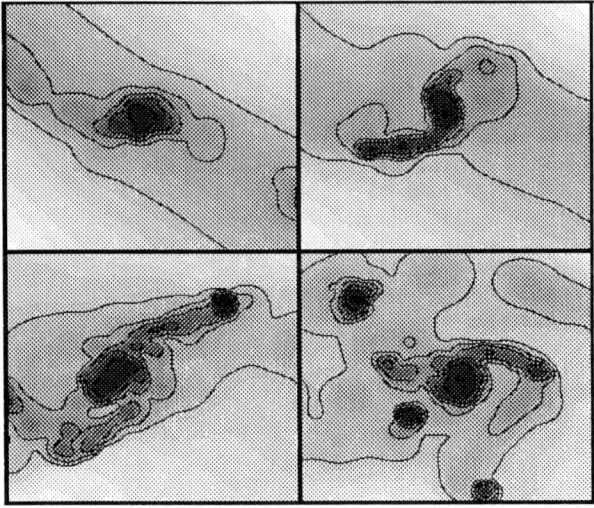

Figure 3. See text for details.

to get complicated as other, less massive, protostellar discs fall toward the binaries. We have yet to run this simulation further to see what effect these interlopers will have.

3. Rotational Fragmentation Induced by Spin-Up

The second instability is rotational and causes fragmentation of the rapidly spinning protostellar discs formed by Prompt Initial Fragmentation. This instability is normally induced by the accretion of material with high specific angular momentum.

If the accretion rate is low (if the clouds have a low initial mass, or if the collision occurs at a relatively small impact parameter) the original protostellar disc develops spiral arms one of which may detach to form a small companion. The companion then grows at the expense of the original protostellar disc, by intercepting the accreting material. The incoming angular momentum is converted into orbital angular momentum and the discs gradually spiral away from one another to a more stable orbit.

If the accretion rate is high (if the clouds have a relatively high mass, or if the collision occurs at a relatively high impact parameter), or if there is a sudden increase in the spin angular momentum of a protostellar disc (after a merger, for instance) the original protostellar disc simply cleaves into two comparable discs. Both discs intercept the accreting material at roughly the same rate, converting the incoming angular momentum into orbital angular momentum so that they also spiral away from one another.

The simulation illustrated in Figures 3a to 3d is one example of the spiral form of rotational fragmentation. Figure 3a shows a single rotating protostellar disc. Figure 3b shows the same disc after a further revolution. A single spiral arm has formed. The free-fall time of this arm is equal to about half a revolution of the central protostellar disc, and the arm collapses to form a companion — Figure 3c.

In a number of simulations a second, or even a third generation of rotational fragmentation can occur. This is particularly evident in the presence of prolonged accretion flows, or if there is a relatively high mass flow onto a disc. This results in the formation of multiple systems — Figure 3d.

REFERENCES

[1] Duquennoy, A. & Mayor, M., 1991, A&A, 248, 485.
[2] Hernquist, L., 1987, ApJS, 64, 715.
[3] Leinert, Ch., Zinnecker, H., Weitzel, N., Christou, J. Ridgeway, S.T., Jameson, R., Hass, M. & Lenzen, R., 1993, to appear in A&A
[4] Monaghan, J.J., 1992, ARAA, 30, 543.
[5] Pringle, J.E., 1989, MNRAS, 239, 361.
[6] Reipurth, B. & Zinnecker, H., 1993, to appear in A&A
[7] Shu, F.H, Adams, F.C. & Lizano, S., 1987, ARAA, 25, 23.

Star Formation via Interaction of Shocks with Molecular Clouds

By H. Pongracic[1,2], J. Chapman [1], M. J. Disney[1], A. H. Nelsoni[1], J. A. Turner[1], A. P. Whitworth[1]

[1] Department of Physics and Astronomy, University of Wales, College of Cardiff, Cardiff, PO Box 913, CF2 3TH, UK

[2] Research Centre for Theoretical Astrophysics, School of Physics, University of Sydney, N.S.W., 2006, Australia

One mechanism whereby regions of molecular clouds can become locally unstable to collapse and form stars is shock compression. Here we investigate the compression arising from a shock (from a supernova remnant for example) overrunning a molecular cloud and determine under what conditions it becomes locally unstable to gravitational collapse. Preliminary results from 3-D hydrodynamic simulations (using Smoothed Particle Hydrodynamics and Tree Code Gravity) are presented. It is found that when the molecular clouds are Jeans stable, radiative cooling is required to instigate collapse. Even with high Mach number collisions clouds which remain isothermal are disrupted by the passage of a shock but there are no regions which are unstable to collapse.

1. Introduction

Stars are observed to form in molecular clouds but the mechanism of formation is not well understood. On the largest scales molecular clouds are supported by supersonic motion or magnetic pressure or both. A mechanism is required whereby regions of molecular clouds can become unstable to collapse and, under appropriate conditions, contract to form protostars. One such mechanism is shock compression and in the interstellar medium this can occur when two or more molecular clouds or subclouds collide with each other or when a shock (from a supernova or expanding HII region) overruns a molecular cloud. The problem of cloud collisions resulting in star formation has been considered by several authors [4,5,6] and references therein. In this paper we consider a shock interacting with a cloud and investigate under what conditions star formation can occur.

The interaction of shocks with molecular clouds has important consequences for the evolution of dynamics of the interstellar medium in addition to being a possible trigger for star formation. Several simulations concerning this interaction have been made to date, but all but one of these simulations have been two-dimensional hydrodynamical calculations and none have included the effects of gravity. The recent 3-D simulation [9] found that the stable vortex rings formed in 2-D simulations were unstable in three dimensions and this led to the cloud fragmenting in all directions.

2. Shock-Cloud Interaction

To simulate the formation of protostars requires a code which can handle a large change in density and spatial scale. To achieve this we use a 3-D code which combines Smoothed

Particle Hydrodynamics (SPH) and Tree Code Gravity (TCG) [6]. This code has been tested on a variety of problems and found to work well. Because of its inherently Lagrangian nature and multifluid capability, as well as not having any geometrical constraints, SPH is a very suitable method for this problem. By using a spatially and temporally varying resolution length, changes in density of up to 8 orders of magnitude above the initial density can be followed.

The resolution in 2-D simulations is ever improving. For example, Bedogni and Woodward [1] used 38,400 grid points in their Piecewise Parabolic Method, and Tenorio-Tagle & Rozyczka [10] used 40,000 grid points in their second-order finite difference scheme, much improved over the some of the original calculations, e.g. [8, 11]. Klein, McKee & Collela [3] used the equivalent of 7×10^6 zones in their adaptive mesh refinement scheme. However, the 2-D simulations may not necessarily be an accurate representation of 3-D phenomena. This was borne out when Stone & Norman [9] made the first 3-D simulations of the shock-cloud interaction and found that the vortex rings which were stable in two dimensions, fragmented in three-dimensions. Although the 3-D simulations can't as yet match the resolution found in 2-D calculations, there is still much that can be learned from lower-resolution 3-D simulations.

None of the previous simulations have included the effects of gravity; all have concentrated on the hydrodynamics. Most have tried to explain the emission observed on the edges of supernovae remnants [7] and/or the appearance of the interstellar medium. Here we specifically include the effects of gravity, as we are trying to determine under which conditions star formation will occur.

The time scales relevant to this problem are the cloud crushing time t_{cc} and the free-fall time t_{ff}. The cloud crushing time is defined to be $t_{cc} = R\chi/v_b$ where R is the radius of the cloud, v_b is the velocity of the supernova blast wave and χ is the density contrast between the cloud and the ambient medium. Rayleigh-Taylor and Kelvin-Helmholtz intstabilities grow on the cloud crushing timescale [3].

The calculations begin with a Jeans-stable self-gravitating molecular cloud, modelled as an isothermal sphere of radius 1 pc and mass 75 $M_\odot$, in pressure balance with a warmer, more diffuse intercloud medium. The cloud is composed of molecular hydrogen with number density < 300 cm^{-3} and a sound speed of ~ 0.6 km s^{-1}. These parameters are typical of those observed in molecular clouds. Provided the molecular cloud is far enough away from the explosion centre the curvature of the shock front can be neglected and we can consider a plane parallel shock impinging on the cloud. The ratio of cloud to intercloud density in these calculations is taken to be 4. In these preliminary calculations we use 36082 particles.

For the cloud considered above and Mach numbers of 4 – 10, $t_{cc} \sim 0.8 \times 10^6 - 0.3 \times 10^6$ years which is comparable with the free-fall time of $\sim 0.6 \times 10^6$ years indicating that gravity is likely to be important in this calculation.

Cooling also plays an important role. Previous calculations of cloud collisions [2,6] have shown that fragmentation and collapse to high density condensations (which can be identified with protostars) occur more readily if molecular cooling is incorporated in the equation of state. The earlier shock-cloud interaction calculations [8,10,11] included explicit cooling, whereas the later ones [1,3,9] used an adiabatic equation of state. In the calculations presented here, we use a piecewise polytropic equation of state (described in [6]), which mimics the heating and cooling processes expected to be important in the interstellar medium. This equation of state is valid in the optically thin limit only.

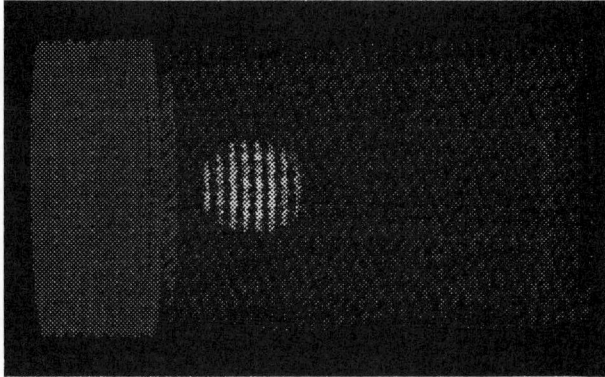

FIGURE 1. 3-D view of particle positions at $t = 0$ on a greyscale according to the density. The material behind the shock on the left has a higher density (shown in white). A stable isothermal cloud is embedded in a lower-density intercloud medium. The scale of the image is 9pc × 6pc × 6pc.

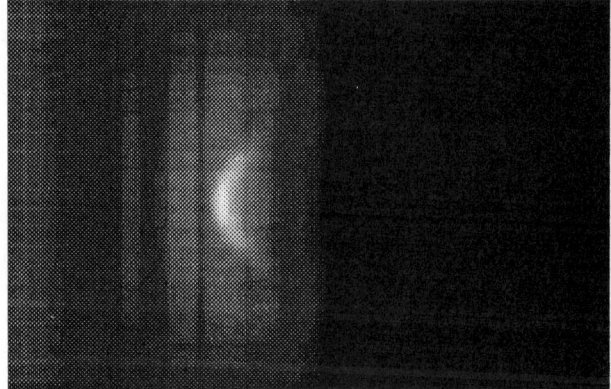

FIGURE 2. A 2-D greyscale representation of the column density in the computational domain. The bow shock is seen at time $t = 1.1$ Myrs. The supernova shock has overtaken the cloud and is still travelling through the lower density intercould material. White represents the maximum density, the minimum is black. Image size: 9pc × 6pc.

3. Results and Discussion

We present here the result of a simulation whose endpoint is a condensation which may evolve into a protostar and discuss the particular conditions under which this occurs.

Figure 1 shows a stationary stable isothermal sphere embedded in an ambient intercloud medium with a Mach 4 shock impinging on it from the left. The supernova shock drives a shock into the cloud material and a bow shock is formed to the left of the cloud as shown in figure 2. The transmitted shock compresses the cloud material and at time t=1.6 $\times 10^6$ years gravity becomes important and a region of the cloud collapses (figure 3). The supernova shock continues past the cloud and into the ambient medium. The condensation contains only cloud material indicating that no intercloud material has been entrained.

The condensation, which is disk-shaped with a diameter of $\sim 10^4$ AU and thickness $\sim 3 \times 10^3$ AU, has a density of $\sim 10^7$ times the density of the material from which it

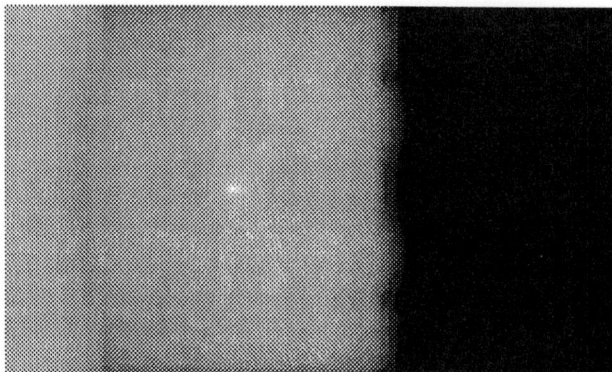

FIGURE 3. Logarithmic greyscale image of the column density at t=1.6 Myrs where a portion of the cloud has collapsed to $\sim 10^7$ times the initial density. The original supernova shock is still travelling through the intercloud material on the right. Maximum density (white) is $2 \times 10^7 \rho_{\text{initial}}$. (a) Image size: 9pc × 6pc. (b) Close-up view of the protostellar object showing an extended disk. Image size: 0.3pc × 0.3pc.

formed. It contains approx 11 $M_\odot$ of material when it forms and continues to accrete material at a rate of $\sim 10^{-6} M_\odot/\text{yr}$. At this stage the disk-like condensation is no longer collapsing. It is not clear whether the object formed is rotating but it is moving to the right as a result of the angular momentum imparted to it. The calculation was stopped when the resolution became compromised and the optically thick regime could no longer be neglected.

The time of formation of the object is ~ 1.6 Myrs (=0.8 t_{ff}) which is twice the cloud crushing time, confirming that gravity acts on a timescale comparable to the development of Rayleigh-Taylor and Kelvin-Helmholtz instabiblities. In previous simulations [3,9] it was shown that these instabilities develop on timescales of 4–10 t_{cc} so that including gravity significantly alters the outcome of the simulation.

We note that cooling is required to instigate a collapse, indicating that some sort of gravo-thermal instability occurs. A simulation run with a Mach 10 shock and isothermal equation of state produced two condensations (with $\rho_{\max} \sim 200 \, \rho_{\text{initial}}$) but these condensations did not collapse any further and the cloud was disrupted by the passage of the shock. It was found that even shocks at modest velocities (e.g. Mach 1.5) will produce a region which collapses if a cooling equation of state is used. However, a Mach 4 collision with cooling, but without gravity produced no potential protostellar object.

In these preliminary simulations only one potential protostellar object is formed when both cooling and gravity are included in the calculations. This may result from the idealised initial conditions that were used. Observations show that many stars occur in binary or multiple systems and that molecular clouds are clumpy and have structure on many scales. Perhaps a shock overtaking a clumpy cloud could result in the formation of multiple objects. This mechanism is currently under investigation.

4. Conclusions

The preliminary simulations presented here indicate that protostellar disk-like structures with dimensions similar to those observed in Young Stellar Objects can form when a shock (from a supernova or expanding HII region) overruns a molecular cloud, but that both molecular cooling and gravity are required for this to occur. The collapse occurs on a similar timescale to the development of Rayleigh-Taylor and Kelvin-Helmholtz instabilities perhaps radically altering the morphology which has been observed in previous simulations.

REFERENCES

[1] Bedogni R. & Woodward P. R. 1990, AA, 231, 481.
[2] Chapman S. J., Pongracic H., Disney M. J., Nelson A. H., Turner J. A. & Whitworth A. P. 1992, Nature, 359, 207.
[3] Klein R. I., McKee C. F. & Colella P. 1989, 10th Santa Cruz Summer Workshop, Springer Verlag, 696.
[4] Lattanzio J. C., Monaghan J. J., Pongracic H., & Schwarz M. P. 1986, MNRAS, 215, 125.
[5] Nagasawa M. $ Miyama S. M. 1987, ProgTheorPhys, 78, 1250.
[6] Pongracic H., Chapman S. J., Davies J. R., Disney M. J., Nelson A. H., & Whitworth A. P. 1992, MNRAS, 256, 291.
[7] Serabyn E., Lacy J. H. & Achtermann J. M. 1992, ApJ, 395, 166.
[8] Sgro A. G. 1975, ApJ, 197, 621.
[9] Stone J. M. & Norman M. L. 1992, ApJ, 390, L17.
[10] Tenorio-Tagle G. & Rozyczka M. 1986, AA, 155, 128.
[11] Woodward P. R. 1976, ApJ, 207, 484.

The Formation of OB Subgroups

By S. J. Chapman, A. S. Bhattal, M. J. Disney, J. A. Turner, A. P. Whitworth

Department of Physics and Astronomy, University of Wales, Cardiff CF2 3TH, UK

We analyze the gravitational stability of shocked interstellar gas layers, and show that, irrespective of the mechanism that generates them, (i) they first fragment gravitationally when the hydrogen column density reaches a value $\sim 6 \times 10^{21}\ cm^{-2}$; and (ii) the resulting fragments are massive, $\geq 7 M_\odot$. These results may help to explain Larson's relations between the masses and sizes of clumps in GMCs, and why massive star formation tends to be self-propagating. We present a numerical simulation which confirms this analysis and produces a cluster of ~ 50 protostars resembling a nascent OB Association. The protostars are all massive, $\geq 5 M_\odot$, and most are in binary or multiple systems.

1. Introduction

There are probably many different modes of star formation. At one extreme is what we might call the quasistatic mode, exemplified by the star formation occurring in quiescent GMCs like Taurus-Auriga. This mode appears to operate rather inefficiently to produce sparsely distributed low-mass stars. At the other extreme is what we might call the dynamical mode of star formation, exemplified by the star formation in Orion, 30 Doradus and presumably also starburst galaxies. This mode appears to operate with high efficiency, to produce stars with a wide range of masses, concentrated in clusters, e.g. [5]. It is unclear whether these regions are dynamically disturbed solely as a consequence of their efficient star formation, or whether it is their dynamic state which engenders a high star formation efficiency. Probably it is both, and consequently the dynamic star formation mode is self-propagating. Self-propagating star formation can be most easily understood if massive protostars condense efficiently out of shocked layers [4]. Such layers result either when two clumps of a GMC collide, or when a shell of cool neutral gas is swept up by an expanding nebula (HII region, stellar wind bubble or supernova remnant) around an evolved massive star.

In Section 2 we present a theoretical analysis of the accumulation and gravitational fragmentation of a shocked interstellar gas layer. In Section 3 we present a numerical simulation of this process. In Sections 4 and 5 we discuss the results, and summarize our conclusions.

2. The accumulation and fragmentation of a shocked interstellar gas layer

Consider gas with density ρ_o flowing into a shock front with velocity v_o, measured normal to the front. Suppose that the shocked gas radiates efficiently so that in effect its sound speed relaxes instantaneously to the value a_s. Define the Mach Number of the shock as $\mathcal{M} = v_o/a_s$. For a collision between two clumps in a GMC, ρ_o is the mean pre-shock density in the clumps. The shocked layer forms where the clumps collide. For collisions at finite impact parameter the layer is oblique. v_o is the speed with which the clumps approach the layer, i.e. the velocity component normal to the layer. To first order, the shear in the layer can be ignored. For a shell swept up by an expanding nebula, ρ_o is the density of the surrounding undisturbed gas. v_o is the speed with which the shell expands, and the shock marks the outer boundary of the shell. To first order, the tangential divergence of the radially expanding shell can be ignored.

The column density of the layer builds up according to $\Sigma \sim \rho_o v_o t = \rho_o a_s \mathcal{M} t$, so that at early times, $t \leq t_{switch} \sim (G\rho_o)^{-1/2}$, confinement of the layer normal to its surface is dominated by ram pressure $P_{ram} \sim \rho_o v_o^2 = \rho_o a_s^2 \mathcal{M}^2$, rather than by self-gravity $G\Sigma^2 \sim G\rho_o^2 a_s^2 \mathcal{M}^2 t^2$. Consequently the layer is not strongly centrally condensed normal to its surface. Its mean density is $\rho_s \sim \rho_o \mathcal{M}^2$, its thickness is $Z \sim \Sigma/\rho_s \sim \mathcal{M}^{-1} a_s t$, and as long as it remains one dimensional (i.e. plane-parallel or spherically symmetric) it has plenty of time to relax to hydrostatic balance. However, a shock-compressed layer is subject to several instabilities involving motions parallel to its surface. These instabilities have been explored by Elmegreen & Elmegreen [3], Vishniac [9], Elmegreen [2] and Lubow & Pringle [8]. As emphasized by Lubow & Pringle, the early short-wavelength dynamical modes are not self gravitating, and are therefore unlikely to condense into protostars, so we shall ignore them. In so doing, we are presuming that they do not disrupt the layer, as suggested by Vishniac [9], but simply engender weak turbulence, which then provides seed perturbations for subsequent gravitational instability. Additional seed perturbations will derive from the pre-existing density structures which the shock overruns.

The fastest-growing self-gravitating mode has a wavelength $L_{fastest}$ and a growth time $t_{fastest}$ given by (e.g. [6]) $t_{fastest} \sim L_{fastest}/a_s \sim a_s/G\Sigma \sim t_{switch}^2/\mathcal{M}t$. The layer will fragment into protostellar condensations when $t_{fastest} \leq t$, i.e. for $t \geq t_{fragment} \sim \mathcal{M}^{-1/2} t_{switch} \sim (G\rho_o \mathcal{M})^{-1/2}$. The separations and masses of the resulting fragments, are given by $L \geq L_{fragment} \sim a_s (G\rho_o \mathcal{M})^{-1/2}$ and $M \geq M_{fragment} \sim a_s^3 (G^3 \rho \mathcal{M})^{-1/2}$. We have analyzed in detail the fragmentation of layers produced both by clump/clump collisions (with the clumps obeying Larson's relations), and by expanding nebulae (HII regions, stellar wind bubbles and supernovae) [10]. We find that the time $t_{fragment}$ it takes for a layer to accumulate and fragment, and the column density of hydrogen $N_{fragment}$ through the layer when it fragments, are only weakly dependent on the input parameters. For a wide range of representative values, we obtain $t_{fragment} \sim 1.0 - 1.6$ $Myrs$ and $N_{fragment} \sim 4 - 8 \times 10^{21}$ cm^{-2}. We suggest that this last result may help to explain Larson's relations [7]. Structures with smaller column-densities are rarely seen because they are poorly defined and short-lived, while larger column-densities are seldom attained (i.e. they are pre-empted by gravitational instability). By contrast, the initial separations $S_{fragment}$ and masses $M_{fragment}$ of fragments when they start to condense out are strongly dependent on the sound speed in the shocked gas a_s: approximately $M_{fragment} \propto a_s^4$ and $S_{fragment} \propto a_s^2$. It follows that the large range of sizes for the structures which apparently subscribe to Larson's relations should be attributed to a spread in the values of the effective sound speed a_s obtaining in the gas from which they form. Larger fragments will be produced when a_s is large, because the gas is hot and/or beacause there is additional support due to turbulence, rotation, magnetic fields and/or cosmic rays. Since a_s has an effective minimum value ~ 0.2 $km\ s^{-1}$, we have $S_{fragment} \geq 0.4 - 0.7$ pc and $M_{fragment} \geq 7 - 24$ $M_\odot$. Therefore shocked interstellar gas layers are likely to spawn massive stars in weakly-bound groups.

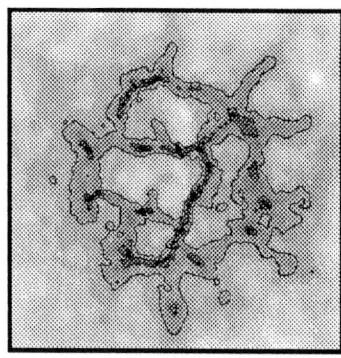

FIGURE 1. See text for details.

3. Numerical simulation of a clump/clump collision

We have simulated collisions between clumps in GMCs for a variety of clump masses M_o, collision speeds v_o, and impact parameters b_o, using our Smoothed Particle Hydrodynamics code, which is well-suited for treating problems in self-gravitating gas dynamics involving large density contrasts [1]. We present here the results of a head-on collision ($b_o = 0$) between two identical clouds (mass $M_o = 750\ M_\odot$; diameter $D_o = 13\ pc$) at speed $v_o = 1.4\ km\ s^{-1}$ (Mach-7); simulations at finite impact parameter produce qualitatively similar results. Before the collision, the clumps are in detailed stable hydrostatic equilibrium. The gas has a barotropic equation of state which corresponds to its having isothermal sound speed $a_o \sim 0.6\ km\ s^{-1}$ for $n_o < 300\ cm^{-3}$ (*i.e.* in the unshocked clouds) and $a_s \sim 0.2\ km\ s^{-1}$ for $n_s > 10^4\ cm^{-3}$ (*i.e.* in the shocked layer); in between these densities it cools approximately according to $a \propto n^{-1/3}$. Figure 1 shows the column-density of hydrogen through the compressed layer, viewed face-on, *i.e.* looking along the collision axis. The frame is 3 pc by 3 pc; the contours are equally spaced logarithmically between $10^{21}\ cm^{-2}$ and $10^{26}\ cm^{-2}$. The layer has condensed into a network of filaments, and the filaments have condensed into strings of beads. Each bead can be resolved into several rotationally-supported protostellar discs, most of which are in binary and multiple systems. Figure 2 shows a close-up of one such system from near the centre of the layer. The frame is now 0.05 pc by 0.05 pc, and the contours are equally spaced logarithmically between $10^{23}\ cm^{-2}$ and $10^{27}\ cm^{-2}$. In total the layer contains ~ 50 protostellar discs having masses in the range $5 - 40\ M_\odot$, diameters in the range $200 - 1000\ AU$, and peak densities in the range $10^{10} - 10^{11}\ cm^{-3}$; the binaries and multiples have separations in the range $1000 - 3000\ AU$. This simulation was performed on a Cray-YMP using 150,000 particles.

4. Discussion

We qualify these results with the following *caveats*. (1) We are not claiming that the substructure in a real GMC is necessarily in the form of stable equilibrium clumps. However, by treating stable equilibrium clumps, we are able to minimize the parameter-space of the initial conditions, *and* we can be certain that the ensuing gravitational instability is caused by the collision. (2) Without the radiative cooling implicit in our equation of state, the collision would not induce gravitational instability. However, we believe that our equation of state is a reasonable representation of the thermodynamics of real protostellar gas. (3) The effective viscosity in our code is much larger than pure molecular viscosity. The viscosity of real protostellar gas is unknown, but there are reasons for believing that

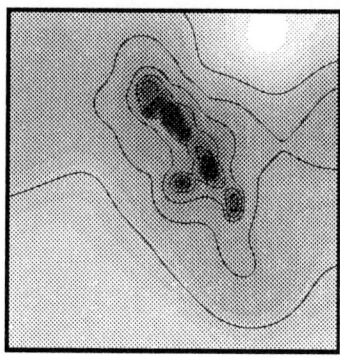

FIGURE 2. See text for details.

it too is much larger than pure molecular viscosity. (4) The code is statistically robust, in the sense that, if we vary the number of particles (over a factor of ~ 30) or their initial positions, the results are unchanged statistically, i.e. the mean numbers, masses, diameters, multiplicities and separations of the protostars formed are the same.

5. Conclusions

By considering the accumulation and gravitational fragmentation of a shocked interstellar gas layer from a very general standpoint, we have been able to show that the protostellar fragments which condense out of it are massive ($M \geq 7$ $M_\odot$). This supports the popular paradigm for self-propagating star formation in which the dense shell of cool neutral gas swept up by the expanding nebulae around the massive stars of one generation spawns the next generation of massive stars. We find that shocked layers should always fragment into protostars when the column-density of hydrogen reaches $\sim 6 \times 10^{21}$ cm^{-2}, and we suggest that this may be part of the explanation for Larson's relations [7]. Our simulations also demonstrate the feasibility of modelling the formation of small clusters of protostellar discs using modern particle methods and supercomputers. With refinements now in hand, it may be possible to follow the evolution of the protostellar discs formed here even further and so predict the statistical properties of the binary systems formed.

REFERENCES

[1] Chapman, S.J., 1992, Ph.D. dissertation (University of Wales)
[2] Elmegreen, B.G., 1989, ApJ, 340, 786
[3] Elmegreen, B.G. & Elmegreen, D.M., 1978, ApJ, 220, 1051
[4] Elmegreen, B.G. & Lada, C.J., 1977, ApJ, 214, 725
[5] Lada, E., 1991, ApJ, 393, L25
[6] Larson, R.B., 1985, MNRAS, 214, 379
[7] Larson, R.B., 1981, MNRAS, 194, 809
[8] Lubow, S.H. & Pringle, J.E., 1993, MNRAS, 263,701
[9] Vishniac, E.T., 1983, ApJ, 274, 152
[10] Whitworth, A.P., Bhattal, A.S., Chapman, S.J., Disney, M.J. & Turner, J.A., submitted to MNRAS

Numerical Simulations of Planetary Growth

By D. M. Kary[1], J. J. Lissauer[2]

[1] Physics Department, University of California, Santa Barbara, CA, 93106, U.S.A.

[2] Astronomy Program, Earth and Space Sciences Department, State University of New York, Stony Brook, NY 11794, USA.

Planet formation is the process by which material in a disk of dust and gas surrounding a (usually young) star is accumulated into bodies of mass 10^{25} to 10^{31} g. Since we only have detailed knowledge of our own Solar System, most of our understanding of planet formation comes from analytic and numerical modelling. The process begins with the formation of a disk around a young star, followed by accumulation of km-sized solid 'planetesimals', and then by the mutual accretion of these planetesimals to form planets. If these planets are large enough and form early enough, they can accrete nebular gas to form massive atmospheres. The process of going from planetesimals to terrestrial-type planets, planetesimal dynamics, can be characterized by three distinct phases: the planetesimal swarm, the mixed body phase, and the mutual accretion of large bodies. In the first of these, the large number of planetesimals indicates that statistical mechanics techniques provide the best modelling. The outcome of such models is frequently the development of a small number of 'runaway' bodies, which grow much faster than the average planetesimal. In the second phase, the behaviour of the mixture of large and small bodies can be modelled by direct numerical integration of the planetesimal motions, either in the restricted three-body limit or (more recently) as an n-body simulation. We present results from a series of simulations in the restricted three-body regime in which we show that a growing planet clears a gap in the planetesimal semimajor axis distribution. Finally, when the population of small planetesimals has been depleted, the mutual accretion of the larger embryos to terrestrial-like planets can be modelled through Monte Carlo techniques; n-body simulations are expected to supercede such Monte Carlo models when adequate computational power becomes available. Due to the long timescales involved, it is possible that the cores of the giant planets never went through this last phase, but rather they became massive enough to capture large atmospheres during an earlier phase of the accretion process.

1. Introduction

The origin of planetary systems is one of the most fundamental problems of astrophysics. Yet, because planets are so difficult to detect at astronomical distances, we have confirmed knowledge of only one planetary system, the Solar System in which we live. Thus, at present, theoretical modelling is the best means by which we can estimate the abundance and diversity of planetary systems in our Galaxy, including those planets which may harbor conditions conducive to the formation and evolution of life. Models of planetary formation are developed using our single example of a planetary system, supplemented by limited astrophysical observations of star forming regions. Data from other planetary systems around both main sequence stars and pulsars [48,76] may eventually provide further constraints.

Numerical studies have played a vital role in enhancing our understanding of the process of planet formation. Due to the breadth of the field, we will only give a brief overview of the process of planet formation in Section 2, with a more detailed discussion of numerical models of the growth of solid planets in Section 3. Included in this section we introduce some new simulations of gap clearing in the planetesimal distribution by a protoplanet. A more analytically oriented review of solid planet growth is presented by Lissauer and Stewart [41]. Giant planet formation is overviewed in Lissauer et al. [39], and a general review of planetary formation appeared in Lissauer [38]. The book *Protostars and Planets III* [35] provides an encyclopedic review of the broader field of star and planet formation.

2. Planet Formation

The nearly circular and coplanar orbits of the planets in the Solar System argue for planetary formation in a flattened disk revolving about the Sun [29,34]. Astrophysical models suggest that such disks are a natural byproduct of star formation from the collapse of rotating molecular cloud cores [8,9,21,60]; however, hydrodynamical calculations have not yet linked the observed states in the interstellar medium (ISM) to a star surrounded by a disk [43,61]. Observational evidence for the presence of disks of Solar System dimensions around pre-main sequence stars has increased substantially in recent years [5]. These observations suggest that the lifetimes of these protoplanetary disks range from $\sim 10^6 - 3 \times 10^7$ years [59,62].

Protoplanetary disks contain a mixture of gas and condensed matter. There is no reason to suspect that the composition of the high specific angular momentum material which collapses to form a protoplanetary disk is different from that of the low angular momentum matter which ends up in the star itself, so the relative abundance of the elements in the solar nebula is believed to have been very similar to that in the Sun; meteoritic evidence appears to confirm this hypothesis for all but the most volatile elements [53]. However, the contrast between the rocky compositions of the terrestrial planets and the ice-rich compositions of most satellites in the outer Solar System supports theoretical arguments that nebula temperatures decreased rapidly with distance from the Sun [4].

Meteoritic evidence also indicates that condensed matter in the solar nebula consisted of a mixture of surviving interstellar grains and solar nebula condensates, with nebula condensates dominating in the meteorite source region (the inner asteroid belt). All of this material seems to have passed through a phase in which it had grain sizes of 0.05 - 100 mm [31]. Thus, planet formation can be thought of as the process by which these grains, along with solar nebula gas, were combined to form bodies with masses of 10^{25} to 10^{30}g.

Models of protoplanetary disks suggest that they are gravitationally unstable if their masses are larger than 1/3 the star's mass [54]. At the opposite extreme, augmenting the masses of the planets in our Solar System with sufficient volatile material to bring them up to solar composition implies a "minimum mass solar nebula" of 0.01 to 0.02 $M_\odot$ [64]. Strictly speaking, this lower bound, though widely used, refers only to the total amount of material (mostly gas which was subsequently accreted by the Sun or expelled from the Solar System) that passed through the disk during its lifetime, rather than the disk's mass at any one instant. Current models which account for ejection of solid bodies from the Solar System and produce planets within the timescale dictated by the observations suggest that the amount of matter in the protoplanetary disk near the beginning of the planet formation epoch was several times larger than the "minimum mass" value, with most of the excess material having resided in the outer parts of the Solar System [51,37].

The dominant paradigm of modern planet formation theory is the "planetesimal hypothesis", which states that planets grow within circumstellar disks via accretion of small solid bodies known as planetesimals [10,19,51]. Sufficiently massive planetary bodies embedded in a gas-rich disk can gravitationally capture large amounts of gas, thereby producing jovian-type planets [7,20,42]. Regular satellites of the giant planets accrete in circumplanetary disks in a manner qualitatively similar to the growth of planets in the solar nebula. Note that in this picture, planet and circumplanetary disk formation is fundamentally different from star and solar nebula formation; planetary growth begins with the accumulation of solid bodies, with the accretion of substantial amounts of gas and possible formation of a disk occurring only if and when the solid planet becomes sufficiently massive [39].

The process of planetary growth is usually divided into several stages. In the first stage, microscopic grains collide and grow via pairwise collisions during settling towards the midplane of the disk. If the disk is laminar, the solids may be able to collapse into a thin enough layer for gravitational instabilities to occur [12,14,50]. If the disk is turbulent, gravitational instabilities are suppressed because the dusty layer remains too thick (and hence insufficiently dense for gravitational collapse [66]). Under such circumstances, continued growth via binary agglomeration depends upon the (currently unknown) sticking and disruption probabilities for collisions between larger grains [67]. The gaseous component of the protoplanetary disk plays an important role in the dynamics of solid bodies in this stage of planetary growth [2,65].

Once solid bodies reach kilometer-size (for parameters believed to be appropriate for the terrestrial region of the solar nebula), gravitational interactions between pairs of solid planetesimals dominate electromagnetic forces, gas drag, and collective gravitational effects. These planetesimals continue to grow via pairwise mergers. The rate of solid body accretion is determined by the size and mass of the largest bodies, the surface density of planetesimals, and the distribution of relative planetesimal velocities. The evolution of the planetesimal size distribution is determined by the gravitationally enhanced collision cross-section, which favors collisions between planetesimals with smaller velocities. Runaway growth of the largest planetesimal in each radial accretion zone appears to be a likely outcome of early planetesimal accretion [75]. The subsequent accumulation of the resulting protoplanets leads to a large degree of radial mixing in the terrestrial planet region, and giant impacts are probable. Growth via binary collisions proceeds until the protoplanets become dynamically isolated from one another. Some planetesimals are never incorporated into planets, and remain as asteroids, comets, and interstellar debris. The origin of the asteroid belt is discussed in Wetherill [70,72,73]. Comets are reviewed in Mumma *et al.* [44].

As a planet grows, the depth of its gravitational potential well increases, so it becomes more capable of trapping and retaining gases from the solar nebula (primarily H_2 and He). Compositional evidence suggests that Earth and smaller bodies formed almost exclusively from condensed matter, while Uranus (whose mass is 14.6 times Earth's) and more massive planets were also able to accumulate gaseous material, although not in sufficient quantities to mirror the Sun's composition. The stage of gaseous accretion (if it occurs) is terminated when either the disk is dissipated [51], or the planet accretes all of the gas within its region of the disk, or gravitational torques from the planet clear a gap in the disk [36].

It is expected that during the formation of their atmospheres each of the giant planets was surrounded by a gas and dust disk. Within this disk a planetesimal accumulation process, analogous to that which formed the solid planets, may have led to the formation of some of the giant planet satellites [33]. Other satellites were probably planetesimals which were captured by the forming planets [49,52]. Earth's Moon is now believed to have originated in a late, giant impact event [18]. A more detailed discussion of satellite origins is given in Stevenson *et al.* [57] and Lissauer *et al.* [39].

The planetary formation process is unlikely to be as purely sequential as lain out above. Grain growth and even the accretion of large planetesimals may well begin during the epoch when a protoplanetary disk is still accreting and redistributing matter. Given that the theories of each of these epochs are still rather primitive, a sequential study of each stage is probably adequate for now. However, to the extent that planetary growth depends on, e.g., the initial size distribution of planetesimals, we must recognize that various processes currently being treated as separate events occur simultaneously and can affect one another.

3. Numerical modelling of planetesimal accumulation

The basic physics underlying the growth of kilometer-sized planetesimals into solid planets involves gravitational encounters (which are very well understood), gas drag (which is fairly well understood, except in a few limited regimes), and physical collisions (whose outcomes are well-understood in some regimes and can be parameterized in other regimes). Thus, in principle, numerical simulation of the growth of planets from planetesimals should be straightforward. However, following the orbits of $\sim 10^{13}$ interacting, kilometer-sized and larger bodies for up to 10^8 orbits is well beyond the capabilities of modern computers. Thus, alternative techniques have been developed to approximate the behaviour of the swarm of planetesimals and the planets which form out of them. In this discussion we will divide the planetesimal accumulation process into three specific phases determined by the distribution of mass in bodies of various sizes. The first phase, corresponding to the earliest period of accretion, is characterized by a large number of small bodies, describable by a simple size distribution function. In the next phase there is a mixture of both small planetesimals and one or a few larger planetary embryos. As well as reviewing previous work, we introduce some new simulations we have performed of this phase. The third phase consists primarily of large mass ($> 10^{25}$ g) bodies. Some models of planetesimal accretion were designed to cover the entire accumulation right up to planet-sized bodies. However, we have included these models in specific phases (usually the earliest one) as a reflection of the where the assumptions made have the greatest validity.

3.1. Phase I. The Planetesimal Swarm

The earliest numerical models of planetesimal accumulation followed from the analytic models developed by Safronov [51]. In its simplest form, this approach describes the distribution of planetesimal masses by a single function and follows the evolution of this distribution through a coagulation relation, which in turn is a function of the mass and velocity distributions and time. The effects of mutual gravitational scatterings between planetesimals on the overall velocity distribution are handled following the formalism developed by Chandrasekhar [11] for the motions of stars in a galaxy. Planetesimals are assumed to have a mean relative velocity (random velocity) approximated by [51] (cf. Lissauer & Stewart [41])

$$v \approx v_K \sqrt{e^2 + \frac{5}{8}i^2}, \qquad (1)$$

where v_K is the local circular Keplarian velocity, and e and i are the rms eccentricity and inclination. The random velocity is a measure of the relative velocity of individual pairs of bodies. In the commonly used 2+2-body approximation (i.e., considering only the gravitational interactions of the two bodies involved during close encounters and only the Sun's gravity at other times), the effective gravitational radius for an encounter is

$$R_{2B} = R_p \sqrt{1 + \left(\frac{v_e}{v}\right)^2}, \qquad (2)$$

where R_p is the sum of the planetesimals' physical radii, and v_e is their mutual surface escape velocity. As v grows, the gravitational cross-section of the bodies decreases. In

Safronov's original analytic work [51], the random velocities of all planetesimals were assumed to grow in proportion to the escape velocity from the largest bodies. Thus, the timescales for accreting the giant planets were much longer than that now believed available for capturing significant amounts of gas, and in the case of Neptune the accretion time was even longer than the age of the Solar System [51].

However, significant simplifications of the velocity distribution were required in order to obtain an analytic solution, so Hayashi *et al.* [19] and Greenberg *et al.* [15] performed numerical integrations of the coagulation equations. Although Hayashi *et al.* basically confirmed the analytic results, Greenberg *et al.* found an acceleration of the accretion rate for large bodies. In this process, which Greenberg *et al.* termed "runaway accretion," the largest bodies double in mass at an ever increasing rate, whereas the smaller bodies grow slowly and eventually are either accreted by the large planetary embryos or ground to dust by high-speed collisions. In [58] it was shown that the runaway growth of the largest planetesimal in each region of the swarm is a consequence of dynamical friction, which dramatically reduces the random velocities of the largest bodies and hence allows them to accrete much more efficiently. If one assumes a minimum mass solar nebula, then runaway accretion leads to roughly lunar-mass 'protoplanets' or 'planetary embryos' in the terrestrial planet region and 1 $M_\oplus$ embryos in the giant planet zone [37]. At this point the embryos have accreted most of the smaller planetesimals, so that further growth must occur through mergers of embryos. This is a much slower process than runaway accretion, and in the outer Solar System it still presents a problem for producing giant planet cores before the solar nebula gas has disappeared (See Subsection 3.3). Other advances in statistical modelling of planetesimal swarms have included multiple semimajor axis bins [45], interactions between zones [55], and more complete velocity evolution equations [3,58,75].

An alternative to the standard statistical approach has recently been developed by Aarseth *et al.* [1]. The model they have developed only looks at a small box within the planetesimal swarm and does full n-body simulations of material within that box (with the Sun's gravity included). The simulations done so far have used 100 to 400 particles. This has allowed Aarseth *et al.* to estimate that runaway accretion should produce lunar sized bodies in the terrestrial region in roughly 10^4 to 10^5 years. Beyond this point, the local box approximation breaks down because large embryos begin to influence bodies well outside the box.

3.2. *Phase II: Large and Small Bodies*

The late stages of runaway accretion are not amenable to statistical models because the runaway body does not fit into the mass distribution function. One statistical model which addresses this is that developed by Spaute *et al.* [56]. This model treats runaway bodies separately from the rest of the swarm. In this regard, the Spaute *et al.* model is a step towards the next phase of planetesimal accumulation, that of mixed large and small bodies.

However, most attempts to study a swarm with a runaway body embedded in it have abandoned statistical models entirely. The approach used by Wetherill & Cox [74], Ida & Nakazawa [25], and Greenzweig & Lissauer [16,17] has been to concentrate on determining the probability that particles with a given set of orbital parameters and negligible mass will hit a finite mass planet in a single encounter. This is done though direct numerical integration of the Newtonian equations of motion in the restricted 3-body regime for a swarm of non-interacting test masses. By only considering the motion of the particles in the restricted three-body limit (particles acted on by the star and planet) this method allows for the study of a large number of particles (typically 10^6 or more) and hence gives very precise statistics on the deviation from pure 2+2-body behaviour in the accretion rate. Greenzweig & Lissauer [16] found that when the random velocities are low enough, the accretion rate

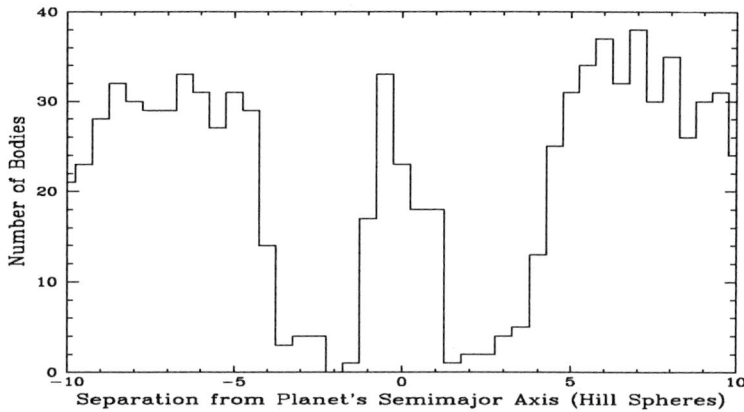

FIGURE 1. The separation in semimajor axes between a 10^{-6} $M_\odot$ protoplanet and 878 planetesimals after ~ 16000 orbits. The planetesimals started out on circular, coplanar orbits distributed to give a uniform surface density disk of material. The protoplanet is on a circular orbit, and the distance scale is measured in terms of the planet's Hill sphere radius. An additional 89 particles fell outside of the region show and 33 particles passed within 10^{-5} R_H of the planet and were stopped.

can be up to twice the two-body accretion rate, before dropping below the two-body rate as the disk becomes two-dimensional and the relative velocity becomes dominated by Kepler shear. Also, Greenzweig & Lissauer [17] found that when the eccentricities of the swarm planetesimals are Rayleigh distributed about a mean, the accretion rate is typically three times that for a swarm of planetesimals which all have the same eccentricity as the mean.

A related procedure is to follow a smaller number of test particles for a longer period of time [19,30]. Although using fewer particles reduces the accuracy of the impact statistics, these simulations provide a clearer picture of the long-term behaviour of the planetesimal swarm near a larger body. Hayashi et al.'s simulations [19] showed that, on timescales of a few synodic periods, the protoplanet forces the planetesimal swarm into a highly non-isotropic distribution in which bodies that start out near the planet are thrown into high eccentricity orbits much further away. Kary et al. [30] concentrated on studying the behaviour of planetesimals which migrate radially through the solar nebula due to gas drag. They found that, contrary to earlier expectations, most planetesimals which migrate toward a protoplanet are not accreted. Instead, many of these planetesimals decay into orbits inside the planet's orbit, after which they can continue to migrate toward the star. Kary et al. also found that the process of trapping particles in resonances outside the planet's orbit could be far more complicated than described in earlier work [68,47].

We have performed a set of simulations of the long term evolution of planetesimals qualitatively similar to those of Hayashi et al., but with a larger number of planetesimals (typically 1000 per simulation) and with more careful examination of the behaviour of bodies which pass within the planet's Hill sphere radius, $R_H \equiv (M_p/(3M_\odot))^{1/3} a_p$, where M_p is the planet's mass and a_p is the planet's semimajor axis. The Hill sphere is a useful scaling relationship in the restricted 3-body problem. We used a Bulirsh-Stoer integrator [16,30] to directly integrate the equations of motion of planetesimals all having the same initial eccentricity in orbits near that of a 10^{-6} $M_\odot$ protoplanet for ~ 16000 orbits. The planetesimals were distributed in a uniform surface density 2-dimensional swarm with random initial apses and longitudes. We found that the planet is very effective at perturbing material which gets close to it, thereby creating a gap in the distribution of planetesimal semimajor axes (Fig. 1). For example, those planetesimals which start out in

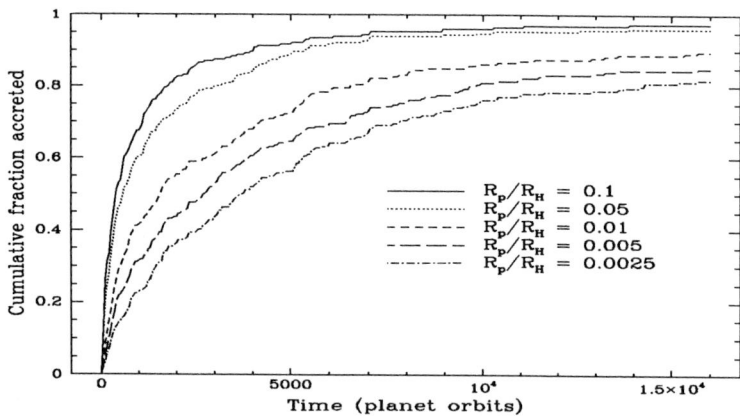

FIGURE 2. Cumulative fraction of material from within the accretion zone (initial planetesimal semimajor axes $a_p - 3.37 R_H < a < a_p - 1.13 R_H$ and $a_p + 1.09 R_H < a < 3.46 R_H$) accreted by planets with various radii, R_p, as a function of time (measured in planet orbits).

circular orbits with semimajor axes between 1.1 and 3.8 Hill spheres from the planet's orbit are either perturbed into high eccentricity orbits with semimajor axes well outside of this region or accreted by the planet on timescales of order 10 synodic periods. The fraction of bodies starting within the accretion zone which are accreted by planets of various sizes is shown in Fig. 2. If the planetesimals start out with some eccentricity, the width of the gap is larger. Bodies which start in orbits closer to the protoplanet are trapped in "horseshoe orbits", while those bodies farther away are not perturbed enough to produce significant orbital migration. Note that the gap in planetesimal semimajor axes doesn't necessarily translate into a gap in the actual positions of the planetesimals, since bodies on eccentric orbits will continue to spend some of their time in the "cleared" region.

More recently full n-body simulations of a protoplanet in a swarm of smaller bodies circling the Sun have been done using hardware specially designed for gravitational n-body simulations [22,23,24]. These simulations use up to 800 particles with various masses to map out the effects of the planetesimal swarm on a much larger protoplanet, and to estimate the change in the protoplanet's accretion rate with time. These simulations have confirmed the effectiveness of dynamical friction as a cause of runaway accretion, and the authors argue that the runaway continues until the protoplanet becomes 50-100 times the mass of the planetesimal swarm bodies. At this point the excitation of the particle eccentricities and inclinations by the growing planet slows the planet's runaway growth (although it continues to grow significantly faster than would be predicted in pre-runaway growth models).

3.3. *Phase III: Embryo Mutual Accretion*

As protoplanetary embryos clear out the smaller planetesimals around them, they enter a new phase of planetary growth with much longer timescales. In this phase, the ∼ lunar-sized (in the terrestrial planet region) embryos perturb each other into crossing orbits from which they can mutually accrete to form a few planets. Because of the long timescales involved (typically 10^7 to 10^8 orbits) there haven't been any attempts to do full n-body simulations of this process; however, Wetherill [69,71,73], Ipatov [27,28], Fernández and Ip [13], and Ip [26] have used a Monte Carlo technique (originally developed by Öpik [46] for studies of asteroid and comet orbital evolution) to model this final stage of solid planet formation. Each planetary embryo's periapse and node are assumed to circulate with a short enough period to be treated as a random variable, and the probability that two embryos in crossing orbits will have a close encounter is tested. If such an encounter occurs, the embryos either impact or mutually perturb each other's orbits, with the outcome determined statistically. This process is continued for all bodies (typically a few hundred) until the remaining embryos are in stable non-crossing orbits. In the terrestrial planet region these are the final planets, while in the outer Solar System they constitute the cores of the giant planets.

One of the most important results of these studies is that the late stages of planet formation (at least in the terrestrial region) are very stochastic. Although the outcome of the accretion process typically consists of 2 to 5 planets inside Jupiter's orbit, the number and mass distribution of these planets depend on the choice of random numbers as well as the initial conditions. In addition, planets are often hit by bodies whose masses are a significant fraction of the planet's final mass. One such giant impact is believed to have resulted in the formation of the Earth's Moon [18].

In the outer Solar System, two additional complications constrain this phase of accumulation. In order for the giant planets to accrete their large atmospheres, they need solid cores of roughly 15 $M_\oplus$. Using a minimum mass solar nebula, the cores one expects to produce through runaway accretion are only ~ 1 $M_\oplus$. Following this, accretion must occur via large-body pairwise accretion, which takes longer than the 10^6 to 10^7 years for which there is significant nebula gas. For Uranus and Neptune, this problem is exasperated by the fact that 1 $M_\oplus$ embryos can easily perturb each other into Saturn-crossing orbits. From there it is easy for these bodies to be ejected out of the Solar System entirely [63,39]. One plausible way around both of these problems is to use a larger than minimum mass solar nebula [37]. This could increase the mass at which runaway accretion is cut off and thereby produce the giant planetary cores without going through large body mutual accretion. However, the axial tilts of Uranus and Neptune indicate that they were also hit by bodies of up to 10% of their final masses [40]. Thus, there must have been some large planetesimals in the Uranus-Neptune region, even if there weren't enough to be dynamically important. Alternatively, it has been suggested that 1 $M_\oplus$ bodies could under certain circumstances have accreted sufficient gas to have large cross-sections and hence high enough collision efficiencies to produce Uranus and Neptune in a minimum mass nebula [39].

The presence of excess solid material in the outer Solar System implies that this material had to have been ejected by the giant planets, a process which would result in considerable orbital evolution of these planets [13,28]. The transfer of energy and angular momentum would result in Jupiter's orbit evolving inward (toward the Sun). In contrast, Uranus and Neptune would evolve outward. This is because the bodies from which they take angular momentum (and hence thrown into Saturn and Jupiter crossing orbits) are much more efficiently ejected from the Solar System than the ones to which they give angular momentum (and hence thrown into higher orbits or directly out of the Solar System) [13,63]. However, there is still considerable debate over key aspects of outer planet formation models, so any conclusions drawn from them should still be regarded as rather speculative (even by the high standards of speculation normally set in planet formation studies).

4. Conclusions

Numerical modelling of planet formation is a young field, with many new techniques still being developed. Computers are only just beginning to become fast enough to allow n-body simulations of parts of the process, and we are still far from being able to model all of planetary accretion this way. Thus, techniques that approximate the behaviour of a swarm of bodies or which only look at a small part of the planet formation process are likely to remain the norm for the foreseeable future.

The stochastic nature of the planet formation process means that it is virtually impossible to make a single model and run it from the beginning of planet formation to the end and expect to get precisely the Solar System we see. Although this stochastic behaviour limits what we can say about the detailed order of events in the formation of our Solar System, it also provides a rich field of study as we explore the range of possible planetary systems (even from virtually identical starting conditions).

This work was supported in part by the NASA Planetary Geology and Geophysics Program under grants NAGW-2061 at UCSB and NAGW-1107 at SUNYSB.

REFERENCES

[1] Aarseth, S. J., Lin, D. N. C. & Palmer, P. L. 1993, ApJ, 403, 351.
[2] Adachi, I., Hayashi, C. & Nakazawa, K. 1976, ProgTheorPhys, 56, 1756.
[3] Barge, P. & Pellat, R. 1991, Icarus, 93, 270.
[4] Barshay, S. S. & Lewis, J. S. 1976, ARAA, 14, 81.
[5] Beckwith, S. V. W. & Sargent, A. I. 1993, *Protostars and Planets III*, ed. E. H. Levy, J. I. Lunine & M. S. Matthews, (Univ. Arizona Press: Tucson), 521.
[6] Black, D. C. & Matthews M. S., Eds. 1985, *Protostars and Planets II*, (Univ. Arizona Press: Tucson).
[7] Bodenheimer, P. & Pollack, J. B. 1986, Icarus, 67, 391.
[8] Cameron, A. G. W. 1962, Icarus, 1, 13.
[9] Cassen, P. M. & Moosman, A. 1981, Icarus, 48, 353.
[10] Chamberlin, T. C. 1905, *Carnegie Institution Year Book #3 for 1904*, (Carnegie Institution: Washington), 195.
[11] Chandrasekhar, S. 1942, *Principles of Stellar Dynamics*, Chicago Univ. Press.
[12] Edgeworth, K. E. 1949, MNRAS, 109, 600.
[13] Fernández, J. A. & Ip, W.-H. 1984, Icarus, 58, 109.
[14] Goldreich, P. & Ward, W. R. 1973, ApJ, 183, 1051.
[15] Greenberg, R., Wacker, J. F., Hartmann, W. L. & Chapman, C. R. 1978, Icarus, 35, 1.
[16] Greenzweig, Y. & Lissauer, J. J. 1990, Icarus, 87, 40.
[17] Greenzweig, Y. & Lissauer, J. J. 1992, Icarus, 100, 440.
[18] Hartmann, W. K. 1986, *Origin of the Moon*, ed. W. K. Hartmann, R. J Phillips & G. J. Taylor, (Lunar & Planetary Inst: Houston), 579.
[19] Hayashi, C., Nakazawa, K. & Adachi, I. 1977, PASJ, 29, 163.
[20] Hayashi, C., Nakazawa, K. & Nakagawa, Y. 1985, *Protostars and Planets II*, ed. D. C. Black & M. S. Matthews, (Univ. Arizona Press: Tucson), 1100.
[21] Hoyle, F., 1960, QJRAS, 1, 28.
[22] Ida, S. & Makino, J. 1992, Icarus, 96, 107.
[23] Ida, S. & Makino, J. 1992, Icarus, 98, 28.
[24] Ida, S. & Makino, J. 1993, Icarus, in press.
[25] Ida, S. & Nakazawa, K. 1989, AA, 224, 303.
[26] Ip, W.-H. 1989, Icarus, 80, 167.
[27] Ipatov, S. 1987, SolSysRes, 21, 129.
[28] Ipatov, S. 1989, SolSysRes, 23, 16.
[29] Kant, I. 1755, *Allegmeine Naturgeschichte und Theorie des Himmels*, Köingsburg und Leipzig, Johann Friederich Petersen. English translation by W. Haste 1968 "Universal Natural History and Theories of the Heavens", In *Kant's Cosmology*, (Greenwood Publ: New York).
[30] Kary, D. M., Lissauer, J. J., & Greenzweig, Y. 1993, Icarus, 106, in press.
[31] Kerridge, J. F. & Anders, E. 1988, *Meteorites and the Early Solar System*, ed. J. F. Kerridge & M. S. Matthews, (Univ. Arizona Press: Tucson), 1149.
[32] Kerridge, J. F. & Matthews, M. S., Eds. 1988, *Meteorites and the Early Solar System*, (Univ. Arizona Press: Tucson).
[33] Korycansky, D. G., Bodenheimer, P. & Pollack, J. B. 1991, Icarus, 92, 234.
[34] Laplace, P. S. 1796, *Exposition du Système du Monde*, English translation, H.H. Harte 1830, "The System of the World", (Dublin Univ. Press: Dublin).
[35] Levy, E. H., Lunine, J. I. & Matthews, M. S., Eds. 1993, *Protostars and Planets III*, (Univ.

Arizona Press: Tucson).
[36] Lin, D. N. C. & Papaloizou, J. 1979, MNRAS, 186, 799.
[37] Lissauer, J. J. 1987, Icarus, 69,249.
[38] Lissauer, J. J. 1993, ARAA, 32, 129.
[39] Lissauer, J. J., Pollack, J. B., Wetherill, G. W. & Stevenson, D. J. 1994, *Neptune and Triton*, ed. D. P. Cruikshank & M. S. Matthews, (Univ. Arizona Press: Tucson), in press.
[40] Lissauer, J. J. & Safronov, V. S. 1991, Icarus, 93, 288.
[41] Lissauer, J. J. & Stewart, G. R. 1993, *Protostars and Planets III*, ed. E. H. Levy, J. I. Lunine & M. S. Matthews, (Univ. Arizona Press: Tucson), 1061.
[42] Mizuno, H. 1980, Prog. Theor. Phys. 64, 544.
[43] Morfill, G. E., Tscharnuter, W., & Volk, H. J. 1985, *Protostars and Planets II*, ed. D. C. Black & M. S. Matthews, (Univ. Arizona Press: Tucson), 493.
[44] Mumma, M. J., Weissman, P. R., & Stern, S. A. 1993, *Protostars and Planets III*, ed. E. H. Levy, J. I. Lunine & M. S. Matthews, (Univ. Arizona Press: Tucson), 1177.
[45] Nakagawa, Y., Hayashi, C. & Nakazawa, K. 1983, Icarus, 54, 361.
[46] Öpik, E. J. 1951, ProcRoyIrishAcad, 54A, 165.
[47] Patterson, C. W. 1987, Icarus, 70, 319.
[48] Phillips, J. A., Thorsett, S. E. & Kulkarni, S. R., Eds. 1993. *Planets Around Pulsars*, Astron. Soc. Pacific Conf. Series, 36.
[49] Pollack, J. B., Burns, J. A. & Tauber, M. E. 1979, Icarus, 37, 587.
[50] Safronov, V. S. 1960, Annd'Astrophys, 23, 901.
[51] Safronov, V. S. 1969, *Evolution of the Protoplanetary Cloud and Formation of the Earth and Planets*, (Nauka Press: Moscow). English translation: NASA TTF-677, 1972.
[52] Sasaki, S. 1990, LunPlanetSci, 49, 284.
[53] Sears, D. W. B. & Dodd, R. T. 1988, *Meteorites and the Early Solar System*, ed. J. F. Kerridge & M. S. Matthews, (Univ. Arizona Press: Tucson), 3.
[54] Shu, F. H., Tremaine, S., Adams, F. C. & Ruden, S. P. 1990, ApJ, 358, 495.
[55] Spaute, D., Lago, B. & Cazenave, A. 1985, Icarus, 64, 139.
[56] Spaute, D., Weidenschilling, S. J., Davis, D. R. & Marzari, F. 1991, Icarus, 92, 147.
[57] Stevenson, D. J., Harris, A. W. & Lunine, J. I. 1986, *Satellites,*, ed. J. A. Bergstrahl & M. S. Matthews, (Univ. Arizona Press: Tucson), 39.
[58] Stewart, G. R. & Wetherill, G. W. 1988, Icarus, 74, 542.
[59] Strom, S. E., Edwards, S. & Strom, K. M. 1989, *The Formation and Evolution of Planetary Systems*, Space Tel. Sc. Inst. Sym Series 3, ed. H. A. Weaver & L. Danly, (Cambridge Univ. Press: Cambridge), 91.
[60] Terebey, S., Shu, F. H., & Cassen, P. 1984, ApJ, 286, 529.
[61] Tscharnuter, W. M. 1987, AA, 188, 55.
[62] Walter, F. M., Brown, A., Mathieu, R. D., Myers, P. C. & Vrba, F. J. 1988, AJ, 96, 297.
[63] Weidenschilling, S. J. 1975, Icarus, 26, 361.
[64] Weidenschilling, S. J. 1977a ApSpSci, 51, 153.
[65] Weidenschilling, S. J. 1977b MNRAS, 180, 57.
[66] Weidenschilling, S. J. 1980, Icarus, 44, 172.
[67] Weidenschilling, S. J. & Cuzzi, J. N. 1993, *Protostars and Planets III*, ed. E. H. Levy, J. I. Lunine & M. S. Matthews, (Univ. Arizona Press: Tucson) 1031.
[68] Weidenschilling, S. J. & Davis, D. R. 1985, Icarus, 62, 16.
[69] Wetherill, G. W. 1980, ARAA, 18, 77.
[70] Wetherill, G. W. 1989, *Asteroids II*, ed. R. P. Binzel, T. Gehrels & M. S. Matthews, (Univ. Arizona Press: Tucson), 661.
[71] Wetherill, G. W. 1990, AnnRevEarthPlanetSci, 18, 205.
[72] Wetherill, G. W. 1991, Science, 253, 535.
[73] Wetherill, G. W. 1992, Icarus, 100, 307.
[74] Wetherill, G. W. & Cox, L. P. 1985, Icarus, 63, 290.
[75] Wetherill, G. W. & Stewart, G. R. 1989, Icarus, 77, 330.
[76] Wolszczan, A. & Frail, D. A. 1992, Nature, 355, 145.